RISK ANALYSIS
FOUNDATIONS, MODELS, AND METHODS

INTERNATIONAL SERIES IN
OPERATIONS RESEARCH & MANAGEMENT SCIENCE
Frederick S. Hillier, Series Editor Stanford University

Vanderbei, R. / *LINEAR PROGRAMMING: Foundations and Extensions*
Jaiswal, N.K. / *MILITARY OPERATIONS RESEARCH: Quantitative Decision Making*
Gal, T. & Greenberg, H. / *ADVANCES IN SENSITIVITY ANALYSIS AND PARAMETRIC PROGRAMMING*
Prabhu, N.U. / *FOUNDATIONS OF QUEUEING THEORY*
Fang, S.-C., Rajasekera, J.R. & Tsao, H.-S.J. / *ENTROPY OPTIMIZATION AND MATHEMATICAL PROGRAMMING*
Yu, G. / *OPERATIONS RESEARCH IN THE AIRLINE INDUSTRY*
Ho, T.-H. & Tang, C. S. / *PRODUCT VARIETY MANAGEMENT*
El-Taha, M. & Stidham , S. / *SAMPLE-PATH ANALYSIS OF QUEUEING SYSTEMS*
Miettinen, K. M. / *NONLINEAR MULTIOBJECTIVE OPTIMIZATION*
Chao, H. & Huntington, H. G. / *DESIGNING COMPETITIVE ELECTRICITY MARKETS*
Weglarz, J. / *PROJECT SCHEDULING: Recent Models, Algorithms & Applications*
Sahin, I. & Polatoglu, H. / *QUALITY, WARRANTY AND PREVENTIVE MAINTENANCE*
Tavares, L. V. / *ADVANCED MODELS FOR PROJECT MANAGEMENT*
Tayur, S., Ganeshan, R. & Magazine, M. / *QUANTITATIVE MODELING FOR SUPPLY CHAIN MANAGEMENT*
Weyant, J./ *ENERGY AND ENVIRONMENTAL POLICY MODELING*
Shanthikumar, J.G. & Sumita, U./*APPLIED PROBABILITY AND STOCHASTIC PROCESSES*
Liu, B. & Esogbue, A.O. / *DECISION CRITERIA AND OPTIMAL INVENTORY PROCESSES*
Gal, T., Stewart, T.J., Hanne, T./ *MULTICRITERIA DECISION MAKING: Advances in MCDM Models, Algorithms, Theory, and Applications*
Fox, B. L./ *STRATEGIES FOR QUASI-MONTE CARLO*
Hall, R.W. / *HANDBOOK OF TRANSPORTATION SCIENCE*
Grassman, W.K./ *COMPUTATIONAL PROBABILITY*
Pomerol, J-C. & Barba-Romero, S. / *MULTICRITERION DECISION IN MANAGEMENT*
Axsäter, S. / *INVENTORY CONTROL*
Wolkowicz, H., Saigal, R., Vandenberghe, L./ *HANDBOOK OF SEMI-DEFINITE PROGRAMMING: Theory, Algorithms, and Applications*
Hobbs, B. F. & Meier, P. / *ENERGY DECISIONS AND THE ENVIRONMENT: A Guide to the Use of Multicriteria Methods*
Dar-El, E./ *HUMAN LEARNING: From Learning Curves to Learning Organizations*
Armstrong, J. S./ *PRINCIPLES OF FORECASTING: A Handbook for Researchers and Practitioners*
Balsamo, S., Personé, V., Onvural, R./ *ANALYSIS OF QUEUEING NETWORKS WITH BLOCKING*
Bouyssou, D. et al/ *EVALUATION AND DECISION MODELS: A Critical Perspective*
Hanne, T./ *INTELLIGENT STRATEGIES FOR META MULTIPLE CRITERIA DECISION MAKING*
Saaty, T. & Vargas, L./ *MODELS, METHODS, CONCEPTS & APPLICATIONS OF THE ANALYTIC HIERARCHY PROCESS*
Chatterjee, K. & Samuelson, W./ *GAME THEORY AND BUSINESS APPLICATIONS*
Hobbs, B. et al/ *THE NEXT GENERATION OF ELECTRIC POWER UNIT COMMITMENT MODELS*
Vanderbei, R.J./ *LINEAR PROGRAMMING: Foundations and Extensions, 2nd Ed.*
Kimms, A./ *MATHEMATICAL PROGRAMMING AND FINANCIAL OBJECTIVES FOR SCHEDULING PROJECTS*
Baptiste, P., Le Pape, C. & Nuijten, W./ *CONSTRAINT-BASED SCHEDULING*
Feinberg, E. & Shwartz, A./ *HANDBOOK OF MARKOV DECISION PROCESSES: Methods and Applications*
Ramík, J. & Vlach, M. / *GENERALIZED CONCAVITY IN FUZZY OPTIMIZATION AND DECISION ANALYSIS*
Song, J. & Yao, D. / *SUPPLY CHAIN STRUCTURES: Coordination, Information and Optimization*
Kozan, E. & Ohuchi, A./ *OPERATIONS RESEARCH/ MANAGEMENT SCIENCE AT WORK*
Bouyssou et al/ *AIDING DECISIONS WITH MULTIPLE CRITERIA: Essays in Honor of Bernard Roy*

RISK ANALYSIS FOUNDATIONS, MODELS, AND METHODS

by

Louis Anthony Cox, Jr.

Cox Associates and University of Colorado

KLUWER ACADEMIC PUBLISHERS
Boston/Dordrecht/London

Distributors for North, Central and South America:
Kluwer Academic Publishers
101 Philip Drive
Assinippi Park
Norwell, Massachusetts 02061 USA
Telephone (781) 871-6600
Fax (781) 871-6528
E-Mail <kluwer@wkap.com>

Distributors for all other countries:
Kluwer Academic Publishers Group
Distribution Centre
Post Office Box 322
3300 AH Dordrecht, THE NETHERLANDS
Telephone 31 78 6392 392
Fax 31 78 6546 474
E-Mail <orderdept@wkap.nl>

Electronic Services <http://www.wkap.nl>

Library of Congress Cataloging-in-Publication Data

A C.I.P. Catalogue record for this book is available from the Library of Congress.

Printed on acid-free paper.

Printed in the United States of America

To Alvin W. Drake and Alexander Haslam, outstanding teachers

CONTENTS

CHAPTER 1
Introduction and Basic Risk Models

CHAPTER 2
Risk Assessment Modeling 43

CHAPTER 5

CHAPTER 6

CHAPTER 7

Preface

The United States spends over $100 billion dollars each year to comply with health and environmental regulations designed to reduce public and occupational health risks. Yet, many of these expenditures produce few or no real health benefits (*Fortune* magazine, September 19, 1994, pp 91-105.) This situation occurs in part for political and group decision-making reasons, in part for psychological reasons, and in part because risk analysts use statistical methods and models that can predict unrealistically large human health risks from economic activities and products, and therefore unrealistically large health benefits from reducing or controlling them.

Operations research and management science (OR/MS) methods can help. They can greatly improve quantitative risk assessments by making risk models more accurate. They can inform and improve regulatory resource allocation decisions, the design and conduct of "toxic tort" litigation, the regulation of insurance markets (including liability insurance), and corporate decisions about how much to invest in risk research and control. They can help businesses define and create responsible production processes and make R&D and marketing decisions about how and whether to develop, test, and market new products. This book explains relevant OR/MS methods and models for achieving these benefits.

Risk analysis also poses many interesting theoretical and applied challenges to operations researchers and management scientists. Continued advances in applied statistics, stochastic processes, model management, simulation modeling, uncertain reasoning, knowledge-based methods, machine learning, and decision analysis will be needed to fully represent and use scientific information to improve risk management decision making.

Thus, quantitative analysis of risk management decisions poses large opportunities and large challenges for members of the OR/MS community.

Both the challenges and the opportunities of applying OR/MS techniques to health risk analysis are likely to remain relevant and interesting to risk analysts and decision makers for years to come. Conversely, the needs of the health risk analysis community for improved methods of risk assessment and risk management are likely to provide a continuing source of fresh and challenging research problems for OR/MS theorists and practitioners. This book seeks to bring operations research and health risk analysis together in a framework that benefits both. Its main goals are to transfer OR/MS insights and methods to health risk analysts and to inform the OR/MS community about exciting and challenging open problems in health risk analysis.

Compared to other recent texts and monographs on quantitative risk analysis, this one emphasizes and exploits OR/MS methods and approaches in applications to health risk modeling and risk management. It surveys modern decision analysis methods and insights as they apply to health risk management decisions and uses them to integrate the scientific (Chapters 1-4) and policy (Chapters 5-8) aspects of cancer risk analysis. It also updates older discussions of risk analysis issues (especially for chemical carcinogens) in several places. Some of the examples present original models and results illustrating the practical application of OR/MS methods to health risk analysis problems.

The topics covered and the level of presentation are designed to make the most important ideas accessible and attractive to the following groups within the risk analysis community:

- Industry toxicologists and risk analysts engaged in quantitative health risk assessments
- Corporate risk managers responsible for making decisions about products and processes based on results of quantitative risk assessments.
- Scientists doing research on quantitative models of human health risks
- Policy makers needing to use risk analysis in thinking about health, safety, and environmental issues.
- Regulators involved in occupational health, environmental, and consumer product regulation (OSHA, EPA, CPSC, FDA, and their stakeholders).
- Litigators involved in toxic torts (especially those in which probabilistic causation of health effects is a key issue)
- Academic researchers teaching courses on health, safety, and environmental risk analysis.

A quantitative background is required for most of the detailed discussions, but summary tables and boxes throughout the text should help casual readers to obtain a quick overview of the central ideas of many topics with a minimum of preparation and effort. Thus, it may be used as a handbook to introduce core analytic methods and technical ideas for risk analysis, perhaps as part of a second course in quantitative methods for risk analysts.

Acknowledgements

I thank David Burmaster and Kimberly Thompson of the Harvard Center for Risk Analysis for many stimulating conversations and for opportunities to teach topics in Chapters 3 and 4; Arcady Novosyolov for reviewing Chapters 4-7; Paolo Ricci for enjoyable conversations and collaborations on most aspects of risk analysis, especially those in Chapters 4 and 8; and those brilliant, devoted, and tireless champions of style templates, Kate Michmerhuizen and Jason McCrea, for their expert help in turning mere text into a final manuscript. Doug Popken collaborated on the simulation model of Campylobacter at the end of Chapter 2. Kathy Macneal provided top-notch assistance in organizing and analysing several data sets. Pat Beatty, Dan Byrd, Dennis Copeland, Peter Craig, Cosmo DiPerna, Larry Frarey, Dave Paley, Mary Paxton, Paul Price, Michael Vaughn, and Jim Wilson provided real-world problems and data, together with thoughtful discussions of what makes them hard and important, that inspired many of the case studies and examples.

Chapter 1

Introduction and Basic Risk Models

1. INTRODUCTION

What is health risk analysis, and what can it potentially be to make it most valuable to public health decision-makers and other health risk managers? This book develops an answer based on the following ideas. A *health risk* describes the potential for a hazard – the source of the risk – to adversely affect the health of those exposed to it. Not all of the possible health outcomes are equally valued. A health risk increases the probabilities of some of the less preferred outcomes, such as shorter lives or more frequent illnesses. Quantitative health risk analysis determines by how much different actions change the potentials for various outcomes to occur and quantifies uncertainty about the answers. It helps to identify actions that lead to more preferred probabilities for consequences.

Representative questions addressed by health risk analysis include the following:

- Does diesel exhaust cause human lung cancer? Do silicone breast implants, extremely low-frequency electromagnetic fields (ELF-EMFs), environmental tobacco smoke (ETS), Alar in apples, plasticizers in children's toys, and benzene in Perrier cause the adverse health effects that have sometimes been ascribed to them? *Hazard identification* addresses such questions.
- How does risk of cancer vary with the concentration and timing of occupational exposures to benzene? For example, are there exposure concentrations below which benzene cannot cause an increased risk of leukemia? How, if at all, do the answers depend on the individual

biochemistries, ages, and diets of those exposed? *Multivariate dose-response models* address these questions.

- How extensively should contaminated soils around an abandoned hazardous waste site be sampled? Which homeowners' properties, if any, should be cleaned, given the sampling results? Answering these questions may involve *exposure assessment, risk characterization*, and *risk management decision-making*.
- How much compensation should a company or a government body such as the Department of Energy pay to a former employee who develops lung cancer if the employee was occupationally exposed to radiation but also had non-occupational risk factors for lung cancer, such as smoking, poor diet, and medical radiation? *Retrospective attribution of risk* to competing causes, including probability-of-causation calculations (Chapter 4), can inform compensation policies in such settings. Other issues, such as whether the occupational exposures were made with the voluntary and informed consent of exposed workers, perhaps in return for hazard pay, may also affect compensation policy decisions.
- Antibiotics administered to farm animals can improve their health but create a risk of developing antibiotic-resistant strains of bacteria that may survive in under-cooked meat and lead to illness in consumers. Illnesses induced (at least in part) by resistant strains might be less responsive to prescribed antibiotics than illnesses from non-resistant strains. Is this risk best managed by banning the use of antibiotics on the farm, by assuring that animals treated with antibiotics are processed to kill bacteria (resistant or not), by encouraging people to use sanitary practices in preparing meat, or by encouraging physicians to consider the possibility of resistance in choosing antibiotics to prescribe? What is the best mix of countermeasures for reducing the risk of illness?

Health risk analysis provides methods and principles for answering such questions. It is often divided into methods for assessing, communicating, and managing health risks. *Health risk assessment* quantitatively estimates the health risks to individuals and to groups from hazardous exposures and from the decisions or activities that create them. It applies specialized models and methods to quantify likely exposures and their resulting health risks. Its goal is to produce information to improve decisions. It does this by relating alternative decisions to their probable consequences and by identifying those decisions that make preferred outcomes more likely. Unlike many areas of expert decision analysis, health risk assessment typically draws on explicit engineering, biomathematical, and statistical consequence models to describe or simulate the causal relations between actions and their probable effects on health.

Health risk *communication* characterizes and presents information about health risks and uncertainties to decision-makers and stakeholders. Health risk *management* applies principles for choosing among alternative decision alternatives or actions that affect exposures, health risks, or their consequences. Risk management is often viewed as a process that maps information obtained via risk assessment to choices of risk management actions. These may include risk acceptance, avoidance, transfer (e.g., insurance), mitigation of consequences, or compensation. From this perspective, risk assessment and risk communication should provide information about exposures and risks to support good risk management decision-making. Chapters 5-7 present formal techniques for making and evaluating risk management decisions.

1.1 Distinguishing Characteristics of Risk Analysis and Relations to Other Disciplines

Several key issues set health risk analysis apart from other fields. These include:

- *Causal modeling of exposures and health effects.* In health risk analysis, opinions and beliefs about the probable consequences of decisions are expressed as explicit quantitative models of exposure and of exposure-response relations. These models usually represent aspects of *causation* and are intended to describe how changes in actions propagate through one or more causal chains to change exposures and health effects. This emphasis on causation contrasts with many statistical and biostatistical models, which emphasize *inference* about the probability of observing an effect, given observed data about an exposure. Causal and statistical relations may be very different, as illustrated at the start of Chapter 4.
- *Vital outcomes.* Consequences of health risk management decisions may include illness or death. Such consequences require special evaluation techniques. For example, how should a decision-maker value death or extreme illness or injury to himself or others in choosing among actions with uncertain consequences? Unlike many economic consequences, some health outcomes cannot be fully compensated or redressed *ex post* by transferring dollars to those who have suffered them. This makes valuing risks that involve potential losses of life and health more difficult than valuing risks that only affect market goods. Moreover, asking people about their willingness to pay (WTP) to remove such risks, or about their willingness to accept (WTA) monetary compensation to bear them, elicits responses that may reflect political and psychological attitudes more strongly than economic tradeoffs and values (see Chapter 5). For example, stated WTP and WTA amounts for avoiding health and

safety risks may reflect beliefs about rights and concerns about fairness and equity in the allocation of risks or costs while being quite insensitive to the probabilities or magnitudes of the consequences involved (Kahneman et al., 1999). Preferences revealed by practical behaviors and choices, including willingness to accept compensation for hazardous occupations, reflect similar concerns (Viscusi et al., 1997). Thus, individual economic values and preference trade-off rates for risks involving potential loss of health or life may be difficult or impossible to define and measure apart from political concerns and cultural attitudes.

- *Inferred risks and consequences.* Risks cannot be directly observed or measured. They refer to potentials. Therefore, they must be inferred from models, knowledge, assumptions, and data. Inferences about human health risks are often drawn from animal studies and/or from chemical structures and experimental results in various assays and biological systems. This calls for conclusions based on analogy, induction from examples, and perhaps deduction about likely biochemical processes – forms of inference that are often challenging compared to routine statistical calculations of conditional probabilities or expected values. Chapter 2 mentions approaches to inference that have been attempted for identifying chemical carcinogens from chemical structures and biological activity data, so far with very imperfect success. Chapter 3 discusses methods and challenges for drawing statistical inferences about hazards and risks from epidemiological data.
- *Multiple stakeholders and decision makers.* Multiple stakeholders and decision makers usually participate in activities that create health risks. For example, exposure to a hazard often arises from the joint decisions or behaviors of those who create or own it and those who are exposed to it. Those who create or own a hazard may differ from those who bear the resulting risk and from those who must regulate or otherwise manage it. Single-person theories of decision making (Chapters 5-7) must be extended to handle risk management decisions with multiple participating or affected parties (Chapter 8).
- *Communication:* Risks and uncertainties must often be communicated to stakeholders who may not have the data, resources, or expertise to accurately quantify and understand risks without assistance. Yet, objective communication of risk information is often difficult or impossible (Chapter 5). How to communicate risk information without misleading or manipulating the audience is a substantial challenge.
- *Perception, comprehension, internalization* and *action:* Stakeholder understanding and perceptions of risk, and their resulting behaviors, may be affected in unexpected ways by attempts to describe the results of formal risk assessments (Chapter 5). Even if risk information is

communicated accurately, behaviors may not change in the ways that models of rational behavior predict. Understanding how stakeholders do and should make decisions based upon risk information is an active area of research.

These aspects make health risk analysis a highly challenging and rewarding field of research and application. It overlaps with and benefits from established areas of operations research – especially applied probability and reliability theory, simulation modeling, prescriptive and psychological decision theories, and public sector operations research, as well as many areas of economics and statistics. Yet, its core problems are sufficiently different and difficult to strain even the best available methods from these related disciplines. The following box lists other academic disciplines that provide core methods and insights for risk analysis.

Core Disciplines Used in Risk Analysis

Risk Analysis = Health risk *assessment*
(including exposure assessment and dose-response modeling)

- Applied probability and statistics/biostatistics/epidemiology
- Uncertainty analysis and uncertain reasoning
- Engineering and scientific modeling, toxicology, biomathematics
  - Effluent quantification and monitoring
  - Environmental fate and transport
  - Exposure modeling, measurement, and monitoring
  - Pharmacokinetic and pharmacodynamic modelling

+

Health risk *management* decision making

- Decision analysis, decision sciences; Psychology of choice
- Fairness theory, risk equity and justice, political economy
- Cost-risk-benefit analysis, value of life, social utility theory
- Economics of information and incentives

+

Health risk *communication*

- Technical tools: Risk profiles, tornado diagrams, risk contours, etc.
- Psychological and cultural aspects of risk communication
- Organizational risk communication
- Societal communication and opinion networks and processes

This book explores the intersection of health risk analysis with these other areas. It describes techniques and models useful for risk analysts and shows how the methods and concepts applied in different parts of the risk assessment, communication, and management process depend on and reinforce each other. It summarizes progress to date and identifies pressing open questions in health risk analysis that benefit from and challenge the disciplines on which it draws. Throughout, special emphasis is given to problems that operations research and management science can help solve.

1.2 The Traditional Health Risk Analysis Framework

The discipline of health, safety, and environmental ("HS&E") risk analysis is traditionally presented as a sequence of steps, as in Table 1. Information produced at each step is passed to subsequent steps. Back-tracking and iterative refinement may occur as more information is gained. Quantitative risk characterization draws together all of the technical information, synthesizing hazard, exposure, and quantitative dose-response information into summaries of risk and uncertainty to inform risk management decisions. These decisions, in turn, may generate actions and information that lead to new information about hazards, exposures, and dose-response relations.

The initial steps, i.e., hazard identification, dose-response modeling, and exposure assessment, typically require the specialized substantive knowledge of engineers, epidemiologists, industrial hygienists, toxicologists, and other scientists and subject-matter experts. Risk quantification and characterization use statistical, probability, engineering, and simulation modeling methods to reduce this information to probabilistic summaries of the likely frequency and severity (and perhaps also the distribution in a population) of adverse health consequences from different risk management alternatives. It is now widely accepted that uncertainties about these components of risk must be included in the risk characterization.

Methods for reducing substantive knowledge to probabilistic summaries of potential adverse consequences and uncertainties are also commonly used in operations research. They are introduced in this chapter and examined more thoroughly in Chapter 2 and 4. Ideally, these summaries can be interpreted and used to guide decisions using the methods in Chapters 5-7, without requiring the specialized subject-matter knowledge that went into them. Risk communication (Chapter 8) applies results from the psychology of risk perception and choice under risk and uncertainty (Chapters 5-7) to frame and describe the risk information in ways that will help stakeholders and decision-makers to understand and internalize it.

Table 1: Steps in Traditional Risk Analysis Framework

Step	**Purpose and Description**	**Relevant information and techniques**
Hazard identification	Identify potential sources of harm or loss. These sources are called *hazards*. Hazard identification identifies possible adverse health effects of activities or exposures ("forward chaining") and possible causes of observed adverse effects ("backward chaining").	• Human data: Epidemiology, clinical and public health statistics. (See Chapter 2.) • Animal tests and bioassays • *In vitro* tests (See Chapter 8.) • Structure-activity patterns, molecular modeling, pattern recognition and statistical classification techniques
Quantitative exposure-response and dose-response modeling	Quantify the magnitude of risk created by exposure of a target to a hazard. Characterize the probable frequency and severity of adverse outcomes or losses caused by exposure to the hazard.	A quantitative risk assessment (QRA) runs multiple exposure scenarios through a *dose-response model* to predict likely health impacts. Statistical, simulation, or biomathematical models of biological processes are used to quantify dose-response relations.
Exposure assessment	Quantify the number of people (or other valued targets) receiving various levels or intensities of exposure to a hazard over time. Relevant exposure metrics may depend on dose-response relations.	Environmental fate and transport models, possibly summed over multiple media (paths) and sources Studies of human activity patterns Biological monitoring of exposed individuals and receptors
Risk characterization	Combine estimated probabilities and severities of losses, together with indications of uncertainty or confidence, to create an overall summary and presentation of risk.	Monte Carlo simulation calculates risks by sampling multiple scenarios. Risk profiles, probability distributions, tornado diagrams, and other trade-off and sensitivity analyses display risk, uncertainty, and variability.
Risk communication	Deals with how to present risk information to stakeholders. Considers how different types of recipients perceive risks and internalize/act on messages about them, in deciding what messages to send via what media.	Psychological theories and models and behavioral/experimental findings on risk perception and effective risk communication. (See Chapters 5 and 8.)
Risk management decision-making	Decide what actions to take to control risks and hazards – i.e., accept, ban, abate, monitor, further research, reduce, transfer, share, mitigate, or compensate.	Risk-cost-benefit analysis, formal decision analysis for groups and individuals, risk quantification and comparison (see Chapters 5-8.)

Thirty years of use have established the value of conceptual frameworks such as Table 1 to risk analysts and researchers, corporate and public risk managers, and regulatory agencies. However, the neat decomposition of risk analysis into well-demarcated sequential steps belies its iterative and exploratory nature.

Most health risks are highly uncertain. Recommendations to begin expensive risk management activities commonly trigger re-analyses of hazards, exposures, and plausible risk magnitudes as those who must bear the costs of risk management question the need for and health benefits from their efforts. Each round of risk characterization and risk management decisions can prompt new research into underlying causal mechanisms of hazard and details of exposure. The results may eventually lead to revised risk estimates and recommendations for risk management. For example, after years of risk research and analysis, the health risks from exposures to air pollutants, industrial chemicals (e.g., benzene, dioxin), electromagnetic radiation, radon, microbial hazards in food, and so forth remain controversial. Thus, in practice, the steps in the risk analysis process are tightly linked and improvements in information at one step may trigger further iterations elsewhere in the process.

To understand each part of the risk analysis framework, it is necessary to understand the other parts. The following sections introduce key risk assessment and risk management models, concepts and terms. They briefly discuss each of the steps in Table 1, emphasizing the types of models used and the information produced at each step. Subsequent chapters examine each step in more detail.

1.3 Defining Risks: Source, Target, Effect, Mechanism

A risk can be defined by answering the following four questions about its causal structure.

1. *What is the <u>source</u> of the risk?* This answers the question "Risk *from* what?" A source of risk is called a hazard. As yet, there is no generally accepted taxonomy or ontology of hazard types, specifying all and only those things that can be viewed as sources of health risks. However, depending on the risk management interventions being considered, the following are often identified or proposed as hazards:
   - *Individual activities, choices, and behaviors.* Legal and regulatory constraints on individual choices are sometimes imposed to reduce risks.
   - *Industrial processes,* such as production, transportation, storage, consumption, and disposal of products and materials.
   - *Economic and industrial activities,* operations, and facilities.
   - *Hazardous substances and materials* ("Hazmats").

- *Flows of energy and materials* in proximity to human populations.
- *Situations, events, and conditions* that arise only occasionally, e.g., earthquakes, tornados, floods, and other "natural hazards".

2. *What or whom is the target that is at risk?* This answers the question, "Risk *to* what or to whom?" Typical answers include individuals, classes of individuals (e.g., those named in a class action suit), and populations or sub-populations such as the elderly, children, asthmatics, or residents in certain neighborhoods.
3. *What is the adverse effect of concern that the source may cause in exposed targets?* This answers the question, "Risk *of* what?" Typical answers for health risks include increased frequency, probability, severity, or duration of undesirable health states; statistical life-shortening; and immediate death. Effects are often modeled as changes in the values of attributes describing health status (Chapters 6 and 7).
4. *By what causal mechanism, does the source increase the probability of the effect in exposed targets?* The question here is whether the relation between exposure to the source and increased risk of adverse health effects is causal. Does a causal path transmit changes in exposures to changes in health? Or is their association non-causal, e.g., because unhealthy persons coincidentally tend to be exposed? An acceptable answer might be quite vague or general (e.g., "A metabolite of the hazardous chemical binds to DNA and causes lesions that can lead to malignancy in affected cells"). In practice the biological mechanism(s) by which an adverse health effect is produced are often unknown. In such settings, hypothesized risks can still be identified and quantified, but there may be uncertainty about whether the resulting risk is real or is only a result of confounding and incomplete data analysis. Chapter 4 discusses various methods for deciding whether a causal relation is likely to be present even when its mechanism is not known.

Although every fully defined risk refers to a definite source, target, effect (or set of effects), and one or more causal mechanisms, the most useful definitions of these components may depend on the risk management decisions to be made. For example, health risks commonly arise from the juxtaposition of hazardous facilities or materials and target populations. Whether the source of risk is best identified as the location of a facility, how it is operated, or private decisions about whether to live near it depends on the purpose of the risk analysis. The answer may differ depending on whether the goal is to help decision makers choose a site for the facility, to advise and inform owners or potential purchasers of nearby residences, or to help local zoning boards select locations for facilities and residences. Similarly, the risks to consumers from a product may depend on both the manufacturer's care in making it and the consumer's care in using it. An employee's safety may depend on his own level of care (e.g., compliance with safety precautions and use of safety equipment such as masks) as well as on his employer's care in the design and conduct of operations. Chapter 8 considers ways to manage risks created by the joint decisions of several parties. The end of Chapter 4 addresses the related question of how to attribute a risk to the multiple risk factors that jointly create it, e.g., decisions by facility owners, neighbors and employees.

2. BASIC QUANTITATIVE RISK MODELS

Once the conceptual building blocks of a risk, i.e., its source-target-effect-mechanism (STEM) components, have been specified, the task of quantifying its magnitude can begin. The most common quantitative measures of risk use applied probability theory to describe how likely are adverse effects in targets, given characteristics of the hazardous exposure and of the targets. This section introduces basic probability models and decision analysis concepts, notations, and terms. These are fundamental for quantifying risk in all areas of applied risk analysis, including engineering safety and reliability, finance, and insurance, as well as health risk analysis.

Quantifying a risk consists of two stages: (a) Select one or more quantitative models to describe the risk; and (b) Estimate the unknowns in the model(s). The following paragraphs introduce mathematical models used to represent risks in many different situations. Chapters 2 to 4 describe techniques for estimating the unknown quantities in these models from data.

Like other areas of operations research, management science, and biostatistics, health risk analysis relies on specialized quantitative modeling techniques. Many of the same modeling issues arise in all of these areas. They include: how to create or *construct* risk models from knowledge, assumptions, and data (Chapter 4); how to *validate* and *refine* the models based on gaps between predicted and observed results (Chapter 3); how to *use* models to improve decisions (Chapters 5-8); and how to *manage* models and decisions over time as new data and knowledge become available. Partial answers to these general modeling questions can be developed independently of the details of the particular models being formulated, validated, used, and managed. To do so, it is convenient to use *directed acyclic graphs* (DAGs) to represent the computational or causal structures of risk models, highlighting input-output relations among their parts while suppressing details.

Example: DAG Structures of Risk Assessment Models

The input-output relations and computational logic of many complex risk models can be clarified by DAG diagrams that hide most details while showing the logical structure of what is calculated from what. For example, a simple health-risk DAG model might be represented by the following graph:

Decision → *Exposure* → *Health Consequences*

Each variable (i.e., *Decision, Exposure,* and *Health Consequences*) in such a graph has several possible values. In a deterministic model, each variable's value is determined by the values of the variables that point into it. In a probabilistic model, the probability distribution for the value of each variable is determined by the values of the variables that point into it. Variables with no inward-pointing arrows (such as *Decision*) are model inputs. The values of its inputs determine the (joint) probability distribution for the values of all other variables in a completely quantified probabilistic DAG model.

Models are typically composed of simpler sub-models, with the simplest ones consisting of probability laws, scientific and engineering laws and formulas, or empirical relations. For example, the preceding model is composed of two sub-models: an exposure model (corresponding to the link: *Decision* → *Exposure*) and a dose-response or exposure-response model (corresponding to the link: *Exposure* → *Health Consequences*.) The probabilistic relation between decisions and health consequences is denoted by:

Pr(Health Consequences | Decision),

where the conditional probability symbol "|" is read "conditioned on" or "given". In this notation, the calculations required to compose the two sub-models may be shown symbolically as:

$$Pr(\textit{Health Consequence} = c \mid \textit{Decision} = a) = \Sigma_x Pr(\textit{Health Consequences} = c \mid \textit{Exposure} = x) Pr(\textit{Exposure} = x \mid \textit{Decision} = a)$$

Here, c indexes possible health consequences, a indexes possible risk management decisions or acts, and x ranges over possible exposures. (Sums are replaced by integrals if the sets of alternative values require it.) This composition equation is equivalent to the diagram:

Decision → *Exposure* → *Health Consequences.*

Constructing exposure models to quantify the *Decision* → *Exposure* link, i.e., $Pr(\textit{Exposure} = x \mid \textit{Decision} = a)$, and constructing exposure-response models to quantify the *Exposure* → *Health* Consequences link, i.e., $Pr(\textit{Health Consequences} = c \mid \textit{Exposure} = x)$ are two of the principle activities required in health risk assessment.

If epidemiological data are available, it may be possible to estimate the end-to-end relation, *Decision* → *Health Consequences* directly from data. Then the estimated relation between decisions and consequences can be compared to the relation implied by the composition of the *Decision* → *Exposure* and *Exposure* → *Health Consequences* links to help validate the models.

2.1 Risk as Probability of a Binary Event

If the adverse effect of interest is binary – for example, occurrence or not of lung cancer before death from other causes – then its *probability of occurrence* is a natural quantitative measure of risk. It may be denoted symbolically as follows:

$p = Pr(E \mid x, z)$ = conditional probability that binary event E occurs in a target with characteristics or attributes z and with exposure x to the hazardous source.

Here,

- E is the binary event of interest, i.e., occurrence of the adverse effect in an exposed target.
- x and z are typically vectors of variables, referred to in statistical modeling as "independent variables" and "covariates", respectively.

x describes relevant aspects of the target's exposure to the source, such as the intensity and duration of exposure. z describes relevant aspects of the target, such as pharmacokinetic and metabolic parameters that determine internal doses from exposure histories, or response parameters that determine individual "frailty" or sensitivity to exposure.

The notation $Pr(E \mid x, z)$ is read as "The conditional probability that E occurs, given x and z." Considered as a function of x alone, with z held fixed, it is called a *quantal dose-response relation* or an *exposure-response relation,* depending on whether x summarizes dose (e.g., in an animal experiment) or exposure to the source. The modifier "quantal" distinguishes dose-response models with binary outcomes from dose-response models with continuous damage measures, survival times, or other non-binary outcomes.

Behind this conditional probability notation is a causal hypothesis that x and z jointly determine the probability of E. This implies that, once x and z are known, the values of their causal predecessors (e.g., the source strength, meteorological conditions, distance to the target, and so forth that led to the exposure history summarized in x) have no further relevance for determining the probability of E. More formally, E is said to be *conditionally independent* of the causal predecessors (if any) of x and z, given x and z, while x and z are said to be *sufficient for* all their predecessors, for purposes of determining the probability of E. Chapter 4 develops this concept further.

Even if the causal ordering of variables is unknown and the conditional independence relations among variables have not been determined, the conditional probability $Pr(E \mid x, z)$ can still be estimated from population data. As discussed in Chapter 2, the estimated numerical value for $Pr(E)$ will change as it is conditioned on different sets of information, but always in such a way that more information tends to bring the estimated value closer to a limiting true value.

The structure of the probability model $Pr(E \mid x, z)$ may be diagrammed as follows:

$$x \rightarrow p \leftarrow z \qquad \text{(A simple DAG model)}$$

In any such DAG graph, the value of a node depends on the values of its immediate parents but is conditionally independent of the values of its more remote ancestors (if any), given the values of its parents.

Much of the probability and statistical modeling discussed in this book, especially in Chapters 3 and 4, takes place in the framework of DAG models displaying hypotheses about the causal relations among variables. The basic idea has been specialized and developed in detail in several contexts, including the theories of Bayesian networks and causal graphs (Jordan, 1999; Jensen, 1996; Pearl 1988) and influence diagrams (Schachter, 1985). Chapter 4 introduces these concepts as they arise in risk analysis modeling.

Example: One-Hit Dose-Response Models

Suppose that a person or an animal inhales for an amount of time, T, a concentration c of a chemical that generates carcinogenic "hits" at an average rate of b hits per year per unit of concentration. "Hits" may be interpreted as interactions between metabolite molecules and cellular DNA or other target macro-molecules that will initiate cancer if the damage done is not repaired before a cell divides. Suppose that a single hit suffices for this purpose. Let the repair efficiency for the exposed individual be denoted by $1 - f$, meaning that a fraction f of all hits go unrepaired, and hence lead to cancer. The fraction f reflects the "frailty" of an individual, i.e., his or her susceptibility to hits. It might depend on individual diet, exercise, enzymology, gender, and other controllable and uncontrollable risk factors. b is interpreted as the *carcinogenic potency* of the chemical.

Under these assumptions, what is the probability that the exposure causes cancer in the exposed individual? Applied probability modeling (e.g., Ross, 1970) shows that the number of unrepaired hits accumulated by time T follows a *Poisson distribution* with intensity $bcfT$. Its probability density function (PDF) is:

$$Pr(k \text{ hits by time } T \mid b, c, f) = e^{-bcfT}(bcfT)^k/k! \qquad \text{(Poisson PDF, intensity } bcfT\text{)}$$

Hence, the probability of no hits by time T is:

$Pr(0 \text{ hits by time } T \mid b, c, f) = e^{-bcfT}$ (recalling that 0! = 1)

and so the probability of at least one hit by time T must be:

$p = 1 - \exp(-bfcT)$. (One-hit dose-response model)

This is called a *one-hit dose-response model* with observed exposure variables $x = (c, T)$ and unobserved parameters $z = (b, f)$. For fixed values of b, f, and T, the one-hit dose-response model reduces to:

$$p = 1 - \exp(-qc).$$

This has a single unknown parameter, defined as $q = bfT$. It is an example of a *reduced parameter*, meaning one that directly connects observed input and output quantities. By contrast, b and f are *structural* model parameters that refer to perhaps unobserved constructs in the causal process mediating between exposure and probability of response. The parameter q may be estimated statistically from data pairs $(c, n/N)$, where N = number of individuals exposed to concentration c and n = number of individuals who respond.

Example: Multistage Dose-Response Model

A frequently used generalization of the one-hit model for chemical carcinogens is the following *multistage dose-response model*:

$p = 1 - \exp[-(q_0 + q_1x + q_2x^2 + \ldots + q_mx^m)]$ (Multistage model)

In this model, x is a numerical summary of dose (most commonly defined as $x = cT$, i.e., the cumulative dose metric). The number of stages, m, and the parameter vector $q = (q_0, q_1, q_2, \ldots, q_m)$ are estimated from data by statistical methods such as maximum-likelihood estimation (Chapter 3). The multistage model is motivated by an underlying biomathematical theory of carcinogenesis in which cells must traverse m stages before becoming fully malignant. Transition rates are assumed to be linear functions of the exposure variable, x. These assumptions imply a relation between x and p that modelers often approximate by the multistage model formula. However, it is a poor approximation, in that the multistage model implies qualitative properties for the dose-response relation different from those implied by the underlying biological motivation (Chiu et al., 1999).

Example: Selling Prices Often Greatly Exceed Buying Prices

This example introduces some of the issues that make risk-cost-benefit analysis and public health risk management such challenging disciplines. Chapters 5-7 give many more examples in the context of a normative framework for decision-making.

Thaler (1992, p. 73) conducted an experiment in which subjects were told to suppose that they had been exposed to a rare fatal disease. They now had probability $p = 0.001$ of a painless death within two weeks, but this risk could be removed by immediately taking a vaccine. They were asked to decide the greatest amount they would be willing to pay for the vaccine to eliminate the 0.001 risk of death from the disease. These same subjects were also asked for the minimum compensation that they would require to participate in a medical experiment that posed a 0.001 probability of a quick and painless death. In other words, they were asked for the "selling price" at which they would be willing to accept an incremental risk of 0.001 of prompt mortality (in effect, selling some security for cash.)

Both willingness-to-pay (WTP) and willingness-to-accept payment (WTA) have often been proposed as ways to assign monetary values to non-market and intangible goods, including risks. In the Thaler experiment, however, *WTA exceeded WTP by more than a factor of 10* for most subjects. Thus, attempts to value a risk reduction or to estimate the perceived magnitude of a risk by expressing it in "equivalent" dollar terms, based on the preferences of those affected, must deal with the fact that WTP and WTA are often very different. WTA typically exceeds WTP by far more than would be predicted by economic models in which buying and selling prices diverge because of wealth effects or loss aversion (Rabin, 2000).

Other experiments have confirmed this pattern in many settings. For example, Viscusi *et al.* (1987) found that 77% of respondents recruited at a mall and a hardware store would refuse to pay *any* positive amount, however small, for a can of insecticide that posed an incremental risk of skin poisoning of 1/10,000. (The baseline was a \$10 can that posed a risk of 15/10,000.) On the other hand, these same subjects would be willing to pay only \$3.78 on average to eliminate the 15/10,000 risk entirely. So, the dollar value of a 1/10,000 incremental risk was generally greater than \$10 for WTA, while the dollar value of the 15/10,000 risk was on average less than \$4.00 for WTP.

These experiments indicate that even simple, binary risks to life and health – i.e., risks that are well described by a single probability number and a known health-loss or mortality outcome – are not easy to evaluate. Thaler (1992) suggests that responsibility for voluntarily assuming an additional risk, as opposed to merely avoiding an existing one, may impose extra psychological costs. Then, the value of accepting or avoiding a risk depends on more than just its probability and consequence, but also on responsibility and potential for blame or regret.

2.2 A Binary Event with Time: Hazard Rate Models

For some binary events, such as death, the key question that a quantitative risk assessment answers is not *how likely* the event is to occur, but *when* it is likely to occur. The most widely used probability tools for answering this question are *survivor functions* and *hazard functions.* Table 2 reviews these and closely related concepts. Each representation contains exactly the same information about when the risk event occurs, meaning that any one of them uniquely determines the others.

Table 2: Equivalent Mathematical Representations of Timing Information

Concept Name	**Mathematical Formula or symbol**	**Meaning**
Random variable	T	Time at which a binary risk event occurs (modeled as a random variable)
Cumulative density function (CDF)	$F_T(t) = Pr(T \leq t)$	Probability that the event occurs by time t
Probability density function (PDF)	• $f_T(t) = Pr(T = t)$ in discrete time • $f_T(t) = dF_T(t)/dt$ if time is continuous and $F_T(t)$ is absolutely continuous	Initial probability density for arrival at time t.
Survival function	$S_T(t) = Pr(T \geq t) = 1 - F_T(t)$ if $F_T(t)$ is absolutely continuous	Probability that event does not occur by t.
Hazard function	• $h_T(t) = f(t)/S(t)$ (continuous time) • $h_T(t) = Pr(T = t \mid T \geq t)$ (discrete time)	Arrival intensity of event at time t, given that it has not arrived before t.
Cumulative hazard function	$H_T(t) = \int_{[0,t]} h(s)ds$ if $F_T(t)$ is absolutely continuous	Cumulative hazard for arrival by time t.
Mean residual life function	$m_T(t) = E(T - t \mid T > t)$	Expected remaining time until arrival, given survival until time t.
Counting process	$N_T(t) = 0$ if $t < T$; else, $N_T(t) = 1$, for all times $t \geq 0$.	Random variable indicating whether arrival has occurred yet for each t.

The vocabulary and set of mathematical concepts in Table 2 is used to describe the probable timing of binary events throughout the applied sub-fields of quantitative risk analysis. The key concepts of *hazard rates* and *hazard functions, probability distributions, survival functions,* and *counting processes* are used by reliability engineers, actuaries, statisticians and biostatisticians, epidemiologists, economists, toxicologists, and clinical pathologists to describe, discuss, and compare risks. The corresponding framework of analysis is routinely applied to describe risks of human mortality, cancer incidence rates in laboratory mice, outages of industrial equipment or processes, accidents and service interruptions in complex engineering systems, progression of diseases ending in death, and failures of consumer products (Lawless, 1982; Andersen et al., 1993). The following discussion introduces some of the basic concepts and techniques.

Let T denote the time at which the binary event E occurs, and assume that the time axis starts at 0 = "now" and runs forward to infinity. (If E never occurs, then T is infinite.) The following function of time represents the probability that the event does not occur before time t. It is called the *survivor function* for T:

$S(t) = Pr(T \geq t)$, for all $t \geq 0$. (Survivor function for T)

In engineering risk assessment applications, where E is the event of system failure, $S(t)$ is also called the "reliability function". For simplicity, we will model T as a random variable with an absolutely continuous probability density function. Its cumulative distribution function (CDF), denoted by $F(t)$, is defined as:

$F(t) = Pr(T \leq t) = 1 - S(t)$, for all $t \geq 0$. (CDF of T)

The instantaneous intensity of occurrence of event E at any time $t \geq 0$, given that the target survives until t without E occurring, is given by the *hazard function, h(t),* formally defined as follows:

$h(t) = lim_{dt \downarrow 0}[Pr(t \leq T < t + dt \mid T \geq t) / dt] \ \forall t \geq 0.$ (Hazard function for t)

Interpreted, $h(t)$ is the expected number of occurrences per unit time of E (i.e., its expected arrival rate) at time t, given that it has not occurred before time t. The integral of $h(t)$ from 0 to t is called the *cumulative hazard* at time t and is denoted $H(t)$. In discrete time, the hazard function is defined as:

$h(t) = Pr(T = t \mid T \geq t).$ (Discrete-time hazard function)

Note that:

$$h(t) = Pr\{(T = t) \ \& \ (T \geq t))\} / Pr(T \geq t) = Pr(T = t)/S(t) = f(t)/S(t).$$

All of these functions may have T added as a subscript to form $S_T(t)$, $F_T(t)$, $h_T(t)$, $S_T(t)$, and so forth, as in Table 2, if there is any doubt about what random variable is being referred to. Where the random variable T is understood from context, the subscript need not (and in this book will not) be shown.

A simplification made in this discussion, but dropped later, is that there is a single event (e.g., death of an individual, failure of a system, detection of a tumor) of interest. It will be called the *risk event*. Therefore, in this section, risk can be characterized without considering the severity or magnitude of consequences associated with the risk event, but only the probability of its occurrence.

For the following examples and discussion, the reader is assumed to be familiar with basic probability theory, random variables, and stochastic processes at the level of Feller (1968), Ross (1970), or Grimmett and Stirzaker (1982).

Example: Basic Probability Calculations for Event with a Deadline

Problem: An event is certain to occur on one of the next five days and is equally likely to occur on any one of them. Find the PDF, CDF, survival function, and hazard function for the time, T, at which the event occurs, treating time as discrete (measured in days).

Solution:

1. The PDF of T is $Pr(T = t) = 1/5 = 0.2$ for $t = 1, 2, 3, 4, 5$. It is 0 otherwise.
2. The CDF of T has values of $F(t) = (0.2, 0.4, 0.6, 0.8., 1.0)$ for days $t = (1, 2, 3, 4, 5)$, a value of 0 for $t < 1$, and a value of 1 for $t > 5$. If $F(t)$ is plotted as a function of continuous time, t, then it will be a step function with steps at 1, 2, 3, 4, and 5 days and horizontal segments joining them.
3. The survival function for T has values of (1, 0.8, 0.6, 0.4, 0.2) on days (1, 2, 3, 4, 5), a value of 1 for $t < 1$, and a value of 0 for $t > 5$. Note that, in this discrete time model, $F(t) = 1 - S(t + 1)$ for $t = 1, 2, \ldots$, rather than $F(t) = 1 - S(t)$ as for continuous-time models.
4. The discrete-time hazard function is $\{h\} = (0.2, 0.25, 0.33, 0.5, 1)$ at the beginnings of days (1, 2, 3, 4, 5), respectively. Here, $\{h\}$ is a notation for a time series of values.

Converting Among Equivalent Mathematical Representations

Let T be a real-valued random variable with an absolutely continuous PDF. Then any of the quantities $F(t)$, $f(t)$, $S(t)$, $h(t)$, $H(t)$, $m(t)$, and $N(t)$ can be derived from any of the others using the following identities:

(a) $f(t) = F'(t) = -S'(t) = S(t)h(t)$, where a prime denotes first derivative

(b) $S(t) = 1 - F(t)$

(c) $h(t) = f(t)/S(t) = F'(t)/[1 - F(t)] = -d[\log S(t)]/dt = H'(t)$ for $S(t) > 0$

(d) $S(t) = \exp[-H(t)]$; conversely, $H(t) = -\log S(t)$

(e) $F(t) = E_F[N(t)]$, where $E_F[N(t)]$ denotes the expected value of $N(t)$ with respect to the probability density implied by F. When the distribution for calculating expectations, F, is understood from the context, then the simpler notation $E[N(t)]$ will be used.

(f) $h(t) = dE_F[N(t) \mid T \geq t] / dt$
$= \lim_{dt \downarrow 0} \{Pr[N(t + dt) = 1 \mid N(t) = 0]/dt\}$

(g) $m(t) = \int_{t \leq x \leq \infty} S(x)dx / S(t)$ and $S(t) = [m(0)/m(t)]\exp[-\int_{[0,\, t]} dx/m(x)]$.

These identities follow from the definitions in Table 2 by simple algebra and calculus; see Lawless (1982), pp. 8-9 and p. 44 or Lancaster (1990), pp. 6-13 for derivations.

The following heuristic interpretations may help to build intuition for these identities.

1. The formula in part (a),

$$f(t) = S(t)h(t),$$

implies that the probability that the risk event occurs between time t and $t + dt$, which is approximately $f(t)dt$, is the product of (i) the probability that it does not occur before time t, which is $S(t)$; and (ii) the conditional probability that it occurs between t and $t + dt$, given that it has not occurred before time t, which is approximately $h(t)dt$. Thus, the hazard rate may be interpreted as the instantaneous "arrival intensity" or "occurrence rate" for the risk event at time t, measured in units of expected occurrences per unit time. This also motivates part (f).

2. The formula in part (d),

$$S(t) = exp[-H(t)]$$

can be derived as the limit of the following discrete-time product identity:

$$S(t) = [1 - h(1)][1 - h(2)] \dots [1 - h(t - 1)].$$

This identity states that survival to the start of period t without occurrence of the risk event is equivalent to non-occurrence of the risk event in all of the previous periods. The construction of a "product integral" leading from this product to the continuous-time identity $S(t) = exp[-H(t)]$ as the length of the time steps goes to zero is given by Andersen et al. (1993), p. 90.

Although the different quantities described in Table 2 provide mathematically equivalent characterizations of risk (i.e., probability of the risk event over time), inasmuch as any one of them can be derived from any other, the hazard rate description may be seen as being most physically fundamental in many situations. Because it is an instantaneous rate, it can change from moment to moment to reflect causal influences such as the momentary intensity of exposure to a hazard. Once the hazard function has been determined from causal influences, the survival function and other descriptions can be obtained by integrating hazard rates to obtain $H(t)$ and then applying the preceding identities.

Example: The Weibull Hazard and Survivor Functions

Problem: Suppose that T is a random variable with hazard function:

$h(t) = (\lambda\beta)(\lambda t)^{\beta-1}$ for $t > 0$, where $\lambda > 0$ and $\beta > 0$ are parameters.

Find the survivor function and PDF of T.

Solution: Identities (c) and (a) above yield the answers:

$S(t) = exp[-H(t)] = exp[-(\lambda t)^{\beta}]$ for $t > 0$

$f(t) = h(t)exp[-H(t)] = (\lambda\beta)(\lambda t)^{\beta-1}exp[-(\lambda t)^{\beta}]$ for $t > 0$.

In this case, T is said to have a *Weibull distribution*.

2.3 Calculating and Interpreting Hazard Functions

The total risk to an individual of developing some undesirable response, such as death or cancer, in a given year *t* may be defined as the *probability* that the individual will develop the response in that year, assuming survival until then. In other words, if he is alive and healthy at the start of year *t,* then the probability that he will have developed the response by the end of year *t* may be defined as his risk of that response in year *t.* (This definition is for dichotomous responses. The more difficult problems of defining risk for a health response that may have varying degrees of severity or that is spread out over time are treated in Chapters 2 and 7.) This probability is also called the individual's (discrete-time) *hazard rate* for the response in year *t.* Hazard rates are used by actuaries, reliability engineers, biostatisticians, and epidemiologists to describe risks. In health risk assessment, hazard rates are used to calculate probabilities of cause-specific deaths or illnesses, to derive survival time distributions, and to quantify total risks over time.

Example: Calculating Hazard Rates and Probabilities of Causation

Problem: An individual is equally likely to die of cancer in any of the next 3 years. He has zero probability of surviving for more than 3 years.

(a) What is his discrete-time hazard rate for death due to cancer in each of the next 3 years if cancer is the only possible cause of death?

(b) Suppose that the same individual is also at risk from other causes (e.g., heart disease, car accidents, suicide) and that the total mortality hazard rate from all sources *other* than his cancer is 0.1 per year. If we learn that he dies in year 2 (but not why), then what would be the probability that the cause of his death is his cancer? This type of retrospective assessment of *probable causes* can be important in litigation when the actual cause of an event (e.g., the occurrence of a certain kind of cancer in someone exposed both to medical radiation and to radiation from nuclear weapons testing fallout) cannot be determined, but the approximate time of the occurrence can be.

(c) Under the same assumption of a 0.1 annual mortality rate from competing (non-cancer) causes, what is the probability that the individual will *not* die of cancer?

(This type of question is important to actuaries in forecasting likely claims experience for a population of policy holders known to be at risk.)

Solution:

(a) Suppose his hazard rate is a in year 1, b in year 2, and c in year 3. By definition, the probability that he will survive year 1 is $(1 - a)$, and his probability of surviving both years 1 and 2 is $(1 - a)(1 - b)$. The probability of surviving years 1, 2, and 3 is zero, by hypothesis, so $c = 1$ (i.e., there is 100% probability of death in year 3, given that he survives to the beginning of year 3). The probability that he will die in year 1 is a, by definition. The probability that he will die in year 2 is the probability that he will survive year 1 and then die in year 2; that is, it is $(1 - a)b$. The probability that he will die in year 3 is the probability that he will survive years 1 and 2; that is, it is $(1 - a)(1 - b)$. Hence, we have the system of equations $a = (1 - a)b = (1 - a)(1 - b) = 1/3$. Solving, we obtain $a = 1/3$, $b = 1/2$, and $c = 1$.

(b) Given that the individual dies in year 2, the probability that the cause of his death is cancer is the ratio of the cancer hazard rate in year 2 to the total hazard rate in year 2; thus, it is $0.5/(0.5 + 0.10) = 83.33\%$. (See Elandt-Johnson and Johnson, 1980, for a general discussion of probabilities of causation.)

(c) The probability that the individual dies of a competing cause before he dies of cancer can be found by calculating the probability that he dies in each year, weighting it by the conditional probability that he dies of a non-cancer cause if he does die in that year (thus obtaining the total probability that he dies of a non-cancer cause in that year), and then summing over all 3 years. The probability of death in year t is given by the product:

$$[1 - h(1)]\,[1 - h(2)]\ldots[1 - h(t - 1)]h(t),$$

where $h(i)$ denotes the total hazard rate in year i. (This is the probability of surviving all the years up through $t - 1$ and then dying in year t.)

The probability of non-cancer death, given death in year t, is $0.1/h(t)$; that is, it is the ratio of the hazard rate for non-cancer death in year t to the total hazard rate in year t. The prospective probability of death due to a cause other than cancer is therefore

$$(0.43)\left(\frac{0.1}{0.43}\right) + (0.57)(0.6)\left(\frac{0.1}{0.6}\right) + (0.57)(0.4)\left(\frac{0.1}{1.1}\right) = 17.8\%$$

(It is assumed that, in year 3, cancer and non-cancer causes of death compete in such a way that there is a chance of $0.1/1.1$ for the individual to die of non-cancer causes before he dies of cancer.)

In summary, discrete-time hazard rates allow survival time probabilities prospective (actuarial) probabilities of deaths or diseases due to specific causes, and retrospective (evidential) probabilities that a health effect was due to a specific cause to be calculated. Similar definitions and calculations can be made for continuous-time risks.

Example: Continuous-Time Hazard Functions

Problem: Suppose that a cancer patient is equally likely to die at any time between 0 and 3 years from now, and that she is certain to die within 3 years. Find her *continuous-time hazard function,* expressing instantaneous risk as a function of time.

Solution: The individual's time of death is uniformly distributed on the interval [0,3]. The probability that she will survive the first t years (where t is between 0 and 3) is initially given by the *survivor function* $S(t) = 1 - t/3$. For $t > 3$, $S(t) = 0$. The initial probability that her death will arrive between t and $t + dt$ is given by $f(t)dt = dt/3$, for sufficiently small dt. Here, $f(t) = 1/3$ is the probability density function for her time of death. Through the use of conditional probability, the probability that her death will arrive in the next time increment dt, given that she has survived until time t, is $f(t)dt/S(t) = dt/(3 - t)$. The ratio $h(t) = f(t)/S(t)$ is defined as the individual's *continuous-time hazard function.* In the present example, $h(t) = 1/(3 - t)$.

The $h(t) = f(t)/S(t)$ definition holds in general for absolutely continuous survivor functions. Intuitively, $h(t)dt$ for small dt is the probability that the individual will die between t and $t + dt$, given that she survives until t. $h(t)$ is a useful and intuitive quantitative measure of instantaneous risk. Continuous-time hazard functions can be used just like discrete-time hazard functions to calculate survival probabilities and prospective and retrospective probabilities of causation.

Defining total risk in a given year as *response probability* makes comparisons of individual risks difficult. For example, we might want to say that one individual is exposed to twice as much individual risk as another (perhaps in support of an argument about how compensation payments ought to be distributed). But probabilities do not support such comparisons. It is not true that a probability of 1 is only twice as great as a probability of 0.5; moreover, it is not clear what would be meant by a probability that is twice as great as a probability of 0.9. The probability scale is not cardinal in this way. Hazard rates provide a way of redefining risk to make such ratio comparisons meaningful. If the probability of response in year t is p, then we may define the transformed value

$$r(p) = -log(1 - p),$$

rather than p itself, as the risk in year t. ((Note: log x denotes the natural logarithm of x.) For small values of p, the values of p and $r(p)$ are numerically very close, with $r(p)$ being only very slightly larger. For example, when $p = 0.01$, $r(p) = 0.01005$. At $p = 0$, $r(p)$ is also zero. For large probabilities, however, $r(p)$ diverges from p more and more, approaching infinity as p approaches 1. For example, $r(0.5) = 0.69$, $r(0.635) = 1$, $r(0.8) = 1.6$, $r(0.9) = 2.3$, $r(0.99) = 4.6$, and $r(0.999) = 6.91$.

Interpretively, *r(p)* is the value of the continuous-time hazard rate, which, if maintained over an interval of unit length, produces a probability of response of p. This hazard rate, rather than the probability that it generates, provides the most natural quantitative measure of risk.

On the $-log(1 - p)$ hazard rate scale, unlike the probability scale, risk ratios (as well as sums and differences) can meaningfully be compared. In other words, *the continuous-time hazard rate scale is fully cardinal.*

Example: Comparing Risk Ratios

Problem: Factory A has a 90% probability of having an accident (a chemical spill) within a year. Factory B has twice as great an accident risk as factory A (so that on average, over the long run, twice as many accidents happen at factory B as at factory A per unit time.) Assume that both factories are in steady state, so that the average number of accidents per unit time remains constant at each factory. What is the probability that factory B will have an accident within a year?

Solution: An accident probability of 0.9 corresponds to a constant annual hazard rate (risk) of $-\log(1 - 0.9) = 2.3$. Twice this is 4.6, corresponding to an annual accident probability of $1 - \exp(-4.6) = 0.99$. Thus, factory B has a 99% probability of an accident within the next year.

Hazard rates are the appropriate measures of risk for use in risk-benefit ratios and other ratios used in risk management decision-making. When probabilities are very small, however, so that a probability and its corresponding hazard rate are numerically almost identical, either can be used for decision-making purposes.

Processes with constant hazard functions for the next arrival of an event play a special role in risk analysis. They are called *Poisson processes.* Poisson processes provide a yardstick for measuring the instantaneous hazard rates of more general processes. The $-log(1 - p)$ measure of risk is dictated by the mathematical properties of Poisson processes. The probability of surviving for T periods with no arrivals in a Poisson process with hazard rate h is e^{-hT} Hence, the probability of at least one response within T periods is $p = 1 - e^{-hT}$. Setting $T = 1$ and solving for h in terms of p in this formula gives $h = -log(1 - p)$.

The general importance of Poisson processes for interpreting quantitative expressions of health risks arises from the fact that any health effect occurring in an individual can be thought of as the first arrival in a random process. The risk to an individual at any moment refers to the probability of an arrival of a response (i.e., an adverse health effect) in the immediate future. In this "waiting-time" interpretation of individual risk, *h(t) is the average arrival rate at time t of responses in a Poisson process with constant hazard rate equal to h(t).* Thus, Poisson processes provide the key for interpreting hazard rates in terms of arrival rates. Specifically, *h(t)* is measured in units of average (i.e., expected) number of responses (or arrivals) per unit time at time *t*. Since *expected number of arrivals per unit time* defines an absolute scale, sums, differences, and ratios of hazard rates can meaningfully be computed, solving the problems that occurred in trying to compare probabilities.

Although it is convenient to interpret risk numbers in terms of expected annual frequencies, or average times until occurrence, intuitions about these quantities can be misleading. For example, if the expected frequency of accidents in a certain type of power plant is 1 in 1000 plant-years of operation, this does *not* imply that there is a 50% chance of an accident within 500 years. Nor does it imply that the probability of going for 1000 years without accident is 50%. (As the next example shows, it is only 37%.) Similarly, statements about the expected number of diseases per person-year or per lifetime of exposure can have counterintuitive implications for the lifetime probability of disease. Suppose that the leukemia hazard rate from exposure to benzene in the workplace were 1 expected case per million person-years of exposure, for example. Then the probability that a randomly selected person will develop cancer from a year of exposure would not be 1×10^{-6}, as might be expected, but 6.3×10^{-5}.

Example: Expected versus Actual Waiting Times in a Poisson Process

Problem: The continuous-time hazard rate for arrival of a catastrophic accident at an industrial plant is $0.01 = 1 \times 10^{-2}$ expected accidents per year. What is the probability that the plant will survive for 50 years without an accident?

Solution: Since the hazard rate is constant over time, this is an example of a Poisson process. The probability of no arrivals in an interval of length L = 50 in a Poisson process with average arrival rate $h = 0.01$ is given by $e^{-hL} = \exp[-(0.01)(50)] = e^{-½} = 0.61$. ($e^{-hL}$ is the survivor function for the Poisson process.) Thus, although the *expected* amount of time until an accident arrives is $1/h$ =100 plant-years, there is a 39% chance that the plant will survive less than half that long without an accident. The probability that no accident will arrive before the *expected* arrival time is only about 37% in any Poisson process. It is thus considerably more likely than not that the actual waiting time to the first arrival will be less than the average (or expected) waiting time.

In summary, Poisson processes provide a useful language for describing and comparing the magnitudes of health and safety risks. Three concepts are critical to the use of Poisson processes in risk assessment. First, the unit of risk is (expected) number of arrivals per unit time. Only for Poisson processes is the expected number of arrivals proportional to elapsed time, making the expected number of arrivals per unit time a well-defined number.

Second, for risk processes with time-varying hazard rates, the amount of risk generated between t and $t + dt$, for sufficiently small dt, namely $h(t)dt$, can be thought of as the amount of risk that would be generated by a Poisson process with average arrival rate $h = h(t)$ operating from t to $t + dt$. Thus, Poisson processes can be used to interpret general (non-constant) hazard functions. Finally, sums, differences, and ratios of risks can meaningfully be interpreted and compared when the Poisson risk scale, that is, expected number of arrivals per unit time, is used as the measure of risk.

Example: Risk Sums

Problem: An 89-year old male is exposed to two sources of risk: heart attacks, which arrive according to a Poisson process with rate $h_1 = 0.1$ arrival per year, and strokes, which arrive according to an independent Poisson process with rate $h_2 = 0.15$ arrival per year. What is the probability that he will survive for a year without experiencing either a heart attack or a stroke?

Solution: The superposition of the two risks is a Poisson process with hazard rate $h_1 + h_2 = 0.25$. The probability of surviving a year with neither sort of arrival is therefore $\exp(-0.25) = 0.78$. In general, the sum of two hazard rates corresponds to a superposition of two Poisson processes. The hazard rate consisting of the sum $h_1 + h_2$ may be thought of as the risk of a health effect generated by the first arrival from either of two competing (or superimposed) Poisson processes, one with average arrival rate h_1 and the other with average arrival rate h_2.

2.4 Hazard Models for Binary Events

To assess the risk to a target from exposure to a hazardous source, it is common to use a *hazard model.* This expresses the hazard function as a function of not only time, but also exposure conditions and covariates describing the target. Such a hazard function model may be symbolized as follows:

$$h(t \mid x, z). \qquad \text{(hazard model)}$$

Diagramatically, its structure is:

$$x \rightarrow h(t \mid x, z) \leftarrow z \qquad \text{(DAG for hazard model)}$$

In words, a hazard model expresses the instantaneous hazard rate at any time as a function of two sets of input quantities: exposure variables and the target's covariates. These may vary over time, as in the examples of age and exposure duration. The symbols x and z are then interpreted as denoting the *histories* (i.e., the time series) of values for the exposure variables and covariates up through and including the time for which the hazard rate is being assessed.

Example: The Proportional Hazards (PH) Model

Biostatistical hazard models often fit parametric hazard functions to available data without first doing any detailed biological modeling. A frequently used statistical hazard model is the *proportional hazards (PH)* model, which may be expressed as follows:

$$h(t \mid x, z) = exp(qx)h(t \mid z). \qquad \text{(Proportional hazards model)}$$

In this model, $h(t \mid z)$ denotes the hazard rate at time t for an unexposed target of "type" z (i.e., having individual characteristics represented by covariate vector z.) It is called the *baseline hazard function,* since it gives the hazard rate for an unexposed individual. This function of time need not be explicitly specified, making the PH model "semi-parametric" (it involves the parameter q but leaves the function $h(t \mid x)$ unspecified). The actual hazard rate for an individual exposed to x is proportional to the baseline hazard rate, with proportionality constant $exp(qx)$. The parameter vector q is usually estimated from survival data in which the values of x and z are recorded for each individual over some interval of time, possibly different for each individual. The occurrence time of the risk event (if it occurs during the observation interval) is also recorded for each individual. If $q = 0$, then exposure has no effect on risk, and this is frequently a hypothesis to be tested.

Example: A Biologically Motivated Hazard Model

Unlike purely statistical models, biologically motivated ones attempt to use biological knowledge to guide the formulation of model equations and to suggest plausible forms for the functional relations among variables in a model. In many recent biomathematical models of cancer causation, the hazard rate for the first occurrence of a malignant cell in an organ at time t, if one has not been created before then, is modeled as follows:

$$h(t) = \mu_2(t)I(t) \qquad \text{(Biologically based hazard model)}$$

where

- $I(t)$ = number of premalignant (or "initiated") cells at time t
- $\mu_2(t)$ = transformation rate (per initiated cell per unit time) of initiated cells to malignant cells.

Both quantities typically depend on the history of exposures and on covariates determining cytotoxic (cell poisoning) responses and cell transformation rates from exposure histories. Even though estimates of these quantities based on data may be uncertain, the model may still be used to help develop theories and to make predictions about how cancer risk is likely to develop over time, provided that additional models are introduced to predict $\mu_2(t)$ and I(t).

The complete causal structure of a biologically-based risk assessment (BBRA) model of chemical carcinogenesis can be summarized in DAG form as follows:

$$\{x\} \rightarrow \{y\} \rightarrow \{\boldsymbol{\mu}, \mathbf{N}\} \rightarrow \{\mu_2, I\} \rightarrow \{h\} \rightarrow \{p\} \quad \text{(BBRA model DAG)}$$

The sub-models that are composed to form the BBRA model are as follows:

Component Models of a Biologically-Based Cancer Risk Model

- **PBPK model:** $\{x\} \rightarrow \{y\}$. The notation is as follows: {} = time series, {x} = time series of administered doses, {y} = time series of the vector of internal metabolites, i.e., {y} is an abbreviation of $\{y(t), t \geq 0\}$ (PBPK output).
- **Cytotoxicity/cell kinetics model:** $\{y\} \rightarrow \{\mathbf{N}\}$, **N** = vector of cell population sizes (including stem cells that can become malignant).
- **Genotoxicity/cytogenetics model:** $\{y\} \rightarrow \{\boldsymbol{\mu}\}$, $\boldsymbol{\mu}$ = vector of cell transformation rates. In stochastic two-stage models of carcinogenesis, $\boldsymbol{\mu} = (\mu_1, \mu_2)$.
- **Cell initiation model:** $\{\mathbf{N}, \boldsymbol{\mu}\} \rightarrow \{I\}$, I(t) = size of "initiated" cell population at time t, $dE[I(t)]/dt = [b(t) - d(t) - \mu_2(t)]I(t) + \mu_1(t)N_1(t)$.
- **Hazard rate model:** $\{\mu_2, I\} \rightarrow \{h\}$, {h} = arrival rate of first malignant cell, $h(t) = \mu_2(t)I(t)$.
- **Time-to-tumor model:** $\{h\} \rightarrow \{p\}$, $p(t) = 1 - \exp[-\int_{[0,t]}h(s)ds]$

If a hazard model is given, then the probability that the event E ever occurs can be obtained immediately as:

$$p = \Pr(E \mid x, z) = 1 - S(\infty) = 1 - \exp[-H(\infty \mid x, z)]$$

where $H(t \mid x, z)$ is the cumulative hazard function for $h(t \mid x, z)$. In this sense, survival models subsume models of binary events without timing.

2.5 Probabilities of Causation for a Binary Event

Using hazard functions to quantify the instantaneous risk of occurrence of an event leads to a useful extension to multiple sources. Suppose that a binary event (e.g., lung cancer, system failure) can be caused by any of several mutually exclusive, non-interacting sources or causes. Let $h_i(t)$ denote the hazard function for occurrence of the event at time t due to cause or source i. Then the total risk of occurrence at time t, which we continue to denote by $h(t)$, is just the sum of these *cause-specific hazard rates*, $h_i(t)$:

$$h(t) = \Sigma_i h_i(t).$$

The probability that the event is caused by source i at time t, if it in fact occurs at time t, is just $h_i(t)/h(t)$. This is sometimes called the *probability of causation*, i.e., the probability that source i caused the event.

The concept of total individual risk for a dichotomous health response at a single moment in time, say t, has now been reasonably well specified. It is the expected number of arrivals per unit time evaluated at moment t, where the first "arrival" corresponds to occurrence of the health effect. In many applications, however, the problem is not to estimate the *total* risk to which an individual is exposed for some health response but to estimate the *incremental contribution* to his or her risk made by some particular cause or source. This is the risk that is said to be *attributable* to the source. For example, in assessing the individual risk of lung cancer attributable to cigarette smoking, one must subtract out, or otherwise adjust for, lung cancers that would have occurred even without cigarette smoking: only the incremental risk above this base line is to be attributed to cigarettes.

Example: Epidemiological Calculation of Attributable Risk

Problem: Suppose that the hazard rate for leukemia among the children in town A is 0.001 case/child-year, while in town B it is 0.002 case/child-year. If the children are thought to be identical, except that those in town B have been exposed to drinking water from a contaminated well, then what would be the quantitative risk of child leukemia attributed to drinking of the contaminated water?

Solution: The attributable risk would be 0.002 - 0.001 = 0.001 case/child-year of exposure. Out of every thousand cases of child leukemia in town B, it is expected that 500 would have occurred even if the drinking water had not been contaminated. Thus, 500/1000 = 0.5 is the *attributable proportion of risk* due to the contaminated well water.

Few concepts in risk analysis have occasioned as much perplexity and debate as that of the attributable risk due to a source. The random arrival model of risks clarifies the key elements of this debate. Suppose that N *sources* or *causal factors* contribute to the risk of a health response in an individual (e.g., cigarette smoking, radiation, and diet in the causation of cancer). Each source can be thought of as firing a random stream of tiny invisible bullets, or "hits", at the exposed individual. (Hit and arrival are synonyms). The average arrival rate or *intensity* of hits from source i at time t is source i's hazard rate at time t, denoted by $h_i(t)$. As usual, the *first* hit or arrival from any source is assumed to correspond to occurrence of the health response. The source of the first hit may be identified as the *cause* of the response. Suppose that a health response does occur at time t. Without knowing which source generated the hit that caused the response, one can still calculate *probabilities of causation* for each of the N potential causes.

Example: Probability of Causation Calculations for Competing Risks

Problem: Suppose that a worker in a uranium mine has a hazard rate for lung cancer of 0.002 expected case/year from all non-occupational sources, and a statistically independent hazard rate of 0.001 case/year from occupational exposure to radiation. If he does get lung cancer, then what is the probability that it was caused by occupational exposure?

Solution: Under the stated assumptions of additive hazard rates from statistically independent occupational and non-occupational sources, the probability of causation for occupational exposure is just 0.001/(0.001 + 0.002) = 33.33%; that is, it is the ratio of the cause-specific hazard rate to the total hazard rate. Note that this formula, which has been proposed for use by the federal government in compensating possible victims of nuclear weapons testing fallout depends critically on the assumption of statistically independent, non-interacting, hazard rates. The biological realism of this assumption is open to question.

To recapitulate, if N potential sources of a health effect are statistically independent, so that the arrival rate from one source is unaffected by the presence of other sources, then the total risk to the individual at time t from all sources is given by the sum $h(t) = h_1(t) + \ldots + h_N(t)$. The contribution to this total made by source i is $h_i(t)$, and this is the amount of risk, measured in expected arrivals per unit time, attributable to source i. Since source i contributes the fraction $h_i(t)/h(t)$ of all expected hits per unit time at time t, this is the probability that source i contributed the hit that caused the observed health response. If the time of this hit is unknown (e.g., because the expression of its effects as an observable health response is delayed by a latency period of uncertain length), then more complicated formulas for probability of causation can be developed.

Example: Relative Risk and Probability of Causation

Problem: Let an individual's age-specific background risk of leukemia be given by a hazard function $h(t)$. According to the *relative risk hypothesis,* occupational exposure to a dose x of a leukemogen (e.g., benzene) or of radiation has the effect of multiplying the hazard function by a constant, $(1 + bx)$, where b is the potency parameter.

(a) If the relative risk model is correct for benzene, and if a 60-year old worker develops leukemia after being exposed to 1 ppm of benzene 8 hours a day for 40 years, then develop a formula for the probability that it was the benzene exposure that caused his leukemia.

(b) How might a court use such information to make compensation decisions in cases where the plaintiff has suffered a probabilistic harm but is unable to offer definite proof of the cause of his injury (e.g., leukemia)? (For this problem, assume that tort litigation rather than Worker's Compensation is being sought as a remedy.)

Solution:
(a) Under the relative risk model, the worker's background age-specific hazard rate and the age at which he develops a response (here, leukemia) are irrelevant. All that matters is the product bx at the time of response. Since leukemia has a relatively short latency period, e.g., 11 years, its time of occurrence can often be estimated accurately. If the time of onset is t and the dose received as of time t is $x(t)$, then the probability that the worker's leukemia was caused by exposure to benzene (assuming competing risks) will be

$$p = bx(t)/[1 + bx(t)].$$

If $x(t)$ and b are known, then probability of causation can be determined. In practice, application of this formula requires a biologically meaningful measure of the dose $x(t)$, estimation of the potency coefficient b, and acceptance of the relative risk model for leukemogenesis.

(b) Suppose that the court uses probabilistic evidence to try to minimize the expected error in its compensation decisions. Specifically, suppose that if the cause of the plaintiff's leukemia (benzene exposure or background) were *known,* then the defendant would be required to pay him a compensation award of C or 0 if benzene exposure or background, respectively, were the cause. When the cause of the plaintiff's injury is uncertain, the court tries to assign a compensation amount that minimizes the expected cost of error. If the probability that benzene exposure caused the leukemia is p, and if the court awards an amount A between 0 and C, then the expected error in compensation is

$$J(A, p)=p(C - A)+(1 - p)A;$$

that is, it is the underpayment if the defendant is guilty, $C - A$, times the probability that he is guilty, p, plus the amount of overcompensation if the defendant is innocent, A, times the probability that he is innocent, $1 - p$. To minimize this expected error, the court should adopt the following rule: *if* $p > 0.5$*, then award full compensation,* $A = C$*. If* $p < 0.5$*, then award zero compensation.* This corresponds to the *preponderance of evidence* standard in tort law. More generally, if every dollar of compensation mistakenly withheld is considered k times as bad as a dollar mistakenly awarded, then the rule becomes: award full compensation if $p > 1/(1 + k)$; otherwise, award nothing.

The above definitions of attributable risk and probability of causation are satisfactory for the random arrival model of causation. In the epidemiological and biostatistics literature, this is called the *competing risk model:* each source is regarded as competing with the rest to be the first to cause a health response. If it holds, then the risk increments contributed by different sources can be distinguished and displayed, at least in principle, as in Table 3.

Table 3. Risks that Increase Probability of Death by One in a Million

Activity	**Cause of Death**
Smoking 1.4 cigarettes	Cancer, heart disease
Drinking 1/2 liter of wine	Cirrhosis of the liver
Traveling 10 miles by bicycle	Accident
Traveling 300 miles by car	Accident
Flying 1000 miles by jet	Accident
Flying 6000 miles by jet	Cancer from cosmic radiation
Living 2 months in average brick building	Cancer caused by natural stone or radioactivity
One chest x-ray taken in a good hospital	Cancer caused by radiation
Living 2 months with a cigarette smoker	Cancer, heart disease
Eating 40 tablespoons of peanut butter	Liver cancer from atlatoxin B
Eating 100 charcoal-broiled steaks	Cancer from benzopyrene
Drinking thirty 12-oz cans of diet soda	Cancer caused by saccharin
Living 20 years near PVC plant	Cancer caused by vinyl chloride (1976 standard)
Living 150 years within 20 miles of a nuclear power plant	Cancer caused by radiation
Risk of accident by living within 5 miles of a Nuclear reactor for 50 years	Cancer caused by radiation
Living 5 years at site boundary of a typical nuclear power plant in the open	

Source: Adapted from R. Wilson, Analyzing the risks of daily life, *Technology Review,* 81, (1979). See also R. Wilson and E. A. C. Crouch, Risk assessment and comparison: an introduction, *Science,* 236, 267-270 (1987).

However, the competing risk model is not applicable if either:

1. Causes are not mutually exclusive. This could happen, for example, if a response is triggered by a combination of multiple hits, possibly of different types or sources, within a certain time of each other.
2. Causes are not statistically independent. This happens if the arrival rate of hits from one source, or their success in doing biological damage (and so their *effective* hit rate), depends on what other sources are present.

If the competing risk model does not hold, then it may not be clear what numbers such as those in Table 3 mean, since part of the risk attributed to a cause may be due to other causes that interact with it.

Example: Interacting Causes

Problem: Suppose that the background hazard rate for lung cancer for a particular individual, in the absence of cigarette smoking or occupational exposure to radiation, is 0.001 expected case per year. If he smokes, this rate increases to 0.003. If he works in a uranium mine, it is 0.002. If he smokes *and* works in a uranium mine, it is 0.006: thus, smoking and exposure to radiation interact multiplicatively in increasing his risk of lung cancer. Assuming that he does both, so that his total risk is 0.006, what is the lung cancer risk due to (i.e., attributable to) smoking?

Solution: Since the risk would be only 0.002 without smoking and is 0.006 with smoking, the attributable risk due to smoking might be calculated as 0.006 - 0.002 = 0.004, and the attributable proportion of risk for smoking as 0.004/0.006 = 67%. A difficulty with this approach is that applying the same formula to the radiation exposure would give (0.006 - 0.003)/0.006 = 50% as its share in causation; the two assigned shares, of 67% and 50%, however, sum to more than 100%. Thus, this approach tends systematically to inflate the proportion of risk attributed to each factor if the factors interact synergistically. An approach that avoids such difficulties notes that smoking increases risk by 0.003 - 0.001 = 0.002 in the absence of radiation exposure and by 0.006 - 0.002 = 0.004 in the presence of radiation exposure. Its average effect, therefore, averaged over the presence and absence of the radiation factor, is to increase risk by (0.002 + 0.004)/2 = 0.003, and this can be defined as the risk attributable to smoking in this example. A similar calculation for radiation again gives 0.003 as its attributable risk; thus, each factor is assigned a 50% share in causation. More generally, if there are N factors, then averaging the incremental contribution to total hazard rate made by the ith factor over all the $N!$ possible orderings of the N factors gives a measure or definition of the risk attributable to factor i. Chapter 4 provides theoretical justification for this measure. For competing risks, averaging risk increments over factor orderings gives the same result as the probability of causation formula for the proportion of risk attributed to any specific factor.

2.6 Risk Models with Non-Binary Consequences

When the risk event is not dichotomous, but instead has multiple levels of severity, hazard functions alone no longer suffice to quantify risk. Models must be developed that account for the probable severity as well as the probable timing of a risk event and that allow multiple events over time. Table 4 summarizes some common risk models. Models 1-4 have already been discussed. Models 5-13 deal with non-binary consequences.

Roughly speaking, the agenda of much of applied risk assessment consists of the following steps:

(a) Determine which of the models in Table 4 (or possibly other models) best represents the risk to be studied.
(b) Estimate the models from available data and knowledge.
(c) Determine how each of the risk management options being considered will change the joint frequency distribution of individual risks.
(d) Decide which frequency distribution of individual risks is most preferred. Choose the corresponding risk management option.

Chapters 2 to 4 discuss topics (a) to (c), while chapters 5-8 address topic (d) for individual and multiple decision makers.

Table 4: Summary of Common Quantitative Representations of Risk

	Risk Representation	Name	Interpretation
1	p	Bernoulli trial	p = probability of a binary event
2	h(t)	Hazard function	h(t) = hazard rate at time t for occurrence of a binary event
3	• h(t | x, z) • S(t | x, z)	• Hazard model • Survival model	• h(t | x, z) = hazard function • S(t | x, z) = prob. of surviving until time t without event occurring, given exposure x and covariates z.
4	L= (p, c; 1- p, 0)	Binary prospect (see Chapter 55)	Prospect yields outcome c with probability p, else 0. Usually, c is measured in natural units such as dollars or life-years.
5	L = (c, p) = $(c_1, p_1; c_2, p_2, \ldots, c_n, p_n)$	Prospect (Chapter 5)	Prospect yields outcome c_j with probability p_j. c and p are vectors.
6	L = (**c**, p) = $(\mathbf{c}_1, p_1; \mathbf{c}_2, p_2, \ldots, \mathbf{c}_n, p_n)$	Multi-attribute prospect (Ch 7)	Prospect yields outcome vector $\mathbf{c}_j$ with probability p_j.
7	L = $(L_1, p_1; L_2, p_2, \ldots, L_n, p_n)$	Compound lottery (Chapter 5)	Lottery L offers lottery L_j as outcome with probability p_j. ("Degenerate" lotteries with probability 1 for specific consequences are special cases.)
8	(h, F)	Compound risk model (See Chapter 3)	h(t) = hazard function for occurrence of risk event. F(c | t) = CDF or risk profile for the magnitude of its consequences if the event occurs at time t.
9	(h, F)	Renewal reward process	The random time ("cycle length") between renewals has hazard function h(t). F(c | t) is the CDF of the reward received from a cycle of duration t.
10	E[h(t) | Info(t)]	Stochastic hazard function	h(t) is a random variable. E[h(t)| Info(t)] is its expected value, conditioned on the information (prior probability measure and observations) in filtration (observed history) Info(t).
11	(h, P, **v**)	State transition model with rewards (Chapter 7)	h is a matrix of state transition probabilities. P = initial state probability vector. **v** = vector of value per unit time for being in each state.
12	[h(t | z), G(z)] = (h, G)	Population risk model (Chapters 2, 3)	h is an individual risk model with inputs z. G(z) is the joint CDF of z in the population.
13	(**p**, **H**)	Population risk model for sets of individuals (Chapter 5)	• **p** is an S-vector of scenario probabilities. • **H** is an S x N array of individual risks specifying the risk to each individual, i = 1, 2, …, N, from each scenario, 1, …, S.

3. HEALTH RISKS FROM HUMAN ACTIVITIES

The inputs to a health risk model are estimated individual or population exposures and individual characteristics or covariates. Its outputs are estimated health risks, typically described by probabilities or as expected transition rates among health states with different values. For purposes of health risk management decision-making and policy-making, such models must be embedded in a larger causal framework that has alternative decision options as inputs and evaluations of their probable consequences as outputs. The consequences of interest typically include not only health outcomes, but also economic impacts of the activities that create exposures, as well as evaluations of the fairness, legitimacy, and performance of the risk management decision process.

The larger causal process addressed by public health risk managers and policy-makers may be schematized as follows. First, a regulator, legislator, court, or other social decision-maker (SDM) specifies a set of *rules* that the individual agents in the society must follow. The agents are consumers, producers, employees, neighbors of production processes – in short, the members of society, cast in stereotyped roles. Of course, both the concept of a single SDM and the idea of well-defined agents playing single, well-defined roles are only convenient fictions, useful for understanding some of the main aspects of health risk management. Rules, in conjunction with monitoring, reporting, and enforcement provisions, map agent actions and outcomes to penalties and payments. Agents take into account the rule-generated consequences of their actions as well as other consequences due to the operation of causal laws, in deciding what to do. The SDM seeks to design and enforce rules – e.g., civil and criminal liability rules, compensation systems, social insurance provisions, and so forth – that will elicit desirable behaviors and promote desirable outcomes, as evaluated by the SDM or by the agents themselves.

Chapter 8 discusses the SDM's rule-design and process-design problems. It also identifies other possible roles for the SDM. These include being a credible information source, arbitrator, contract enforcer, funder of socially worthwhile projects when private incentives do not suffice (e.g., because they do not capture positive externalities), insurer of last resort, and representative of the interests and preferences of multiple stake-holders in social choice processes. Social risk management choices usually require balancing the multiple, partly conflicting interests of different stakeholder groups. Siting decisions for hazardous waste sites, hazardous production and storage facilities, HAZMAT transportation routes, or health care facilities bring the interests of different parties into especially sharp relief. Chapters 7 and 8 present principles and techniques for making such difficult decisions, as well as for deciding about which public health projects to fund (if any) when their health benefits are uncertain and when different parties may receive their costs, risks, and benefits.

3.1 Risk Management Decision-Support Sub-Models

Models that support social risk management decision-making are typically composed of several sub-models linked so that the outputs from one are inputs to others. The most common sub-models are as follows:

Sub-Models Commonly Used in Risk Management Decision-Making

- ***Risk management decision models.*** These map predicted probable health and other consequences of SDM decisions into choices of which decisions to make. (See Chapters 5-8).
- ***Behavioral decision models*** map SDM decisions and other information into decisions and/or behaviors of affected individuals.
- ***Exposure models*** map individual decisions and behaviors to resulting exposures (Chapter 2)
- ***Dose-response models*** map exposures to probable health consequences. (See Chapters 3 and 4)
- ***Health impact models*** map probable health consequences, together with information about individual health care, insurance, litigation, and other decisions, into a set of health, financial, and other impacts on individuals and society.

They are described next. To a first approximation, each of these sub-models provides information to the next, with the cycle being closed when dose-response and health impact models are used to inform social risk management decisions.

Behavioral and prescriptive decision models

Behavioral decision models map the rules and information provided by the SDM into agent acts, corresponding to individual choices and behaviors by consumers, producers, distributors, and other affected agents. For example, an SDM's rule mandating a speed limit of 65 miles per hour on interstate highways will elicit a frequency distribution of speeds at which individual drivers actually drive. For brevity, we will represent behavioral models by the mnemonic diagram:

SDM acts $\rightarrow$ *agent acts.*

In reality, agents will typically consider information and consequences other than those specified by the SDM in deciding what to do. Moreover, the link between SDM acts (e.g., providing information, setting standards and constraints) and resulting behaviors is complex. It involves aspects of risk perception, assimilation and internal coding of risk information; formation of beliefs, preferences, and intentions; attitudes toward the source of risk information as well as toward the type of risk involved (e.g., voluntary vs. involuntary, natural vs. technological, and so forth). Chapter 2 reviews key portions of the psychology of risk perception and risk communication, as well as behavioral decision-making under risk. From the SDM's perspective, however, what matters most is the conditional frequency distribution of agent choices or behaviors, given the SDM's decision. The abbreviation *SDM acts* $\rightarrow$ *agent acts* will symbolize this relation.

In addition to behavioral decision models, a rich theory of normative, or prescriptive, decision making can help to clarify what choices *should* be made from among competing risk management alternatives. Chapters 5 to 7 discuss some of the best-developed prescriptive decision theories for health risks, as well as formal definitions and models of risk to which they lead.

Chapter 8 examines decision theories for groups of individuals. Such theories cover decisions made through processes in which multiple agents participate and possibly interact. Table 5 sketches a partial taxonomy of decision settings that suggests the richness of the area. In this table, decision-makers are those who choose actions or behaviors that create health hazards, while stakeholders are those who bear (or care about) the health or other consequences of the decisions. These coincide in some decision settings, but not in all of them. The taxonomy is only partial since the different categories are not mutually exclusive and are not completely exhaustive. For example, distributed hierachical decision-making may involve many local decisions that can fruitfully be analyzed using some of the other models in the table.

A full discussion of each of the decision settings in Table 5 would encompass theories and evidence from economics, law, ethics, and management science that stray far from the main themes of this book. Chapters 5 through 7 introduce behavioral and normative decision theories for individual decision-making about risky prospects. Chapter 8 surveys key ideas from theories and models of decision-making about risks with multiple decision-makers and stakeholders. They emphasize empirical findings and prescriptive theories about how decisions are or should be made under risk, leaving other questions of economic efficiency and evaluation of decision processes to the large literatures on each decision setting in Table 5.

Table 5: A Partial Taxonomy of Risk Management Decision Settings

Decision Setting	**Frameworks theories, and models**	**Examples**	**Comments**
Private individual risk choices	Behavioral and normative decision theories	Recreation, diet, active smoking	Single decision-maker (d.m.) bears consequences of own decisions.
Private decisions with externalities	Tort law, Coase Theorem (Chapter 8)	One polluter, mutiple neighbors	Others bear some of the consequences of the single d.m.'s acts.
Centralized decisions	SDM decision theory, social utility theory	Regulatory agencies (EPA, OSHA, CPSC)	Social decision maker (SDM) sets policies that affect many agents.
Private binary transactions	Contract and tort law, principal- agent models, bilateral bargaining/ negotiation models and game theory	Consumption, occupation, home-buying	Two parties (seller and buyer, employer and employee) engage in transactions that may create hazards (e.g., from consumer products, occupational hazards).
Decentralized risk decisions by many individuals	Political economy models, theories of evolutionary games and equilibria	Wood burning, vaccination of children, safe sex behaviors	Individual behaviors affect multiple other individuals, depending on how they choose to behave.
Collective decisions via political processes	N-person game theory, political economy models, collective choice and social utility theories Cost-risk-benefit analysis, economic theories of regulation	Siting and zoning processes for hazardous facilities Regulation of exposures to environmental tobacco smoke or to chemicals	Multiple interest groups must agree to reach a decision. Subgroups and coalitions may block or delay agreement if they have enough power. Stakeholders, decision-makers, and other participants in the process may overlap.
Small group decisions	Club theory, theory of syndicates, team theory	Deciding whom to admit to an insurance pool.	Several parties must decide what to do and how to decide.
Arbitrated agreements	Law and economics game theory models; conflict resolution and mediation models	Decision to litigate. Settlement out of court, with threat of court trial as no-agreement default.	Two parties may either reach agreement themselves or ask a third part to choose an outcome for them.
Regulated industry	Optimal monitoring and enforcement models; hierarchical control models	Inspection and grading of foods, standards and certification processes	Regulator's actions for specific cases and companies help to establish a reputation that may affect the actions of other regulated companies in the same industry.
Hierarchical and distributed decision-making	Decentralized team theory, mathematical organization theory	Implementation of national environmental audit and clean-up programs through regional and local administrations	Each decision-maker in the hierarchy may have some local information and scope of decision-making authority and be evaluated and funded based on contributions to higher level goals.

Risk management decisions made by individual agents include all of the following: choosing what levels of care to take in producing or in using a consumer product; where to locate and how to operate hazardous facilities; levels of care to use in recreational activities; use of health care and medical services to screen for and treat health damage; how much and what kinds of insurance policies (against health damage and liability suits) to purchase; whether to bring suit for injuries when damage occurs, and so forth. All of these are included in the category of agent acts. However, some individual decisions and behaviors, especially those determining the locations of emissions sources and of exposed individuals and the amounts of emissions produced, provide direct inputs for exposure models.

Exposure models

Exposure models map certain agent decisions, such as design choices for facilities, heights of stacks on power plants, investments in scrubber or other cleaning technologies, and so forth, into resulting exposure patterns for individuals and populations. These models typically describe the dispersion and transport of hazardous materials through different media – indoor and outdoor air, surface water and ground water, soil, food and water – between the source(s) and the target receptor(s). Physical, chemical, and engineering sub-models link the source strength and the positions of target receptors, perhaps integrated over time, to predict quantitative exposures received by the targets. In addition, exposure models must model the distribution over time of human populations among locations and activities resulting in exposures. We will use the abbreviation

agent acts $\rightarrow$ *exposures*

to symbolize the input-output relation for the exposure model.

Dose-response models

Dose-response models map individual exposures and covariates to individual risks (e.g., probabilities or hazard functions) of adverse health responses. Dose-response relations are often described by individual risk models such as those in Table 3. A generic dose-response model may be denoted by:

$h(t \mid x, z)$ (Individual dose-response model)

or represented via a diagram as:

$x \rightarrow h \leftarrow z.$

Here, x is an exposure metric or an exposure history (also denoted by $\{x\}$) and z is a vector of parameters and covariates. Together, they determine the risk, h, to an exposed individual. In words, the mnemonic diagram for dose-response relations is:

exposures → individual risks ← individual covariates.

A dose-response model may be evaluated as a whole by statistical methods from data on (x, z, c) triples. Here, x and z are the individual exposure and covariate variables just mentioned and c is an observed individual health consequence variable, e.g., summarizing age at death and what adverse conditions were observed to be present at death. Chapter 3 discusses statistical dose-response models and estimation methods.

Alternatively, a dose-response function can often be decomposed into two main components, represented diagrammatically as follows:

$x \rightarrow y \leftarrow z_1$ (Pharmacokinetic model)
$y \rightarrow h \leftarrow z_2$ (Pharmacodynamic model)

The notation in these diagrams is as follows:

- y is a metric or vector of internal doses or biologically effective doses.
- z_1 = vector of pharmacokinetic parameters. The relation between x and y is described by a pharmacokinetic model. The vector z_1 specifies the parameters of this model and any covariates affecting pharmacokinetics.
- z_2 = vector of pharmacodynamic parameters. The relation between y and h is described by a pharmacodynamic model. The vector z_2 specifies its parameters and any covariates affecting pharmacodynamics.
- As usual, h indicates individual risk, e.g., an age-specific hazard function for mortality. Any of the more specific individual risk models in Table 4 may be substituted for the generic h.

Pharmacokinetic and pharmacodynamic sub-models may be composed to form the whole dose-response relation, as follows:

$$\begin{array}{l} \mathrm{x} \rightarrow \mathrm{y} \rightarrow \mathrm{h} \leftarrow \mathrm{z}_2 \\ \quad\;\;\, \uparrow \\ \quad\;\;\, \mathrm{z}_1 \end{array}$$ (Dose-response model DAG)

In words, the pharmacokinetic component calculates the value of y from inputs x and z_1. The pharmacodynamic component calculates the value of h from inputs y and z_2. Thus, the model is built from the two sub-models via composition of the input-output functions calculated by the sub-models.

Health Impact Models

Health impact models describe how the financial and other consequences of the health damages predicted by the exposure and dose-response models are managed, mitigated, and distributed among individuals by institutions such as insurance, liability, and health care. While often important for policy making, health impact models fall largely outside the main scope of risk analysis, which focuses instead on quantifying probable health effects. Many of the modeling methods and principles in this book are relevant for health impact modeling and related topics in welfare economics, risk-cost-benefit analysis, and microeconomics. For a fuller development of these topics, see Johansson, 1995.

Risk Management Decision Models

Risk management decision models complete the risk management cycle by mapping available information about probable consequences of acts into choices of which acts to take. Acts available to social decision-makers (SDMs, e.g., regulators or administrators) usually include collecting additional information to reduce uncertainty about exposures and risks, as well as opportunities to disseminate existing information and warnings and to require or constrain activities by private agents. Risk management decision models can be used to quantify the expected value of additional information for improving decision-making, and hence can help to set research priorities. They also prescribe interim decisions to be made unless and until additional information becomes available.

Formal decision models are idealizations that do not describe how real agencies and organizations function. Even in the best public health risk management organizations, subjective judgments, practical experience, and political constraints and processes may drive different decisions from what a formal decision analysis would recommend. Other, less benign, processes are easily imagined. For example, a bureaucratic decision process might seek to maximize agency size and funding by publishing hazard and risk messages that scare the public. Even if such behavior is not deliberate, it may provide a competitive advantage to organizations competing for public dollars and approval, so that the agencies that end up with budgets to address risks will tend to be those that have successfully mobilized public concern and resources. A less discouraging process that still diverges from many normative decision models might involve agencies that simply respond to public preferences, perceptions, and pressures, without seeking to evaluate or reconcile them. In many cases, analysis produces insights or identifies options that lead to improved decisions, as judged by preferences for probable consequences, than those resulting from less formal approaches. These insights can be useful even in realistic decision-processes that differ substantially from those in formal decision models.

Chapters 5 to 7 review formal decision analysis principles and methods that are useful for both public (SDM) decision making and private (agent) decision making about how to manage health risks. It also presents evidence on the value of such methods in experimental and real settings.

In summary, the stylized framework for health risk management outlined in this section consists of the following cycle.

1. *The SDM communicates risk information and posts rules* that require or constrain individual activities. Chapter 8 addresses the problem of designing processes and rules from the SDM's perspective.
2. *Individual agents respond to the SDM's acts* and to other risk-related information by adopting behaviors that affect exposure and risk. Chapters 5-7 describe behavioral and normative models for individual decision making under risk.
3. *Agent actions lead to exposures.* Exposures cause health risks and consequences. Exposure modeling is a sub-field of its own, not treated extensively in this book. (Chapter 2 illustrates techniques used in exposure modeling and Chapter 3 presents an example of the use of statistical sampling in estimating the spatial distribution of soil contamination.)
4. *Dose-response models describe the relation between exposures and health risks.* Chapter 3 discusses statistical models of dose-response relations. Chapter 2 examines pharmacokinetic and pharmacodynamic sub-models that complement purely statistical modeling.
5. *Risk analysts produce information about the health risks and consequences of exposures.* They work with exposure information and dose-response models to synthesize risk characterizations. Such information can inform and improve both agent decisions and SDM decisions, but only if it is communicated effectively to decision-makers. The psychology and principles of risk perception, decision-making, and communication, discussed in Chapters 5-8, can be applied to help ensure that technical analyses lead to presentations that improve decisions. New decisions taken by the SDM in light of new risk information trigger further rounds of agent decisions, health consequences, information collection, analysis, and presentation.

A simplified mnemonic diagram for this iterative process is as follows:

SDM actions → agent behaviors → exposures → risks → health consequences → SDM actions

Each arrow in the diagram represents one of the models described above. Each model calculates outputs at the arrow's head from inputs that point into it (and perhaps other inputs, not shown).

This account ignores many practical complexities and modifying factors encountered in real policy-making and implementation. Yet, it provides a useful blueprint for understanding the parts of risk analysis and how they fit together. The following chapters fill in the details and show how the blueprint is applied in practice.

Chapter 2

Risk Assessment Modeling

1. INTRODUCTION

Risk assessment models relate actions to their probable consequences. Ideally, a prospective risk assessment model should identify the potential consequences of alternative risk management actions and quantify their probabilities if each action is taken. In addition, a thorough quantitative risk assessment (QRA) must characterize uncertainty about these probabilistic predictions and their sensitivities to different assumptions and modeling choices. Predictions of probable consequences and their uncertainties provide the crucial information needed to make well-informed choices among different actions.

At a minimum, a prospective QRA model should provide a well-documented procedure to (a) Compute or estimate probabilities of adverse effects from descriptions of targets exposed to hazardous sources; and (b) Characterize uncertainties about these probabilities. A retrospective risk assessment model turns the prediction problem around. Given that an adverse consequence has occurred, it seeks to determine how much of the responsibility for causing it should be attributed to particular sources or actions that made its occurrence more probable.

This chapter introduces quantitative techniques useful for both prospective and retrospective risk modeling. It also provides a case study showing how risk modeling techniques can be applied to a particularly challenging area: assessing the health risks from food-borne microbial pathogens. This example clarifies how uncertainty analysis and assumption-based ("contingent") analysis can be used to provide useful guidance to inform current risk management decisions, even when key scientific questions have not yet been answered.

Example: A Prospective Risk Assessment Model for Cancer Risks

A widely used paradigm for risk assessment of chemical carcinogens is the following. The dose-response relation between lifetime cumulative exposure to a carcinogen, x, and resulting lifetime probability of tumor, p, is assumed to be (at least approximately) described by the one-hit model:

$$p(x) = 1 - e^{-qx}. \qquad \text{(One-hit dose-response model)}$$

where q is interpreted as the effective potency of the carcinogen (see Chapter 1). For values of qx less than about 0.1, $p(x)$ is numerically well-approximated by:

$$p(x) = qx. \qquad \text{(Linear dose-response relation)}$$

Recall that the potency parameter q can be interpreted as the average number of cancer-causing "hits" (e.g., of metabolite molecules on DNA targets) per unit of cumulative exposure per lifetime, assuming (not necessarily realistically) that a single hit can cause cancer.

Now, suppose that exposure to the carcinogen occurs in food and that a population of people ingests an estimated total annual amount of carcinogen, X, that is approximately normally distributed with mean 1×10^7 mg/yr and standard deviation of 5×10^5 mg/yr. The estimated value of q is log-normally distributed with geometric mean 2.1×10^{-6} cancers per mg and 95% confidence interval $[9.8 \times 10^{-7}, 4.5 \times 10^{-6}]$. This confidence interval is intended to express uncertainty about the dose-response relation. Then the prospective risk model consisting of the one-hit dose-response model and the normally distributed exposure estimate predicts that a plausible upper bound on the cancer cases that will be prevented per year by removing the carcinogen entirely from the food supply will be:

Plausible upper bound of health benefit =
$(4.5 \times 10^{-6})(1E7 + 1.96*5E5) = 49.4$ cancers/yr.

Note, however, that the product of the 95% upper confidence limit (UCL) on the potency estimate (i.e., 4.5E-6 cancers per mg) and the 95% UCL on the population exposure estimate (i.e., 1.1E7) does *not* give the 95% UCL on their product, but instead will typically greatly exceed it. Moreover, theories of rational decision-making, as developed in Chapters 5-7, in general require the whole probability distribution of consequences as input, rather than just a plausible upper bound. Monte Carlo uncertainty analysis, discussed later in this chapter, solves both problems by quantifying the probability distribution for the output:

Expected number of cases in population = $1 - e^{-qX}$

from input probability distributions for q and X.

Now, suppose that it were to be discovered the assumptions of the one-hit model do not correctly describe the relation between exposure and cancer risk – for example, because the true biological relation involves an exposure threshold below which there is zero excess risk. Then the probability distribution of qX, as well as the confidence interval calculated for q in the example, may be meaningless. They lack *construct validity* if the concepts being estimated, such as the potency parameter q, do not correspond to anything in the real world. It is common for such *model uncertainty* to overwhelm the uncertainty due to sampling variability captured in traditional statistical confidence intervals. Chapter 3 discusses methods to deal constructively with model uncertainty.

Finally, note that the political desirability of a ban on the carcinogen may depend not only on the probability distribution of qX, but also on whether the carcinogen occurs naturally in the food or results from unnatural sources such as man-made pesticides or additives. Chapters 3-5 describe other factors that drive perceptions of the value of reducing risks. Thus, risk management decision-making must consider more than just the numerical aspects of a risk.

The following box summarizes several motivations for risk assessment modeling.

Why Use Risk Modeling?

1. Predict probable consequences of alternative risk management actions.
   - Propagate alternative feasible input combinations through the model.
   - Compare resulting risks and uncertainties about consequences.
   - Choose the input combination yielding the most preferred consequence probabilities. (See Chapter 4 on influence diagrams and Chapter 5.)

2. Provide framework for discussion and conflict resolution.
   "Rational" conflicts arise from differences in:
   - *Values* and preferences for outcomes, *beliefs* about likely consequences of actions, and *attitudes* toward risk (see Chapters 5 and 6)
   - Risk assessment modeling addresses the "beliefs" component.

3. *Consolidate*/clarify relevant scientific and engineering knowledge.
   - Make assumptions explicit and testable (e.g., via causal graphs, Chapter 4)

4. *Explore* possibilities; generate new scientific hypotheses

If preferences for different sets of consequence probabilities and their uncertainties can be determined, then prospective risk assessment models can potentially be used to optimize choices of controllable inputs to achieve the most-preferred probabilities for outputs. The inputs to a risk model are typically divided into *decision variables* that can be chosen directly by a decision-maker from within some range or constraint set, and *state variables* (or "scenario variables") that cannot be controlled directly. Chapters 5-7 discuss criteria for how to choose controllable inputs to obtain a most-preferred probability distribution of outputs (i.e., consequences), assuming that probabilities of consequences can be predicted for different feasible input combinations of the decision variables (usually interpreted as risk management alternatives). The central task of prospective risk assessment models is to provide this predictive information.

Risk models help to consolidate, document, and make explicit the knowledge and beliefs about how variables affect each other needed to calculate consequence probabilities for different decisions. Hence, in addition to supporting individual decision-making, they can potentially provide a basis for reasoned resolution of conflicts that arise from disagreements about the probable consequences of actions. They also provide a framework for scientific progress in testing and refining the hypotheses embodied in the modeling assumptions.

Example: Expert Resolution

A large literature in management science and operations research addresses the problem of how best to combine the opinions of different experts when their opinions are expressed as probabilities for events. Many proposals are based on apparently reasonable axioms, such as the unanimity axiom that if all of the experts agree, then the combined estimate should agree with them. But a more knowledge-based perspective raises counterexamples such as the following. Suppose that a parallel reliability system works (or "succeeds") if and only if at least one of its two sub-systems, A and B, works. Suppose that it is common knowledge that the prior probability that each sub-system works is 0.5. A decision-maker hires two experts, one to inspect sub-system A and the other to inspect sub-system B. Each expert is able to determine with certainty whether the sub-system she inspects will work. If each expert reports that her posterior probability that the system as a whole will work (after completing her inspection) is 0.5, then should their expert opinions be combined using the unanimity axiom? In other words, should the decision-maker accept 0.5 as their combined estimate of the probability that the system will work? Obviously not. The only way that each expert will have a posterior success probability of 0.5 for the system is if she observed that the sub-system she inspected will not work. Therefore, the correct probability of system success, given the two experts' reports, is 0.

A prospective risk model in this context might be represented by a look-up table such as the one in Table 1.

Table 1: A System Reliability System State Table

System State	Sub-system A status (Inspected by Expert 1)	Sub-system B status (Inspected by Expert 2)	Whole System Status	A's report	B's report
0	0 = Not working	0	0	0.5	0.5
1	0	1	1	0.5	1
2	1 = working	0	1	1	0.5
3	1	1	1	1	1

This table also shows the reports of the two experts for each system state. The DAG is:

A's and B's reports $\leftarrow$ *Status of sub-systems A and B* $\rightarrow$ *System State* $\rightarrow$ *Whole System Status*

The inference task is to estimate the probability of the *Whole System Status* variable from the experts' reports and knowledge of the model. Given such an explicit model, the way to combine any pair of expert reports is clear. But if an explicit model were not used, then a combination procedure that satisfies unanimity might seem reasonable. More generally, how to resolve disagreements among experts about probabilities – and even how best to use estimates that agree – may be difficult or impossible to determine unless the reasons for disagreement (or agreement) are clear. Reference to an explicit risk model can provide the required clarity. Although look-up tables are unwieldy for large models, equally explicit representations, such as influence diagrams (see Chapter 4), are appropriate even for large models.

Example: Coherent Health Risk Research Portfolios

Suppose that several investigators propose to study a causal process leading from exposure to a non-genotoxic chemical carcinogen (one that does not damage DNA) to increased risk of kidney tumors in male rats. Assume that the causal model DAG for the process being studied is as follows:

$$\{x\} \rightarrow \{y\} \rightarrow \{N\} \rightarrow p$$

where

- x = dose level (e.g., in mg of chemical per kg of animal body weight)
- y = internal dose received by relevant kidney cells
- N = size of kidney cell population at risk of malignant transformation
- p = lifetime risk (i.e., probability) of tumor
- {} indicates a time series variable (see Chapter 1).

Suppose that four different teams of investigators propose to collect experimental data and reduce uncertainties about the following component relationships:

- Team A: $\{x\} \to p$. This emphasis might correspond to a standard National Toxicology Program (NTP) two-year bioassay study, for example. The inputs are exposure conditions and the output is lifetime incidence of tumor for animals in each exposure group.
- Team B: $\{x\} \to \{y\}$. This team proposes to study the pharmacokinetics and metabolism of the chemical in male rats, perhaps leading to a simulation model predicting internal doses in kidney cells from given exposure patterns.
- Team C: $\{y\} \to \{p\}$. Proposals to examine the relation between internal doses (as revealed by various lab tests and biological markers) and tumor probabilities have been relatively uncommon, but are becoming more popular as the need for biological realism in risk assessment has become more widely recognized.
- Team D: $\{x\} \to \{N\} \to p$. This team proposes to examine the effects of exposures on cell turn-over rates and target cell population sizes and proliferation, as well as the correlation between these cell kinetics parameters and cancer risk.

From the funding organization's point of view, B and C together form a *coherent portfolio* of research to reduce uncertainty about the overall ($\{x\} \to p$) relation, since the components they study can be composed to give the overall relation of interest.

$$(x \to p) = (x \to y) \bullet (y \to p)$$

where "•" denotes composition of functions and the (hopefully suggestive) notation $(y \to p)$, for example, denotes the function mapping values of y to values of p. (Curly braces on time series variables are omitted for convenience.) Similarly, team D has proposed a coherent decomposition of the dose-response relation as:

$$(x \to p) = (x \to N) \bullet (N \to p)$$

By introducing new data on y and N, respectively, these efforts can potentially reduce uncertainty about the $(x \to p)$ relation that would be estimated solely from observations on tumor incidence rates in different exposure groups. Thus, they can complement the holistic approach proposed by Team A. However, there is no way to integrate the results of projects B and A unless C is funded. And, there is no way to integrate results from either of B or C with results from D, i.e., the relations studied in these projects cannot be composed to form the $(x \to p)$ relation. The project subsets {A, B}, {A, C}, {B, D}, and {C, D} are examples of incoherent research portfolios. Similarly, a project that studied the relation $(y \to N)$ would not be immediately useful when combined with any one of these four studies, even though it could be valuable as part of a coherent set of projects corresponding to the decomposition

$$(x \rightarrow p) = (x \rightarrow y) \bullet (y \rightarrow N) \bullet (N \rightarrow p).$$

Much of the existing theory of optimal statistical design of experiments deals with how to allocate observations among different exposure groups to minimize errors in estimated relations. For example, if it can be assumed that internal dose is proportional to administered dose over the range of interest, so that:

$$y = kx,$$

then choosing (a) The values of x at which to measure y; and (b) How many measurements to take at each selected x value (i.e., how big to make the corresponding exposure group) in order to minimize the expected width of the 95% confidence interval for k are standard questions for the design of experiments (Silvey, 1981). Making these choices to also help answer questions such as whether the $(x \rightarrow y)$ relation is linear, concave, convex, or s-shaped is a less standard design problem. Figuring out how best to decompose networks of causal relations into effective research portfolios, especially when taking into account the costs of collecting data and the statistical error characteristics of observations made on different variables, has received comparatively little attention in the statistical design literature. This appears to be a promising area for application of operations research methods.

1.1 Approaches to Quantitative Risk Assessment (QRA): Probability, Statistical, and Engineering Models

Quantitative risk assessment (QRA) models have two main technical aspects. One is a general framework of *conditional probability calculations* for structuring, defining and calculating individual and population risks from available data. This framework uses probability identities to structure and combine the different data elements contributing to a QRA model. An example is the following formula for the probabilistic exposure-response relation of an individual with uncertain pharmacokinetic parameters:

$$Pr(response = r \mid exposure = x) =$$
$$\Sigma_y Pr(response = r \mid internal\ dose = y) * Pr(internal\ dose = y \mid exposure = x).$$

It may be abbreviated by suppressing the explicit indices, yielding:

$$Pr(response \mid exposure) =$$
$$\Sigma_{internal\ dose} Pr(response \mid internal\ dose) * Pr(internal\ dose \mid exposure).$$

or, in symbols.

$$Pr(r \mid x) = \Sigma_y Pr(r \mid y)Pr(y \mid x).$$

We will often use such abbreviations. Similarly, the function composition notation introduced in the previous examples, e.g.,

$$(x \rightarrow p) = (x \rightarrow y)\bullet(y \rightarrow p),$$

can be extended to include composition of probability relations via the identity:

$$Pr(p \mid x) = \Sigma_y Pr(p \mid y)Pr(y \mid x).$$

If exposure is uncertain, then the overall response probability is just the expected response probability with respect to what is known:

$$p = E(p) = \Sigma_x Pr(p \mid x)Pr(x) = \Sigma_x \Sigma_y Pr(p \mid y)Pr(y \mid x)Pr(x).$$

(As usual, sums must be replaced by integrals and probabilities must be replaced by probability density functions if the number of levels being considered is continuous instead of discrete.) While such identities follow from probability theory, they are devoid of empirical content, except that they must be consistent with underlying DAG relations, such as:

exposure → *internal dose* → *response.*

The second aspect of risk modeling is substantive. It deals with how to use knowledge, data, and assumptions to estimate the key empirical relations used as components of the risk model. These components include the internal dose-response sub-model:

Pr(response | internal dose) or Pr(response = r | internal dose = y),

the pharmacokinetics sub-model:

Pr(internal dose | exposure), or Pr(internal dose = y | exposure = x),

and the exposure sub-model:

Pr(x) = Pr(exposure = x)

for all combinations of *r*, *y* and *x* values of interest. Different modeling strategies are available to quantify these relations. Statistical and engineering approaches are the most prominent. Once the component relations have been estimated, simulation software or DAG methods can be used to combine networks of statistical and engineering formulas into coherent models.

As discussed further in Chapter 3, a traditional *statistical risk modeling* strategy postulates specific parametric risk models, such as the following:

$$Pr(response = 1 \mid internal\ dose = y) = 1 - e^{-qy} \ for\ y \geq 0$$

for a binary (0-1) response. It then estimates unknown model parameters, e.g., q, from available data via maximum-likelihood estimation (MLE) and other statistical estimation techniques. Several modern computational statistical methods avoid specific parametric assumptions by fitting *non-parametric* relations to the available data. Important computational methods include classification trees, discussed later in this chapter, and a variety of algorithms for fitting smooth response surfaces to data. Among the best developed methods and algorithms for interpolating smooth relations among observed data points are interpolating splines, iteratively weighted least squares ("loess"), wavelet and orthonormal basis series expansions of unknown functions, and kernel smoothers. The Multivariate Adaptive Regression Splines (MARS) technique combines ideas from classification trees and smooth interpolation approaches. These algorithms are described in many texts (e.g., Cherkassky and Mulier, 1998). Software implementing them is available in repositories such as http://lib.stat.cmu.edu/ and in commercial packages such as S-PLUS™, SAS™, and Statistica™. Such statistical tools are an important part of a risk assessor's tool kit.

Statistical methods can fit relations to input-output data for a whole system without having to understand anything about its internal workings. By contrast, *engineering models* attempt to describe, predict, and simulate the behaviors of systems, including those leading to individual or population health responses, by *composing the behaviors of sub-systems.* The behaviors of the primitive (i.e., not further decomposed) sub-systems used as basic building blocks are modeled via scientific and engineering laws and/or described by empirical relations. Physical laws such as conservation of mass and flow balance equations can help to structure the overall description of the process being modeled.

Quantitative descriptions of the sequence of sub-processes leading from source emissions to health effects usually reflect chemical laws governing the transformation, transport, and deposition of chemicals (e.g., Henry's law; volatilization, deposition, and absorption rate laws; equilibrium partition coefficients, and so forth.) Engineering models often use a mix of theoretical and empirical relations to describe dispersion and flow of chemicals through ground water, surface water, soil, and air (Conway, 1982; Paustenbach, 1989). They may also include applied probability laws and formulas describing stochastic phenomena such as waiting times between events and the evolution of random diffusion and transport processes in heterogeneous media.

We shall call all such compositional models *engineering models,* although many of them have been developed in related fields such as toxicology, biometrics, industrial hygiene, and exposure assessment. Some of the types of engineering models most commonly used in health risk assessment modeling are as follows.

Examples of Engineering Models Used in Risk Assessment

1. *Fault trees, event trees,* and *influence diagrams* modeling operations or process failures leading to releases of hazardous materials (Barlow, 1998).
   - Techniques: Discrete-event models and stochastic simulation of fault trees.
   - Influence diagram algorithms
   - Applications: Accident and safety models, e.g., compound Poisson process models for frequency and severity of releases, losses, or damage

2. *Source models* describing releases from facilities into the environment.
   - Techniques: Chemical engineering models of leaks, spills, and discharges

3. Pollutant *fate and transport models* describing flow from source to targets
   - Engineering techniques: Continuous simulation models
   - Dispersion models (e.g., Gaussian Plume, Gaussian Puff) for air, surface water, ground water, soil
   - Statistical models: Regression models, time series analysis
   - Simulation models: Models of seepage, soil absorption, volatilization, etc.

4. *Human lifetime exposure models* describing exposures of human populations to substances in the environment
   - Monte-Carlo simulation of the frequency distribution of individual behaviors (e.g., mobility, activity, and use patterns) in the population

5. *Physiologically-based pharmacokinetic (PBPK) models.* These describe the uptake by animals or humans of substances from the environment and the distribution and metabolism of substances following inhalation, ingestion, injection, or dermal absorption

6. *Pharmacodynamic models* and other biologically-based and biologically motivated health risk models of the interactions between substances and target macromolecules, cells, or cell populations (e.g., multistage and MVK dose-response models)

7. *Ecosystem compartmental flow models* (for pollutants, population dynamics)

References: Barlow, 1998 for topic 1; Paustenbach, 1989 and Conway, 1982 for topics 2-7.

Statistical and engineering approaches to quantifying input-output relations are complementary. Several DAG modeling frameworks, including influence diagrams (Barlow, 1998), can combine elements of both. The compositional modeling characteristic of the engineering approach is captured by the input-output relations in the DAG, while the conditional probability distribution of each variable (i.e., node), given the values of its predecessors, may be determined by statistical regression modeling, non-parametric modeling, discrete-event simulation, computational mathematical models, or other techniques.

Section 2 describes the general framework of conditional probability calculations used to quantify risks for individuals and populations. The following sections then introduce techniques for risk assessment modeling. Statistical approaches are deferred until Chapter 3.

2. CONDITIONAL PROBABILITY FRAMEWORK FOR RISK CALCULATIONS

The actions of greatest interest in health risk assessment usually affect exposures of individuals and populations to known or suspected hazards. The causal graph of most frequent interest, shown in words on the first line and in symbols on the second, is:

Act → *Population exposure* → *Individual exposures* → *Individual risks*
↑
individual covariates

$$a \to F_x(x \mid a) \to x \to \boxed{p = f(x, z)} \leftarrow z$$

Note that the conditional symbol "|" is used to indicate conditioning on an *act,* as well as conditioning on an *event.* Acts may not be events in the formal probability-theory sense of measurable subsets of a sample space, but the conditioning notation is still suggestive and useful. The interpretation of this diagram is that choice of a particular act a from a set of possible risk management alternatives A induces a population frequency distribution of individual exposures. This distribution is represented by the cumulative distribution function (CDF) of exposure for a randomly selected individual:

$$F_x(x \mid a) = Pr(exposure \leq x \mid act = a) \text{ for } a \in A \text{ and for } x \geq 0.$$

For example, building a power plant exhaust stack at different heights will affect the number of people exposed to various levels of average annual concentrations of its pollutants in inhaled air. Similarly, posting and enforcing a certain speed limit on a stretch of highway will affect the number of people per year who travel along it at various speeds. Individual exposures, x, are drawn from the exposure distribution induced by a. Together with any individual characteristics that affect risk, summarized in the covariate vector z, the individual exposures determine individual risks, modeled here as probability p of a binary adverse outcome, such as a cancer death or an automobile accident fatality.

The key challenge for prospective risk assessment modeling is to quantify the causal relations between (a) Acts and exposure distributions; (b) Population exposure distributions; and (c) Resulting health risks. The key challenge for retrospective risk assessment modeling is to quantify the causal relation between individual health risks or adverse health outcomes and the exposures (or acts) that may have contributed to then. Even given a known risk model $p = f(x, z)$ with known inputs x and z, deciding how much of the blame for an adverse response should be attributed to exposure x (or to individual components of x, if x contains multiple variables) requires some normative principles for risk attribution and risk accounting. Although a full treatment of attribution of risk in the presence of joint causes is deferred to Chapter 4, some simple approaches based on probability of causation are discussed later in this chapter.

2.1 Calculating Average Individual Risk When Individual Behaviors May Respond to Acts

Of course, the distribution of exposures in the population determined by the chosen act often depends on the behaviors of the individuals in the population. Then, arrows from z (which includes individual behaviors) to x (and from a to z, if individual behaviors are partly determined by the choice of risk management strategy) must be added to the above DAG. Still, individual behaviors (and covariates) can be "marginalized out" of the quantitative probabilistic relation between acts and risks if it is desired to focus on the act-consequence relation. This is done by summing (or integrating) risks *conditioned* on behaviors over all behaviors, weighted by their frequencies.

More explicitly, let the covariate vector z subsume indicators of relevant individual behaviors. Then the probability density function of exposure for a randomly selected individual, given act a, is:

$Pr(x \mid a) = \Sigma_z Pr(x, z \mid a) Pr(z \mid a)$. (Exposure PDF with z marginalized out)

If individual's exposure depends on the behaviors and other characteristics in *z*, as well as on act *a*, then the following substitution may be made:

$$Pr(x, z \mid a) = Pr(x \mid z, a) Pr(z \mid a).$$

The conditional probability $Pr(z \mid a)$ can incorporate a *behavioral model* allowing for the possibility that the individual's behavior (or other risk-affecting characteristics) change based on the chosen risk management act, *a*. If *x* and *z* help to determine each other, however, then it is easier to use the joint distribution:

$$Pr(x, z \mid a).$$

This could be the case, for example, if pollution levels around a power plant affect the decisions of people about where to live, which in turn affect their exposures to the pollution. In this case, *z* might include information on location, and *a* affects the exposure at each location, as well as the distribution of people among locations (and hence the joint distribution of *z* and *x* values in the population.)

For a randomly selected individual, the risk associated with act *a* (before learning the individual's particular characteristics, say z_i) is:

$E(p \mid a) = \Sigma_{(x, z)} Pr(response \mid x, z) Pr(x, z \mid a)$. (Risk for random individual)

If the DAG model is: $a \rightarrow x \leftarrow z$, perhaps augmented with an arrow from *a* to *z*, then this formula can be factored as:

$E(p \mid a) = \Sigma_x \Sigma z f(x, z) Pr(x \mid z, a) Pr(z \mid a)$. (Risk when x does not affect z)

If individual characteristics and behaviors affecting risk are statistically independent of exposure *x* and act *a*, then this equation simplifies to:

$$E(p \mid a) = \Sigma_x \Sigma_z f(x, z) Pr(x \mid a) Pr(z).$$

If the risk-relevant characteristics and the exposure are known for a particular individual, then that individual's risk is of course just $p = f(x, z)$.

Summing over multiple (x, z) combinations can be tedious. More importantly, the relevant (x, z) combinations for predicting risks may not be known *a priori*. Fortunately, modern data-mining software can identify (x, z)

combinations that are significantly associated with individual risks in large data sets, and can perform risk calculations on sample data. In particular, *classification tree* programs (e.g., Breiman et al., 1984; Biggs et al., 1991) organize risk calculations based on large data sets in a convenient format.

Example: Calculating Individual Risks via Classification Trees

The following classification tree, produced by the commercial program KnowledgeSeeker™ (Biggs et al., 1991), shows how the annual lung cancer death rate in a hypothetical randomly sampled cohort of 30-year old white females varies with different combinations of factors. Each node in this tree contains three numbers. The bottom number is the number of individuals (out of a total of 98,069 in the sample) who match the description corresponding to that node. This description is found by tracing the path from the top of the tree (its root node) to the node being examined. The other two numbers give the percentage of women in the subset described by the node who died of lung cancer during a one-year follow-up period (middle number in the node) and the percentage who survived (top number).

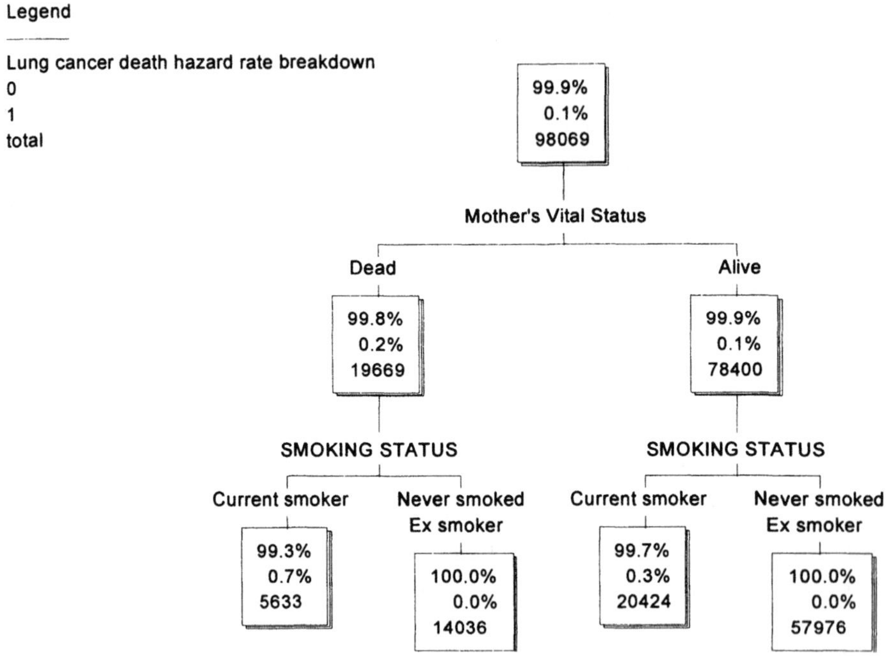

For example, 19,669 out of the 98,069 thirty-year old women in the sample had mothers who were still alive at the start of the 1-year observation period. Of these women, 0.1% died of lung cancer during the next year. In the subset of these women that were current smokers (size = 20,424), the sample value of the lung cancer mortality hazard rate was 0.3% instead of 0.1%. By contrast, among women whose mothers had already died, current smokers have a hazard rate of 0.7%. The individual risk of lung cancer associated with current smoking, calculated with respect to the information in this tree only, is 0.0%, 0.3%, or 0.7%, depending on which leaf node describes the individual. Thus, an individual's risk depends on what is know about her. Calculated without conditioning on any specific information about mother's vital status or the individual's smoking status, the average individual risk (hazard rate for lung cancer) is just 0.1 (top node).

In this example, x may be interpreted as the smoking exposure variable, with ordered categorical values of Current, Ex, and Never, while z is the mother's vital status. The act variable a coincides with x, i.e., the decision whether to be a current smoker. The prospective risk model inferred from the data is:

$E(h \mid a, z) = 0$ if a = never smoker or ex-smoker
$= 0.3\%$ if a = current smoker and z = mother still alive
$= 0.7\%$ if a = current smoker and z = mother not still alive

where h denotes the hazard rate for lung cancer death within one year. The average individual risk for a current smoker is given by the formula:

$$E(h \mid a = \textit{Current smoker}) = \Sigma_{(x,\, z)} Pr(\textit{response} \mid x, z) Pr(x, z \mid a)$$
$$= (0.007)[5{,}633/(5{,}633 + 20{,}424)] + (0.003)[20{,}424/(5{,}633 + 20{,}424)] = 0.004.$$

This can also be seen at a glance from a classification tree in which the act (a = current smoker) is conditioned on first:

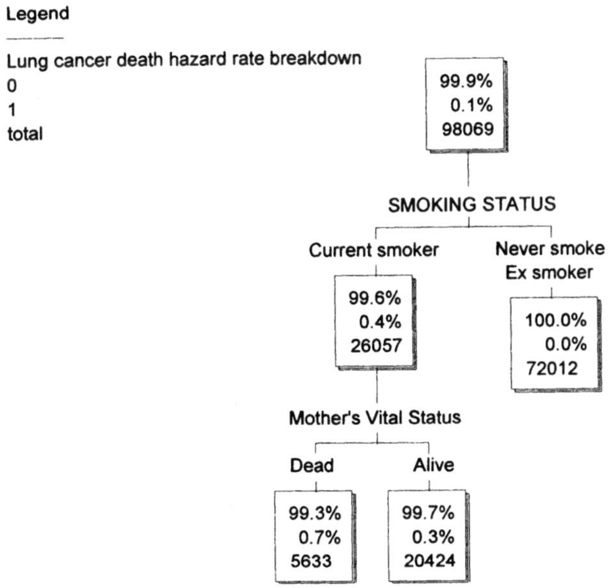

Thus, using classification trees to condition on what is known automatically produces sample estimates of conditional individual risks, making manual calculations unnecessary.

The risk information in the above trees could be shown in a 2 x 2 table. However, the classification tree program has the following advantages:

(a) It *automatically* selects category boundaries for variables of any type (e.g., nominal, ordered categorical, or continuous) to optimize an objective function (e.g., to maximize expected reduction in classification entropy or error, depending on the algorithm used.) For example, it automatically pooled the ex-smoker and non-smoker categories after determining that risk does not differ significantly between them in this population.

(b) The classification tree format extends easily to more than 2 variables and to variables of different types (e.g., continuous as well as discrete).

(c) The tree shows how individual risks change as one conditions on more or less information. For example, if information about individual medical histories of lung disease (e.g., chronic bronchitis) were added to the data set, then the individual risk (with respect to all available information) for a woman whose mother is still alive might change as indicated in the following refined tree:

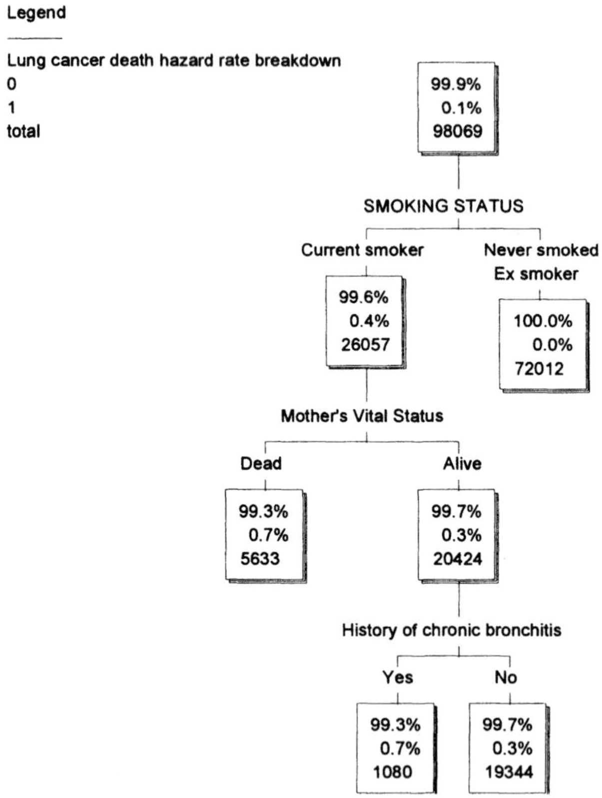

(d) When constructed carefully, classification trees sometimes have direct interpretations as *causal models,* showing how individual risks in a population will change if certain acts (branches at nodes) are taken. This important point is developed in subsequent examples and is placed on a rigorous technical and philosophical foundation by Shafer (1996). In the above tree, a straightforward causal interpretation would imply that a decision to not smoke would reduce individual risk of lung cancer at age 30 to approximately 0.0% from approximately 0.7% (among 30-year old women with a history of chronic bronchitis and mother still living). Whether this causal interpretation is justified requires additional analysis, as discussed in Chapter 4.

As illustrated in this small example, classification tree analysis provides a model-free (non-parametric) approach to analyzing multivariate data. It is also used to "mine" large databases for useful knowledge (Fisher and Lenz, 1996). In general, classification tree algorithms search for *combinations* of values of independent variables that best predict the value of the dependent variable. Ability to predict is measured by *splitting criteria* such as the entropy, variance, or statistical significance of differences among the conditional frequency distributions of values for the dependent variable, conditioned on the answers to questions asked. Questions are represented by partitioning the possible values of variables into subsets and asking which subset the value for a particular individual belongs to. Based on the answer, a new question is asked. The result is the classification tree, with nodes representing questions and branches at each node representing possible answers. Internal nodes are also called *splits*, while leaf nodes ("tips" of the tree) represent probabilistic classifications or predictions of the value of the dependent variable.

Many tree-growing algorithms stop growing a tree when no additional questions can be found that will improve its ability to predict the value of the dependent variable, according to the selected splitting criterion. Alternatively, trees may be grown larger than needed and then pruned back until the estimated error rate is minimized (Breiman *et al.*, 1984). Different splitting criteria and pruning or stopping criteria lead to different specific classification tree algorithms. The classification trees in this chapter were created using the specific algorithm of Biggs *et al.* (1991). This algorithm uses estimated statistical significance (based on F- and chi square statistics for continuous and categorical variables, respectively), with a Bonferroni adjustment to correct for multiple comparisons in choosing class boundaries, for its splitting and stopping criteria. Each tip of the tree, corresponding to a unique branch or path through it, represents a sequence of questions and answers. The conditional frequency distribution of the value of the dependent variable, based on the questions and answers leading to the tip, constitutes the classification tree's (probabilistic) prediction at that tip.

Although tree-growing algorithms have become fairly sophisticated, e.g., by incorporating refinements to help find robust splits using multiple subsamples of data, the final trees are often easy to understand and interpret.

Tree-structured estimates of risk are often particularly appropriate for health risk analysis applications. Estimates of a dependent risk variable being conditioned on more and more information as one moves down a branch reflects the natural information structure of many risk assessment settings in which the information to be used to estimate risk must be decided as part of the assessment process. We will use classification trees as models for understanding the relation between risk estimates and information.

2.2 Population Risks Modeled by Conditional Probabilities

For a population, the risk associated with act *a* is determined by the set of individual risks that *a* induces in the population. The expected fraction of the population with individual risks of r or more (for r between 0 and 1) is:

$$1 - F_{p|a}(r) = (1/N)\Sigma_i \Pr[f(x_i, z_i) > r \mid a] \text{ for } 0 \leq r \leq 1. \quad \text{(Population Risk Profile)}$$

where N is the number of individuals in the population and *i* indexes the individuals. In practice, the conditional probability

$$\Pr[f(x_i, z_i) > r \mid a]$$

must often be estimated by numerical techniques such as Monte Carlo simulation or Discrete Event Simulation, discussed later in this chapter.

The population risk profile just defined does not necessarily contain all the information that is relevant for choosing among alternative risk management acts.

Example: Population Risk Profiles Omit Risk Equity Considerations

Consider the five risk management options A-E shown in the table below for a society consisting of two individuals. There are two possible states, 1, and 2, each with probability 0.5. If state *s* occurs and act *a* has been chosen (for *s* = 1, 2 and *a* = A, B,..., E) then the mortality risks to individuals 1 and 2 are as shown in the "State 1" and "State 2" columns. For example, if State 1 occurs and act A has been chosen, then individual 1 dies, while individual 2 does not. The remaining columns – risk to Individuals 1 and 2, probabilities on 0, 1, and 2 deaths, and expected number of deaths – are calculated from the data in the State 1 and State 2 columns.

Acts A and B have identical risk profiles – one person dies with probability 1 – but very different risk equity properties. Act B always sacrifices individual 1, while act A assigns equal risks to the two individuals. Similarly, acts D and E have identical risk profiles. But for act D, both individuals have identical risks, while for act E, individual 1 always has a risk that is 0.5 greater than individual 2's. (This might arise if the individuals are neighbors of a facility that can have either a major accident or a minor one, with 1 living closer to it than 2. A minor accident never affects 2 but gives a risk of 0.5 to 1. A major accident always kills 1 and gives a risk of 0.5 to 2.)

Table 2: Risks for Different Acts and States. (n = Number of Deaths.)

Act	**State 1 risks for individual (1, 2)**	**State 2 risks for individual (1, 2)**	Risk to Ind. 1	Risk to Ind. 2	Pr(n = 0)	Pr(n = 1)	Pr(n = 2)	E(n)
A	(1, 0)	(0, 1)	0.5	0.5	0	1	0	1
B	(1, 0)	(1, 0)	1	0	0	1	0	1
C	(1, 1)	(0, 0)	0.5	0.5	0.5	0	0.5	1
D	(0.5, 0.5)	(0.5, 0.5)	0.5	0.5	0.25	0.5	0.25	1
E	(0.5, 0)	(1, 0.5)	0.75	0.25	0.25	0.5	0.25	1
Probs:	0.5	0.5						

All five options imply the same expected number of deaths (i.e., 1), but distribute risks to the two individuals in different ways. Past risk analysis literature seeks to compare options such as A-E in terms of their risk profiles, represented by the columns for Pr(n = 0), Pr(n = 1), and Pr(n = 2) in this table (Boiney, 1995; Fishburn and Sarin, 1991). Typical insights are that societal aversion to catastrophic (multiple-fatality) outcomes would make A preferable to C and D, while aversion to identifiable (rather than anonymous statistical) individual mortality risks and preference for equitable risk distributions would make A preferable to B. D is preferable to E on grounds of equitable distribution of individual risks. D might be preferred to A if the state is learned before consequences occur, since it does not lead to identified individual fatalities in advance of the event. On the other hand, A might be preferred to D because it lacks D's potential for catastrophic mortality, i.e., death of both individuals. The most important lesson from such examples is that *the probability distribution for number of fatalities does not contain all relevant information for choosing among acts.* Fatality risk profiles are not necessarily adequate to inform decisions. Instead, as discussed in Chapter 5, it may be necessary to consider probabilities of different subsets of fatalities in a population in order to adequately account for risk equity and catastrophe-aversion concerns.

Despite the importance of subset-valued fatality risks for general analyses of population risks, many public health risks have special characteristics that justify a simpler approach. In particular, health responses in individuals are often conditionally statistically independent of each other, given the individual exposures. Suppose that each person described by the pair (x, z) has the same independent probability of response and the same conditional probability distribution for the severity of response, e.g., life lost, quality-adjusted life-years (QALYs) lost, etc. Each (x, z) pair may be interpreted as a single cell of a multi-way contingency table, if risks can be adequately characterized using a finite number of such (x, z) combinations. Then in a large population with many individuals described by each (x, z) pair, it may be reasonable to care only about the *sum* of the health values (i.e., life lengths, QALYs, utilities, etc.) over all cells. This is justified if the measure of value has been defined so that each incremental unit is equally valued, as might be the case for QALYs or for individual utilities of health states; see Chapter 6. Invoking a Central Limit Theorem then suggests that the total health value for the population will be approximately normally distributed, with mean and variance given by the sums of the means and variances of the values from all cells *(x, z)*.

In this framework, defining population risk in terms of the (approximately) normal distribution of total health value associated with each act *a* in *A* provides enough information to choose among acts. Indeed, the theory of single-attribute expected-utility decision-making described in Chapter 6 suggests the following specific formula for making such choices: Act A is preferred to act B if and only if

$$E(V \mid A) - (k/2)Var(V \mid A) > E(V \mid B) - (k/2)Var(V \mid B),$$

where $(V \mid a)$ denotes the approximately normally distributed total value (e.g., total QALYs in the population) if act *a* is chosen (for $a = A, B$). As usual, $E(V \mid A)$ and $Var(V \mid A)$ denote the mean and variance of the random variable $(V \mid A)$. In this formula, *k* denotes a coefficient of risk aversion (see Chapter 6) indicating how willing a decision-maker is to trade-off expected gain for reduction in variance. The risk-adjusted "certainty-equivalent" value associated with act *a* may be defined as:

$$CE(a) = \Sigma_{(x,\, z)}N(x, z \mid a)\{E[V(x, z)] - (k/2)Var[V(x, z)]\},$$

where $N(x, z \mid a)$ denotes the number of individuals in cell *(x, z)* if act *a* is taken, assuming that this is known or can be modeled.

Example: Classification Tree Calculations of Population Risks

Population risks can be calculated using classification tree software, which also automatically constructs a finite number of cells *(x, z)* with statistically significantly different conditional distributions of risk. For example, consider a hypothetical population of white males in which the average age at death is 66.75 years, with conditional means based on diet and smoking behaviors as shown in the following tree. The expected life-years in this population in the absence of any interventions is 66.75 per person, with standard deviation 8.1 years (not shown). The conditional sample mean and standard deviation at all nodes are automatically computed by the program, although standard deviations are shown here only for the leaf nodes.

Now, suppose that the splits in the tree below correspond to *causal* impacts, so that choosing to go down a particular branch at a node gives one the corresponding conditional distribution of life length. Then a public health program that changed all behaviors to a "healthy life style" could extend average lifespan to 70.32 years with standard deviation 7.51 years. (In this illustration, "healthy life style" is defined as no beer, fewer than 3 cups of caffeinated coffee per day, no smoking, and consumption of raw vegetables at least once per week. The KnowledgeSeeker™ classification tree program uses the notation [0, 1) for a continuous variable like beer consumption to denote the interval that contains 0 but not 1, while [1, 24] is the interval from 1 to 24, inclusive.) Conversely, making all unhealthy choices could reduce average life to 58.9 years with standard deviation 6.97 years.

The means, standard deviations, and node population sizes produced by the classification tree program provide the information needed to calculate *CE(a)* for any act *a* (i.e., for any leaf of the tree, if all variables are considered decision variables) via the following certainty equivalent formula (see Chapter 6):

$$CE(a) = \Sigma_{(x, z)}N(x, z \mid a)\{E[V(x, z)] - (k/2)Var[V(x, z)]\}.$$

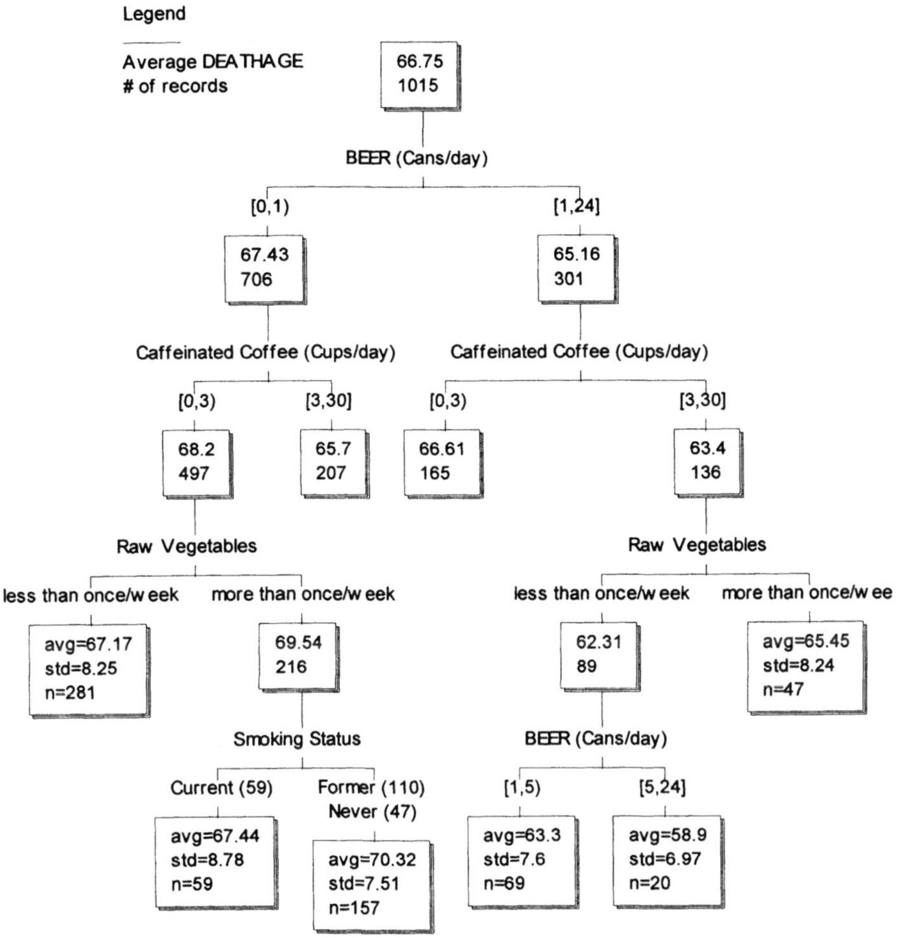

Here, the summation index *(x, z)* now indexes the 6 leaf nodes of this tree; $N(x, z \mid a)$ is the number of people who will be described by leaf node *(x, z)* if act *a* is taken; $V(x, z)$ is the normal distribution with mean $N(x, z \mid a)*avg(x, z)$ and a variance of $N(x, z \mid a)*[std(x, z)]^2$ for leaf node (*x*, *z*) where "*avg(x, z)*" and "*std(x, z)*" are the conditional average per person (the top number shown in the leaf node) and the standard deviation calculated by the classification tree program. More generally, if the effect of a public health campaign can be modeled as inducing a new frequency distribution of the population among the leaves in the tree, then the certainty equivalents of the leaves can be weighted by the new frequency distribution to evaluate the desirability of the campaign.

2.3 Trees, Risk, and Martingales

Classification trees show that the risk for an individual or a population generally depends on the information that is used (i.e., conditioned on) in calculating it. It is easy to imagine that there might always be additional variables that could be introduced that might change an estimated risk if they were included. This raises the disturbing possibility that risk has no objective meaning – that it might be a concept corresponding to nothing objective, represented by a host of different estimates, all equally valid (or invalid), depending on what information one happens to use in calculating them. This threat can be laid to rest by showing how to form useful definitions and stable estimates of risk. To do so, it is essential to answer questions such as the following. Can a risk estimate based on one set of information ever be objectively better than an estimate based on a different set of information? Similarly, is there any sense in which risk estimates tend to converge to a true value as more information is collected? The answer to both questions is yes.

The following tree suggests part of the solution to the problem.

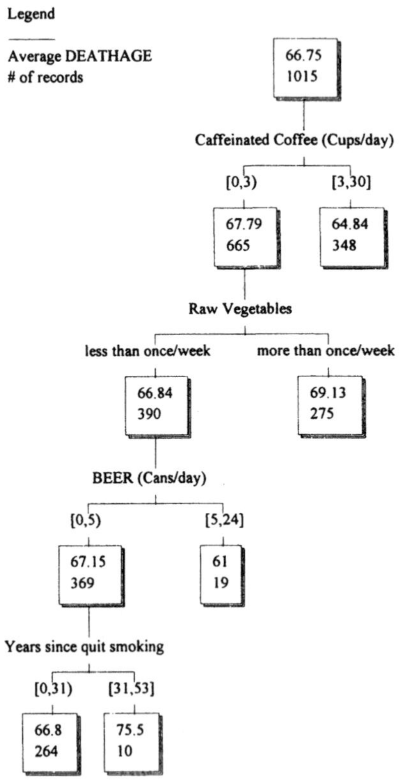

It shows the conditional mean of life length conditioned on more facts as one moves down the tree, using the same hypothetical data set as in the previous example. The mean changes as information is accumulated, increasing from 66.75 years at the top of the tree to 75.5 years at the bottom for an individual or group for whom all facts are as shown along the path leading to the bottom right leaf node. (Note that never smoking is identified as less beneficial than being a former smoker who has quit for at least 31 years. This presumably would not reflect a causal health benefit from past smoking, but rather a different causal mechanism – perhaps a tendency among those who have successfully quit smoking to have healthier lifestyles, or perhaps just the fact that those who have quit smoking for at least 31 years must survive long enough to attain this survival time. Such interpretive ambiguities often enliven epidemiological risk discussions, as discussed further in Chapter 3.)

Although the mean can increase or decrease as new facts are gathered, it satisfies the following property (known as the *martingale property* in the theory of stochastic processes). Let M_0, M_1, M_2, ..., denote the sequence of mean values at the successive nodes 0, 1, 2, …, along the path describing an individual or population in a classification tree. For example, in the above tree, $M_0 = 66.75$, $M_1 = 67.79$, and $M_3 = 67.15$ years. Now, *before* learning the result of the next piece of information to be collected about the individual or population for whom risk is being quantified, it is always true that:

$$E(M_{n+1} \mid M_0, M_1, M_2, \ldots M_n) = M_n \qquad \text{(Martingale property)}$$

For example, starting from node 1 along the above path, and before learning about a particular individual's consumption of raw vegetables, we have:

$$E(M_2 \mid \text{node } 1) = (390/665)(66.84) + (275/665)(69.13) = 67.79 = M_1.$$

Once an additional fact is gathered (e.g. raw vegetable consumption is less than once a week), then the actual value of M_2 (either less than or greater than M_1) will be known. But its expected value, starting from node 1, is just M_1. Technically, the sequence of conditional means forms a *Doob martingale* (Ross, 1996, p. 297). Now, even if there are an infinite number of relevant facts that can be introduced that would modify the conditional mean value, the Martingale Convergence Theorem (Ross, 1996, p. 319) implies that, with probability 1, *the sequence $\{M_n\}$ approaches a finite limit* as $n \to \infty$. This limit is just the conditional expectation of the dependent variable (life-years, in this example) given the entire sequence of facts.

Thus, it is *not* the case that different, mutually inconsistent estimates of risk can be achieved by introducing new facts and conditioning on them until a desired answer is produced. Indeed, if the maximum size of each adjustment $|M_{i+1} - M_i|$ is sufficiently small, then the theory of large deviations for martingales can be used to show that the probability that new facts will cause the conditional mean to vary by much from its current value is small. Azuma's inequality (*ibid*) places a quantitative upper bound on the probability of large deviations.

Next, consider trees of finite depth. For any two random variables V and I, we have the conditional variance formula (Ross, 1996, p. 51):

$$Var(V) = E[Var(V \mid I)] + Var(E(V \mid I)]. \qquad \text{(Conditional variance formula)}$$

Therefore, letting V be the dependent variable (e.g., life-years, probability of lung cancer in the next year, probability of eventual death with lung cancer, etc.) and letting I be the information about which branch at a classification tree node describes an individual, we have:

$$E[Var(V \mid I)] < Var(V) \qquad \text{(Average variance decreases with information)}$$

(This assumes that each node branches to children with different mean values, so that $Var(E(V \mid I)] > 0$, as will be the case in practice.) The decrease of variance on average as new information is gained can be seen in the above tree. Each node with children has at least one child with a standard deviation (and hence variance) smaller than its own. The branching probabilities are such that the *average of the variances of the children is always less than the variance of the parent.* In this sense, "deeper is better", i.e., conditioning on more information reduces the expected variance around the conditional mean.

For small sample sizes, the distinction between true and sample variances becomes important, and care must be taken not to grow a tree deeper than the data warrant. Over-growing trees may be prevented in practice using statistical cross-validation tests and the pruning approach detailed in Breiman et al., 1984. But expected variance decreases with increasing depth in a tree holds as long as sample sizes are not too small.

The idea that "deeper is better" in trees, i.e., that conditioning on more information leads to better estimates, can be formalized in several ways not involving variance. For example, the *entropy* of the conditional distribution for the dependent variable decreases on average as one conditions on more information (Breiman et al., 1984). Moreover, the *expected utility* (see Chapter 5) of the best decision made with respect to all available information increases on average as one conditions on more information (Laffont, 1989, Chapter 4). In this very important sense, risk estimates from nodes further down in a tree are to be preferred to the less-informed risk estimates in the nodes above them, i.e., to their parents or ancestors.

2.4 Value of Information in Risk Management Decisions

To better understand the value of a more-informed risk estimate, suppose that a decision-maker is considering giving up a pleasurable or beneficial activity that creates a probability h of an adverse outcome within a year and that otherwise has no adverse effects. If I is the set of all information available when the decision must be made, then an optimal decision rule has the following form (see e.g., DeGroot, 1970):

Abandon the activity if and only if $E(h \mid I) > h^*$. (Optimal decision rule)

In other words, the activity should be abandoned if and only if the estimated risk is greater than some maximum acceptable risk level, h^*. This maximum acceptable risk level is an increasing function of the benefit, enjoyment, or utility derived from the activity. It is a decreasing function of the severity of the adverse outcome.

The value of h at a classification tree node that has several children is a weighted average of the conditional (child-specific) values at the child nodes. Suppose that the expected value of h at that node is above h^*, so that the optimal decision if no further information is gained is to abandon the activity. (A symmetric argument applies if $h < h^*$ at the node.) If it is possible to resolve the uncertainty at the node and learn which of its children holds before making a decision, then one of two things will happen. Either the new, more informed value will still be above h^*, so that the decision to abandon the activity remains; or else the new value of h, say $E(h \mid I')$, will be less than h^* and the optimal decision changes. If the decision changes, it will be because the new information has shown that the old decision would have a lower value than previously expected. Therefore, *the new information has value whenever it leads to a changed decision.* If it changes the best decision, then it has allowed a decision to be avoided that one would no longer want to make in light of the new information.

For this reason, *more-informed decisions are always expected to lead to greater expected utilities than are less-informed decisions.* The new information will change the best decision – the one that would have been made without it – based on less information, if and only if doing so is expected to be advantageous. The technique of *pre-posterior analysis* in Bayesian decision theory (Raiffa, 1968; DeGroot, 1970; Laffont, 1989; Barlow, 1998) calculates the expected reward with and without the new information, and places an economic value on the information based on the difference in reward that it is expected to make.

The following example shows how value-of-information (VOI) concepts can be used in a simplified product safety testing program.

Example: Comparing Information Values from Different Tests

Setting. A manufacturer is considering producing a product such as a vaccine, a cold remedy, or a consumer appliance, but is uncertain about whether it might occasionally cause harm (e.g., undesired side-effects) in some members of the consuming population. The manufacturer assigns utilities or payoffs to the four possible act-state pairs (from the binary produce/don't produce decision variable and the binary harm/no harm state variable) as follows:

Manufacturer's Payoff Table. (Numbers are utilities).

State: / **Act:**	Occasional Harm = s_1	No Harm = s_2
Produce Product = a_1	-2	2
Do not Produce = a_2	0	0

The number in row i, column j, for $i = 1, 2$ and $j = 1, 2$ which may be denoted u_{ij}, is the reward if act a_i is taken and the state turns out to be s_j. Assume that the manufacturer wishes to take the action yielding the largest expected value of the payoff numbers, i.e., the greatest expected utility.

These four payoff numbers could be re-scaled to run from 0 to 1, instead of from –2 to 2, as in Chapter 5's discussion of utilities, without changing the analysis. Also, the binary variables could be replaced with much larger act and state sets, e.g., to reflect different levels of production, conditions for use, and degrees of harm. But this 2 x 2 example suffices to illustrate value-of-information comparisons and calculations, and the results can be extended to larger examples in an obvious way. Chapter 5 explains how to define and construct utilities for outcomes so that choosing the act with greatest expected utility is the unique course of action consistent with certain axioms of rational decision-making.

Now, suppose that the decision-maker can obtain imperfect information about the probable safety of the product, e.g., from laboratory tests or clinical trials. For simplicity, consider a test that can only give results of positive or negative ("Yes" or "No") and that has the following performance characteristics:

TEST A Characteristics	Occasional Harm = s_1	No Harm = s_2
Pr(Test A = Yes):	0.8	0.4
Pr(Test A = No):	0.2	0.6

If the true state is that the product causes no harm, then this test has a 40% chance of reporting "Yes" (a false positive) and a 60% chance of reporting "No". If the product does cause occasional harm, then the test has an 80% chance of reporting "Yes" and a 20% chance of reporting "No".

(a) How much should the manufacturer be willing to pay for the information provided by this imperfect test before deciding whether to produce the potentially hazardous product, if the two possible states (No Harm and Occasional Harm) are *a priori* considered equally likely?

(b) Suppose that the probabilities of the two possible states are unknown and considered equally likely to be anywhere between 0 and 1. How would this change the answer in part (a)?

(c) Suppose that the manufacturer can choose between Test A, described above, and Test B with the following characteristics. Which test is preferable, and how does the answer depend on the prior probabilities of harm and on the utilities assigned to the different outcomes?

TEST B Characteristics	Occasional Harm = s_1	No Harm = s_2
Pr(Test B = Yes)	0.5	0.7
Pr(Test B = No)	0.5	0.3

Solution.

(a) Test A may be regarded as a random variable, A, whose value is correlated with the true but unknown state variable. It gives "Yes" with probability

$$\Pr(A = \text{Yes}) = [\Pr(s_1)\Pr(A = \text{Yes} \mid s_1) + \Pr(s_2)\Pr(A = \text{Yes} \mid s_2)]$$
$$= (0.5)(0.8) + (0.5)(0.4) = 0.6.$$

It gives "No" with probability 0.4. If it gives "Yes", then the posterior probability of harm is:

$$\Pr(s_1 \mid A = \text{Yes}) = \Pr(A = \text{Yes} \mid s_1)\Pr(s_1)/\Pr(A = \text{yes}) = (0.8)(0.5)/(0.6) = 2/3$$

by the definition of conditional probability. If it gives "No", then the posterior probability of harm is:

$$\Pr(s_1 \mid A = \text{No}) = \Pr(A = \text{No} \mid s_1)\Pr(s_1)/\Pr(A = \text{No}) = (0.2)(0.5)/(0.4) = 0.25.$$

If the test result is A = "Yes", then the expected payoff from producing is:

$$EU(a_1 \mid A = \text{Yes}) = (0.67)(-2) + (0.33)(2) = -0.67.$$

If A = "No", then the expected payoff from producing the product is:

$$EU(a_1 \mid A = \text{No}) = (0.25)(-2) + (0.75)(2) = 1.$$

Hence, *the decision-maker should produce if and only if test A says "No"*. The expected payoff using this optimal decision rule is

$$\text{EU with Test A} = \text{EU(best action if A = No)}\Pr(A = \text{No})$$
$$+ \text{EU(best action if A = Yes)}\Pr(A = \text{Yes}) = (1)(0.4) + (0)(0.6) = 0.4.$$

By contrast, if the decision is made without any testing, then the expected value is 0 and an optimal decision is not to produce. Hence the manufacturer should be willing to pay for the test information, since it increases the expected utility of the optimal decision. If all payoffs are measured in units of millions of dollars and if the decision-maker is risk neutral (meaning that utilities are proportional to dollars, see Chapter 6), then the economic value of the test can be expressed in dollar terms as (0.4)($1M) = $400,000.

(b) If the probability of state s_1 is uniformly distributed between 0 and 1, then its expected probability (with respect to this information or assumption) is just $E(p_1) = 0.5$. But an expected probability is a probability. Hence, the analysis with these unknown state probabilities is identical to the analysis in part (a). As discussed in Chapter 6, this mathematical equivalence of a known probability of 0.5 and an uncertain probability with a uniform distribution does not guarantee that the two situations will be *perceived* as being equivalent. Many decision-makers display an aversion to unknown or uncertain probabilities, sometimes called *ambiguity-aversion* (Raiffa 1968; Camerer, 1995).

(c) The obvious way to answer question (c) is to repeat the preceding calculations for Test B and to carry out sensitivity analyses to show how the best choice of experiment varies with the state probabilities and with the utilities. A more insightful approach recognizes that **B** = **AG**, where **B** and **A** are the 2 x 2 matrices representing tests B and A, respectively (i.e., the 2 x 2 matrices of state-to-signal probabilities for the two signals "Yes" and "No"). **G** is a "garbling matrix" that can be interpreted as a probabilistic mapping of the test A result to a signal of "Yes" or "No" for Test B. More specifically, in this example, **A** = [0.8, 0.4; 0.2, 0.6], **B** = [0.5, 0.7 ; 0.5, 0.3], and **G** = (0.2, 0.8 ; 0.8, 0.2) (using the notational convention that semicolons separate rows of the matrix and rounding all numbers to one significant digit). Test B may be interpreted as a measurement in which test A is first performed, and then its answer is randomly "garbled", i.e., changed, with probability 0.2. From this perspective, it is now intuitive that test A is preferable to test B, since B is just a garbling of A. Indeed, *all* decision-makers will obtain greater (or equal) expected utility from a test A than from a test B, for *any* payoff matrix and *any* prior probabilities, if and only if B is a garbling of A (Laffont, 1989). In other words, for any tests A and B mapping the same set of states to sets of signals, A is preferred to B under all conditions if and only if there is a Markov matrix **G** such that **B** = **AG**. Here, **A** and **B** are the matrices of conditional signal probabilities, given states, representing tests A and B, respectively.

Example: A Risk Research Stopping Problem

Suppose that the probability of an adverse health effect from a new product, conditioned on initially available information, is 0.5. This probability can be revised by collecting and conditioning on additional information. Suppose that new information is obtained by performing tests that answer specific questions. Each test produces one bit of information, i.e., the information from learning the answer to a single binary question with *a priori* equally likely answers. Each bit of information obtained is equally likely to increase or decrease the conditional probability of harm (based on all information received so far) by 0.1, until the truth (probability = 0 or 1) is discovered.

This idealized model of the research process is a Gambler's Ruin problem (Ross, 1996) and a martingale. The expected amount of information that must be collected to discover the truth, starting from probability p, is $10p(10 - 10p)$ bits. The theory of ruin problems can be used to determine how much research is optimal and when to stop because the expected cost of continuing enough to change the current best decision (with respect to the available information) outweighs the expected benefits from reduced decision errors.

3. BASIC ENGINEERING MODELING TECHNIQUES

Many areas of applied mathematics, statistics, and engineering are used to construct the detailed exposure models, dose-response models, and other substantive sub-models used in health risk assessment. Four modeling techniques that are especially common in health risk assessment are:

- *Compartmental flow models.* Compartmental models and other *continuous simulation* models use systems of ordinary differential equations (ODEs) and algebraic formulas to determine the rates of flows of substances among compartments over time. They are widely used to model exposure processes, pharmacokinetics, and metabolism of inhaled or ingested chemicals.
- *Monte Carlo uncertainty models* give numerical approximations for the probability distributions of functions of random variables. Thus, inputs described by random variables are propagated through the model to obtain probability distributions of outputs
- *Stochastic transition models* and other *applied probability models* describe intensities of probabilistic transitions (of populations, individuals, cells, etc.) among distinct states over time. Often, the first passage of a system into a particular state (e.g., death of an individual, malignancy of a stem cell, or extinction of a population) is the risk event of interest, and stochastic models are used to calculate the probability distribution of this first passage time.
- *Discrete-event simulation models* (DES) provide approximate numerical solutions for stochastic transition models. Monte Carlo simulation of stochastic processes is used to obtain numerical estimates of event probabilities and first passage times.

The following sections introduce and illustrate these techniques. Mastering them makes much of the technical risk assessment literature accessible.

3.1 Compartmental Flow Continuous Simulation Models

Compartmental flow models are used in exposure and risk assessment modeling to describe the migration of substances among environmental and biological compartments following a release or exposure. They typically use mass balance equations (tracking what goes into a compartment, what goes out, and what stays in the compartment) and empirically derived or estimated equilibrium partition coefficients to describe how substances partition themselves among compartments over time. Many environmental fate and transport models use compartmental flow modeling concepts, perhaps augmented with descriptions of physical processes such as diffusion of contaminants in a moving stream of air or water.

Compartmental flow modeling methodology has been used for decades for both environmental fate modeling and classical pharmacokinetic modeling. The mathematical frameworks are identical. In both cases, linear time-invariant (constant-coefficient) ordinary differential equations (ODEs) provide the best-known modeling approach. In this framework, concentrations of chemical(s) in n different ecological or physiological compartments at any time, t, are represented by an n-vector $y(t)$. Fractional flow rates per unit time among compartments are summarized in an $n \times n$ matrix, A, with elements that respect mass balance. The system of linear time-invariant (LTI) ODEs describing the evolution of concentrations over time is:

$$d\mathbf{y}(t)/dt = \mathbf{A}\mathbf{y}(t) + \mathbf{u}(t), \qquad \text{(LTI compartmental flow model)}$$

where $\boldsymbol{u}(t)$ represents exogenous inputs to the compartments (e.g., from new source emissions) at time t. The symbolic solution to this system of equations is:

$$\mathbf{y}(t) = e^{\mathbf{A}t}\mathbf{y}(0) + \int_{[0,\,t]} e^{\mathbf{A}(t-s)}\mathbf{u}(s)ds,$$

where e^{At} and $e^{A(t-s)}$ are matrix exponentials. In practice, quantitative solutions can easily be determined by numerical integration. This is accomplished by *continuous simulation* software packages. General-purpose mathematical and scientific software such as MATLAB™ also provide fast numerical solutions to such systems.

Estimates of the flow rate parameters in **A** can be made from sufficiently long and rich observations on the input-output histories $\{\mathbf{u}(t), \mathbf{y}(t), t \geq 0\}$, in conjunction with simple assumptions about observation errors. Pre-packaged statistical estimation routines are available in professional nonlinear regression packages such as NONMEM™. For simple input patterns, explicit formulas for **y**(t) can be obtained by solving the above equation symbolically using the symbolic mathematical software engines in packages such as MAPLE™, MATHEMATICA™, or MATLAB™. If the linear compartmental flow model is modified to include nonlinear rate coefficients, e.g., inter-compartmental flow rates that depend on the current value of **y**(t), perhaps reflecting transfer mechanisms that become saturated at high flow levels, then symbolic solutions may not be achievable, but numerical solutions can still be generated via continuous simulation.

Example: A Passive Smoking Exposure Model for the Workplace

Problem: Suppose that smokers in an office generate *S* micrograms per hour of nicotine entering the office air, starting at 9:00 AM each work day. If the dilution volume of the office space is *V* cubic meters and the effective air exchange ratio for tobacco smoke is *q* air changes per hour (including both ventilation and removal of smoke by absorption on surfaces, etc.), then how does the concentration of nicotine in office air change during the day? (Assume for simplicity that the source strength, *S*, remains constant throughout the day.)

Solution: Denote the concentration of nicotine in office air at time *t* by *y(t)*. This is a one-compartment model, with office air being the only compartment for which concentration is to be modeled. (To model the concentration of nicotine in the blood of a non-smoker in the office, it would be necessary to consider additional compartments. Such models are explored for benzene in the next example.)

The concentration in the office air increases due to a contribution of new smoke of (S micrograms per hour/V cubic meters). At the same time, it is being decreased by the introduction of clean air at a fractional rate of *q* air changes per hour. So, the net change in concentration at time *t* is given by the following ODE:

$$dy(t)/dt = (S/V) - qy(t). \qquad \text{(Mass balance ODE for smoke)}$$

This is a special case of the LTI model

$$dy(t)/dt = \mathbf{A}\mathbf{y}(t) + \mathbf{u}(t)$$

with all quantities scalar, A = -q, and u(t) = S/V. Its solution is:

$$y(t) = e^{At}y(0) + \int_{[0,\,t]}e^{A(t-s)}u(s)ds = e^{-q}ty(0) + (S/V)\int_{[0,\,t]}e^{-q(t-s)}ds.$$

Assuming that y(0) = 0 (since smoke dissipates over night, so that each day starts with zero concentration), this time-varying solution becomes:

$$y(t) = (S/qV)(1 - e^{-qt}) \text{ for } t \geq 0 \qquad \text{(Concentration history in office air).}$$

In steady state, inflows and outflows balance, i.e.,

$$dy(t)/dt = (S/V) - qy(t) = 0,$$

or,

$$y^* = S/qV. \qquad \text{(Steady-state concentration in office air)}$$

Repace et al. (1998) used essentially this model with some algebraic substitutions and elaborations (such as source strength = emissions rate per smoker times number of smokers) to estimate workplace concentrations of nicotine in air due to environmental tobacco smoke (ETS). (They also assumed without justification that adverse health effects are proportional to concentrations, even at low concentrations, yielding the sensational conclusion that 400 lung cancer deaths and 4000 heart disease deaths per year among non-smokers might be caused by ETS in offices. However, this portion of the risk assessment seems speculative, as it fails to distinguish between statistical association and causation; see Chapters 3 and 4.)

3.2 Applications to Pharmacokinetic Models

Compartmental models are frequently used in risk assessment to model pharmacokinetics and metabolism of chemicals within exposed people or animals. The input-output structure of classical compartmental models of pharmacokinetics and metabolism, as well as of modern physiologically-based pharmacokinetic (PBPK) models, may be expressed very simply in DAG notation as:

$$\{x(t)\} \rightarrow \{\mathbf{y}(t)\} \qquad \text{(PBPK model DAG).}$$

Here curly braces {} enclose time series and the notation is as follows:

- {x(t)} denotes the time series of administered doses or exposures up to and including time t.

- {**y**(t)} denotes a vector giving the concentrations of the administered compound and all of its metabolites in each compartment in the model, up to and including time t.

Since exposures up to time t determine physiological compartment concentrations of a substance and its metabolites up through time t for *all* t, the causal graph can be further abbreviated by dropping *t*, as follows:

$$\{x\} \rightarrow \{\mathbf{y}\}.$$

This notation should be interpreted as indicating that the history of x up to any time determines the history of **y** up through that same time.

Population pharmacokinetic models are developed from data collected from multiple individuals, e.g., sampled blood-level and tissue-level concentrations over time, as well as individual covariates such as sex, age, and weight. Many of these models emphasize the dependence of {**y**} on individual physiological parameters and covariates. The frequency distribution of these quantities in a population may be estimated from data and can help to understand and predict the health impact in a heterogeneous population of different dosing regimens or exposure histories. In addition, inter-individual variability in PK parameters can be characterized from population PK data. A DAG model for a population PK model is:

$$\{x\} \rightarrow \{\mathbf{y}\} \leftarrow \{\mathbf{z}\}$$

where **z** is the vector of individual parameters and covariates that mediates the effect of {x} on {y}. It is also sometimes convenient to use the following more compact notation to express the same relation among the variables:

$$(\{x\}, \{\mathbf{z}\}) \rightarrow \{\mathbf{y}\}.$$

In words, the values of x and **z** up through any time t jointly determine the history of **y** up through time t.

Monte Carlo uncertainty analysis, discussed in the next section, provides ways to estimate the frequency distribution of {**y**} in the population from the joint population frequency distribution of {x} and {**z**}.

Finally, the current state of the art in biologically-motivated dose-response modeling combines pharmacokinetics with *pharmacodynamic* (PD) models describing the interactions of chemicals with cells. Examples include models of the interactions of genotoxic chemicals or metabolites with DNA; models of cell surface receptor-mediated stimulation of inappropriate cell division or cell death; and models of cytotoxic (cell-poisoning) damage and compensating cell proliferation or other altered cell kinetics in response to such damage. Such models promise to increase the amount of biological knowledge and realism brought to bear in quantitative modeling of dose-response relations.

Example: Compartmental Pharmacokinetic (PK) and PBPK Models

Suppose that an animal or person is exposed to 10 ppm of benzene in an inhalation chamber for 6 hours. How might the concentration of benzene in its blood stream change during this exposure? Compartmental flow models can answer such questions.

A classical compartmental pharmacokinetic (PK) modeling approach postulates that benzene passes from chamber air into various internal compartments, first reaching the blood via the lungs and then partitioning between between the blood and one or more deeper compartments (e.g., fat, muscle, organs, etc.) The flows of benzene among these compartments might be diagrammed as follows if all non-blood physiological compartments are aggregated into one:

Chamber air $\Leftrightarrow$ *blood* $\Leftrightarrow$ *deep compartment* $\Rightarrow$ *excretion.*

Here, thick arrows indicate flows of material among compartments, to avoid confusion with the information arrows in DAG diagrams. These flows may be described quantitatively by ordinary differential equations (ODEs). For example, a simple linear, constant-coefficient ODE for concentration in blood in this system is:

d[blood concentration]/dt =
k_1[Chamber air concentration – k_2(blood concentration)]
– k_3[(blood concentration – k_4(deep compartment concentration)]

The k_i are micro-constants that reflect compartment volumes and fractional transfer rates per unit time, e.g., in response to concentration gradients between adjacent compartments. These constants may be estimated from inhalation experiment data by non-linear regression programs. Programs such as NONMEM™ (http://pkpd.icon.palo-alto.med.va.gov/) automate much of the required statistical and mathematical modeling. In addition, pharmacokinetic modeling programs can test hypotheses about the number of compartments required and their interconnection patterns. For example, they can test whether the "deep compartment" must be disaggregated into multiple other compartments to adequately fit time course data from measurements of concentrations in chamber air and animal blood. Once the micro-constants and the structure of the compartmental model have been estimated, the flows among compartments can easily be simulated. In the current case with linear, constant-coefficient ODEs, the amounts in different compartments over time can also be solved for analytically.

An alternative, more biologically based modeling approach simulates the distribution of chemicals into different organ and tissue groups as blood circulates, as well as simulating the pharmacokinetic and metabolic processes that convert administered doses of chemicals to internal doses of metabolites in different physiological compartments. Current *physiologically-based pharmacokinetic* (PBPK) models accomplish this simulation by representing the flow of a chemical from compartment *j* to compartment *i* at time *t* (for non-metabolizing compartments) by the equation:

$$f_{ij}(t) = k_{ij}[C_j(t) - C_i(t)/P_i].$$

Here, $C_i(t)$ is the concentration in compartment *i* at time t. P_i is an *equilibrium partition coefficient,* such that the concentrations in compartments *i* and *j* are in equilibrium (i.e., no net flow occurs among them to further equalize concentrations) when $C_j(t) = C_i(t)/P_i$. k_{ij} is a *rate constant* that reflects compartment volume and the rate of flow in response to a concentration gradient. The rate of change in the concentration of a chemical in compartment *i* at time *t* is thus

$$dC_i(t)/dt = \Sigma_j[f_{ij}(t) - f_{ji}(t)],$$

i.e., it is driven by the sum over all compartments *j* of the flow into *i* from *j* minus the flow out of *i* into *j*. (If *j* is not adjacent to *i*, the flow between them is zero. The physical flows could be expressed in units of mass per unit time, e.g., fractions of a mole per minute. Then, dividing by compartment volume gives the change in concentration per unit time.)

If the chemical is metabolized in compartment *i* – as benzene is in liver and bone marrow, for example – then the expression for $dC_i(t)/dt$ must be modified. The most common revision is to include a non-linear *Michaelis-Menten* term

$$[-V_{max}C_i(t)/P_i]/[K_m + C_i(t)/P_i]$$

on the right-hand side:

$$dC_i(t)/dt = \Sigma_j[f_{ij}(t) - f_{ji}(t)] - [V_{max}C_i(t)/P_i]/[K_m + C_i(t)/P_i]$$

to describe saturable metabolism. Here V_{max} and K_m are the *Michaelis-Menten parameters* for the aqueous component of the compartment. The Michaelis-Menten formula is fundamental in enzymology. It is derived from mechanistic considerations of available enzyme molecules binding to available (non-bound) substrate molecules.

Given the values of the pharmacokinetic parameters k_{ij} and P_i and the metabolic parameters V_{max} and K_m, the system of ODEs can easily be integrated numerically to solve for the time series $\{C_i(t),\ t \geq 0\}$. The solutions can be compared to experimental observations to validate the model or to indicate a need for model refinements and corrections.

Example: Uncertainty Analysis of Linear Dynamic Models

It is sometimes argued that the use of increasingly complex "biologically-based" risk assessment (BBRA) models to capture increasing mechanistic understanding of carcinogenic processes may run into a practical barrier that cannot be overcome in the near term: the need for unrealistically large amounts of data about pharmacokinetic and pharmacodynamic parameters. This example shows that, for a class of dynamical models widely used in biologically-based risk assessments, it is unnecessary to estimate the values of the individual parameters.

Consider the mapping from administered dose history (a scalar time series) to the resulting history of metabolite concentrations in different physiological compartments of a physiologically-based pharmacokinetic (PBPK) model (a vector time series). If metabolism and pharmacokinetics are linear over the exposure range of interest (meaning that doubling the magnitude of the input at each moment will double the magnitude of each output at each moment) and if the rate coefficients remain constant over time, then the PBPK model can be represented as:

$$dy(t)/dt = \mathbf{A}y(t) + \mathbf{b}x(t)$$

where **A** is a matrix of intercompartmental flow rate coefficients and **b** is an allocation vector describing how the scalar dose administered at time t, denoted by x(t), is distributed among compartments. (For example, x(t) might reflect injection into the arterial blood or a muscle tissue group, concentration inhaled into alveolar air, or amount ingested by oral gavage.) Classical compartmental flow models of pharmacokinetics are often explicitly described in this form (Mazumdar, 1989, Chapter 5; Seber and Wild, 1989, Chapter 8). Ramsey-Andersen type PBPK models (Riggs, 1963; Ramsey and Andersen, 1984) may also be expressed in this form over dose ranges where nonlinearities (e.g., due to Michaelis-Menten saturable kinetics) are small enough to be ignored. This is realistic for many pesticides and volatile organic compounds over a wide range of administered concentrations. For example, the Ramsey-Andersen PBPK equation for tissue uptake by a nonmetabolizing tissue is

$$dC_i(t)/dt = (Q_i/V_i)[C_{art}(t) - C_i(t)/P]$$

where

$C_i(t)$ = concentration of chemical in tissue group i at time t

Q_i = blood flow (volume per unit time) to tissue group i

V_i = tissue distribution volume

P = tissue:blood partition coefficient

$C_{art}(t)$ = concentration of chemical in arterial blood.

This is clearly of the form

$$dy(t)/dt = ay(t) + bx(t)$$

upon making the identifications

$$y(t) = C_i(t), a = (-Q_i/PV_i),\ b = (Q_i/V_i), x(t) = C_{art}(t).$$

Thus, the entire system of (linear) PBPK equations in the Ramsey-Andersen model can be expressed as a special case of the form $d\mathbf{y}(t)/dt = \mathbf{Ay}(t) + \mathbf{b}x(t)$.

Suppose that the effect of internal dose, as measured by the concentration of some metabolite in a tissue or susceptible cell population, is to increase the rate of initiating mutations (per susceptible stem cell per unit time), but with the size of the "normal" stem cell population at risk of initiation remaining approximately constant Then it is plausible that the risk due to exposure should depend only on the area-under-curve (AUC) of the internal dose time series. Let

$$z(t) = \mathbf{cy}(t)$$

be the relevant measure of internal dose (a scalar), where **c** is a vector of weights. For example, z(t) might be a component of **y**(t), an aggregate (i.e., a nonnegative linear combination) of several components of y(t), or a partition coefficient times a component of **y**(t). Let AUC(z) denote the total AUC for the internal dose:

$$AUC(z) = \int_{0 \leq t \leq \infty} z(t)dt.$$

Assume that tumor risk depends on the internal dose history {z(t)} only through AUC(z) and that dosing for any finite period leads to responses (e.g., concentrations of metabolites in compartments) that all eventually subside and approach zero after dosing stops. Thus, all compartments eventually drain (perhaps through other compartments) to the external environment. (This is realistic for the pharmacokinetics of many chemicals. However, it would not hold for chemicals that become permanently sequestered in a tissue, e.g., fat or bone. In applications of this concept to pharmacokinetic or other biomathematical models, the "relaxation time" for the system, i.e., the time for responses to return almost to zero. should be significantly less than the lifetime of the animal. Then *the relation between AUC(z) and AUC(x) is as follows:*

AUC(z) = k * AUC(x) for some constant k.

For, the system dynamics for a general linear constant-coefficient system can be represented by the following vector differential equation, where **A** has full rank (Wolovich, 1994):

$$dy(t)/dt = \mathbf{A}\mathbf{y}(t) + \mathbf{B}\mathbf{x}(t) \qquad \text{(state dynamics equation)}$$
$$\mathbf{z}(t) = \mathbf{C}\mathbf{y}(t) \qquad \text{(output equation)}$$

If the original system has a flow rate matrix **A** that does not have full rank, then an observationally equivalent system with the same input-output behavior and **A** of full rank can be constructed, e.g., by deleting irrelevant states (compartments not affected by the input) and aggregating redundant states (e.g., those connected to the rest of the system by single input and output flows) (Wolovich, 1994, p. 85; Cobelli et al., 1979).] For a globally asymptotically stable system with initial condition **y**(0) = **0**, the limiting behavior is by definition such that **y**(t) → 0 as t → ∞. Integrating both sides of the state dynamics equation from time t = 0 [defined as the first time at which **x**(t) > 0] to t = ∞ gives **y**(∞) - **y**(0) = **0** = **A***AUC(**y**) + **B***AUC(**x**), where AUC(**x**), for example, denotes the integral of **x**(t) from t = 0 to t = ∞. Rearranging gives AUC(y) = $-\mathbf{A}^{-1}\mathbf{B}$*AUC(x). Premultiply by **C** to obtain:

$$AUC(\mathbf{z}) = -\mathbf{C}\mathbf{A}^{-1}\mathbf{B}*AUC(\mathbf{x}).$$

Define $\mathbf{k} = -\mathbf{C}\mathbf{A}^{-1}\mathbf{B}$. Specializing to scalar x(t) and z(t) proves the above result.

The significance of this result for complex biologically-based risk assessment (BBRA) models is as follows. *If risk is determined by the AUC of some metabolite in some compartment of a globally stable linear constant-coefficient compartmental flow model, then it is determined by the AUC of the administered dose time series (i.e., by total administered dose) and a single other number, k.* Data uncertainties about the values of the numerous biological constants typically involved in complex BBRA models (e.g., the partition coefficients and metabolic rate parameters in PBPK models) are irrelevant as long as AUC(x) and AUC(z), or their ratio, can be estimated from data. Once k is known, it can be used to predict values of AUC(z) (and hence risk) for any value of AUC(x), even if the relevant metabolite(s) and target cell populations are not known.

This result illustrates a general principle: for many complex models, the number of degrees of freedom in the input-output relation is much smaller than the number of input quantities that must be specified to initialize the model. Moreover, sensitivity analyses of complex simulation models reveal that uncertainty about the input-output behavior of a model (e.g., as quantified by the coefficient of variation of an input:output ratio) may be much less than the uncertainties (coefficients of variation) for its input quantities. As an example, a sensitivity analysis of the steady-state ratio of benzene metabolized in the liver to benzene concentration in inhaled air in a PBPK model for B6C3F1 mice (AIHC, 1993, p. 66) found that:

- A 100-fold variation in K_m/V_{max} around its nominal level (i.e., multiplying or dividing by 10) changes the steady-state input-output ratio by less than 2%.
- A 100-fold variation in Q_{alv}/N, where Q_{alv} = alveolar ventilation rate and N = blood:air partition coefficient, produces only about a 4-fold variation in the steady-state input-output ratio. (In this sense, uncertainty about input-output behavior is only about 1/25 as great as uncertainty about the value of Q_{alv}/N.)
- A 10-fold increase in the nominal value of the liver blood flow rate, Q_{liver}, increases the steady-state input-output ratio by only about 25%, from 0.8 to 1.0, but a 10-fold decrease in Q_{liver} would reduce it almost to zero. Fortunately, this parameter value is known quite well, so that its estimated value is unlikely to be high by a factor of 10.

In addition, the steady-state input-output ratio at low concentrations depend on the original PBPK model parameters only through the three "reduced parameters" K_m/V_{max}, Q_{alv}/N, and Q_{liver} (Bogen and McKone, 1988). Thus, uncertainty about the ratio depends only on uncertainty about these three quantities, rather than on uncertainties in all of the original parameters. (Thus, it would be inappropriate to estimate uncertainties by a Monte-Carlo analysis that sampled K_m, V_{max}, Q_{alv}, N, and Q_{liver} independently.) Further, although both K_m and V_{max} were individually quite uncertain, their ratio is closely approximated by a measured quantity – the rate of low-dose metabolism – and is therefore far less uncertain than either its numerator or its denominator. These points apply to many other PBPK models. Hetrick et al., 1991, reached similar conclusions via Monte-Carlo analysis of PBPK models for methylene chloride, methylchloroform, and styrene.

3.3 Monte Carlo Uncertainty Analysis

Monte Carlo uncertainty analysis is used to propagate probabilistic inputs through formulas or models to obtain approximate probability distributions for their outputs. The basic idea is simple, although it can be implemented with great sophistication. Suppose that a known model is given, say,

$$y = f(x),$$
$$x = \text{input},\ y = \text{output}$$

along with a probability distribution function F, where $F(a) = \Pr(x \leq a)$, for the value of the input x. Although the probability distribution of y could then be derived by the calculus technique of change of variables, it is often simpler to approximate it by (a) Randomly drawing multiple values of x from F, (b) Calculating $y_i = f(x_i)$ for each sampled input value x_i; and (c) Forming the sample distribution function of the resulting y_i values. This sample distribution approximates the true probability distribution of y induced by the probability distribution of x. It is called a *Monte Carlo* approximation.

Two advances greatly extend the power of simple Monte Carlo uncertainty analysis. One is that it can easily be applied to probabilistic formulas or sub-models, e.g., to nodes in DAG models that map combinations of input values to conditional probability distributions over output values. If the deterministic model formula

$$y = f(x)$$

is replaced with a probabilistic model

$$Pr(y \mid x)$$

then the Monte Carlo uncertainty analysis simply draws values of y from $Pr(y \mid x)$ for any given x value, instead of calculating it deterministically. The other major extension applies to cases in which x is a vector of inputs and it is hard to sample directly from F. Then, in many cases, the components of x can be sampled in a cycle from appropriate conditional distributions that are far easier to sample from than is the joint distribution F. Repeating many cycles eventually leads to sampled values of the components that approximate successive draws from their joint distribution. *Gibbs sampling* and other forms of Markov Chain Monte Carlo (MCMC) sampling, discussed further in Chapter 3, exploit this idea.

Sophisticated implementations of basic Monte Carlo uncertainty analysis include techniques familiar from Monte Carlo simulation applications in other areas of operations research. These include variance reduction methods (e.g., antithetic variables), Latin hypercube sampling, and graphic user interfaces that allow DAG networks of formulas to be assembled and that then propagate probabilistic inputs through the entire network of formulas. Commercial software products such as Analytica™, discussed in Chapter 4, package Monte Carlo uncertainty analysis techniques into convenient software for modeling and uncertainty analysis.

Propagation of distributions via Monte Carlo simulation plays a second major role in health risk assessment, in addition to facilitating probabilistic uncertainty analysis. Often, the goal of analysis is to estimate the population frequency distribution of exposures or risks. For example, the model

$$y = f(x)$$

could represent a dose-response model, with y = probability of adverse health effect and x = a summary of exposure. If the frequency distribution of x in a population is known, then the frequency distribution of y in the population can be obtained by the same Monte Carlo sampling process outlined above. Thus, Monte Carlo analysis can be used to quantify both uncertainty, when x represents a random variable, and variability, when x has a distribution of values within a population (see e.g., Price et al., 1996).

Example: Monte Carlo Uncertainty Analysis in a Dose-Response Model

Suppose that the probability of an adverse health response in a consumer who ingests x colony-forming units (CFUs) of a pathogenic bacterium in contaminated food during a meal is:

$$\Pr(\text{illness} \mid \text{exposure} = \text{x CFUs}) = 1 - \exp(-\lambda\alpha x) \text{ for } x \geq 0.$$

Here, λ is interpreted as a potency parameter (a property of the bacterium) and α as a susceptibility or "frailty" parameter (a property of the consumer). The DAG diagram for this situation is:

exposure x → response probability p ← susceptibility α

↑

potency λ

According to this (simplified) model, the risk to an individual of illness from ingesting this bacterium depends on the three quantities x, λ, and α, all of which may be uncertain. Uncertainty about α for a randomly selected individual may reflect inter-individual variability of α values within the population. It may also reflect limited ability to accurately estimate α from available data, even if it can be assumed that all individuals in the population have the same value of α.. Uncertainty arising from the fact that a parameter has a frequency distribution of different values within a population is often referred to as *variability.* It cannot be eliminated by additional sampling.

Uncertainty due to limited ability to estimate the value of a parameter from available data (or to estimate its frequency distribution of values in a population, if it has more than one value) is often just called uncertainty. Uncertainty about x and λ for an individual is called *epistemic uncertainty* if it is thought that true values exist but are simply not known to the analyst. By contrast, the uncertainty about whether a response will occur, even if (x, λ, α) are all known for an individual, is called *stochastic uncertainty.* A stochastic uncertainty arises if the relation between inputs and outputs is probabilistic rather than deterministic. Physical sources of stochastic uncertainty include the random binding of competing signaling molecules to cell surface receptors over time, sensitivity of physical systems to initial conditions to a degree that cannot be resolved by measurements, and random diffusion of chemicals or substances in biological systems. Finally, model uncertainty refers to uncertainty about whether the relation Pr(illness | exposure = x CFUs) = 1 – exp(-λαx) is correct. Doubt about whether the functional form of the model is correct and whether all relevant variables have been included are types of model uncertainty.

A Monte Carlo uncertainty analysis of risk to an individual in this example works by sampling values of the three quantities (x, λ, α) from their joint probability density function (PDF), assuming that it is known. (If the quantities are statistically independent, then the Monte Carlo program can sample each quantity from its own marginal PDF. In general, the PDFs of inputs must be estimated from data using statistical methods or modeling.) For each sampled triple of values for (x, λ, α) the occurrence or non-occurrence of the response is simulated as the value of a Bernoulli random variable having value 1 with probability 1 – exp(-λαx) and value 0 otherwise. Thus, the Monte Carlo approach can simulate stochastic as well as epistemic uncertainties. Averaging the value of this binary indicator variable over many repetitions approximates the average (i.e., unconditional) individual response probability induced by the joint PDF of the three inputs.

Monte Carlo simulation can also be used to calculate a frequency distribution of individual risks in a population of individuals from the population frequency distributions of x, λ and α. Thus, it is able to handle variability as well as stochastic and epistemic uncertainties. Each sampled triple of (x, λ, α) values for an individual generates a corresponding individual risk of 1 – exp(-λαx). Hence, the joint PDF of (x, λ, α) values in the population determines a probability distribution for individual risks. The applied probability technique of change-of-variables can be used to find this frequency distribution analytically in simple cases, but Monte Carlo simulation can be used even when exact mathematical analysis is difficult or impossible.

Example: Higher-Order Monte Carlo Uncertainty Analysis

A subtle but important point is that the joint PDF of the inputs can be used to determine a probability distribution for population frequency distributions of individual risks. To illustrate the concept, consider a bivariate risk model in which each individual has a constant exposure level and a constant response threshold. An individual with exposure x and response threshold t responds if and only if $x > t$. Suppose that individual response thresholds are uniformly distributed between 0 and 1, i.e., $t \sim U[0, 1]$ and that exposure levels are also independently distributed between 0 and 1, i.e., $x \sim U(0, 1]$. Then what is the frequency distribution of individual risks in a population of size N?

The answer is that the actual frequency distribution of individual risks is uncertain, but its probability distribution can be determined. This example is simple enough so that either an analytic solution or Monte Carlo uncertainty analysis is practical. For a Monte Carlo analysis, one might first assign to each of N hypothetical individuals values of t drawn from U[0, 1]. Then, the probability that individual i with threshold t_i will have an adverse response to his or her random exposure is $1 - t_i$. After all N individuals have been assigned response levels, a frequency distribution of their individual risks, i.e., of $1 - t_i$, can be calculated. Repeating the entire exercise many times allows one to calculate the probability distribution of the frequency distribution – for example, the fraction of runs for which the cumulative frequency distribution of individual risks, evaluated at any particular quintile, lies between any two values of interest. Thus, Monte Carlo analysis can be used to calculate probabilities of population frequency distributions of individual risks from probability distributions of input quantities.

Analytically, the total number of responses, R, has a binomial distribution with parameters N and $p = 0.5$ (since $\Pr(x > t) = 0.5$ for each individual.) For example, the probability that all N individuals will have zero risk is $\Pr(R = 0) = 0.5^N$. Notice that, in this example, the *unconditional* risk for each individual (i.e., the risk not conditioned on information about x or t) is 0.5, while the *conditional* risk for an individual with response threshold t is $1 - t$. The risk for an individual with exposure x and threshold t is either 0 or 1. Thus, individual risk must be defined with respect to what is known about the individual.

3.4 Applied Probability and Stochastic Transition Models

Applied probability models are used in all areas of applied risk analysis to describe and predict how changes in model inputs affect the probabilities of different outcomes. Stochastic process models describing the probable time evolution of dynamic systems, such as cell populations, infected human populations, or individual patients, have been used to model epidemics, cancer development, and disease progression, respectively. These models can be used to quantify the probability distribution of the time until the first occurrence of a specific undesirable health state in an individual or in a population, conditioned on the history of observations made so far. They can also be used to estimate the probable amounts of time that will be spent in different health states.

Applied probability analysis typically determines the mathematical form of the model relating input parameters, such as transition rates, to observed health consequences. The values of the parameters must be estimated from data by appropriate statistical methods. Chapter 3 introduces relevant statistical principles and techniques.

The logic of stochastic process models is often similar to that for compartmental flow models (see e.g., Matis and Kiffe, 2000). The main difference is that discrete individuals (e.g., cells or people) make probabilistic transitions among states at discrete moments, instead of continuous quantities flowing deterministically among compartments in continuous time. However, the framework of a set of distinct states (or compartments), dynamics defined by transition rates (or flow rates) among them, and determination of probable state trajectories (or deterministic state trajectories) from initial conditions and rate parameters applies to both. While continuous simulation programs can be used to numerically solve compartmental flow models for the amounts in different compartments over time, discrete-event simulation programs can be used to numerically simulate the transitions of individuals among states over time.

Example: SIR and Reed-Frost Models for Epidemics

Stochastic models and discrete-event simulation (DES) are often used in health risk assessment to model infectious diseases in a population: One of the simplest models distinguishes among only three states, with allowed transitions as follows:

Susceptible ⇒ Infected ⇒ Removed. *(SIR transition model)*

In this "SIR" model, individuals can make one-way transitions from the Susceptible to the Infected compartment, and from the Infected to the Removed (e.g., recovered and immune) compartment. Each such transition is a discrete event for a specific individual, making it natural to use DES methods to study how the process evolves.

Suppose that a population of size N is distributed among these compartments. Let S(t), I(t), and R(t) denote the number of individuals in the Susceptible, Infected, and Removed compartments, respectively, at the start of discrete time period t. The time periods are chosen so that one period is the length of the duration of infectiousness. (This simplification can be relaxed, but is adequate for purposes of illustration.) The members of the population are assumed to mix completely and randomly. Let p denote the probability of effective contact between any infected and any susceptible person in one time period. Then the evolution of the infected population is described by the stochastic difference equation:

$$\Pr[S(t+1) = k \mid S(t), I(t)] = {}_{S(t)}C_k(1-p)^{kI(t)}[1-(1-p)^{I(t)}]^{S(t)-k} \quad \text{(Reed-Frost Model)}$$

where ${}_{S(t)}C_k$ denotes the combination of S(t) individuals taken k at a time. In words, this is the probability that exactly k out of the S(t) susceptibles avoid effective contact by any of the I(t) infectives, while the rest do not.

This simple Markov model, called the Reed-Frost model, cannot be solved analytically. But it can easily be simulated in any discrete-event simulation modeling package. Doing so and studying the state trajectories from a few hundred runs for each of various parameter combinations suggests some interesting qualitative conclusions, such as the existence of a threshold value for p above which an epidemic is highly likely.

The SIR model can also be approximated as a continuous compartmental flow model. The flow diagram is still:

$$S \Rightarrow I \Rightarrow R$$

The flows are described by the following system of ODEs:

$$dS/dt = -aIS$$

$$dI/dt = aIS - bI \quad \text{Continuous Kermack-McKendrick Model}$$

$$dR/dt = bI$$

Here, a and b are flow rate parameters describing the rate of new infection per susceptible-infected pair per unit time and the recovery or removal rate, respectively. These equations give a continuous-time, deterministic approximation to the discrete-event stochastic system for large populations. (This is a continuous version of the *Kermack-McKendrick* model of epidemics; see Hoppensteadt and Peskin, 1992.) This system of ODEs reveals that there is indeed a population threshold size for S beyond which epidemics will spread once started. Since

$$dI/dt = I(aS - b),$$

it follows that $dI/dt > 0$, i.e., the pool of infectives grows over time, if and only if $S(0) > b/a$. Thus, the threshold size for S required to produce an epidemic (defined as $dI/dt > 0$) is:

$$S^* = b/a.$$

For initial susceptible population sizes larger than this threshold, an initial infected population pool will grow, reach a maximum, and then decline to zero, leaving the final susceptible population smaller than its initial size, S(0), by an amount that depends on how much larger S(0) was than S*.

Although such analytic results are elegant and suggestive for some real epidemics, deterministic modeling of epidemics cannot realistically express the role of probabilistic events early in the initial development of an epidemic, when the infected pool may be very small. SIR models have been fit to data from a variety of real epidemics, but the fits are sometimes poor. The mathematical modeling and analysis have therefore been extended to include important realistic complexities (Grenfell et al., 1995; Hethcote, 2000). These include age-structured populations with age-dependent infection and recovery/removal parameters; incomplete mixing of populations; geographically distributed and heterogeneous populations; and seasonal or other periodic (e.g., school year based) forcing of infection rates. An additional compartment for "Exposed" (but not yet Infective) individuals is often added, forming an "SEIR" compartmental model. The possibility of chaotic trajectories (and hence limited ability to predict the timing of outbreaks) has been explored in both SIR and SEIR models of childhood epidemics of measles and chickenpox. SIR and SEIR models have been used in planning vaccination programs and other risk management strategies.

For recommended further reading, including statistical methods for estimating the parameters of stochastic epidemic models from data, see Andersson and Britton (2001).

Example: A Stochastic Model of Carcinogenesis

The age-specific hazard rate for occurrence of a tumor at time *t* in an individual who has survived until *t* without tumor can be calculated from biological quantities instead of being estimated from population survival data on age-at-first-tumor. In a simple multistage model of carcinogenesis, stem cells are hypothesized to make transitions among successive states, corresponding to increasing sets of transformations along the path to malignancy. The state transition diagram for a three-state model, with states of "Normal", "Initiated", and "Malignant" for each stem cell, is as follows:

N(t) = Normal stem cells $\Rightarrow$ I(t) = Initiated stem cells $\Rightarrow$ M(t) = malignant stem cells

This model is often applied to situations in which a single chemical carcinogen is thought to act on different stages for an extended period. Let T denote the duration of exposure. The probability that at least one malignant cell is formed by time T may be derived as follows.

i. The number of initiated cells created from time t = 0 to time t = T, denoted by I, is a Poisson random variable with mean $\mu_1 NT$. Here, μ_1 = transition rate of stem cells from the Normal to the Initiated state. It is measured in units of expected transitions (interpreted as somatically heritable transformations) per cell per unit time. Below, the analysis is extended to the case in which this transformation rate depends linearly on the dose rate, so that $\mu_1 = a_1 + b_1x$.

ii. Conditioned on the value of I, the arrival times of the initiated cells are independently and uniformly distributed over the interval from 0 to T (Ross, 1983, p. 37). The probability that an initiated cell created at time t < T has not undergone a second (malignant) transition by time T is $\exp[-\mu_2(T - t)]$. Here, μ_2 = transition rate of cells from the Initiated to the Malignant state. Below, the analysis is extended to the case in which this transformation rate also depends linearly on the dose rate, so that $\mu_2 = a_2 + b_2x$. Integrating over all times t in (0, T] with respect to the uniform density 1/T shows that the total probability that an initiated cell created between 0 and T survives until time T without becoming malignant is given by the following quantity:

$$s = [1 - \exp(-\mu_2 T)]/\mu_2 T = \Pr(\text{initiated cell does not become malignant}).$$

iii. The probability that none of the I initiated cells becomes malignant is s^I. Therefore, the conditional tumor probability that at least one cell becomes malignant is

$$P = 1 - s^I.$$

iv. The unconditional tumor probability (averaged over the conditional probabilities for all values of I weighted by their Poisson probabilities) is:

$$E(P) = 1 - E(s^I) = 1 - \exp[-(1 - s)\mu_1 N] \qquad \text{(Exact two-stage model for risk)}$$

where the expectation follows directly from the moment generating function for a Poisson random variable. This formula has a physical interpretation as:

tumor probability = 1 - exp[-(expected number of malignant cells formed)].

This completes the derivation of the tumor probability.

To apply these formulas to risk assessment, it is necessary to include dose explicitly and to develop methods for estimating the unknown quantities from data. Dose may be included as follows. Define the dose-dependent cumulative hazard at time T as

$$H(x) = (1- s)\mu_1(x)NT.$$

The formula for s and algebraic rearrangement yield

$$\begin{aligned} H(x) &= [\mu_1(x)NT][\mu_2(x)T - 1 + \exp(-\mu_2(x)T)]/\mu_2(x)T \\ &= N[\mu_1(x)/\mu_2(x)]\{\mu_2(x)T - 1 + \exp[-\mu_2(x)T]\} \\ &= N[(a_1 + b_1x)/(a_2 + b_2x_2)]\{(a_2 + b_2x)T - 1 + \exp[-(a_2 + b_2x)T]\}. \end{aligned}$$

Now, recall that $H(x) = -\ln[1 - P(x)]$, where $P(x)$ denotes the tumor probability at dose x. Equating these two formulas for H(x) gives the regression model

$$H(x) = N[(a_1 + b_1x)/(a_2 + b_2x)]\{(a_2 + b_2x)T - 1 + \exp[-(a_2 + b_2x)T]\}.$$

H(x) may be either convex or concave, depending on the (nonnegative) values of its parameters.

To fit this regression model to data from animal dose-response experiments, the observed quantities x and P(x) (and hence $H(x) = -\ln[1 - P(x)]$) can be tabulated from experimental data. The observed values of P(x), namely

$$P^*(x) = n(x)/N(x)$$

where

n(x) = number of responding animals exposed to x
N(x) = total number of animals exposed to x,

contain binomial sampling error. From the data points $P^*(x)$ for several dose groups, x, it is possible to estimate the parameters of the regression model.

It may appear that there are six unknown parameters to be estimated, namely (N, a_1, b_1, a_2, b_2, T). However, N can be absorbed into the constants a_1 and b_1, since it multiplies them, and only the products Na_1 and Nb_1 can be identified from the data (e.g., P(x) depends on N and a_1 only through their product.) Similarly, any fixed T can be absorbed into a_2 and b_2 (and into a_1 and b_1 in the first term, on multiplying by T/T). These simplifications reduce the quantal regression model to the following four-parameter family:

$H(x) = [(a_1 + b_1x)/(a_2 + b_2x)]\{(a_2 + b_2x) - 1 + \exp[-(a_2 + b_2x)]\}$. (Regression model)

The four parameters of this model are scaled versions of the corresponding original parameters. They can be estimated by any of several nonlinear regression techniques (Seber and Wild, 1989), e.g., using a maximum-likelihood or a least squares criterion.

This example illustrates how applied probability modeling based on a hypothesized underlying transition process can be used to specify the functional forms of parametric risk models. These parametric models may then be fit to data using statistical techniques, as discussed further in Chapter 3. Different underlying biological models, e.g., based on different assumptions about how exposure affects the kinetics of cell proliferation and death (not considered in the above model), lead to different parametric families of models. Statistical data analysis can then help to evaluate the relative plausibility of different hypothesized models if the parametric models based on them imply significantly different observed values of experimental data.

Finally, there are many more general cancer risk assessment models in which stem cells make probabilistic transitions among more than two stages, with cell divisions and deaths taking place at different rates in different stages. Many of these models with constant exposures can be described or approximated as Markov processes with a state vector $\mathbf{y}(t)$ giving the number of cells in each state at time t. The vector of expected number of cells in each state then evolves according to the linear system of equations:

$$dEy(t)/dt = \mathbf{A}E[\mathbf{y}(t)],$$

where $\mathbf{A}$ is the matrix of transition rates (possibly dose-dependent) and E denotes expected value. As implied by our discussion of compartmental flow models, this has the symbolic solution:

$$Ey(t) = e^{\mathbf{A}t}\mathbf{y}(0)$$

when $\mathbf{A}$ is constant. Formulas for the variance also follow readily from the theory of Markov processes. For specific structures of the transition matrix, e.g., if transitions occur only in one direction along a linear sequence of states, this symbolic expression can be evaluated (e.g., using the MAPLE symbolic math tool box in MATLAB™) yielding explicit formulas for the expected number of cells in each compartment as a function of time – including the expected number of cells that have reached malignancy. Seber and Wild (1989) discuss algorithms for estimating the transition matrix $\mathbf{A}$ in such models from data on $\mathbf{y}$ collected at different times. Matis and Kiffe (2000) provide a compartmental treatment of stochastic population models applied primarily to populations of animals, people, and ecosystems instead of cells, but the mathematical framework is the same.

4. INTRODUCTION TO EXPOSURE ASSESSMENT

An individual's risk of an adverse health effect due to exposure to environmental hazards is usually modeled as being determined by two things: a dose-response function and an exposure history. The dose-response function (perhaps represented by a biologically-based simulation model or by a parametric statistical model) maps the exposure history to a conditional probability or hazard function for occurrence of the effect. Exposure history, in turn, depends on (a) The locations and strengths of emission sources; (b) Physical and chemical laws and meteorological, soil, and water transport processes governing dispersion, deposition, and transformation of substances in the environment; and (c) The locations and activities of exposed individuals over time.

Exposure assessment is a large and rich discipline by itself, with its own models, technical methods, software tools, texts, training courses, and journals. This section only briefly introduces key concepts and methods that are essential for many risk assessment applications. Exposure assessment methods can provide crucial information to risk assessment modelers, but its technical details lie beyond the scope of this treatment.

Most exposure assessment models treat each individual as a system – in essence, a box with inputs and outputs – that absorbs substances from the environment and eliminates them or their metabolites at rates determined by dose, biochemistry, and activity levels. An individual makes transitions among different micro-environments (such as home, car, office, etc.) throughout the day, absorbing ambient substances from each. Transitions among these micro-environments generate exposure histories for individuals. Population exposures are frequency distributions of individual exposure histories.

Key exposure modeling concepts for health risk assessment include the following:

- Environmental *fate and transport models* describe how emissions from a source move through air, surface water, ground water, soil, and perhaps biota, perhaps accumulating in some compartments or undergoing chemical transformations along the way. Compartmental flow modeling is often useful in this context, although other engineering models (e.g., of dispersion of an effluent plume in air or in water under various conditions) are also essential.

- *Human activity models* describe how much time people spend at various locations (especially, home, transportation, work, and recreation) and how fast they breathe, intake food and water, and so forth at those locations throughout the day, week, and year. Much effort has gone into analyzing the statistical patterns of various activities, both on short time scales (e.g., showering; see Burmaster, 1998), medium term (e.g., daily activities by age groups; see e.g., Funk et al., 1998) and long ones (e.g., fishing habits over a lifetime; see Price et al., 1998.)
- *Mobility models* describe the length of stay of people in households, occupational locations, and other locations where they may be exposed to a particular combination of sources and pathways. These models can be extended to include frequency distributions of activity patterns (e.g., proportion of time spent outside) within specific subpopulations identified by characteristics such as age, sex, education, number of children, and so forth (Field et al., 1998).
- *Exposure monitoring* devices, including personal monitoring badges and micro-environment monitors (e.g., air sampling equipment with built in graded filtration and analysis capabilities, air flow controls, and so forth) provide measurements that can replace or augment model-based predictions of exposure levels.
- *Biomarkers of exposure* use data from blood, tissue, urine, saliva, hair, or other biological samples, in conjunction with PK (or PBPK) and perhaps PD models, can also be use to estimate actual exposures received.

Example: Environmental Fate and Transport Modeling Software

Fate and transport models describe how emissions from different types of sources (e.g., point sources, stationary area sources, and mobile sources on land or water) spread through air, surface water, ground water, soil, airborne dust, and the food chain. They predict concentrations of chemicals and their products, e.g., from chemical reactions, dispersion, degradation, and aging occurring between emission and arrival at a receptor location, at different times and locations around the source(s). Necessary inputs are the characteristics of the emissions and of the environmental pathways being considered.

For airborne contaminants, the most widely used predictive models are based on the *Gaussian plume* model of dispersion. These models require as inputs descriptions of source emissions release rates (e.g., grams per second), source geometry (e.g., point source, area source of known shape and size, line source, etc.), release height, and temperature and stack gas velocity of effluent plumes. They also require wind direction and speed, terrain type, and meteorological stability class in order to estimate how the emissions will disperse. Analogous models for soil and groundwater contamination require other input parameters that affect volatilization of chemical from soils, sorption of chemical into the soil matrix, soil porosity and density, water flow characteristics, and chemical parameters such as the octanol-water partition coefficient.

Data for fate and transport modeling are readily available for many chemicals (see e.g., http://www.tds-tds.com/fs_efd.htm), as is environmental fate and transport simulation and modeling software for use in exposure assessment. The U.S. EPA's Center for Exposure Assessment Modeling (CEAM) Software Products web page provides software packages for predicting pollution concentrations in air, surface water, ground water, soil, and different parts of the food chain. The underlying models in many cases draw on compartmental flow and continuous simulation modeling methodology. They implement a range of engineering models, from simple compartmental flow models to sophisticated numerical simulation models with many empirically derived parameter values built in. (See http://www.lakes-environmental.com/lakeepa1.html and http://www.ess.co.at/GAIA/models.html).

Detailed information, documentation, and user's guides for the methods and algorithms used in exposure models can be downloaded from easily accessed web sources. Many of the models are accompanied by validation information and test data sets. The fate and transport component of exposure modeling can be considered relatively well developed, providing risk assessors with a variety of reasonably high quality, well documented predictive tools for modeling the exposure consequences of environmental releases. Human mobility models, activity models, and related data are being added to software and data repositories.

5. A CASE STUDY: SIMULATING FOOD SAFETY[1]

Quantitative risk assessment (QRA) of microbial hazards and food safety poses challenges not encountered in chemical or radiation QRAs. The ability of microorganisms to reproduce under favorable circumstances, even after partial decontamination, invalidates the mass-balance calculations often used for chemical exposure and QRA modeling calculations. Evolution of resistant strains under selection pressures from antibiotics and spread of infections from one victim to another have no parallels in the world of chemical risk assessment. And dose-response relations for microorganisms are often unusual, with smaller doses of bacterial pathogens sometimes creating a higher probability of illness than larger doses (Teunis et al., 1999).

The modeling techniques presented in this chapter often suffice to identify effective risk-management strategies and to eliminate ineffective ones, even in complex situations with high uncertainty and variability in individual exposures and dose-response relations. This section examines how a discrete-event simulation model of human exposures to a common food-borne bacterium, *Campylobacter jejuni,* can be used to estimate the probable human health impacts of alternative risk management strategies. Simulation reveals such large differences in expected health impacts between the most and least effective risk management strategies that policy

[1] This section is based on Cox LA and Popken DA, 2001. *A Dynamic Simulation Model of Campylobacter Illnesses,* Report to Animal Health Institute, Cox Associates, Denver, CO.

recommendations become obvious even without the types of refined decision analysis modeling discussed in Chapters 5 through 7.

5.1 Background: The Potential Human Health Hazard

Campylobacter jejuni (CP) is the most commonly diagnosed cause of bacterial gastroenteritis in the United States. In undercooked chicken, non-poultry meats and seafood as well as in raw milk and water, restaurant lettuce, and other contaminated foods or water consumed without adequate cooking, CP can cause infectious diarrhea lasting for a week or more (Friedman et al., 2000). The probability of sporadic CP infection has recently been modeled as a function of CP contamination levels, measured in colony forming units (CFUs) of CP in ingested food. Both an approximate Beta-Poisson dose-response model and an exact hypergeometric dose-response model (Teunis and Havelaar, 2000) give similar maximum-likelihood estimates of the dose-response function describing risk of illness as a function of CFUs ingested, based on data from experiments in 111 healthy young adult volunteers (Black et al., 1988).

Although erythromycin is usually the indicated treatment for CP illness, physicians sometimes prescribe ciprofloxacin instead. Ciprofloxacin is a member of the fluoroquinolone (FQ) family of antibiotics. It is a broad-spectrum antibiotic and hence is often used for empiric treatment. If a patient has been infected (at least in part) by an FQ-resistant strain of CP, then it seems plausible that the effectiveness of an FQ prescription might be diminished, potentially leading to extra days of illness.

Since 1995, veterinarians in the U.S. have also used another FQ, enrofloxacin, to combat fatal infectious respiratory disease in roughly 1% to 2% of chicken broiler flocks. This use creates selection pressures for mutations (most notably in the DNA gyrase gene, *gyrA*) favoring development of FQ-resistant CP strains in chickens. Consumption of undercooked chicken (and other undercooked meats) outside the home is a significant risk factor for *Campylobacter*iosis in humans (Friedman et al., 2000). Therefore, this veterinary use of FQ creates a potential risk of extra days of illness in humans with FQ-resistant CP treated with FQ.

The risk is only potential, because it is not yet known whether excess illness-days actually do result from FQ treatment failure or impairment in this context. Some studies indicate that resistant strains are associated with longer illnesses (e.g., Smith et al., 1999), but it is not clear whether FQ resistance *causes* excess illness days or is simply non-causally *associated* with excess illness days. For example, both longer illness durations and resistance to FQ might be attributes of some strains of CP, even if resistance does not cause excess illness-days. Yet other evidence suggests that FQ most effectively *reduces* illness-days among patients infected with strains that would be classified as resistant (Marano et al., 2000), emphasizing that "resistance" is a matter of degree (Minimum Inhibatory Concentration or MIC) rather than a dichotomous attribute. Piddock (1999), noted that "Preliminary data showed that only 1/39 patients with ciprofloxacin-resistant *Campylobacter* enteritis did not respond to ciprofloxacin therapy, suggesting that many bacteria classified as resistant by laboratory tests may nevertheless be amenable to antimicrobial therapy." (Definitions of "did not respond" and "amenable" were not given, however.) Despite these ambiguities, the fact that FQ treatment effectiveness in humans *might* be diminished by veterinary use of FQ invites risk assessment and risk management.

Many investigators have suggested that the need for risk management intervention is already urgent, and is becoming more so. They believe that veterinary use of enrofloxacin in chickens is a probable major cause of the trend of increasing FQ-resistance among human CP cases, documented in many countries between 1990 and 2000 (e.g., Smith et al., 1999). Yet, this interpretation of historical trends is undermined by the following considerations:

- Human resistance rates to ciprofloxacin in the U.S. were increasing several years before enrofloxacin was introduced for veterinary use (data in Smith et al., 1999). Indeed, by 1989, sme physicians were already concerned that over-prescription of ciprofloxacin to humans might eventually lead to growth in resistance (e.g., Hooper and Wolfson, 1989).
- Resistance rates have increased at roughly comparable rates in countries with and without widespread use of enrofloxacin in chickens (e.g., Gaudreau and Gilbert, 1998).
- The association between FQ use in farm animals and frequency of FQ-resistance in CP isolates from the animals appears to be weak or non-existent (Piddock et al., 2000).

Although several authors have tried to argue that human use of FQ is unlikely to be a major driver of FQ-resistant CP strains in humans, while veterinary use probably is (e.g., Smith et al., 1999), the evidence that veterinary use plays any significant role in human FQ-resistant CP seems to be too limited to be very compelling.

Thus, a risk manager might reasonably conclude that, as in many other practical applications, the available epidemiological data and scientific understanding do not single out a clear cause for an observed worrisome trend in risk numbers – in this case, the increasing resistance of human CP isolates to FQ. Nor are the health consequences of that trend well understood, since resistance does not necessarily imply ineffective treatment (Piddock, 1999). Even the significance of chicken consumption as a risk factor for CP can be questioned, as there is some evidence that chicken prepared and eaten at home appears to be a protective factor against CP infection (e.g., Friedman et al., 2000). But if veterinary use *is* the cause of increasing resistance in humans, or at least a major contributor to it, and if resistance does lead to excess illness-days or other adverse health consequences, then risk management interventions should be considered.

The probable human health consequences of different risk management interventions are necessarily uncertain in this setting. Yet, uncertainty about probable consequences does not imply that useful comparisons among risk management options cannot be made or that actions based on such comparisons should not be taken. (This is the constructive content of the "Precautionary Principle" sometimes advocated by those anxious to avoid decision paralysis in the absence of adequate scientific knowledge.) To the contrary, most of applied decision analysis, as developed in Chapters 5-7, deals with situations like this one in which the probable consequences of different decisions are highly uncertain.

How should the potential risks of excess illness-days from FQ use in chickens be managed, given the uncertainties just discussed? The answer depends in part on whose risk management decisions are to be addressed. Chicken farmers, processors, retailers, preparers, consumers, physicians, and regulators such as USDA and FDA can all take actions to reduce or eliminate health risks from CP infections, including FQ-resistant strains of CP. To decide which actions are likely to prove most effective in creating health benefits, a risk assessment model is needed that will show how different actions affect the hypothesized risks.

5.2 Risk Management Setting: Many Decisions Affect Risk

Risk management of many food-borne microorganisms, including CP, is complicated by the fact that many parties and decisions affect the final risks experienced by consumers. A consumer's risk of chicken-borne CP-induced illness is determined by the microbial loads consumed in servings of chicken. The microbial load of CP CFUs in chicken reaching the kitchen or the table depends on at least the following decisions:

- *Farmer's care.* Sanitation and animal husbandry practices at the farm affect the microbial load of CP among chickens leaving the farm. This load is quantified by the frequency distribution of CP CFUs on broilers sent from the farm to slaughter. It depends not only on care taken at the farm but also on season and on contamination of chicken flocks by wildlife, by run-off of water from adjacent areas, and by other sources.
- *Transporter's level of care.* Microbial load is typically greatly amplified during transportation from the farm to the processing plant and subsequent holding prior to slaughter.

- *Processor's level of care.* Microbial load can be either further increased or greatly reduced by processing. Choices of filtration systems, antimicrobial treatments, use of chiller fluids, and so forth can greatly affect the ratio of outgoing to incoming microbial load. Processing can potentially reduce incoming microbial load by orders of magnitude, although cross-contamination from shared chiller fluid can also introduce CFUs onto previously uncontaminated chicken. Cooking can eliminate CP and other pathogens.
- *Preparer's level of care.* Proper washing, freezing, and cooking of chicken can eliminate CP from chicken portions before they are eaten. Improper storage and possible cross-contamination of other foods on the premises where the chicken is prepared and consumed can amplify the microbial loads ingested. Recent epidemiological evidence suggests that CP risks from chicken consumption may be primarily associated with non fast-food restaurants and other out-of-home settings, while preparation and consumption of chicken at home appears to be a protective factor rather than a risk factor (Ikram et al., 1994; Adak et al., 1995; Eberhart-Phillips et al., 1997, Friedman et al., 2000).

The interaction of these stages is mainly multiplicative, i.e.,

final microbial load =
*initial load * transportFactor * processFactor * storageAndPrepFactor.*

The initial microbial load (defined as CFUs per chicken leaving the farm) is modified by the three factors shown. (If attention is confined to FQ-resistant strains, then the "initial load" in this equation refers to initial load of FQ-resistant CFUs. In reality, cross-contamination of batches of chicken at processing deviates from the purely multiplicative model, but this does not affect the following points.) Sufficient care in eliminating microbial load at any stage can reduce or eliminate the microbial load that finally reaches consumers. Therefore, *the incremental health benefit from additional care taken at any stage decreases with care exercised at other stages.*

From the standpoint of centralized risk-management decision-making, this concatenated-process structure leads to the problem of how to optimally allocate resources across stages to achieve the greatest health benefit for consumers at the end of the pipeline. From the standpoint of decentralized decision-making, it raises the question of how much care players at each stage should take, based on their expectations about how much care players at other states will take. That actions at multiple stages interact to determine the final risk also raises the risk management question of what level of care regulators responsible for food safety should require parties at each stage to take. Economic efficiency requires that the least costly intervention (or combination of interventions) be used to achieve any chosen level of reduction in microbial load. We will approach the question of what intervention(s) to implement indirectly, by comparing the expected health benefits that can be achieved by different risk management options. Quantifying the costs to different parties of implementing the options is beyond our scope, but quantifying their health benefits provides the risk-related information needed to make economic cost-risk-benefit or purely health-based decisions.

The most important and creative step in risk-management problems is often to generate a sufficiently rich set of promising alternatives to evaluate and choose among. The following example set is adequate for purposes of illustration, but other opportunities may turn out to be preferable. (For example, it might be possible to use FQ as part of a competitive exclusion program to kill infectious intestinal bacteria before deliberately introducing innocuous strains to repopulate the chickens.) The examples to be analysed of risk management options addressing different links in the chain leading from farm to fork to excess days of illness are the following:

Example Risk Management Options to Be Compared

1. Eliminate FQ use at the farm (at least while it is still effective in humans.)
2. Optimize withdrawal period for FQ prior to bringing chickens to market
3. Track FQ-treated flocks from farm to processing plant. At processing, send them to cooking or other further processing (e.g., use in soup or frozen pre-cooked meals) that eliminate CP CFUs
4. Implement processing changes to reduce risk:
   a. Reduce cross-contamination at processing plant by 10%. For example, process FQ-exposed chickens at the end of each shift to avoid contaminating subsequent batches with FQ-resistant strains of CP.
   b. Reduce cross-contamination by 100%. (This puts an upper bound on the impacts achievable by measures designed to reduce cross-contamination.)
   c. Increase CFU-reduction by 10%. This can be done by changes in processing. This case study will focus on what the human health impacts of such a change would be, rather than on the technology options for achieving it.
   d. Increase CFU-reduction by 100%
   e. Both (10%), i.e., a and c
   f. Both (100%), i.e., b and d
5. Change restaurant storage and preparation practices to increase food safety. This as modeled by a 10% reduction in the microbial load multiplier ("storageAndPrepFactor"). For example, make sure that undercooked chicken and other meats are kept adequately refrigerated and are properly stored, washed, and cooked before being served.
6. Change physician prescription practices to reduce over-prescription of FQs in CP (and other) cases, modeled by a 10% reduction in frequency of FQ prescriptions to CP patients.

These options are illustrative rather than exhaustive. Rather than seeking to define or anticipate a full range of detailed technologically possible options, those listed above postulate plausible degrees of risk-reduction at different stages to permit an evaluation of which ones will most effect public health. Quantifying the sensitivity of health benefits to plausible risk control measures at different stages can help to determine where technical creativity in defining specific risk-reduction (or contamination-reduction) measures is most likely to pay off in improved health outcomes.

The risk assessment simulation model presented next quantifies the health impacts of these different interventions under the *assumptions* that chicken-borne CFUs of CP increase the probability of CP illness and that FQ-resistant strains of CP reduce the effectiveness of FQ prescriptions. Hence, the model can provide the health risk information needed for cost-effectiveness or cost-benefit analysis of risk management options conditioned on these assumptions, while leaving open for further research the question of whether the assumptions are correct. Such conditional risk assessment and risk management allows available data and knowledge, in conjunction with assumptions that cover any crucial gaps, to be used to evaluate risk management options. Parties not pleased with results may have a positive incentive to collect information to replace the conditioning assumptions with facts and data.

5.3 Methods and Data: Overview of Simulation Model

This section presents a discrete-event simulation model of the microbial load of CP reaching consumers via ingested chickens. Microbial load is quantified in terms of colony-forming units (CFUs) in servings of chicken. The model randomly samples the initial number of CFUs on each chicken when it leaves the farm from an estimated empirical distribution of CFUs. Then, it simulates the probabilistic amplification and reduction of microbial load at successive stages from farm to table. Individual chickens and chicken servings are tracked as discrete objects throughout the simulation. Each serving consumed may therefore have its own individual history of contamination, CFU increases and decreases, and preparation prior to consumption. Individual consumers are also modeled as discrete objects. Each individual has characteristics (especially age, as well as ethnicity) that affect the statistical chicken consumption and the dose-response relation.

Ingested microbial load enters a non-linear dose-response model that predicts resulting probabilities of infection and illness. Ill patients are assumed to seek care with some probability, and physicians are assumed to prescribe fluoroquinolone (FQ) antibiotics, instead of erythromycin or some other treatment, with some probability. The probability that a CP-infected patient harbors an FQ-resistant strain of CP, generated by use of FQs at the farm to combat illness in chickens, is multiplied by the probability that a doctor will prescribe FQ to that CP-infected patient to obtain the probability of a potentially ineffective prescription. Ineffective prescriptions, in turn, are assumed to create excess days of illness. Thus, the whole model estimates the probable excess days of illness from FQ-resistant CP per 100,000 people per year, based on assumed chicken consumption, factors affecting microbial loads, prevalence of FQ-resistant strains of CP, care-seeking behavior by ill patients, and physician behaviors in prescribing FQ. This model can quantify the impacts on both ineffective prescriptions and total cases of CP illnesses per 100,000 people per year of interventions that change one or more of these inputs. Hence, it provides a basis for evaluating the health benefits of alternative risk management options.

In summary, discrete-event simulation provides a framework for integrating information about FQ usage in chicken flocks, animal husbandry practices, food refrigeration, transportation, storage, processing, and preparation practices, and CP diagnosis and FQ prescription practices. It is flexible enough to describe how all these factors, as well as individual covariates such as age (possibly a surrogate for acquired immunity to CP and for other risk factors and protective factors), interact to affect risk in humans. The simulation approach can support both quantitative risk assessment and risk management decision and policy-making. The following paragraphs summarize the approach, discuss data and assumptions used to implement it, and summarizes results illustrating the types of insights that DES modeling can provide to support risk management decisions.

Model Processes for Chickens: Simulating Microbial Loads

Figure 1 shows the main processes that chickens pass through before being consumed. Each stage contains dynamics that affect its outgoing microbial load of CP. Table 3 summarizes the parameter estimates and data sources used to quantify microbial load and/or prevalence of FQ resistant strains at each stage. These parameter estimates are discussed next.

Chicken broilers mature on the farm for about 8 weeks (Doye et al., 1982), where they develop CFU loads both internally in the cecum and externally on the chicken surface. The extent of CP contamination varies by season, being lower in the spring, and higher in the summer (Friedman et al., 2000b, Stern 1995b). It can be affected by changes in animal husbandry practices. Stern et al. (1995a) reported that 12.1% of 200 chicken exteriors (from 5 farms) were found to be infected with *Campylobacter spp.* at the farm, while after holding and transport, 56% of the chicken exteriors were infected. Jones, et al. (1991) found that 5 of 25 samples (20%) had *Campylobacter jejuni* on the carcass surface just prior to processing. Thus, preprocessing incidence rates between 20% and 56%, with an approximate midpoint of 0.38, may be realistic, depending on the farms and practices considered. We will use this range to express plausible variability in pre-processing microbial loads, while recognizing that individual farms may fall outside this range. As discussed in the section on uncertainty analysis, capturing a plausible range of variability is probably adequate for purposes of analysis: a more refined estimate based on more extensive research would probably change the results of the analysis very little.

The poultry intestine appears to be the primary source of infection (Stern, 1995b; Musgrove, et al., 1993). The intestinal infection level presents risk in that the contents can spread to other birds during maturation, transport and processing via cross-contamination. For example, Sterns' study (1995a) found that, among 10 batches of 10 birds, 9 batches (90%) had some amount of cecal infections although only 6 batches had any surface contamination. If a 90% cecal colonization incidence produced a 38% surface contamination incidence, then a point estimate of the relation between them would be:

Pre-processing surface contamination incidence
= 0.42 x (cecal colonization incidence)

Uncertainty about the multiplier 0.42 might be modeled as uniformly distributed between 0.22 and 0.62, corresponding to surface contamination incidence rates of .20 and .56 respectively. Cecal colonization probability is modeled as a binomial probability with $p = .90$. Here and in several other steps we exploit the fact that assuming a single binomial probability (as opposed to a mixture of different binomial probabilities having the same mean, which might be more realistic but which requires more data) *maximizes* the variance of the simulated relationship (Feller, 1968, p. 231.) Thus, using a fixed binomial parameter helps to bound the true variability in results, as discussed further in the section on uncertainty analysis.

Two studies by USDA-ARS, one in 1998 and one in 1999 (FDA-CVM, 2000b), each reported that 9.4% (12/128 and 45/481) of *Campylobacter jejuni* isolates tested positive for ciprofloxacin resistance (based on the E-test, a commercial test kit by AB Biodisk of Sweden, with a Minimum Inhibitory Concentration (MIC) of 4 mcg/ml.) These samples were obtained from across the country, and should be considered a rough nationwide average. In these studies, the broilers were taken from the drip line after the chiller tank (providing an opportunity for cross contamination). This parameter is modeled as a binomial probability with $p = .094$.

Figure 1: Simulation Model Processes for Chickens

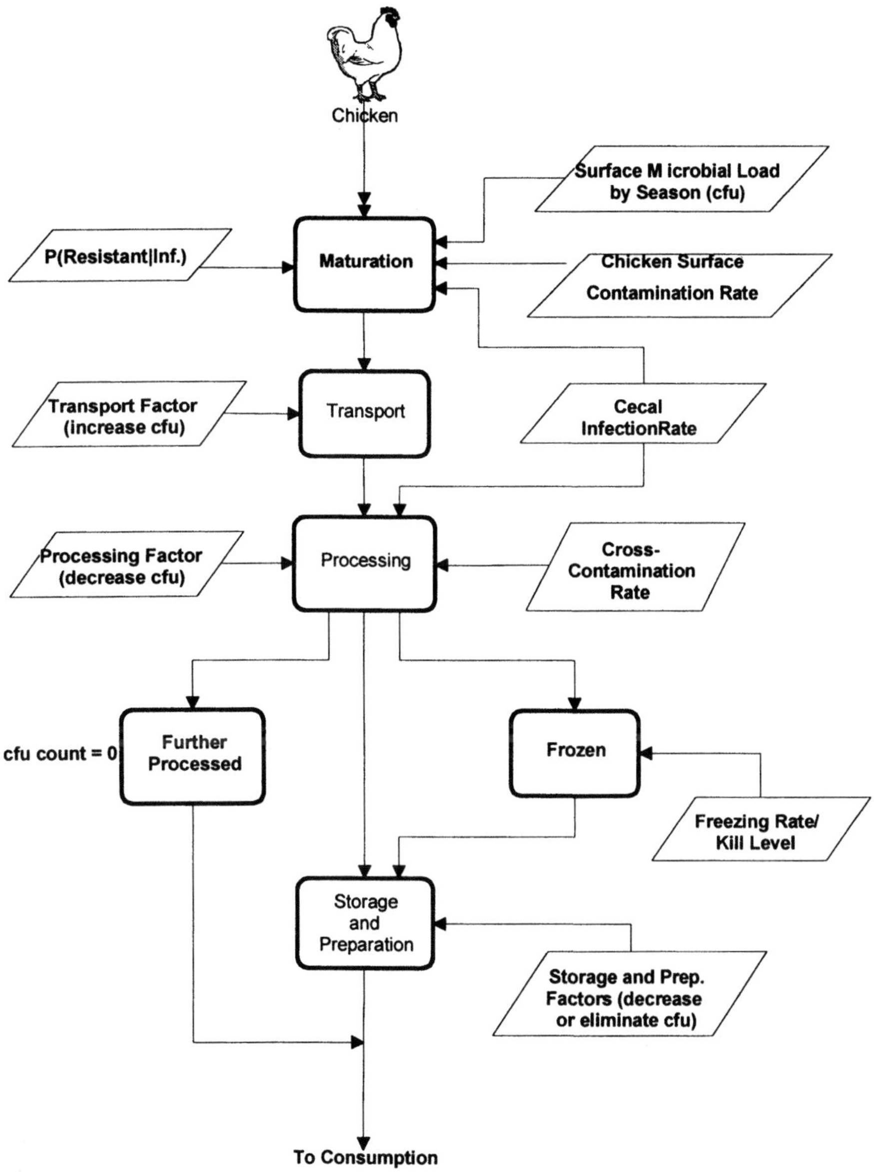

Table 3: Summary of Input Parameter Values for Chicken Processes

Data Input	Values in Simulation Model	Notes
Seasonality of pre-processing incidence	Winter, Spring, Summer, Fall multipliers = 0.82, 0.63, 1.43, 1.13, respectively.	Friedman et al., 2000b; Stern 1995b
Pre-processing incidence of surface contamination	Surface contamination multiplier x cecal colonization incidence. Surface contamination multiplier: U[0.22, 0.62], μ= 0.42. Cecal colonization incidence: binomial probability, p = 0.9.	Stern et al, 1995a ; Jones, et al. 1991
Pr(Resistant infection)	Binomial probability, p = .094.	FDA-CVM, 2000b
Surface microbial load on chickens	Triangular distribution for $\log_{10}$ of values at farm: T(0, 2.98, 6.38)	(Stern, et al., 1995a)
Transportation Factor	Triangular distribution for $\log_{10}$ of factor values: T(1.32, 2.73, 4.24).	(Stern, et al., 1995a)
Processing Factor	Triangular distribution for $\log_{10}$ of factor: T(1.0, 2.23, 3.0) by which rinsing, scalding, etc. decrease surface microbial load.	Stern (1995a, b). Izat, et al. (1988) Lillard (1990), Mead (1995)
Proportion further processed	Binomial probability, p =.4678. CFU count reduced to 0.	Describes prepared and frozen chicken foods,
Frozen Chicken Factor	Select non-further processed chickens with binomial probability p=.163 to freeze. Reduces CFU count by 100 on selected chickens.	(Mead, et al. 1995)
Post-processing Surface Contamination Incidence	0.735 ~= (cross-contamination multiplier) x (surface multiplier) x (cecal colonization incidence) 1.934 x .4222 x .90 (using means)	Cross contamination multiplier is uniformly distributed U(1.368,2.5) (FDA-CVM 2000b)
Post processing incidence without retail infection	Binomial probability, p = .302, for chickens showing infection after processing but not at retail outlet	(USDA, 2000)
Storage and Preparation Factor	Implied value (~1E-5) was estimated by model calibration	No data are available to fully quantify this factor

The level of *Campylobacter* infection (CFUs) found on the surface of chickens at the farm provides the first step in predicting the eventual number of illnesses in humans. The distribution of CFU/chicken at the farm ranges from 0 to $10^{6.38}$, with a weighted-average value of $10^{2.98}$ (Stern, et al., 1995a). A triangular distribution for the $\log_{10}$ of the value captures these three points: T(0, 2.98, 6.38).

Transportation of chickens from the farm to holding pens and subsequent slaughter increases the microbial load on the surface of the carcass by a weighted-average multiplier of 537 ($\log_{10}$ = 2.73) (Stern, 1995a). In the referenced study, the factor increases observed over 12 measurable batches

(batches for which *Campylobacter* was detectable both before and after transport) ranged from a low of 20.89 to a high of 17,378.01. This leads to a Triangular distribution for the $\log_{10}$ of the factor values as follows: T(1.32, 2.73, 4.24), where again the goal is only to capture a plausible range of variability for purposes of eventually simulating expected excess illness-days per year in humans. The log-triangular distribution is almost certainly not correct, but, as discussed later, the exact from of the distribution need not be used for purposes of accurately estimating model outputs.

Following transportation, rinsing, scalding, and other processing steps significantly *decrease* the microbial load on the carcass surface. Mead, et al. (1995) reports that, during processing, CFU/g on skin samples were reduced between 10 and 1000 fold. Effects were measured both before and after a series of hygiene control measures were instituted. Before the changes, the average reduction factor over 5 sample flocks was 100. After the changes, the average reduction factor over 10 sample flocks was 218.8. The overall minimum flock average reduction factor was 10, the overall maximum was 1000, and the overall average was 168.53 ($\log_{10} = 2.23$), similar to an average factor reduction of $10^{2.21} = 162.2$ based on data in Stern (1995a, b). Izat, et al. (1988) report reduction factors ranging from 22 to 110. Lillard (1990) reports an average reduction factor of $10^{2.61}$. Variability in the processing reduction factor may be approximated by a Triangular distribution for the $\log_{10}$ of the factor values as follows: T(1.0, 2.23, 3.0).

A large portion of processed chicken is "further-processed" for use in prepared and frozen foods. In most cases, processing includes heating or freezing processes that effectively eliminate or neutralize CP risks. The proportion of further processed chicken has increased steadily in recent years, from 2% in 1962 to 41% in 1998 (USDA, 1988; DPI, 2000). Fitting a quadratic polynomial to the "further-processed" percentage *vs.* time for data available through 1998 projects a further-processed value of 46.78% for the year 2000 followed by: 2001: 49.26%, 2002: 51.81%, 2003: 54.42%. Although it is risky to project a current trend too far into the future, it is plausible that further increases will tend to decrease overall and FQ-resistant CP risks over the next several years. The simulation model selects chickens for further processing using a binomial probability of 0.4678 (the year 2000 estimate) and sets their post-processing CFU count to zero.

Some processed whole and cut-up chicken is marketed in frozen form, with CFU counts on frozen retail chicken being at least 100 times less than for fresh chicken (Wesley, 1998). This is simulated by reducing the CFU count by a factor of 100 after processing for all frozen chicken. Although Lee et al. (1998) report that *Campylobacter jejuni* can remain viable after freezing and replicate to former levels under the right conditions, Curtis et al. (1995) report that at even modest levels of chilling (68°F), *Campylobacter jejuni* levels in raw chicken steadily decline. Reed (1994) reports that growth of *Campylobacter jejuni* is inhibited below 86°F, consistent with Jacob-Reitsma (2000) who further states that - "actual multiplication during handling or storage at room temperature in the kitchen will not readily occur in moderate climates." Approximately 8.66% of processed chicken currently falls into the frozen category (USDA, 2000), or 16.27% of the chicken that is not further processed. The simulation selects frozen, non-further processed chickens randomly using a binomial probability with $p = 0.1627$.

Cross-contamination of broilers by CP during processing can elevate the incidence of surface CP from perhaps 0.38 ± 0.18 (see above) just prior to processing ranges to post-processing levels of 52% (Jones et al., 1991), 62% (Harris et al., 1986), 86.25% (Stern, 1995b, averaged over seasons and tests), or 88.2% (FDA-CVM 2000b for a nationwide sample of broiler carcass rinse samples.) Based on post-processing incidence rates ranging from 52% to 95%, with an approximate midpoint of 0.735, almost two times (1.934) the estimated preprocessing average incidence of .38, the cross-contamination factor is simulated as a uniform distribution, U[1.368, 2.5] with mean 1.934.

Storage and preparation effects are a function of refrigeration, food handling/safety, and preparation techniques after processing, but before consumption. Stern (1995b) found that 10 days of refrigeration at 4°C reduced detectable *Campylobacter* spp. incidence from 86.25% to 27.5%. On the other hand, *Campylobacter* was shown to survive, albeit at much reduced levels, after 49 days in cooked minced beef stored at 2°C (Curtis, et. al, 1995). Refrigeration and freezing appear to be effective at reducing the CFU count below human infective levels (approximately 500 CFU), but not at completely eliminating the organism. Subsequent improper storage and preparation procedures could allow infections to return to high levels (Lee, et al., 1998). The most effective way to eliminate the risk is through thorough cooking. Koidis and Doyle (1983) showed that cooking meats at 58-60° C inactivates 10^6 *Campylobacter* per gram of meat.

The simulation model distinguishes between elimination and reduction of CP. Estimates of the prevalence of CP on chicken at retail stores include: 31.6% (Jones et al., 1991); 24% (Harris, et al., 1986), 88% (Smith, et al., 1999), 69% (Willis and Murray, 1997) and 44% (Rossiter et al., 2000). The decrease in prevalence from post-processing levels (mean = 73.5%) to this set of values (average = 51.3%) may be due to the refrigeration effects noted by Stern and to lack of appropriate air mixture required for survival of CP. An average retail incidence of 51.3% represents a 30.2% reduction in the average incidence. The simulation model therefore uses a binomial probability of $p = 0.302$ to select chickens showing infection at post-processing but not at the retail outlet. More elaborate (e.g., re-sampling) statistical approaches could be taken to estimate a distribution for this ratio, but a single binomial parameter is used in the model and uncertainty about it is modeled via sensitivity analyses. For chickens whose CP microbial load is not eliminated, a factor for average reduction in CFU count due to storage and preparation is estimated from historical CP incidence rates by a model-calibration calculation discussed below.

Model Processes for Human: Estimating Population Frequencies of Exposure, Dose-Response, Treatment, and Illness-Days

The simulation model includes a sub-model of human consumption of chicken and subsequent events. Each individual is first assigned age and ethnicity attributes by sampling from demographic marginal frequency distributions. Consumers eat chickens at rates (servings per week) estimated from historical data broken down by ethnic group. Each consumer becomes ill or not according to an age-specific probabilistic dose-response model.

Figure 2: Model Processes for Humans

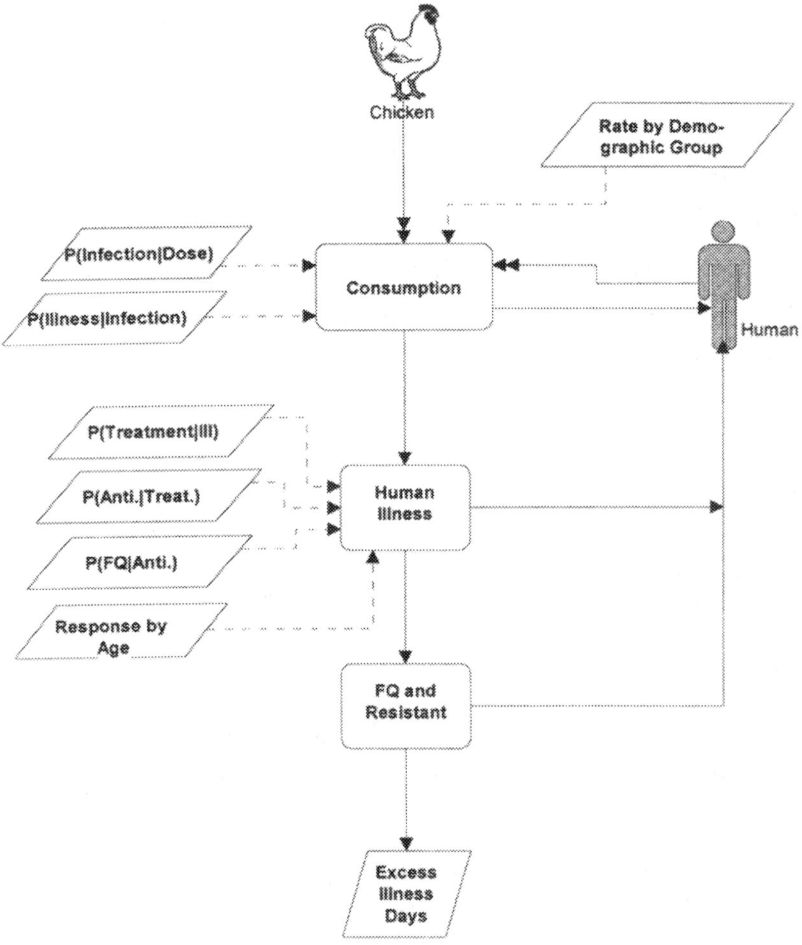

Figure 2 outlines the basic processes. The estimated average amount of chicken consumed per capita in 1999 was 78.5 lbs./year (USDA, 1999), corresponding to about 0.4069 chickens per week (based on average chicken weights derived from USDA, 2000). Each chicken is assumed to provide fours servings, with the CFU count per simulated chicken being divided by 4 to obtain CFUs per serving. The dose-response model described below is then applied to each serving. Per-capita consumption rate can be broken down by subpopulations to indicate variability among ethnic groups, although this does not affect aggregate risk calculations. For example, consumption rate multipliers by ethnic group for black, white, and other are (0.93, 1.42, 1.2) with relative frequencies in the population of (0.823, 0.128, 0.049), respectively (derived from USDA, 1993). The simulation assigns simulated humans an ethnic attribute using the above proportions. For each group, a weekly consumption rate is obtained by multiplying 0.4069 by the appropriate multiplier value. If the joint distribution of ethnicity, age, and consumption were available, it could be used instead of these marginal distributions.

The CDC-FoodNet (1999) reports CP illness rates by age, as follows:

Table 4: CP Incidence Rates and Population Frequencies by Age Group

	Estimated 1998 CP Incidence and Population by Age Group							
Age Groups	< 1	1-9	10-19	20-29	30-39	40-49	50-59	60+
reported cases per 100000	50.4	22.2	10.5	26.1	22	19.2	18.2	14.9
case rate multiplier relative to young adult	2.11 (0)	.930	.440	1.000	1.000	.804	.762	.624
% of US population	.0144 *	0.13*	.1435	.1341	.1583	.1507	.1041	.1652

*Census data provides population estimates for persons under 5 and 5 to 9. We have estimated the breakout between < 1 and 1 to 9 by assuming that 10% of 0-9 are < 1.

** Rows do not sum exactly to 1.0 due to rounding.

Simulated individuals are assigned ages sampled from the population frequency distribution of ages shown at the bottom of Table 4. The data in this table are also used to estimate dose-response probabilities for members of each age group from dose-response probabilities predicted by a model fit to experimental data for young adults (Black et al., 1988, discussed below). The dose-response probability for members of each age group is obtained by multiplying the dose-response probability for the "young adult" category (assumed to be ages 20-39) by the case rate multiplier in Table 4 to allow the model to produce correct age-specific relative illness rates.

A crucial parameter for risk assessment is the number of CP cases per year attributable to consumption of contaminated chicken This number will be used to calibrate unknown parameters within the simulation model – specifically, to estimate the "storageAndPreparation" factor that multiplies the post-processing microbial load. (If this factor were known from direct measurement *and* could be separately estimated from population attributable risk data, then a statistical technique such as generalized least squares could be used to combine the two estimates. But no direct estimate from measurements is currently available.) The estimated number of cases of campylobacteriosis in the US in 1999 was 1.38 million (FDA-CVM, 2001) out of a population of 272,691,000 (USCB, 2000b). However, approximately 27.5% of reported cases are for children under the age of 1 (CDC-FoodNet, 1999), who comprise about 1.44% of the population (USCB, 2000c). We exclude these cases from the chicken-consuming population, as infants consume very little unprocessed chicken, especially outside the home, and are probably at minimal risk of chicken-borne CP via such consumption or cross-contamination from uncooked chicken. This gives an annual CP illness rate in the potential chicken-eating population of:

Annual Campylobacter rate among consumers at least 1 year old
$= (1.38/272.691)*(1 - 0.275) = (0.005)(1 - 0.275) = 0.0037$

The estimated percentage of *Campylobacter spp.* associated with poultry consumption (among non-infants) has historically been cited at levels between 50% and 70% (Fedorka-Cray, et al., 1999), although more recent studies suggest that the chicken-attributable fraction may be much lower.

Friedman et al. (2000a) reports a matched odds ratio for eating undercooked poultry of only 1.9 (7% among cases vs. 4% among controls), in contrast to estimated odds ratios of 9.1 and 7.6 found in two small studies from the 1980s used by FDA-CVM (2000) to estimate a chicken-attributable fraction of nearly 0.6. Eating chicken or turkey cooked outside the home was also a significant risk factor (mOR = 2.4) while chicken or turkey eaten in the home was protective (mOR = .5). Risk factors with matched odds-ratios greater than that for eating undercooked poultry included eating non-poultry meat cooked outside the home (mOR = 2.4), drinking raw milk (mOR = 3.5), living on or visiting a farm (mOR = 2.1), contact with farm animals (mOR = 2.2), and contact with puppies (mOR=2.0). Since the prevalence of some of these factors is greater than that for eating undercooked chicken, it seems that FDA-CVM's attribution of nearly 60% of all CP cases to chicken requires revision to be consistent with these more recent data.

The Friedman et al. (2000a) study data can be used to suggest a plausible upper bound on chicken-attributable risk of CP illness. The only two statistically significant chicken-related risk factors are (a) eating undercooked poultry and (b) eating chicken or turkey cooked outside of the home. The fractions of CP cases engaged in these activities in the week prior to illness were 7% and 46% respectively, compared to corresponding control rates of 4% and 28%, respectively. If *all* CP illneses above the control group level are attributed to chicken, then 7% - 4% = 3% and 46% - 28% = 18% for a total of 21% of cases, would be attributed to chicken. (In reality, some of these patients may also have engaged in other risk activities, such as eating non-chicken meat outside the home. But for bounding purposes we attribute all excess cases to chicken.) This calculation suggests that not more than 21% of CP cases are likely to be attributable to chicken.

This number could be much lower if the effects of confounders, i.e., risk factors that cause CP illness and that are positively associated with chicken consumption, were identified and removed. For example, analysis of CDC population survey data shows that increasing age, which is significantly negatively associated with CP risk among adults (see Table 4), is also significantly associated with average chicken consumption, frequency of dining out in restaurants, children in day care, pets in the home, and other risk variables identified in the literature as risk factors for CP. To quantify and correct for the effects of such factors and to isolate the contribution of chicken consumption, it is essential to use individual-level data to quantify the joint occurrence rates of combinations of these risk factors. However, as of this writing, the CDC has elected not to release its individual-level data on risk factors and CP incidence.

Because the fraction of CP cases attributed to chicken is a key parameter for risk assessment, it is worth examining other approaches for calculating it. Univariate population attributable risks (PARs) may be approximated from the summary data reported by Friedman et al. using the traditional formula:

$$PAR = p(RR - 1)/[1 + p(RR - 1)] \qquad \text{(Univariate attributable risk)}$$

Here, p = population exposure rate (i.e., percent of people exposed to a risk factor) and RR = relative risk of campylobacteriosis among exposed as opposed to unexposed people. For the data summary provided by Friedman et al. (2000b), and approximating the relative risks (not reported) by the odds ratios, this calculation yields the results in Table 5. Although only multivariate calculations based on individual-level data will give correct results, these approximate univariate calculations suggest that eating poultry or non-poultry meat cooked outside the home may be approximately equally dangerous. Indeed, it is tempting to speculate that it may not be consumption of chicken *per se* that carries the risk, but rather consumption of improperly prepared foods (chicken or other) in some restaurants that may be a primary driver of domestically acquired sporadic CP risks in the population. That poultry cooked at home appears to be a protective factor bolsters this conjecture. Chapter 4 considers this hypothesis further and confirms it using causal graph modeling.

Table 5: Univariate Population Attributable Risks for Several Factors

Risk Factor	Cases	Controls	OR	P	PAR*
Undercooked poultry	0.07	0.04	1.9	.056	.05
Poultry not cooked at home	0.46	0.28	2.4	.375	.34
Other (non-poultry) meat not cooked at home	0.51	0.34	2.2	.429	.34
Raw seafood	0.05	0.03	1.8	.041	.03
Raw milk	0.02	0.01	3.5	.015	.04
Farm (living on or visiting)	0.16	0.09	2.1	.127	.12
Farm animals contact	0.11	0.05	2.2	.082	.09
Puppies contact	0.11	0.06	2.0	.086	.08
Poultry cooked at home	0.53	0.69	0.5	.606	-.43

Data source: http://www.cdc.gov/foodnet/pub/iceid/2000/friedman_c.htm

* Rough approximation, unadjusted for confounders and multiple risk factors and using matched OR to approximate relative risk

For illustrative purposes, 0.60 will be used as a baseline value for the chicken-attributable fraction in the simulation model, as it is the approximate midpoint of the historically cited values. Results will be reported separately for the revised fraction of 0.21 computed above. This will be called the "most plausible" case since the recent Friedman data use many more cases and controls than earlier studies, its subjects span multiple geographic regions and demographic groups, and the data are very recent and exclude foreign travel-related cases that could otherwise bias the results.

These two attributable-risk numbers give an estimated annual chicken-attributable Campylobacter illness rate of:

$$0.60*0.005*(1 - .275) = .002202 \qquad \text{(Base case)}$$
$$0.21*0.005*(1 - .275) = .007709 \qquad \text{(Plausible case),}$$

or approximately 220 chicken-attributable campylobacteriosis cases per 100,000 person-years in the base model and (0.21/0.60)*220 = 77.1 cases per 100,000 person-years in the "most plausible" case. This value corresponds to a "StorageAndPreparation" CFU reduction multiplier of approximately 1.1E-5, compared to a value of approximately 2.1E-5 for the base case. (As discussed later, uncertainty about these specific parameter values can be relatively large without greatly affecting uncertainty about the product of the various factors that determine final microbial load and risk.)

Microbial load reaching individuals via servings of chicken enters a non-linear dose-response model that determines individual probabilities of infection and illness. Teunis, et al., (1999) developed the following dose-response model for *Campylobacter jejuni* infection and illness based on human experimental data from Black, et al. (1988) for doses ranging from 8 x 10^2 to 2 x 10^9 in healthy young adults:

- Pr(Infection | Dose) $= 1 - (1 + D/\beta)^{-\alpha}$; D = dose (CFU), $\beta = 7.9$, $\alpha = .15$
- P(illness | infection) $=1 - (1 + \eta/D)^{-r}$, $\eta = 4.4*10^7$, r = .048

The minimum infective dose of *Campylobacter jejuni* observed to cause illness in the Black et al. study was 800 CFU (for a non-chicken medium). Other research suggests that it may be as low as 500 CFU (Robinson, 1981). The simulation model assumes that any dose below 500 CFU per has zero probability of producing an illness.

The fraction of consumers with campylobacteriosis who are treated with FQ is calculated, using the approach of FDA-CVM (2000b), as the product of three terms: (a) probability of being treated, given CP illness; (b) probability that the treatment includes prescribing an antibiotic; and (c) probability that the antibiotic prescribed is FQ. FDA-CVM estimates the third term as P(FQ | antibiotic) =0 .5550. The first two terms are calculated as weighted averages of probabilities conditioned on each of three Campylobacter case types (invasive infections, enteric (intestinal) infections with bloody diarrhea, and enteric infections without bloody diarrhea. The final result is:

Pr(FQ | CP illness) =
Pr(treatment | ill)*Pr(antibiotic | treatment)*Pr(FQ | antibiotic)
= 0.274*0.482*0.555 = 0.073

The next step is to estimate the proportion of FQ-treated CP cases that are infected with FQ-resistant strains of CP. A 1998 CDC FoodNet Campylobacter Case Control Study obtained Campylobacter isolates (FDA-CVM, 2000) from 580 individuals (*C. jejuni,* 92.4%; *C. Coli,* 2.7%; C. "other", 4.8%). Of these, 409 remained after removing cases associated with foreign travel or prior human use of FQ. Of the remaining cases, 6.4% (26 of 409) tested positive for ciprofloxacin resistance using the E-test previously described. A second FoodNet case-control study (Kassenborg, et al., 2000, Marano, et al., 2000) using 1998 and 1999 data found an unadjusted ciprofloxacin-resistant rate of 11% (94 of 858) among CP isolates tested. After removing cases associated with recent foreign travel, the resistance rate falls to 6.5% (37 out of 568). (Among those *not* likely to have acquired resistance from chicken, the rate was much higher -- 34.6% (27 out of 78). There were also two resistant cases associated with prior human use of FQ, however, we could not determine if either one also involved foreign travel. Thus, the rate could be slightly lower. Resistance testing was again defined via the E-test.) Pooling the results from both studies, i.e., 6.4% and 6.5%, gives a combined estimate of 6.4% (63/977) for the FQ resistance fraction among CP isolates from human patients.

This resistance rate based on sampled isolates admittedly provides at best only a rough estimate for the FQ resistance rate among cases of chicken-derived campylobacteriosis consumption. CP infections may be acquired from non-chicken sources that may also harbor FQ-resistant strains, such as beef consumption (FQ was approved for beef cattle in 1998), pets, or water (Koenraad, et al., 1995). The effects of these sources and how they might interact in people are not yet understood. However, since the mean FQ-resistance rate in chicken isolates (9.4%) is higher than the 6.4% resistance rate in isolates from human CP patients (6.4%), it seem clear that a person may become ill from a CP-infected chicken with isolates that would be classified as FQ-resistant at the time of slaughter, but then not test positive for resistance.

This difference in FQ resistance rates between chickens and humans is probably *not* due to differences in the survivability characteristics of resistant vs. non-resistant strain (Curtis et al., 1995), nor to competition on the carcass with non-resistant forms, as refrigeration virtually halts replication. The most plausible explanation is that the amount of resistant organisms required for chicken carcasses to test positive for resistance is often less than that required to cause a resistant illness in humans (Piddock, 1999). A chicken carcass may contain a mixture of *Campylobacter jejuni* strains (Chuma et al., 1997; Lillard, 1990), with resistant organisms constituting only a small fraction of the total. If the resistance rate in CP-infected chickens is R, then we assume that the appropriate resistance rate in humans is R x (.064/.094) = R x 0.68.

Children under age 18 are rarely prescribed FQ because of the risk of irreversible cartilage erosions (Talsma et al., 1999; Li and Brainard, 2000; Morgan D, 1997). The model therefore allocates FQ-resistant cases taking FQ only to individuals 18 and older (about 74% of the population in 1998 based on USCB, 2000c). The allocation is proportional to their age-specific relative Campylobacter incidence rates in Table 4.

The stochastic portion of the simulation ends after the human dose-response modeling, providing a probabilistically determined value for the number of chicken-associated cases of CP illnesses for each age cohort. Multiplying the number of CP illnesses by Pr(FQ | CP illness) = 0.073 and then by Pr(Resistant strain | CP illness) = 0.064 yields the expected number of FQ-resistant cases prescribed FQ. The methodological justification for focusing on the expected number of cases is that the probability distribution for the number of cases is approximately Poisson, and so is completely characterized by its mean. The conditional joint frequency distribution of the severity and duration of each case could easily be sampled from as a final step (perhaps conditioned on the infecting CFU load, individual age, treatment regimen, and other covariates) if it were known. However, useful estimates for this effects distribution are not yet available.

Table 6: Summary of Parameter Values for Human Sub-Model

Data Input Description	**Value**	**Notes**
Chicken-attributable CP cases	60% (Baseline) of all CP cases 21% (Plausible conservative estimate, confounders not controlled)	21% = rough estimate from Friedman et al. (2000) data.
Total Annual CP illness rate	0.005 = [Cases/year (1.38 M) from FDA-CVM 2001] / 1999 Population.	
Chicken-attributable CP Cases/yr.	220.24 cases/100,000 person-years = 0.005*.60 *(1-.275) (baseline model). (0.21/0.60)*220.24 = 77.02 cases/100,000 (most plausible)	(1-.275) = proportion $\geq$ 1 year old (CDC-FoodNet. 1999b)
Weekly Consumption	0.41 x Demographic multiplier	Servings of chicken per week per person week (derived from USDA, 2000).
Adjustment Factors of Consumption Rate	Group — Multiplier (Avg. = 1.0) — % of population White — .93 — 82.3% Black — 1.42 — 12.8% Other — 1.2 — 4.9%	(USCB, 2000a)
P(Infection | Dose)	$1-(1+D/\beta)^{-\alpha}$, D = dose in CFU, $\beta = 7.9$; $\alpha = .15$	(Black, et al. 1988, Teunis, et al., 1999)
P(illness | infection)	$1-(1+\eta/D)^{-r}$ $\eta = 4.4*10^7$; r = .048	(Black, et al. 1988, Teunis, et al., 1999)
Response Rate by Age	See Table 4.	
P(treat | ill)	0.274 = 0.33* Pr(bloody diarrhea, enteric | CP illness) + 0.20* P(non-bloody diarrhea, enteric | CP) + 1* P(invasive illness | CP Illness)	(FDA-CVM, 2000b)
P(antibiotic | treatment)	0.48 = Pr(treat with antibiotic) = 0 .844 if submit stool sample, 0.38 if not, 1 for invasive cases.	Proportion who submit sample = 0.26 for bloody, 0.15 for non-bloody. (CDC; FDA-CVM, 2000b)
P(FQ | antibiotic)	0.55 = Probability that antibiotic is FQ.	(CDC-FoodNet surveillance data, FDA-CVM, 2000b).
Pr(CP case is FQ-resistant | from chicken)	0.064 = average of two case control studies with values of 0.064 and 0.065	FDA-CVM , 2000a pg. 3-12; Kassenborg, et. al., 2000

An important methodological point is that *it is unnecessary* to know the consequence distribution of illness events, for purposes of evaluating and comparing risk management option, *if* preferences decrease in the expected number of cases (e.g., if twice as many expected cases per unit time are considered twice as bad, independent of the precise marginal distribution of adverse health consequences once a case occurs.) Indeed, Poisson (and compound Poisson) processes for arrivals of undesirable events can be unambiguously ordered by first-order stochastic dominance (see Chapter 6) based only on their intensities, even without knowing their other characteristics or the values assigned to consequences of different severities. Table 6 summarizes the baseline (i.e., pre-intervention, pre-sensitivity analysis) parameter values for the human sub-model.

The entire simulation model, composed of the chicken and human sub-models, was implemented in Visual C++. It can be downloaded from www.cox-associates.com/AHI. A typical run of the simulation spans one simulated year and is run for 100,000 iterations (i.e., for 100,000 simulated individuals for 1 year). A typical single simulation run requires about 30 seconds on a Windows NT 4 Workstation with a 433 MhZ processor.

5.4 Results: Baseline and Sensitivity Analyses of Options

Table 7 summarizes output values of various metrics for both the "most plausible" case (chicken-attributable fraction =0 .21) and the "baseline" case (chicken-attributable fraction = 0.60). These values assume an overall FQ resistance rate of 17.6% of CP cases, and a 6.4% resistance rate for CP cases acquired via chicken consumption.

Table 7: Pre-Intervention Metrics for [Most Plausible; Baseline] Scenario*

Metrics [Most Plausible; Baseline]	**Rate (Cases per 100K per yr)**	**Total cases (in 270M)**	**Expected Excess Illness Days**
CP – all sources	506.1	1,380,000	
FQ resistant CP – all sources	89.1	242,896	
FQ resistant CP prescribed FQ – all sources	6.5	17,780	35,560
CP – chicken associated	[77.1; 220.2]	[210,206; 600,587]	
FQ resistant CP – chicken associated	[4.9; 14.1]	[13,453; 34,438]	
FQ resistant CP prescribed FQ – chicken associated	[0.36; 1.03]	[985; 2,814]	[1,970; 5,627]

* Baseline scenario assumes that 60% of all CP cases are attributable to chicken; Most Plausible scenario assumes that this number is 21%. "Expsected Excess Illness Days" assume 2 days of excess illness per FQ prescription to a person with an FQ-resistant strain of CP.

The simulation model allows sensitivity analyses to be run to study the effects of changes in the parameter values. For example, ineffective FQ prescription rates and total chicken-associated CP rates per 100,000 people per year can be plotted against: 1) CFU per chicken leaving the farm, 2) CFU increase due to transport, 3) CFU decrease due to processing, 4) cross-contamination multiplier, and 5) cecal CP contamination rate in farm chickens. Varying CFU per chicken leaving the farm shows that almost no cases occur for initial loads less than about $10^{4.85}$. Thus, human infections are only possible (all else held equal) if a given chicken has an initial microbial load significantly higher than the current mean ($10^{3.09}$).

Reducing the initial CFU load or the transport (CFU increase) factor, or decreasing microbial load during processing, by about 3.2-fold, would virtually eliminate risk of chicken-borne campylobacteriosis. This required change is the same for the infection, transport, and process factors because they have a symmetric (multiplicative) effect on final microbial load.

Figure 3 shows the sensitivity to the transport factor of the rate of ineffective prescriptions (per 100,000 person-years, assuming that any detectable level of FQ-resistant CP causes treatment failure). Note that the risk associated with the current mean value of the transport factor is far higher than would be the case if all chickens experienced this mean value. This reflects the fact that the right tail (above 500 CFUs) of the population frequency distribution of final microbial loads reaching consumers drives the production of CP cases. The variability in CFU loads among chickens is responsible for almost all chicken-attributable CP cases, as the average level of contamination is too small to cause illness.

Finally, define the cross-contamination rate as the average post-processing incidence of infected chicken carcasses (surface) divided by the average preprocessing incidence. A value of 1 corresponds to no cross-contamination. A value of 2 implies one additional infected chicken for every infected chicken going into processing. Virtually all chickens become infected at a value of about 3.6.

Figure 3: Health risk (ineffective prescription rate) vs. transport factor

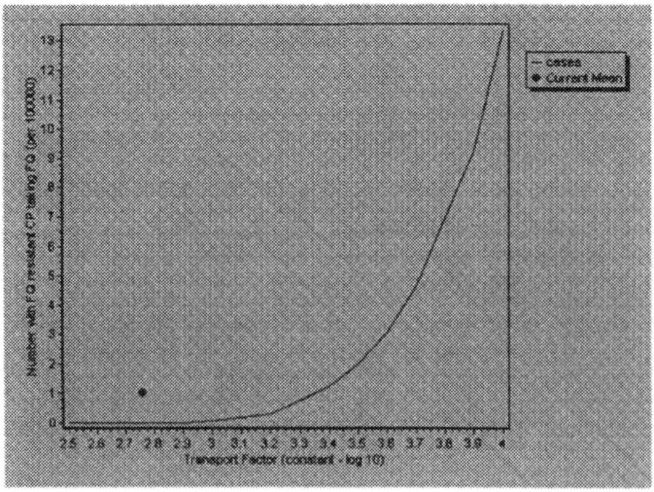

Below 2.5, the risk level is approximately linear in the cross-contamination factor, but risk approaches a plateau as the factor increase from 2.5 to about 3.6. Risk is approximately linear in cecal CP contamination rate.

The simulation model can be used to evaluate the effects of proposed risk management interventions throughout the supply chain. Table 8a below summarizes the results for the "baseline" case (chicken attributable fraction = 0.60) while Table 8b summarizes results from the "most plausible" case (chicken attributable fraction = 0.21). For scenarios requiring runs of the simulation model, results are the average of 10 simulation runs of 100,000 persons for one year (1 million person-years). Table 8 assumes that campylobacteriosis creates 6 days of illness on average.

Attribution for FQ-resistant cases is somewhat more complicated. The key question, as discussed earlier, is whether FQ resistance *causes* additional days of illness or is merely associated with them. One extreme is to assume that *any* FQ-resistant case of CP causes additional days of illness. At the other extreme, only FQ-resistant cases prescribed FQ would cause excess days of illness. Since we have determined that the probability of being given FQ, given that one is ill, is 0.073, the minimum attribution will be 0.073 times the maximum attribution. Table 8 reports corresponding minimum and maximum values for average illness-days reduced, assuming an average of two excess-days of illness.

The risk management options evaluated in Table 8 are as follows:

1. *Eliminate FQ at the farm.* This does not change the incidence of CP incidence but reduces the number of cases of FQ resistance. Of 110.8 FQ-resistant cases per 100,000 person-years (Table 7), 17.54 (or 6.14 in the more plausible case) may be due to consumption of chicken that has FQ- resistant CP. Eliminating FQ use on the farm might eliminate most of these, although some FQ-resistant chicken isolates could stem from other environmental sources (Piddock et al., 2000). However, there is currently no available quantitative attribution of FQ resistance in chickens to its various possible sources. Table 8 assumes that the remaining quantity of FQ resistant chickens is zero.
2. *Optimize FQ withdrawal.* Withdrawing FQ some number of days prior to slaughter can achieve some of the benefits of a ban. If the proportion of chickens with FQ-resistant CP isolates decays exponentially after FQ has been withdrawn with a half-life of 20 days then *FQ proportion(t)* = $e^{-0.035t}$ and the numbers in Table 8 for this scenario follow. The true half-life is not known but is probably more than 2 weeks (Jacob-Reitsma, 1994a).
3. *Track FQ-exposed batches of chickens* and process them to eliminate all CFUs. This option has a similar effect to (1) above but also essentially eliminates all resistant strains for a portion of incoming

batches/farms, whereas (1), for the same batches/farms, would allow resistant strains derived from environmental sources. area serviced by a given processor all apply FQ.

4. *Processing changes.* Processing changes will not change the rate of FQ resistance as a proportion of all CP cases, but will reduce the number of cases of CP illness. This in turn reduces the number of FQ resistant cases. A side benefit that we have not attempted to quantify is a reduction in incidence of illnesses from other food-borne pathogens such as Salmonella. Technical options include: use of chlorinated water sprays (Mead, et al. 1995), increased chlorination of chiller tank water (White, et al., 1997; Yang, et al., 2000), and increased scalding (Yang et al., 2000), electrification of chiller tank water (Li and Slavik, 1996), or irradiation (Thayer, 1995). Rather than enumerating these options, Table 8 focuses on specific numerical targets (e.g., an additional 10% reduction in outgoing microbial load). To model a greater reduction in load during processing, the processing factor distribution is changed from T(1.0/P, 2.23/P, 3.0/P) to T[1.0/P+log10(k), 2.23/P+log10(k), 3.0/P + log10(k)]. For 10% case, k =1.1 and for 100% case k = 2.0. P = .94 for the base case and .94375 for the most plausible case. Other reductions in microbial load during processing are modeled similarly.
5. *Restaurant/Consumer behavior changes.* Changes in consumer or restaurant storage and handling practices can reduce CP illnesses, and thereby the number of FQ-resistant cases. Options include keeping chicken chilled until use, thorough cooking, maintaining clean food preparation surfaces, and consistent use of disinfectants (White, et al., 1997; Josephson, et al. 1997). These changes are effected in the simulation through a 10% decrease in the food preparation (reduction) factor, from 2.5E-5 to 2.25E-5 in the base case and 1.343E-5 to 1.209E-5 in the most plausible case. It is most plausible to ascribe these changes to restaurants, as chicken consumed outside the house is a risk factor, while chicken consumed at home appears not to be (e.g., Ikram et al., 1994; Adak et al., 1995; Eberhart-Phillips et al., 1997, Friedman et al., 2000a.)
6. *Physician behavior changes.* The final change evaluated is a 10% reduction in the rate at which FQ is prescribed by physicans.

Table 8a reveals a key policy-relevant finding. Risk management measures that reduce total CP illness rates prevent dramatically more illness-days per year than measures that specifically reduce chicken-associated FQ rates. This is true even under the extreme assumption that FQ resistance *causes* increased illness days for *all* FQ-resistant cases.

Table 8a: Health Benefits per 100,000 Person-Years: Base Case

	Cases Reduced				Illness Days Reduced[1]		
Risk Management Option	**CP (%)**	**CP (cases)**[1]	**FQ resistant CP**[2]	**FQ resistant CP prescribed FQ**[3]	**CP**[4]	**FQ (Min/Max)**[5]	**Total**
1.Eliminate FQ	0	0	17.54	1.284	0	2.57 / 35.0	**2.6**
2. Optimize withdrawal							
10 days	0	0	5.25	.38	0	0.76 / 10.3	**0.76**
20 days	0	0	8.77	.64	0	1.28 / 17.5	**1.3**
30 days	0	0	11.33	.83	0	1.66/ 22.6	**1.7**
3. Track FQ	0	0	17.54	1.28	0	2.57 / 35.0	**2.6**
4. Processing changes.							
↓cross-cont. (10%)	3.7%	10.3	.66	.048	61.98	.0975	**62.1**
↓cross-cont. (100%)	48.2%	132	8.45	.618	792	/1.32	**793.2**
↑CFU-reduction (10%)	14.7%	40.3	2.58	.189	242.0	1.24/16.9	**242.4**
↑CFU-reduct. (100%)	71.5%	196	12.54	.918	1176	0.38/ 5.2	**1177.8**
both (10%)	15.1%	41.3	2.65	.194	248.0	1.84/25.1	**248.4**
both (100%)	85.4%	234	14.98	1.10	1404	0.39/5.3	**1406.2**
						2.2/30.0	
5. Restaurant changes.	14.6%	40	2.56	.190	240	0.375/ 5.1	**240.4**
6. Physician behavior changes	0	0	0	.424 [6]	0	.848	**0.85**

[1] Assumes 6 days per CP illness, 2 days excess days if ineffective FQ treatment
Nonzero values are from 10 or more runs of the simulation

[2] CP > 0 $\rightarrow$ CP x .064, CP = 0 $\rightarrow$ 274 x .064 x Proportion of FQ relative to original

[3] .0732 x FQ resistant CP

[4] 6 x CP

[5] min = 2 x FQ resistant CP prescribed FQ, max = 2 x FQ resistant CP

[5] .10 x (the number of cases of FQ resistant CP prescribed FQ - all sources)

Table 8b: Health Benefits per 100,000 Person-Years: Plausible Case

	Cases Reduced				Illness Days Reduced[7]		
Risk Management Option	CP (%)	CP (cases)[8]	FQ resistant CP[9]	FQ resistant CP prescribed FQ[10]	CP[11]	FQ (Min/Max)[12]	Total
1. *Eliminate FQ at farm*	0	0	6.14	0.45	0	.90 / 12.3	**.90**
2. *Optimize withdrawal period.*							
10 days	0	0	1.8	0.135	0	0.27/ 3.7	**0.27**
20 days	0	0	3.1	0.225	0	0.45/ 6.1	**0.45**
30 days	0	0	4.0	0.29	0	0.581/ 7.9	**0.58**
3. *Track FQ batches.*	0	0	6.1	0.45	0	0.90 / 12.3	**0.90**
4. *Processing changes.*							
↓*cross-contam.* (10%)	5.5%	5.3	0.34	0.025	31.8	0.05 / .68	**31.85**
↓cross-contam. (100%)	46.4%	44.5	2.8	0.21	267	0.42/5.7	**267.4**
↑CFU-reduction (10%)	12.8%	12.3	0.79	0.058	242.0	0.11/ 16	**73.9**
↑CFU-reduction (100%)	80.1%	76.8	4.9	0.36	460.8	0.72/9.8	**461.2**
both (10%)	20%	19.2	1.2	0.09	115.2	0.18/2.46	**115.4**
both (100%)	88%	84.4	5.4	0.395	506.4	0.79/10.8	**507.2**
5. *Restaurant changes.*	18.6%	17.8	1.14	.083	106.8	9.17/ 2.3	**107.07**
6. *Physicians prescribe FQ 10% less frequently*	0	0	0	0.42[13]	0	0.848	**0.85**

Table 8b shows similar results, although the reductions are smaller, since they start from a smaller base. Thus, although the assumed rate for chicken-attributable risk varies nearly three-fold, from 0.21 (Table 8b) to 0.60 (Table 8a), it does not affect the key conclusion that *interventions that focus on reducing Campylobacter are much more effective in reducing total expected illness-days than interventions that focus solely on FQ resistance.*

In slightly more detail, totally eliminating FQ at the farm is predicted to create less than 1% of the public health benefit, measured as reduction in expected illness-days per capita per year, that would result from incremental

[7] Assumes that all excess days in a resistant case are attributed to resistant organisms from a causal point of view

[8] Nonzero values are from 10 or more runs of the simulation

[9] CP > 0 → CP x .064, CP = 0 → 95.9 x .064 x Proportion of FQ relative to original

[10] .0732 x FQ resistant CP

[11] 6 x CP

[12] min = 2 x FQ resistant CP prescribed FQ, max = 2 x FQ resistant CP

[13] .10 x (the number of cases of FQ resistant CP prescribed FQ - all sources)

improvements in processing that reduce outgoing CFUs and cross-contamination by 10% each. Banning veterinary use of FQ is also predicted to create less than 1% of the health benefit that would be achieved by improving restaurant sanitation by 10%, even if only the benefits from reduced CFUs in chicken are considered. Changing physician's prescription behavior is expected to have relatively little impact on public health, as only a relatively small fraction of consumers affected end up experiencing ineffective prescriptions.

Despite the many assumptions and uncertainties in the modeling, discussed further in the next section, it appears that the different risk-management options in Table 8 can be confidently divided into two subsets based on their probable effectiveness in reducing illness-days per year. Options 1-3 and 6 are the least effective, while options 4 (in various versions) and 5 are much more effective.

Of course, even the least effective options, as judged by predicted health impacts, might still be worth implementing if their costs are sufficiently small, or if they are the only options available to those making the decisions. For example, a regulatory body such as FDA might decide that it is only empowered to focus on FQ-resistance related risks, and thus choose to ignore risk management interventions that do not address them, even if the ignored options could create larger health benefits. Similarly, a decision-maker might be required to consider only the health benefits of initiatives that involve chicken-associated microbial load of a specific pathogen, like CP, as a causal mechanism, ignoring effects on other pathogens and foods.

Such restrictions may lead to decisions that are sub-optimal from the perspective of reducing expected illness-days as much as possible for the resources spent. But they often map to organizational goals and mandates more easily than more comprehensive risk-cost-benefit analyses. Table 8 and the underlying simulation model provide information that is potentially useful for a wide variety of decision processes, even if the most obvious recommendations (e.g., to focus on options 4 and 5) cannot be implemented or require additional constraints and pragmatic considerations to be taken into account. For the results in Table 8 to be useful, however, they must be credible. Technical credibility requires realistically confronting the many uncertainties introduced by the modeling process.

5.5 Uncertainty Analysis and Discussion

How certain are the conclusions in Table 8? The sensitivity analyses already reported provide a partial answer, suggesting that the main results are robust to even fairly large changes in assumed parameter values. This section examines the quantitative uncertainty in estimates of the health impacts of different risk management strategies. Major sources of uncertainty to be addressed include sampling variability, parameter uncertainty and variability, and model uncertainty.

Potential Challenges to the Results: Major Uncertain Modeling Assumptions and Simplifications

As mentioned in the introduction, the true health risks associated with FQ-resistant CP are currently unknown. Even the hypothesis that campylobacteriosis patients treated with FQ will not recover as fast if they have FQ-resistant CP isolates as they would otherwise remains an unproved, though plausible, conjecture. Moreover, the discrete-event simulation model introduced here to help support decisions is based on several crucial assumptions that may be wrong. These include the assumptions that:

- Chicken products cause 60% (Table 8a) or 21% (Table 8b) of CP cases.
- Any amount of FQ-resistant CP, no matter how slight, causes complete treatment failure in patients prescribed FQ.
- Excess illness-days results from any FQ-resistant strain (high extremes in Table 8) or else only from FQ-resistant strains in patients administered FQ (low extremes in Table 8.)
- 6.4% of all CP cases are FQ-resistant
- The effect of storage and preparation on microbial loads is exactly the value required to make the model's predictions match historical rates of CP illness attributed to chicken consumption. (This results from the fact that this parameter was not estimated from independent data, but rather was calibrated so that the model's outputs would agree with historical estimates of chicken-attributable risk for CP illnesses.)

Even if these parameter values are correct on average, there is certainly variability in them that has been ignored, e.g., among geographic regions and from patient to patient. Key components of the model, such as the dose-response relation, have been represented by relatively simple sub-models that ignore complexities such as the effects of a consumer's immune status on susceptibility to CP. Variability that has hitherto been ignored, as well as uncertainties about parameter values, must be considered before the quality of the conclusions in Table 8 can be appraised.

In addition to uncertainties and sources of variability that have not yet been modeled, the discrete-event simulation model contains deliberate simplifications that are known to be incorrect. Examples include (a) Modeling the health impact of ineffective FQ treatment as being a constant (two excess days of diarrhea) for all patients; and (b) Assuming that all individuals in the same age group have the same dose-response relation. Before recommending decisions, the *scope* of the analysis must also be considered, i.e., have all and only the relevant consequences been addressed?

In this case, it could be sensibly argued that the metrics reported in Tables 7 and 8, such as illness-days prevented per 100,000 person-years, should be defined more carefully, e.g., by excluding all people who do not consume chicken products, or who consume only frozen and processed chicken products with negligible or zero microbial load. Changing the denominator with respect to which risk rates are calculated would certainly change the numbers reported, although not necessarily their ratios.

Given all these limitations, how can the simulation model's conclusions reveal anything of practical value to decision-making and risk management? Half the answer is that it produces comparisons and insights that are *robust* to the uncertainties, simplifications, and scoping decisions incorporated into the model. In other words, the main conclusions do not change no matter how some of the open uncertainties and debatable assumptions are resolved. For example, even if the true number of excess illness-days following an ineffective FQ treatment has a complicated probability distribution that depends in unknown ways on possibly unobserved individual characteristics, none of this would change the fact that fewer ineffective treatments cause less damage, and so are preferable. As discussed below, refining this part of the model to make it more detailed and realistic would not change the essential comparisons among the options in Table 8. More generally, it is important to identify conclusions that are insensitive to the uncertainties and approximations in assumptions and scope defined by the model.

The second half of the answer is that the model is a *contingent* one. It compares different risk management options *conditioned* on the policy-relevant assumption that there is a health risk to be managed, with the causes and characteristics ascribed to it in the model. This is more than a logical dodge to circumvent uncertainty. It allows effective and ineffective options to be distinguished under the stated assumptions of the model so that *if* an intervention is selected for implementation, its potential health benefits will have been characterized and compared to those from other options. While acknowledging that scientific uncertainties remain about whether the risks exist and behave as hypothesized, simulation can still be used fruitfully to compare options for managing them in case they turn out to be real.

The following paragraphs address these points in more detail. The use of complex assumption-based risk models is widespread in health risk analysis. Understanding exactly what such models show and how they can be useful despite their explicit and implicit limitations is crucial for both risk analysts and risk managers.

Modeling Sampling Variability by Poisson Counts

Sporadic cases of campylobacteriosis arise as statistically independent, rare events in a large population, at least if multiple illnesses within a household are treated as single events. The number of cases per year may therefore be approximated as a Poisson-distributed random variable, with equal mean and variance, for any fixed set of input conditions (Feller, 1968). Even if different sub-populations have different rates and even if the occurrences of CP illnesses are censored by various binomial probabilities, the final counts per year will still be Poisson-distributed. Then, approximate upper and lower 95% confidence intervals for the numbers in Tables 6 and 8 can be derived by simply adding and subtracting 1.96 times the square root of the mean value to each mean value shown. Although more sophisticated bounds can be formed for Poisson-distributed count data, this simple procedure gives a useful idea of the plausible range of sampling variability. For example, the plausible upper-bound estimate of 985 cases per year in Table 7 would have an approximate 95% Poisson confidence interval of [923, 1047] based on Poisson sampling variability alone.

Under mild mathematical assumptions (Feller, 1968), the number of cases per year for *any* fixed input conditions (and for any type of case – CP illness, CP illness due to chicken-borne CP, CP illness due at least 10% to chicken-borne-CP and treated with FQ, etc.) should be approximately Poisson-distributed. This conclusion does *not* depend on uncertainties about the individual parameters that contribute to the probable number of cases, such as the transportation factor, processing factor, storage and preparation factor, which may have relatively large uncertainties. (Indeed, if risk depends on a product of such factors and if the product is calibrated to match historical data, as detailed previously, then uncertainty about the product can be much less than the uncertainties about its individual components. They may be wrong but the estimated product may still be right.) Therefore, instead of composing posterior probability distributions for the individual parameters to estimate a probability distribution for their product, we will rely on the Poisson confidence limits for the final results to indicate the likely extent of sampling variability. When individual-level data on exposures and responses (e.g., the data of Friedman et al., 2000) are made available, the assumption of Poisson variation in counts of sporadic CP cases can be statistically tested. Meanwhile, the "independent rare event" nature of the data motivate the Poisson assumption as a starting point for uncertainty analysis.

Some sources of sampling variation that might be important in other contexts have been omitted from the simulation model. Most conspicuously, variability in the number of excess days of illness per case of CP (or per ineffective FQ prescription to a CP case) is ignored. This is not because such variability is uninteresting or unimportant – for example, it may inform patient decisions about whether to seek treatment and physician decisions about whether to prescribe an antibiotic – but because all of the risk management decisions considered in this paper affect excess illness-days only through the expected number of cases (of CP illness, ineffective treatments, etc.) generated per year. In effect, what can potentially be controlled by the risk management alternatives is the probable number of cases per year. This random number is completely characterized by its expected value, assuming the Poisson model. Smaller expected values are always preferred. (Technically, the family of Poisson probability distributions can be totally ordered by first-order stochastic dominance based only on their means; see Chapter 6 and Ross, 1996).

The expected number of cases per year under different risk management scenarios summarizes everything that can be directly controlled by decision-making. Consequences in terms of illness-days have additional uncertainty (which could be modeled, for example, via a *compound* Poisson process), but this additional uncertainty is superfluous for judging the relative health benefits from different risk management decisions. These depend only on the ratios of the expected numbers of cases for the different decisions. Therefore, we truncate the modeling with the calculation of the expected number of cases for each risk management decision. Modeling the sampling variability in illness-day consequences does not affect the relative health benefits estimated for different risk management decisions.

Approximate Modeling of Parameter Variability and Its Effects

An advantage of discrete-event simulation methodology is that it can simulate realistic variability in the values of key parameters and variables, such as the microbial loads of CP on individual servings of chicken. The triangular distributions selected in Table 3 to represent variability in key parameters and variables (e.g., initial surface microbial loads and transportation and processing factors) are intended only to reflect the approximate variability in the aggregate data that have been published on these factors. Although values for individual chickens or servings of chicken might fall outside the ranges shown, accurate simulation does not require accurately capturing all possible values and their probabilities. Rough approximations to the frequency distributions describing inter-individual variability in the population are often all that is needed for the distributions of their products (or other smooth functions) to be approximately correct.

As an example, suppose that the microbial load reaching an individual consumer in a particular serving of chicken is modeled as a product of many factors, each sampled randomly and independently from its own distribution of values. These factors could represent the initial microbial load, its gradual amplification during successive hours of transportation, its reduction or amplification at each successive stage of processing, storage, distribution, and preparation, etc. The distribution of this product will be approximately log-normal (under the reasonably mild conditions of a Central Limit Theorem) with a distribution that depends *only* on the geometric means and standard deviations of the factors entering the product. Hence, *the distributions of the individual factors need not be modeled very accurately in order for the simulated distribution of their product to be approximately correct.* In this example, only the geometric means and standard deviations need to be approximately correct.

Our modeling of parameter variability has also relied on simple (mainly binomial and triangular) distributions to approximate the true but unknown population frequency distributions of individual parameters. Without access to original data, it is necessary to use published means, medians, ranges, and similar statistics to construct these approximate distributions. Yet, simulation results based on them may still be quite satisfactory, as sensitivity analysis suggests that the population frequency distribution of microbial loads reaching individuals is quite insensitive to the detailed frequency distributions of individual parameters.

As mentioned previously, the use of single binomial parameters where data do not support a more refined (mixture model) description tends to *maximize* the variability captured in the simulation modeling (Feller, 1968). This is also true of the dose-response relation, where the (unrealistic) assumption that all individuals in the same age group have the same dose-response relation tends to maximize simulated inter-individual variability in responses at any given exposure level. It is not always understood that using single average probabilities in simulation (or analytic) models instead of more accurate and detailed conditional probabilities leads to more, not less, simulated uncertainty. For example, it creates wider simulated confidence bands and measures of run-to-run variability. This fact may be exploited to obtain conservative confidence bands for results from multiple runs.

The Poisson uncertainty analysis framework makes this simulation-based approach to confidence band estimation unnecessary. In the Poisson framework, CP cases are viewed as arising from a superposition of multiple Poisson processes, representing many parallel versions of the process in Figures 1 and 2 with different parameter values. These versions may represent flows from different farms through different processing plants and retailers to different outlets (e.g., restaurants) and consumers. Even without knowing how many versions there are and what the version-specific parameter values are, such a superposition leads to an overall Poisson process for CP cases, since the sum of Poisson processes in a Poisson process. Therefore, the mean number of cases per year is all that finally matters.

How this mean value varies for different risk management options is then the key question. Table 8 addresses it by considering the impacts of options defined in terms of percentage reductions in microbial loads. The final microbial loads enter a dose-response model for each age group. This may be viewed as an average of many unknown individual-specific dose-response functions, making some aggregation error inevitable. While in principle, mixture distribution modeling might be used to attempt to resolve the aggregate dose-response function into several components, the required data are not available. The model has already been calibrated to accurately describe the expected number of cases given historical conditions, and no individual-level data are currently available from which to attempt to estimate the frequency distribution of individual dose-response functions in the population.

Sensitivity analysis provides a partial solution to the problem of unknown and variable dose-response relations. The dominant qualitative features of the modeled dose-response relation are that (a) Risks are low or zero for sufficiently small doses (e.g., less than 500 CFUs in the simulation model); and (b) Illness probability increases rapidly as a function of dose, reaching an approximate plateau of about 0.2 among 20-39 year olds for CFU levels of about 1,000 to 10,000 CFUs. For older age groups, this approximate plateau is lower; it is about 0.14 for people 60 and older. *Any* dose-response relation with these qualitative features tends to produce similar expected numbers of CP cases from given population frequency distributions of microbial loads. Therefore, although individual dose-response relations are unknown and may be quite variable, the average number of CP cases produced by any given frequency distribution of microbial loads is not highly sensitive to this uncertainty. Therefore, the change in expected CP cases per year from different options affecting the microbial load distribution received by consumers can be predicted reasonably well even without further modeling of individual dose-response functions – provided that the assumed aggregate relationship is roughly accurate.

Model Uncertainty and Results of Simple Approximate Calculations

Model uncertainty, i.e., uncertainty about the quantitative causal relations among inputs and outputs of the simulation model, dominate parameter and sampling uncertainties in this study. As previously mentioned, it is not known what impact FQ use in chickens has on development of FQ-resistance in humans. It is not known whether or under what conditions FQ-resistant isolates in humans imply that treatment with FQ will have diminished effectiveness, and the loss of effectiveness (if any) has yet to be estimated. The proportion of all CP cases that is caused by chicken consumption is unknown, but is probably much lower than previous estimates (which do not control for confounders or other risk factors). The impact on human health of banning FQ use in chickens could well be zero or negative, depending on how these and other uncertainties are resolved.

In this setting of huge model uncertainties, it may make little sense to develop more detailed models of parameter uncertainty or variability than those already described. For example, using Bayesian priors in lieu of data could create an illusion that uncertainty has been usefully addressed, but only by introducing a new, potentially obscuring, layer of assumptions. On the other hand, simple sanity-check calculations based on a different logical approach can test – and hence potentially build confidence in – the contingent validity (given its stated assumptions) of the model's predictions.

For example, Table 7 shows an average of 0.36 ineffective prescription cases per 100,000 people per year due to FQ use in chicken, based on simulation runs. To check whether this estimate is roughly consistent with what one should expect, the same quantity can be estimated by a transparent "top-down" arithmetic approach using documented aggregate relationships and assumptions. Instead of composing multiple "bottom-up" relations (e.g., microbial load on chickens leaving the farm, amplification or reduction during transport, processing, storage and preparation) and attempting to estimate their parameters from data, one can simply multiply the following aggregate components:

- Annual Campylobacteriosis rate = 0.005 (= 1.38M Cases per year, from FDA-CVM 2001, divided by 1999 US Population.)
- Proportion of Campylobacteriosis cases in non-infants = 0.725
- Proportion of Campylobacteriosis cases attributable to chicken among non-infants = 0.21 (suggested plausible upper bound without correcting for effects of confounders.)
- Proportion of Campylobacter cases subsequently prescribed fluoroquinolones = 0.073 (based on FDA-CVM 2000b methodology, discussed previously.)
- Proportion of Campylobacter-infected cases with FQ-resistant strain = 0.065 (based on data of Kassenborg et al., 2000).

Doing so gives:

Approximate risk rate = 0.005 x 0.725 x 0.21 x 0.073 x .065
= 4×10^{-6} expected cases of patients with FQ-resistant CP prescribed CP per person-year.

This estimate is close to Table 6's estimate of 3.6×10^{-6} from a million simulation runs of the "bottom-up" simulation model. The simulation model has many advantages, e.g., its estimates may have smaller bias than the results based on a product of mean values and it allows extensive sensitivity and what-if analyses. But its detailed sampling and many iterations necessarily make its numerical results harder to immediately trace and verify. Thus, it is reassuring that the top-down and bottom-up approaches produce similar estimates, given their shared modeling assumptions.

Model uncertainty becomes much more important when the health effects of changing inputs to reduce microbial loads are to be predicted. To illustrate, Figure 4 plots predicted reductions in CP risk from different reductions in *average* microbial loads, for two different statistical models of the population *variability* in individual microbial loads surrounding the average levels. The lower curve represents a log-normal model of the population frequency distribution of microbial loads. This distribution is motivated by the structure of Figures 1 and 2, in which final microbial load is driven by a product of factors, each with its own variability distribution. (Cross-contamination during processing complicates the model structure, but the multiplicative specification provides a useful conceptual simplification and approximation to the full simulation model.) The upper curve in Figure 4 represents a log-exponential model, suggested by FDA-CVM (2001b), for unspecified reasons, as a possible model for inter-individual variability. Both statistical models were estimated by fitting their parameters to historical data, and both go through the current data point in the upper left-hand corner (i.e., 0 reduction in average microbial load or in CP risk below current levels.) Yet, they make utterly different predictions about the human health benefits from reducing average microbial load of CP in chicken by 50%. The log-normal model predicts that this would essentially eliminate chicken-borne CP risks, while the log-exponential model predicts that such risks would be only slightly reduced. Not surprisingly, the simulation model produces a curve similar to that for the log-normal distribution.

Figure 4: Risk impacts of reducing average microbial load, as predicted by log-normal and log-exponential models of heterogeneity in microbial load

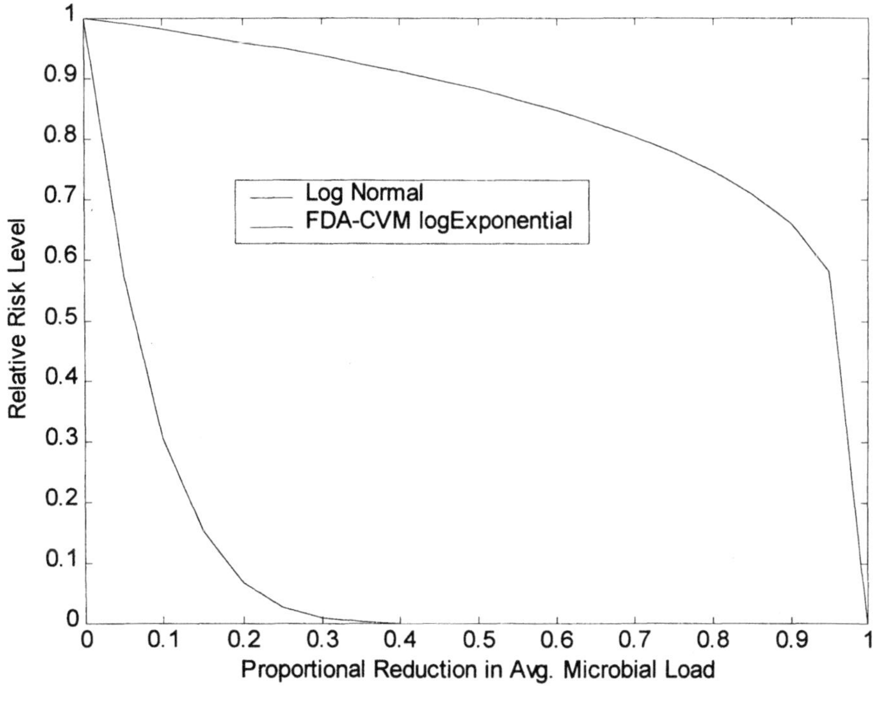

In summary, better and more confident risk modeling will require obtaining better descriptions of the physical distribution of microbial loads. It will also require clarifying the causal relation between ingested microbial loads (of FQ-resistant, FQ-susceptible, or mixed) strains of CP and adverse health responses among patients receiving different treatments. Until this basic descriptive information is available, the predicted quantitative impacts of interventions will be highly sensitive to modeling assumptions. These sensitivities, illustrated by the two very different curves in Figure 4, make careful analysis of parameter uncertainties and sampling variability premature. They are small by comparison to the modeling uncertainties.

However, despite these uncertainties about the *absolute* impacts on human health of different risk management interventions, the *relative* impacts previously shown in Table 8 appear to be quite robust. Under a wide variety of parameter estimates, it remains true that reducing microbial loads of CP at processing is predicted to be a relatively effective strategy for reducing adverse human health impacts of CP in chickens. FQ susceptibility or resistance makes little difference to health impacts once the decision is made to reduce microbial loads.

Robustness to Scoping Decisions

The decision impacts modeled in Table 8 are restricted in scope. They do not consider possible benefits from eliminating FQ in chickens due to lowered FQ-resistance in other farm animals, or to lowered FQ-resistance in pathogens other than Campylobacter. Yet, these impacts might be considered a part of any comprehensive risk-cost-benefit comparison. However, broadening the scope of the assessment to include impacts on other animals and pathogens is unlikely to change the comparative ranking of options in Table 8. There seems to be little or no empirical association between FQ use in the farm environment development of FQ-resistance in farm animals (Piddock et al., 2000) If other microbial pathogens are brought into the scope of the analysis, the comparative advantage of risk measures that reduce microbial loads compared to those that reduce FQ resistance is likely to increase, rather than decrease. Changes in processing or in restaurant food safety practices that reduce probability of illness due to CP can also reduce probability of illness due to Salmonella, Shigella, and E. coli (e.g., Whyte et al., 2001). Other changes in scope, e.g., to consider possible human health benefits from the use of FQ to kill bacteria in chickens, perhaps as part of a competitive exclusion program, would clearly not weaken – but might strengthen – the conclusions from Table 8.

A final aspect of the predicted risk rates in Table 8 is that their ratios are not expected to change as the denominator describing the population at risk changes. For example, suppose that the population generating chicken-associated CP cases were not defined as the whole U.S. population in relevant age groups, but rather were refined to include only those exposed to CP in undercooked chicken. Then, even though the fraction of the population contributing observed CP cases would be smaller, and so the absolute risk rates per capita in this population would be larger, the ratios of the rates would remain unchanged.

5.6 Conclusions

This case study for *Campylobacter jejuni* has shown how a discrete-event simulation model can be used to integrate available data elements, together with policy-relevant assumptions where major scientific uncertainties remain unresolved, to inform risk management decisions. The model quantifies the probable impacts on public health of different options for managing the risks from chicken-borne *Campylobacter jejuni* despite the large uncertainties about the existence and magnitudes of the hypothesized hazards. The basic logic of the model is that of a series process. A sufficient dose (CFUs) to cause campylobacteriosis reaches a consumer via a chicken serving if and only if the amount leaving the farm is non-zero *and* it is amplified during transportation *and* it is not reduced to safe levels (or eliminated completely) by processing and preparation. The probability of this event is found by simulating the many combinations of initial microbial loads and their subsequent histories and counting the fraction that lead to illnesses. The process is not a pure series system because of the possibility of cross-contamination during processing, but this is easily incorporated in the simulation model. Individual attributes that affect exposure and dose-response can also be included if their frequency distributions in the population are known. In the current model, this capability is illustrated by the inclusion of ethnicity (affecting chicken-consumption and hence exposure) and age (affecting the dose-response relation). Importantly, uncertainty analysis suggests that the fact that many relevant individual covariates are not known, and hence cannot be explicitly modeled, probably does *not* greatly change the results of the simulation model. In theory, excluding such conditioning information can lead to overly wide simulated confidence bands by replacing mixture distributions with homogeneous proportions or mean values (Feller, 1968, p. 231). But it does not greatly affect the simulated mean values of the output variables, such as expected number of cases per year. For outputs that are expected to have Poisson distributions, accurate simulation of mean values is sufficient.

The major conclusion from the analysis is that risk management measures that reduce microbial loads are likely to produce relatively great public health benefits, as measured by reduced illness-days per year, compared to other interventions. Such measures should ideally be introduced during processing (e.g., Whyte et al., 2001) or at the point of consumption (e.g., in restaurants) to create the greatest health benefits. Measures taken on the farm may be positioned too early in the process to have maximal impact, since they occur prior to amplification of CFUs during transportation and cross-contamination during processing. Measures taken after consumption are too late to prevent illnesses. Therefore, focusing resources on reducing microbial loads during processing, storage, and preparation is expected to produce at least 100 times greater health benefits than measures such as banning fluoroquinolones from veterinary use or from physician prescriptions.

The estimated relative health benefits from different risk management interventions appear to be robust to many of the uncertainties, simplifications, and scoping choices incorporated into the simulation model. The model's results are also consistent with the results of simple top-down calculations that do not attempt to model the detailed sub-processes affecting microbial loads from farm to fork, but instead use simple aggregate ratios and proportions to estimate risk. Therefore, contingent on the assumptions that a risk exists and that it is based on microbial loads and dose-response relations in approximately the ways assumed, it appears that the comparisons of options in Table 8 are likely to remain valid as better information is collected and released.

To significantly improve the predictive validity and credibility of the simulation model, it will be necessary to improve scientific understanding of several areas that are currently being modeled based on assumptions. The most important, for purposes of accurate risk assessment is to quantify how the frequency distribution of microbial loads measured at different points in the process will be shifted by proposed countermeasures. For example, how much do better farm practices (e.g., improved animal husbandry and hygiene practices) affect the *right tail* of the microbial load distribution (e.g., above 500 CFUs) received by consumers? Data that address such questions could be used to substantially increase the realism of the risk assessment model.

A second major area is to better understand the multivariate dose-response relation characterizing risk of campylobacteriosis as a function of age, gender, diet, food preparation habits, venue of meals (e.g., home vs. restaurant consumption) and possibly other factors, as well as of CFUs ingested in chicken. Better information on this front can help to test key hypotheses taken as fixed in the current model, such as the role of chicken in causing (or preventing) campylobacteriosis both at home and away from home. It may also clarify the roles of confounders – a key advance needed in the epidemiology of chicken-associated campylobacteriosis.

Finally, the health consequence portion of the model could be refined and improved by studying the severity and duration of illness as a multivariate function of the dose of CFUs, individual covariates, composition of the infecting strains (e.g., what proportion are FQ-resistant), and choice of treatments. The data required for such modeling, as well as for studying the roles of confounders, have been collected and are now being analyzed by the CDC (e.g., Friedman et al., 2000b). When released, they may provide the empirical basis for improving on the present simulation model by replacing key assumptions with data.

These advances could certainly increase the realism and credibility of the model. However, to be useful for informing current decisions, the current incomplete and partly speculative knowledge base must be used. As in much applied risk assessment, the data required for better modeling are not yet available. Regulators and chicken industry decision-makers are already proposing and making changes in industry practices and processes to protect and promote human health. The insights that quantitative risk assessment can give must be timely if they are to be used. The framework of contingent modeling analysis and recommendations illustrated here allows current knowledge to be applied to identify the relative orders of magnitude of health benefits expected from different risk management interventions.

The main conclusions from this analysis (see Table 8) are conditioned on currently available data and policy-driven assumptions. They are robust to many of the other modeling assumptions and limitations imposed by current knowledge gaps. However, the contingent analysis framework leaves open the possibility that better scientific information will lead to revised assumptions about the existence and nature of the hypothesized food safety hazards. If it does, such revisions, in turn, can lead to revised and improved comparisons and recommendations of risk management alternatives. This theme is pursued further via causal graph modeling in Chapter 4.

Chapter 3

Statistical Risk Modeling

1. INTRODUCTION

Statistical risk modeling provides a holistic approach to describing and predicting the probabilities of adverse health responses based on individual exposures and covariates. Rather than attempting to describe or simulate the detailed causal and biological processes that lead from exposure to response probabilities within an individual, statistical risk modeling relies on observed data relating exposures, covariates, and responses at the level of the whole organism or a population of organisms. It seeks to describe this multivariate relation parsimoniously, often as a sum of a stable, predictable relation and a random component.

This task is challenging because it is typically not known which aspects of exposure and covariates are relevant to response, what mathematical form the relation among these variables and response probabilities may take, how unobserved variables affect the observed relation, or which observed responses are attributable to an underlying dose-response relation instead of to randomness or to the effects of unobserved variables. This chapter introduces statistical modeling techniques that have proved useful in clarifying and overcoming these obstacles.

Table 1 outlines the principle steps in traditional statistical risk assessment. The main idea is to *assume* a parametric family of models – perhaps motivated by biological reasoning or biomathematical modeling – and then to *estimate* its unknown parameters from data via techniques such as maximum likelihood estimation (see Table 6). Although often useful, this approach makes the validity of risk predictions contingent on the validity of the selected model and its parameter estimates. This chapter reviews techniques for implementing and improving upon the traditional framework.

Table 1: Steps in Traditional Statistical Risk Modeling

1. *Observe/collect* data set, e.g., $D = \{(x_i, z_i, y_i), i = 1, ..., N\}$
- x_i = exposure input for case i (usually an animal or a person)
- z_i = other covariates for case i; y_i = output (i.e., response) from case i.
- N = number of observations
  e.g., $y_i = 1$ if risk event happens to case i; else $y_i = 0$ or, y_i = time of observed response in case i (if response occurs)
  Data sources used in statistical health risk assessment models include:
  - Survey research (e.g., epidemiology surveys)
  - Archival data (e.g., death certificates); Clinical trials and records
  - Designed experiments (e.g., animal dose-response experiments)

2. *Assume/select* a family of statistical risk models:
Model M: $\Pr(\text{Response} = y \mid x, z) = f(y, x, z\,;\,q)$,
- q = vector of model parameters (to be estimated from data)
- x = vector of exposure or dose variables (e.g., concentration, duration)
- z = vector of relevant covariates (e.g., age, sex, body weight, smoking status)
- f = model's functional form
- y = value of response variable

3. *Estimate* model's unknown parameters and/or functional form from data.
Alternative approaches to estimating parameters (q* = parameter estimate):
- Select q* to maximize $\Pr(D \mid f, q)$ (Maximum Likelihood Estimate, MLE)
- Select q* to minimize a measure of average difference between expected (model-predicted) and observed y values
- Condition *probabilities* of q values on observed data, D, i.e., compute the posterior $\Pr(q \mid D)$ or its mode (Bayesian approach)

Estimate the model function, f, from data if possible.
- Nonparametric function approximation methods (e.g., splines, neural nets, classification trees, multiple adaptive regression splines (MARS), wavelets, orthonormal series expansions with coefficients estimated from data, loess, kernel smoothers. See Cherkassky and Mulier, 1998, for these algorithms).
- Use mathematical modeling (Chapter 2) to suggest/constrain functional forms.

4. *Predict/extrapolate* p for new (x, z) values using the prediction formula:
- $p^*(x, z) = f^*(x, z\,;\,q^*)$, where * denotes estimated/predicted values
- Bayesian approach: Sample values of q from posterior $\Pr(q \mid D)$

5. *Test, evaluate,* and *refine* the risk model. Do observed risks match predictions?
- If not, refine model form, f, or parameter estimates, q*
- If so, design new experiments to test predictions for full range of (x, z). Feed back new data to Step 1.

6. *Interpret the results* for risk managers. How big and how certain is the risk at different levels of exposure?

2. STATISTICAL DOSE-RESPONSE MODELING

The risk from a hazard depends on the extent to which individuals are exposed to it and on how large a risk each exposure history creates. Exposure assessment, introduced in Chapter 2, addresses the first issue. Quantitative dose-response modeling addresses the second. Instead of only identifying whether there is a causal relation between exposure and risk of response, as in hazard identification, dose-response modeling undertakes the additional task of quantifying the relation between them. A quantitative dose-response model predicts risk to an individual as a function of exposure and perhaps other variables (covariates).

The main steps or phases in traditional dose-response modeling, as outlined in Table 1, are as follows:

1. Select the data for dose-response or exposure-response relations to be modeled and obtain the data. Needed data elements typically include exposure estimates, responses, and covariate information on individual cases and controls. This step includes choosing what variables to record, the protocols and coding schemes to be used in recording them, and how to address non-responses and other missing data. It also requires defining the variables (e.g., dose/exposure metrics and response metrics) and coding schemes to be used in recording the data.
2. Select one or more statistical dose-response model forms to use in interpreting the data and in testing hypotheses about risk.
3. Estimate the parameters of the model(s) in Step 2 from the data obtained in Step 1. This is also called *fitting the model* to the data. The estimated parameter values, together with the model form, determine the predicted risk given dose (or exposure) and covariates.
4. Evaluate the model. Are its predictions trustworthy and accurate? To what degree? Quantify and explain using goodness-of-fit statistics and graphical model diagnostics. Perform and report results of model cross-validation, Bayesian model-averaging, tests of alternative model forms and other techniques for quantifying modeling uncertainties (see Section 3). Feed the results back to step 2 to assure that selected models appear to be appropriate. If necessary, return to steps 1-3 to refine the model. (If this is done, then use appropriate modeling techniques to correct for model-selection and over-fitting biases, discussed subsequently.)
5. Interpret the results of the modeling in terms of its implications for risk (e.g., is there a significant exposure-response relation?) Characterize remaining uncertainties.

Each step is described next, emphasizing the statistical challenges and difficulties for each. Later sections discuss methods to overcome these difficulties.

2.1 Define Exposure and Response Variables, Collect Data

Collecting and recording dose-response data from laboratory animal experiments or from human epidemiological studies requires an investigator to make several decisions. Unless proper care is taken in statistical design and analysis, these decisions can unintentionally affect the conclusions from the study. For example, the data must include a summary of dose or exposure for each individual (or for each group of individuals, if exposures are indistinguishable within groups). But, in the absence of a biological understanding of how dose affects response, the appropriate dose metric to use in summarizing dose histories may be unknown. Indeed, it is possible that no adequate dose metric may exist, i.e., that no method for summarizing exposure histories by single numbers provides enough information to uniquely determine risk.

Example: Failures of the C x T Dose Metric

For example, the "C x T" rule is often assumed (Finkelstein, 1995). It hypothesizes that equal cumulative exposures (e.g., equal concentration times time, or C x T amounts) create equal risks. This rule represents each exposure history by its area-under-curve (AUC) for purposes of risk calculation. Yet, the C x T assumption fails dramatically for many common chemicals (Checkoway and Rice, 1992; Smith, 1992; Vacek, 1997). Examples include dose-response relations for the following classes of chemicals (Calabrese, 1999; Cox, 1999):

- *Aneugens* (Aaderma et al., 1998). These include leukemogens and carcinogens such as benzene that cause aneuploidy (typically resulting from improper chromosomal segregation or disjunction at mitosis), e.g., by generating reactive metabolites that bind to key molecules such as tubulin, disrupting proper formation of the mitotic spindle. Such chemicals may have sub-linear, threshold-like dose-response functions because it takes at least a certain concentration of metabolites to effectively disrupt spindle formation.
- *Cytotoxic proliferators.* Chemicals such as arachlor, chloroform, diesel exhaust, and formaldehyde cause cell injury and cytotoxicity (i.e., cell poisoning) at high concentrations, but not at low ones. Sufficiently high concentrations lead to cell damage that then stimulates compensating proliferation (e.g., regenerative hyperplasia) of cells in target organs or tissues. If such cell proliferation plays a key role in producing excess tumors, as it appears to for these chemicals, then enough chemical must be present to cause proliferation before risk of cancer from this mechanism will be increased.
- *Receptor-mediated carcinogens* such as dioxin (TCDD) send inappropriate growth or regulatory signals to a cell's nucleus by binding to cell receptors that trigger signaling. Enough chemical must be present to trigger the inappropriate signals. Similarly, chemical carcinogens that act via the peroxisome proliferator superfamily of nuclear receptors are expected to exhibit dose-response thresholds. These chemicals, which include the plasticizers DEHP and DINP, some herbicides and organic solvents, and carcinogenic metabolites of

trichloroethylene (TCE), cause liver cancer in rodents by regulating gene expression in peroxisomes via steroid hormone receptors – specifically, the peroxisome proliferator-activated receptors (PPAR). Humans express far lower levels of the key receptor PPARalpha in liver than do rodents and do not appear to be at increased risk of liver cancer from peroxisome proliferators.

For these and many other chemicals, relatively high dose-rates cause disproportionately severe carcinogenic and cytotoxic responses. Therefore, data collection protocols that assign the same exposure category or metric to individuals having the same cumulative exposures, but different histories of dose-rates, may introduce biases before statistical analysis begins. To avoid exposure metric uncertainty and biases, an investigator may treat duration and intensity of exposure as two separate explanatory variables in a multivariate risk model (Thomas, 1987; Finkelstein, 1995), perhaps with peak exposures being separated from average exposures (Checkoway and Rice, 1992) or various functions of concentration and duration being used (Vacek, 1997) to summarize intensity.

Experiment design choices also affect the results recorded for animal dose-response studies. Such choices include the spacing of doses, the sample sizes of dose groups, the start and stop times (and ages) for dosing of animals, the length of follow-up before terminal sacrifice (if any), and the tests and assays performed to detect effects. Dose-response data must summarize the response observed (if any) for each individual or each dose group. How the dose and response variables are defined, measured and coded can affect the discovery of dose-response relations.

In epidemiology, which covariates are included and how they are coded, (e.g., choice of category boundaries for representing continuous variables such as age by discrete categories) can sometimes make the difference between finding and not finding an association between exposure and response (Brenner and Loomis, 1994). Choices made in defining and coding exposure metrics, covariates, and response variables are at least as important in epidemiology as in animal experiments. For both, failing to model these choices may create spurious exposure-response relations or hide true ones.

Example: Variable-Coding Bias Can Affect Apparent Relations

Suppose that the observed lung cancer incidence rates among a sample of workers exposed "infrequently", "moderately", and "heavily" to a substance are 0.12, 0.06, and 0.14, respectively. For simplicity, assume that all three groups are the same size. An analyst seeking patterns in the data using a regression model with binary explanatory variables must dichotomize these three data points into "not high" and "high" levels of exposure for purposes of analysis. If the lower two exposure levels are assigned to "not high" and the highest level to "high", then the average lung cancer incidence rates will be 0.09 (from {0.12, 0.06}) in the "not-high" group and 0.14 in the "high" group. Arrayed this way, the data seem consistent with a hypothesis of increased risk in the highest exposure group. On the other hand, without prior beliefs about what patterns should be expected in the data, it might be equally plausible to group the data by assigning the lowest exposure level to "not high" and the remaining two groups to "high". Then the average lung cancer risks will be 0.12 in the "not high" exposure group and 0.10 in the "high" group, removing the appearance of a positive association between exposure and risk. This simple example generalizes: *How exposure (and other) variables are coded in a model can affect its results.* In this case, the assignment of "high" and "not-high" designations to different subsets of exposure groups can reverse the conclusions that seem to follow from the data. Decisions about how to code data are often made before statistical modeling begins, but they should be treated as an explicit part of the modeling process.

Example: Incomplete Adjustment for a Continuous Confounder

It is a fairly common practice in applied epidemiology to use binary codes to represent age groups or other contiguous intervals of values for continuous variables. For example, a regression model might contain dummy variables coded 1 for people in an indicated age range and 0 otherwise, with one such variable for each age interval (e.g., < 20 years old, 20-39, 40-64, 65-84, and 85 and over.) Including such dummy variables in a statistical risk model, along with the exposure and response variables of interest, is often said to "control for the effect" or "adjust for the effect" of the coded variable (age, in this example). But in reality, representing wide intervals by dummy variables may only *partly* control for the effect of the variable being coded. For example, suppose that both exposure and risk of response increase with age (so that age is a confounder) but that risk of response depends *only* on age and not on exposure. The relevant causal graph model is:

Risk of response ← *Age* → *Exposure*

If age is positively associated with both exposure levels and risk of response within each age group represented by the dummy variables (e.g., if lung cancer risk and average exposure to a substance increase with age for people in each age group, e.g., between 65 and 84 years old), then this residual confounding (i.e., the confounding that remains after conditioning on age codes) may create a significant positive statistical association between exposure and risk. This non-causal association persists despite the "adjusting" or "controlling" for age by dummy-variable coding. Such spurious (i.e., non-causal) associations reflect the way the data were coded for analysis, rather than any true (causal) effect of exposure on risk. To avoid such residual confounding, it would be prudent to treat age as a continuous variable rather than using dummy-coding to attempt to control for its effects.

Potential Selection and Information Biases in Epidemiological Studies

Investigator choices that may affect the results of epidemiological studies include study design (e.g., retrospective case-control, prospective cohort, or cross-sectional); sample definition and selection; information-elicitation protocols (e.g., design of surveys and code books; training of interviewers);

matching of cases to controls (if any); and operational definitions of exposures and responses. Case-control designs match "cases", i.e., subjects who have an adverse response, with "controls" who don't. Statistical analysis – e.g., analysis using logistic regression models or other exposure-response models discussed Section 2.2 – can then identify statistically significant differences in exposures and other covariates between cases and controls. These may help to explain the different health outcomes. By contrast, prospective cohort studies track individuals over time to see whether those with different exposure histories develop adverse health responses at different rates. Cross-sectional studies examine exposure-response associations at a moment in time and are often harder to interpret causally than prospective cohort studies. Many other designs are possible, but case-control and prospective cohort designs are most common.

The validity of risk estimates from any epidemiological study can be threatened by *selection biases,* in which the subjects included in a study do not represent the population for which inferences are to be drawn; and *information biases,* in which information collected on subjects is biased (Choi and Noseworthy, 1992). Table 2 lists biases of each type and recommended countermeasures for preventing or reducing them. In any epidemiological investigation, the investigator should be aware of these potential biases and should control for their effects in the design, conduct, and data analysis for the study.

Selection biases include hospital admission bias and volunteer/non-response bias, as well as artifacts of differential survival or differential availability of subjects having different exposure and response histories. *Hospital admission bias* occurs if hospitals tend to admit subjects known to have been exposed to a health hazard and those known or suspected to have a certain health effect. If investigators compare hospitalized cases (defined by known effects) against non-hospitalized controls (or controls at different hospitals), such comparisons can reveal an association between exposures and effects among cases compared to controls, even if there is no causal relation between them. The admission process tends to inflate prevalence of exposure in the hospitalized population.

Volunteer bias and *non-response bias* can arise if those who volunteer (or who do not refuse) to provide information are not representative of the whole population for which inferences are to be drawn. For example, if subjects who are concerned that they may have been exposed or that they are developing adverse health effects are especially likely to be included as respondents, then the exposure-response relation estimated from respondent data may differ from what the relation in an unbiased random sample.

Other selection biases can occur if subjects with adverse health effects are more (or less) likely than other subjects to respond – perhaps because they are unequally available to respond, as for fatal effects.

Table 2: Some Common Selection and Information Biases

Selection Biases	Recommended Countermeasures
Hospital admission bias (as well as referral and exclusion biases)	• If possible, use prospective cohort design (Feinstein, 1988) • "Use population-based cases and population-based controls" (Choi and Noseworthy, 1992)
Selective attrition/survival Differential follow-up loss	• Use a well-specified cohort (Feinstein, 1988) • "Include non-surviving subjects in the study through proxy interviews" (Choi and Noseworthy, 1992) • Use transition data and compare to counter-factual survival curves to quantify causal impacts when exposure affects attrition rates (Robins, 1987)
Detection/surveillance bias	• Match cases to controls (or exposed to unexposed subjects) based on similar levels of diagnostic scrutiny and/or cause of admission.
Membership bias (e.g., lifestyle bias, socioeconomic history)	• In cohort studies, use multiple comparison cohorts. • Hard to control in case-control studies.
Self-selection bias Response/volunteer bias	• Achieve response rate of at least 80% by repeated efforts if necessary • Compare respondents with sample of non-respondents
Information Biases	
Intra-interviewer bias	• Keep interviewers blind to study hypotheses and subject classifications
Inter-interviewer bias	• Use same interviewer for study and comparison groups
Questionnaire bias	• Use dummy questions to mask study goals • Avoid leading questions/ leading response options
Diagnostic suspicion Exposure suspicion	• Hard to prevent in case-control studies. In cohort studies, make diagnosis blind to exposure; exposure assessment blind to diagnosis
Recall bias • Differential • Non-differential	• Conduct sensitivity analysis of recall bias during analysis. (Barry, 1996; Greenland, 1996) • Use proxy measures of exposure to check for recall bias. (Basso, 1997) • Use high-quality data based on ongoing direct examination of subjects. (Feinstein, 1988) • Apply bias-correction algorithms

References: Lichtenstein et al., 1987; Feinstein, 1988 and comments in Savitz et al., 1990; Gail, 1991; Shore, et al.,1992.

Example: Differential Non-Response Rates Can Create Non-Causal Associations Between Exposures and Response Rates

A population-based case-control study (Brownson et al., 1992) collected data on self-reported exposure to environmental tobacco smoke (ETS) from 618 lung cancer cases and 1,402 population-based controls. In this study, 25% of eligible controls refused to participate in either a screening interview or the full interview. By contrast, only 1% of eligible cases refused to participate. If potential controls living in ETS households are less (or more) likely than those in non-ETS households to participate in the health survey, then the prevalence of ETS exposure will be reduced (or increased, respectively) among participating controls compared to its prevalence in the general population. Hence, it may also be less among participating controls than among participating cases, creating an apparent exposure-response association, even if the prevalence of ETS exposure among cases is the same as in the general population. Thus, survey-participation rates that are associated with exposure histories and with response status (case *vs.* control) can potentially create non-causal statistical associations between exposures and responses.

In case-control studies, in which each case with an adverse response is matched with one or more controls not having the response, inappropriate matching may lead to non-causal exposure-response associations. For example, suppose that lung cancer patients in a hospital ward are matched with members of an orthopedic ward. If both cases and controls are then asked about their past exposure histories, then it seems plausible that the lung cancer cases may be more inclined to think about and recall past inhalation exposures than the controls (Nilsson, 1996). This illustrates how selection of subjects can lead to a bias in the information obtained from the subjects – in this example, *recall bias,* which is discussed further shortly – suggesting that selection and information biases need not be disjoint.

Under information biases, *diagnostic suspicion bias* arises if subjects diagnosed with an adverse health effect of interest are examined more thoroughly for exposure than are control subjects not similarly diagnosed. Conversely, if subjects known to be exposed are examined more carefully than other subjects for signs of the health effect, then an association between exposure and effect may be created by the diagnostic process alone.

Diagnostic information may also be biased if the diagnostician's knowledge of exposure affects the recorded diagnosis. For example, Sterling et al. (1992) report that, among subjects with identical conditions at death, "Lung cancer as an underlying cause is recorded with a much smaller relative frequency if the decedent is known to be a never-smoker and with a much larger relative frequency when the decedent is known to be a smoker." More generally, cause-of-death information can be highly subjective and variable (Messite and Stellman, 1996). As in other areas of human judgment and decision-making, medical judgments about causes can lead to selective discounting or over-weighting of evidence based on pre-existing hypotheses (Ahn and Bailenson, 1996) and to confident but mistaken diagnoses that tend to make recorded evidence confirm what diagnosticians expect to find (Dubeau, 1986; Harber and Shusterman, 1996; Ben-Shakhar et al., 1999).

Recall bias occurs if remembered exposures (or health effects) collected via a survey differ systematically from the true values. If recalled exposures also differ systematically between subjects with and without health effects (or among subjects with different levels of other covariates), even when they have the same true exposures, then the recall bias is called *differential;* otherwise, it is *non-differential.* Something akin to recall bias can result if cases tend to have exposure estimates provided by surrogates (e.g., next of kin) while controls more often provide their own exposure estimates, if the accuracy of estimates differs between subjects and their surrogates.

As an example of differential recall bias, suppose that exposures that occurred long ago are recalled less accurately than more recent ones. Then subjects with larger cumulative exposures based on more years of exposure may tend to have larger error variances for their cumulative exposure estimates (Greenland, 1993). The distribution of errors varies with the quantity being estimated, invalidating a standard assumption often used in simple regression models to justify formulas for confidence intervals.

Example: Recall Biases May Increase Over Time

The idea that errors in exposure estimates or exposure classifications may increase with elapsed time is supported by some studies of particular carcinogens, such as cigarette smoke. For example, a study of the accuracy of recalled and reported exposures to cigarette smoke in South Germany (Heller et al., 1998) concluded that "Data from the reinterviews revealed that reported smoking status confirmed by cotinine measurement in 1987-1988 conflicted in a number of cases with the data obtained in 1984-1985 using the same procedure. For example, 0.1% of those who stated they were current regular smokers, 4.3% of those who stated they were current occasional smokers, and 17.6% of those who stated they were ex-smokers in 1984-1985 claimed in 1987-1988 to have never smoked... Altogether the true proportion of ex-smokers among self-reported never smokers was about 9.7% (17.8% in men and 6.7% in women)."

The wording of questions on survey instruments can prompt recall biases. For example, asking "Were you ever exposed to X?" may elicit fewer positive responses than asking a conjunction of sub-questions such as "Were you ever exposed to X at work? At home? In social settings? As a child? As a young adult?" (Rottenstreich and Tversky, 1997). Asking "How often have you been exposed to X" and suggesting a set of responses may elicit a yet different response. Such *elicitation biases* can powerfully affect what is recalled and reported. Chapter 5 discusses other types of elicitation biases that can affect perceptions and judgments as well as recall.

Recall bias is a fairly common threat in case-control studies, and its effects should be assessed and corrected for in the analysis if they cannot be avoided during data collection. Its effects can be strong and sometimes surprising. For example, self-reported estimates of exposures (or of other covariates) that are correlated with true values may bias risk estimates unpredictably, i.e., the naïve (biased) estimates may lie either above or below the true risk values (Wacholder, 1995). Even a modest degree of differential recall bias can potentially explain away some statistically significant and (biologically plausible) associations, such as the one between environmental tobacco smoke (ETS) and lung cancer risk in some studies

(Barry, 1996). Thus, it is important to avoid or correct for such biases before interpreting the results of epidemiological risk assessments.

If the probability distribution of measurement errors for exposure estimates – including the effects of recall biases – is known or assumed, perhaps based on a special-purpose study, then the biases that these errors create in risk estimates often can be estimated and removed using relatively simple corrections (e.g., Armstrong et al., 1989). In practice, the distribution of errors is usually unknown. Special-purpose studies to quantify it (by collecting additional data on true values for a sub-set of study subjects or for an additional sample of subjects) are typically not available, and even reasonable assumptions may not be known. In this case, sensitivity analyses should be used to determine whether recall bias or other measurement errors could plausibly explain away observed associations (Greenland, 1996).

Finally, using a cohort design instead of a case-control design, with exposures measured in advance of responses, can avoid the threat of recall bias entirely.

Measurement Errors in Explanatory Variables

The recorded values of estimated exposures and other covariate values may contain errors for various reasons, of which imperfect recall is only one. Use of surrogate respondents; imputation of exposure estimates from job categories, job-exposure matrices (Hawkes and Wilkins, 1997) or industrial hygiene data collected from other workers; sampling variability in industrial hygiene data; and many other sources contribute to errors in exposure estimates for individuals.

Statisticians use the term "measurement error" to describe all of the following:

- Errors in exposure *estimates,* for continuous measures of exposure;
- Errors in exposure status *classification,* for binary or categorical measures of exposure;
- Sampling errors in exposure *measurements,* in situations (e.g., occupational hygiene settings) where exposure is actually measured.

Statistical risk models that ignore any of these forms of measurement error in exposure, confounders, or other covariates can produce incorrect conclusions about the relations between exposure and health effects. The following warning (Kuha, 1994) typifies findings by many methodologists: "Exposure measurement error is common in epidemiologic studies. ... If the possibility of measurement error is ignored and a model is fitted using the erroneous covariate values, the estimates of the exposure-disease association will be biased. In models with multiple covariates this bias can be either positive or negative because of residual confounding. Even qualitative conclusions about the true effects, based on the idea of attenuation, can thus be false in such cases." Similar effects occur when measurement errors in confounders or other explanatory variables are not explicitly modeled (Marshall and Hastrup, 1996; Wacholder, 1995).

In simple linear regression models with one explanatory variable, errors in recorded data for the explanatory variable attenuate estimated effects and can increase false negatives, i.e., failures to detect exposure-effect relations. The attenuation can be corrected for in simple linear regression models with known measurement error variance (e.g., Armstrong, 1990). However, attenuation does not necessarily hold in models with multiple explanatory variables unless strong assumptions (such as jointly normally distributed variables) are made (Greenland, 1993). Substantial research throughout the 1990s established that in general, multivariate statistical risk models that do not explicitly represent uncertainties in exposure estimates (or in other explanatory variables, such as risk factors, covariates, and confounders) can yield regression coefficients and resulting risk estimates that are biased either above or below the correct values, making the risk estimates meaningless or misleading. (See e.g., Aickin and Ritenbaugh, 1996 for applications to multiple linear regression models; Carroll, 1989 for generalized linear models; Armstrong, 1990, Rosner et al., 1990, Schmid and Rosner, 1993, and Kuha, 1994 for logistic regression; and Nakamura and Akazawa, 1994 for the proportional hazards model.)

Example: Exposure Misclassification in ETS Risk Analysis

Measurements of tobacco metabolites in non-smokers (Anderson et al., 2001) and common sense both suggest a possible causal relation between exposure to environmental tobacco smoke (ETS) and lung cancer risk. Such a relation is often considered established (e.g., Haustein, 2000). However, to quantify the magnitude of this relation (and even to confirm that a causal relation exists), it is essential to correct for exposure misclassification errors. For example, suppose that a self-reported non-smoker is more likely actually to be a smoker if she is married to a smoker than if she is married to a non-smoker. Then lung cancer due to active smoking may be higher in self-reported non-smokers married to smokers – a group studied in many epidemiological investigations of ETS-lung cancer associations – that in control groups of self-reported non-smokers not married to smokers. This could create a spurious (non-causal) association between ETS and lung cancer among self-reported non-smokers that would obscure any causal association unless its effects are estimated and removed.

Similarly, one study reported that "The misclassification rate (proportion of reported nonsmokers with [serum] cotinine levels of at least 14 ng/mL) was 4.2% and was significantly higher among subjects who were Black, had a high school education or less, or were reported former smokers" (Wagenknecht et al., 1992). High school education or less is significantly associated with lung cancer risk, even after adjusting for age, smoking, and aspects of diet (e.g., van Loon et al., 1995). It is also associated with greater probability of being married to a smoker (e.g., Kawachi and Colditz, 1996). Therefore, lung cancer due to active smoking may be higher in self-reported non-smokers married to smokers because of the confounding effects of education, as in the following causal graph:

smoker misclassified as non-smoker ← *no college education* → *smoking spouse*
↓ ↓ ↓
excess lung cancer risk ← *(other risk factors, e.g., diet)* *ETS exposure*

Clearly, the causal relation between ETS and lung cancer risk can only be isolated after correcting for misclassification of smoking status.

Past studies of the effects of smoker misclassification on risk estimates have reached mutually inconsistent conclusions. Tobacco industry commentators have observed, with some apparent cogency, that (a) Measured cotinine levels in saliva (an indicator of tobacco smoking) indicate that some people who classify themselves as non-smokers are active smokers; (b) There is substantial differential classification error across exposure groups; and (c) "Misclassification bias alone is likely to explain any lung cancer risk elevation observed in the U.S. epidemiology of environmental tobacco smoke exposure among nonsmoking women" (Ogden et al., 1997; see also Levois and Switzer, 1998 and the review of older literature in Nilsson, 1996). To the contrary, health regulators in many countries have argued that misclassification bias is unlikely to greatly affect risk estimates for ETS. For example, Wells et al. (1998) state that "True smoking status, either occasional or regular, was determined for individual current smokers in 3 existing studies of nonsmokers by inspecting the cotinine levels of body fluids. ...The misclassification rates for the important category of female smokers misclassified as never smokers were, respectively, 0.8%, 6.0%, 2.8%, and 15.3% for majority regular, majority occasional, US minority regular, and US minority occasional smokers. Misclassification rates for males were mostly somewhat higher. The new information supports EPA's conclusion that smoker misclassification bias is small." Such analyses generally treat measured cotinine levels (which indicate *current* smoking exposure) as indicators of true *ever*-smoking behavior. Thus, they may under-estimate true misclassification rates. How much of the statistical association often reported between ETS and lung cancer risk is due to such errors is a topic of on-going debate, with estimates ranging from 100% to nearly 0%.

Numerous methods have been developed to estimate, control and correct for effects of measurement errors in explanatory variables. Many of these are based on the advances in statistics discussed in Section 3. To anticipate that discussion, some especially prominent strategies are listed in the following box. Most of these approaches require a small validation study and/or modeling assumptions to estimate error models. Section 3 introduces some of the key technical innovations such as the bootstrap, Data Augmentation, Gibbs sampling and MCMC methods. These are also now widely used to address many other statistical challenges in addition to explicit modeling and correction of the effects of measurement errors in the explanatory variables.

Methods to Correct Biases Due to Errors in Explanatory Variables

- *Perform sensitivity analyses* of the risk estimates to exposure measurement errors, misclassification, and incorrectly smeasured (or unmeasured, i.e., omitted) confounders (Greenland, 1996).
- *Bootstrap resampling* to estimate and correct for measurement error biases (Huakka, 1995; Kim and Zeleniuch-Jacquotte, 1997).
- *Data Augmentation* algorithms and simple corrections to remove bias from exposure measurement errors (Kuha, 1997). Data Augmentation algorithms (see Table 6) estimate conditional probability distributions of unobserved data elements, such as true exposures, and model parameters that depend on them, given the observed data values.
- *Bias correction algorithms* ("regression calibration") for specific risk models. These methods estimate the bias due to error (e.g., via a Taylor series expansion) and then subtract it from naïve (biased) risk estimates to obtain unbiased ones. (See e.g., Aickin and Ritenbaugh, 1996 for multiple linear regression models; Nakamura and Akazawa, 1994 for the proportional hazards model; Armstrong et al., 1989 for a multivariate normal discriminant analysis model; Spiegelman and Valanis, 1998 and Rosner et al., 1989 for logistic regression models; Kuchenhoff and Carroll, 1997 for linear and logistic regression models with thresholds; Knuiman et al., 1998 for univariate logistic, linear, and Cox models; and Bashir and Duffy, 1997 for additional methods of bias correction.)
- *SIMEX* (Simulation-Extrapolation) adds *simulated* measurement error to observations, extrapolates the bias for *total* measurement error, then subtracts this bias from risk estimates (Marcus and Elias, 1998).
- *Bayesian modeling* of uncertainties in exposure (and other explanatory variables) calculates risk conditioned on different true values of the variables, weighted by the conditional probabilities of these values given the observed (measured) values. (See Müller and Roeder, 1997 for Gibbs sampling and MCMC methods; Schmid and Rosner, 1997 for logistic regression. Section 3 and Chapter 4 discuss Bayesian methods.)

Example: A simple attenuation bias-correction procedure (Kim and Zeleniuch-Jacquotte, 1997)

This example illustrates an approach to correcting for measurement error biases in a simple, univariate setting with a well-defined model for the measurement error. It also introduces the bootstrap technique.

Consider a population that is stratified according to some criterion (e.g., age, income, etc.) Suppose that:

a) True exposures are normally distributed in each stratum
b) Measurement error is additive and is normally distributed in each stratum

c) Repeated measures of exposure are available for each individual
d) The relation between exposure and probability of response is given by a logistic regression model (i.e., there is no uncertainty about the correct model form.)

Then to correct for attenuation bias due to measurement error, adjust the measured (error-full) exposure estimates and estimate coefficients in the logistic regression model as follows:

1. Estimate measurement variance and population exposure variance from mixed effects ANOVA of repeated measurements in each stratum.
2. Multiply each individual's measured exposure by the following correction factor:

$$\frac{\textit{estimated variance from exposure}}{\textit{estimated exposure variance} + \textit{estimated measurement error variance}}$$

3. Estimate logistic regression coefficients by fitting conditional logistic regression models to the transformed (corrected) average exposures calculated in Step 2.
4. *Bootstrap* steps 1-3 to obtain confidence intervals for the coefficients. The basic idea of the bootstrap is as follows:
   - *Generate many random samples* by re-sampling (i.e., drawing sample values at random) from the empirical cumulative frequency distribution of the data. This empirical CDF is constructed from the full sample.
   - For each random sample, *fit the model* (e.g., estimate its parameters and effects) and assess model diagnostics.
   - *Quantify the sample frequency distribution of model parameters,* effects estimates, and diagnostics based on these many "resampled" data sets.

The result indicates the magnitude of bias expected when the original data are used to fit a model and to estimate effects and diagnostics.

Example: The Data-Augmentation Approach to Measurement Error

Data Augmentation is a computational Bayesian algorithm that seeks to estimate the joint posterior distribution of missing data and model unknowns. Let $\underline{X}$ denote measured exposure and X denote true exposure. Suppose that individual responses (e.g., death or not) are measured without error, and denote by $\underline{Y}$ the response variable. Finally, suppose that the probability distribution of $\underline{Y}$ values for any individual depends only X and a vector of unobserved parameters, β, for that individual. The situation might thus be diagrammed as follows:

$$\underline{X} \leftarrow X \rightarrow \underline{Y} \leftarrow \beta$$

In conditional probability notation, the model consists of a *measurement error sub-model* $\underline{X} \leftarrow X$, corresponding to the probability relation $\Pr(\underline{X} \mid X)$ (i.e., the conditional probability of observing each measured value of exposure given the true value, obtained from a validation study, or from repeated measurements); an *exposure-response model* $\Pr(\underline{Y} \mid X, \beta)$; and an *exposure model* Pr(X).

The basic idea of the Data Augmentation algorithm is to treat true values of all quantities (exposures, covariates, and parameters) that are measured with error, or that are not measured at all, as *missing data.* For example, in the above diagram (which does not include covariates), the true values of X and β are treated as missing data. Their values are to be estimated by conditioning on the observations (X̲, Y̲) for each individual (and assuming a specific parametric model for Pr(Y̲ | X, β). The estimation procedure is iterative. It proceeds as follows:

Data Augmentation Algorithm: The Basic Idea

a) Estimate missing data values (the X values) by sampling (or "imputing") their values from the current posterior PDF of X, Pr(X | X̲, Y̲, $\beta^{(i)}$), calculated by Bayes' Rule and assuming that the current estimates of the parameters, $\beta^{(i)}$, are correct.
b) Re-estimate the uncertain parameter values, β, (thus obtaining the updated estimates $\beta^{(i+1)}$), by sampling from its current posterior distribution based on the "complete" data set formed from the known data (e.g., Y̲) and the imputed data for all missing values (e.g., for X). (Schafer, 1997; Kuha, 1997).
c) Iterate steps (a) and (b) to obtain an estimated joint posterior distribution for the parameters and missing data values.

The main required inputs for this approach are parametric models for all conditional relations. These may have to be developed from results of validation study or repeated measurements. Such models may be wrong.

Some of the main strengths of the Data Augmentation approach are:

- It applies to continuous, discrete, and mixed independent variables.
- *Multiple imputations* made during the iterations lead to better estimates than "single-imputation" methods such as regression-calibration.
- It can be used with vague priors. With a flat prior, it yields the likelihood function (and MLE) and information matrix.

For full discussion of the algorithm, its properties and extensions, see Schafer, 1997.

Although this discussion has emphasized measurement error in exposure variables, similar considerations hold for measurement errors in other explanatory variables. As noted by Wacholder (1995), "For many variables measured in epidemiology, particularly those based on self-report, however, errors are often correlated with the true value... Errors in a polytomous confounder or errors correlated with the true value of a continuous confounder may produce unpredictable bias." Similarly, Marshall and Hastrup (1996) found that "Error in the measurement of a confounder... could bias estimates of other study variables... even with statistical adjustment for the confounder as measured. With dichotomous independent variables, 15% measurement error in the form of misclassification could lead the null study variable to appear to alter risk by as much as 50%.... These results support the need for every epidemiologic inquiry to include evaluations of measurement error in each variable considered."

The approaches listed above provide specific methods for following this recommendation. Sensitivity analysis is perhaps the most universally applicable, requiring no detailed knowledge of or assumptions about the probability distribution of measurement error. It is useful for determining how much error would have to be present to reverse study conclusions and to explain away reported associations among variables. The other methods generally require more information or more detailed assumptions, but can provide quantitative probabilistic estimates for the effects of exposure and other explanatory variables on risk even when the measured values of some or all of the variables contain errors.

Omitted Explanatory Variables and Missing Data

In practice, it is common for measurements of some variables to be missing for some individuals. Worse, one or more relevant variables may be completely omitted during data collection. For example, some occupational studies of lung cancer must contend with omission of smoking status from the worker records. From a data analysis perspective, omitted explanatory variables are an extreme form of missing data, and technical methods are available for dealing with both. However, unless such methods are used, failure to measure causally relevant variables and to include them in the modeling can lead to false conclusions. Failure to correctly model missing data can also lead to biased and incorrect risk estimates.

Example: Omitted Confounders Undermine Causal Interpretations

The nature of the dose-response relation (if any) between alcohol consumption and increased risks of lung cancer, breast cancer, and other cancers has been much debated. How measurement errors are modeled (Schmid and Rosser, 1993) and how well confounding by smoking can be corrected (Carpenter et al., 1998) greatly affect the outcomes of epidemiological investigations of alcohol-cancer associations. Omitted confounders can play a crucial role in interpreting observed associations, which typically have relative risks of less than 2. Unmeasured confounders and incomplete adjustment for known confounders such as cigarette smoking provide potential non-causal explanations for observed alcohol-cancer associations. For example, suppose that a cluster of factors generically called "Unhealthy lifestyle" are both positively related to alcohol consumption and cancer risk, consistent with the following diagram:

Model A: Alcohol consumption ← Unhealthy lifestyle → Cancer risk.

In words, this model postulates that the probability that a randomly selected individual has high alcohol consumption is greater among individuals with unhealthy lifestyle. The risk of cancer is also greater among these individuals, but alcohol consumption does not directly increase cancer risk. (Recall that data can be used to test such causal theories against rival theories, such as that in Model B below, by testing their implied conditional independence and composition relations.)

Model B: Unhealthy lifestyle → Alcohol consumption→ Cancer risk.

For example, cancer risk is conditionally independent of alcohol consumption, given lifestyle, in Model B but not in Model A. Chapter 4 discusses such tests further.)

Now, assuming for purposes of illustration that Model A is correct, a data collection effort that omits data on lifestyle will show that cancer risk is positively associated with alcohol consumption. But if this association is entirely due to the omitted confounder of lifestyle, then the observed association would *not* imply that changing alcohol consumption would change cancer risk. Thus, the risk management implications of an observed statistical association may be unclear if it is possible that a confounder has been omitted from the data set. Indeed, even if theories A and B are both partly correct and there is a direct effect of alcohol consumption on cancer risk as well as the spurious association between them created by the confounder "unhealthy lifestyle", the effects of the confounder must still be removed in order to reveal the direct effect. Chapter 4 discusses these issues further.

As might be expected, simplistic methods for treating missing data and unmeasured variables – such as using only subjects with complete data in the analysis, treating "missing" as a separate category or value for variables, or replacing missing values with their sample based on available values – do not work very well in practice. These stratagems can lead to large biases in risk estimates for real data sets with missing data (Vach and Blettner, 1991; Greenland and Finkle, 1995).

More promising approaches for dealing with missing data are as follows:

- Replace missing values with their *conditional expected values* given the values of observed variables (e.g., Paik and Tsai, 1997 for the proportional hazards model). In the proportional hazards model, effects of covariate data that are missing at random also can be estimated and corrected using extensions of the EM algorithm (Section 3, see Table 7) and related computational algorithms (Herring and Ibrahim, 2001).
- Estimate posterior means or *posterior probability distributions* of missing values, conditioned on all observed data, by applying iterative Bayesian algorithms (e.g., the EM and Data Augmentation algorithms in Section 3, see Tables 5 and 6).
- Apply *multiple imputation* methods (Schafer, 1997). These repeatedly replace missing values with "imputed" values (sampled via Monte Carlo simulation), analyze each resulting complete data set, and characterize the frequency distribution of results. (See the case study in Section 5.)

Techniques for dealing with unmeasured variables include using auxiliary information to correct for biases (Marais and Wecker, 1998); regresson-based sensitivity analysis and bias-correction approaches (Lin et al., 1998); model-free sensitivity analyses for possible effects of omitted covariates (Greenland, 1996); and testing for the presence of unobserved ("latent") variables. Rough sensitivity analyses may also be based on analytic results such as that, for a confounder to explain an R-fold relative risk, it must be at least R times more prevalent among the exposed than among the unexposed subjects (Lin et al., 1998 citing Cornfield et al., 1959).

Finite mixture distribution modeling, discussed later in this chapter, provides relatively sophisticated techniques for detecting and quantifying the effects of unmeasured variables. The key idea is to represent unobserved sources of non-random *inter-individual heterogeneity* in observed exposure-response patterns by a new variable, often called an individual's "type". These sources presumably reflect the influences of unobserved variables that mediate observed exposure-response relations. Individuals are hypothesized to belong to different types (the different components of the finite mixture distribution) based on the values of unobserved covariates. The number of types and the conditional probabilities of observed individual data given each type are then estimated from the available data using techniques for missing data, such as the EM algorithm (Table 7). Similar techniques have been used in linear statistical models for latent variables (i.e., unobserved variables) and in Hidden Markov Models (HMMs) that extend techniques for unobserved variables to the states of a Markov process. An individual's state affects both the probabilities of observed measurements and the probabilities of next states, and inferences about both may be based on observations collected over time.

In general, it is good advice that "Potential biases due to unmeasured confounders, classification errors, and selection bias need to be addressed in any thorough discussion of study results" (Greenland, 1996). To do so, it is often possible to apply the techniques mentioned in this section. Current research in statistics and biostatistics is extending the reach of these basic ideas to more general and realistic settings as practitioners become more aware of the need to address missing data and unmeasured variables explicitly in their analyses and statistical risk models.

As a rough quantitative guideline based on experience,it is often considered that omitted confounders may provide plausible non-causal explanations for relative risks of about two or less. For example, Lin et al. (1998) p. 960, note that "Observational studies are subject to potential biases induced by unmeasured or unknown confounders. Many epidemiologists use a rule of thumb that the actual relative risk may be off by a factor of 2 because of unrecognized confounders. This comports well with [our] results." This rule of thumb suggests that relative risks of less than about 2 typically cannot be interpreted as clear evidence of a causal relation, since they might well be explained instead by unrecognized confounders.

2.2 Select a Model Form for the Dose-Response Relation

Once exposure and response variables have been defined and recorded in a data set, the traditional next step is to select a dose-response *model form* to describe the relation between them. This is a family of mathematical functions relating dose (and perhaps other covariates) to probability of response or to other measures of risk, such as the survivor function or hazard function for occurrence of an adverse event. The model form is usually represented by a parametric family of models, in which case parameter values must be estimated from the data collected.

Table 3 lists some of the dose-response models used to analyze and interpret experimental and epidemiological data. There are several types of dose-response models, with the main divisions being quantal (denoted by p, standing for probability, in the first column of Table 3) *vs.* survival models (denoted by h); parametric *vs.* non-parametric or semi-parametric models; and deterministic *vs.* stochastic models. Quantal dose-response models predict the expected fraction of exposed subjects (animals or people) that will develop an adverse response within a specified amount of time if they are exposed to a given dose level.

Table 3 indicates quantal models with a type of "p", since they predict response probabilities. Examples include the probit, logit, one-hit, and multistage models. The probit and logit models both provide families of smooth s-shaped curves to fit to data, but lack direct biological rationale. The one-hit and multistage models have interpretations in terms of possible biological mechanisms, i.e., interactions of metabolite molecules with DNA or other cellular targets (referred to generically as "hits".)

Survival models, also called time-to-tumor models, are represented by hazard functions, or, equivalently, by survivor functions. They are indicated in Table 3 by "h". These models indicate how risk evolves over time. Examples are the proportional hazards and Weibull survival models.

Deterministic dose-response models represent response probabilities or hazard functions as deterministic functions of dose. By contrast, stochastic models represent them as random variables with probability distributions that depend on dose and possibly other information (e.g., the number of initiated cells, in the Moolgavkar *et al.* two-stage stochastic model.) Initially identical subjects with identical dose histories may develop different response probabilities over time in a stochastic dose-response model. Only the Moolgavkar et al. model in Table 3 represents a stochastic dose-response relation, as signified by the $E[h \mid I(t)]$ in column 1. $I(t)$ may be interpreted as the number of "initiated" or pre-malignant cells at time t.

Table 3: Some Important Dose-Response Models

Name/Type	**Model Form**	**Comments**
One-hit, p	$p = 1 - \exp(-qx)$	Assumes that one "hit" can cause a response
Multistage, p	$p = 1 - \exp[-(q_0 + q_1x + q_2x^2 + \ldots + q_mx^m)]$ (Approximate form.)	Assumes that *m* transitions are needed to cause cancer.
Weibull quantal, p	$p = 1 - \exp(-ax^b)$	Nonlinear even at low doses if $b \neq 1$
Probit, p, G	$p = \Phi(qx)$, Φ = unit normal CDF, q = a scalar potency factor	An individual responds if and only if dose x exceeds her tolerance threshold. Individual tolerance thresholds are normally distributed in the population.
Multivariate logistic regression, p, G	• $p = \exp(\mathbf{qx})/[1 + \exp(\mathbf{qx})]$ • $\ln[p/(1 - p)] = q_0 + \mathbf{qx}$, $\mathbf{q}$ = parameter vector, $\mathbf{x}$ = vector of explanatory variables	Logit and probit models are both generalized linear (GLIM) models
Linear relative risk p	$p = (1 + qx)p^*$ for $p < 1$, p^* = background risk	Lacks biological motivation
Multivariate Proportional Hazards (PH), h	$h(t \mid \mathbf{x}, \mathbf{z}) = \exp(\mathbf{qx})h(t \mid \mathbf{z})$, where $h(t \mid \mathbf{z})$ = background hazard function for an individual with covariates $\mathbf{z}$.	Efficient estimation of parameter vector *q* requires special techniqes (e.g., partial likelihood).
Moolgavkar Venzon Knudson ("MVK") two-stage stochastic model of carcinogenesis $E[h \mid I(t)]$	$h(t) = \mu_2(t)E[I(t) \mid M(t) = 0]$, where $E[I(t) \mid M(t) = 0] = (\mu_1N/\alpha_2)(\exp(\delta_2t - 1)/[1 - A + (B - 1)\exp(\delta_2t)]$. A and B are known functions of α_2, β_2, and μ_2. • $M(t)$ = malignant cells at time t • $I(t)$ = initiated cells at t • N = number of normal stem cells • μ_1 = transformation rate of normal to initiated cells • $\mu_2(t)$ = transformation rate of initiated to malignant cells at t • α_2, β_2 = cell division and death rates for initiated cells.	This is a stochastic hazard function model. The exact solution shown here is derived in Hoogenveen *et al.*, 1999. It typically leads to lower hazard functions than earlier, approximate solutions.
Weibull survival model h	$h(t \mid a, b) = (a/b)(t/b)^{a-1}$ Parameters *a* and *b* are scale and shape parameters	Exposure contracts the time axis, moving any adverse effects closer in time.

In other stochastic hazard rate models, I(t) may be interpreted more generically as the *information* generated by all observations up through time t. The hazard rate at time t is conditioned on I(t) and can change as more information – including the passage of time – is obtained. In this more general use, the increasing information sets $\{I(t), t \geq 0\}$ constitute a *filtration* with respect to which the stochastic h(t) values are evaluated.

The other models in Table 3 are deterministic. However, those with types denoted by (p, G) refer to *mixture distribution models,* in which some unobserved characteristic (e.g., a response threshold) follows a frequency distribution (e.g., a normal distribution for the probit model) in the population. Mixture distribution models can represent population variability and inter-individual heterogeneity due to unobserved causes or covariates in dose-response relations. They can also model the effects of hidden or latent variables (such as "frailty") that may not have a well-understood physiological interpretation but that lead to more descriptive models. Mixture distributions are discussed further in Section 4.

Multistage and Logistic Dose-Response Models

Among the models in Table 3, the most commonly used ones are the multistage model for analysis of animal experiment cancer bioassay data and the logistic regression model for analysis and interpretation of epidemiological data. For years, the standard cancer risk model used by EPA and OSHA in interpreting animal tumor data has been the "linearized" multistage (LMS) model. This is a multistage model with non-standard adjustments made in calculating its upper confidence limits to force them to be straight lines with positive slopes at low doses. It has been the default model for regulatory risk assessment of chemical carcinogens, i.e., the one recommended for use when the correct model is not known.

The low-dose slope of the upper confidence limit (calculated via the linearization procedure) is variously referred to as the "q_1*" (read "q one star") value, the *carcinogenic potency* factor, the *slope factor*, and the *unit risk* for a chemical. It is the value reported in EPA's Integrated Risk Information Systems (IRIS) database (http://www.epa.gov/ngispgm3/iris/) and other databases for chemical cancer risks. However, as noted in Table 3, the multistage formula is only an approximate expression of the underlying biomathematical assumptions. An exact formula derived from the same assumptions leads to risk predictions that are significantly smaller at low doses (Chiu et al., 1999). Moreover, the linearity of the 95% upper confidence limit is often unjustified by data or by correct statistical methodology. For example, if the true dose-response function is a multistage model (and also a Weibull model) with $p(x) = 1 - exp(-qx^b)$, where b is any integer greater than 1, then the slope of $p(x)$ approaches zero at $x = 0$, and the slopes of all correctly calculated confidence bands also approach zero at $x = 0$. Despite this, the calculation of unit risks is deeply embedded in regulatory risk assessment methodology.

Logistic regression is ubiquitously used to analyze population data on exposures and cancer risks. It is used routinely to adjust for the effects of potential confounders, and to estimate odds ratios. The odds ratio (OR) for a dichotomous risk factor is the ratio of the odds of developing the adverse effect, i.e., the ratio of p/(1-p) values among people with and without exposure to that risk factor. For a continuous risk factor, OR is the odds ratio associated with a unit increase in the factor. A "risk factor" is an independent variable in the logistic regression model. In both cases, the logistic regression model implies that OR = $exp(q_i)$ for factor i.

The logistic regression model permits odds ratios to be calculated for each risk factor. It allows statistical hypothesis-testing of whether the effect of each factor on risk differs significantly from no effect, i.e., an odds ratio significantly different from 1 – assuming that the logistic regression model form is correct. Although it lacks biomathematical motivation, logistic regression is convenient, familiar, and available enough in risk assessment software to assure its continued use for the foreseeable future.

Logistic regression and other dose-response models, as they are implemented in standard software packages, must be modified to correctly process missing data, uncertain or erroneous values of independent variables, omitted covariates, uncertainty about which risk factors are relevant, uncertainty about model forms, and causal relations among multiple variables. Appropriate techniques are available for dealing with all of these complexities, as discussed in Section 3.

Although the multistage and logistic regression risk models have been most widely used in assessing cancer risks and other health risks due to exposures to various risk factors, usage is gradually changing. For animal bioassay data analysis, biologically motivated dose-response models, such as the Moolgavkar *et al.* model in Table 3, are gaining wider acceptance. The need to use model forms that reflect relevant biology is undeniable, and biomathematical modeling is starting to catch up with the need.

Non-Parametric and Semi-Parametric Models

An important idea for avoiding uncertainties about the correct model form is to avoid selecting any specific parametric family of models, and instead to use non-parametric ("model-free") methods. For example, rather than assuming that excess risk is proportional to exposure, it is less constraining to assume only that there is a smooth, but otherwise unknown, function relating exposure to risk. The shape of the relation can then be estimated directly from the data instead of being assumed. Examples of such model-free approaches include the following:

- Replace the assumption of a linear relation between explanatory variables and the log-odds ratio in logistic regression models with a weaker assumption of an unknown smooth relation (Harrell et al. (1988).
- Apply nonparametric classification and regression tree (CART) algorithms (Breiman et al., 1984) to approximate unknown multivariate relations with multi-dimensional rectangular histograms based on the data (Ye, 1998; Denison et al. 1998).
- Invoke the principle of parsimony (shorter is better) to select among a broad class of smooth curves that can be expressed via wavelet decompositions (Antoniadis et al., 1997). The "minimum description length" criterion selects the model that has the shortest joint description of the data and the model itself. Pragmatically, this involves estimating the largest coefficients in the wavelet decomposition representing an unknown curve.

To fit smooth curves to data without assuming any specific underlying parametric model form relating between them, one might consider the class of all "smooth" curves (e.g., with continuous derivatives up to some order), or sub-classes such as all smooth monotonic curves, all smooth s-shaped curves, and so forth. In such a non-parametric class of curves, the member that best fits the data according to some criterion is called a *nonparametric estimate* of the dose-response model. Non-parametric approaches to curve-fitting include expanding the unknown dose-response relation as a convergent power series or wavelet series and estimating its leading coefficients from data. There are now many algorithms to fit curves and surfaces (such as splines or loess algorithms) in which the penalty weight assigned to rapid changes ("roughness") is a user-tunable parameter. These and other non-parametric strategies (Eframovich, 1999; Tarter and Locke, 1993) typically require more sophisticated and computationally intensive iterative fitting algorithms than parametric estimation (e.g., nonparametric maximum likelihood (NPMLE) algorithms for survival models; isotonic, kernel, and generalized additive regression algorithms, etc. for quantal models). Table 3 does not show non-parametric curve-fitting and data-smoothing models, as there are no model forms to display for them.

Example: A Bayesian Non-Parametric Approach to Dose-Response Curve-Fitting Using Monte-Carlo Uncertainty Analysis

Regulatory risk assessments for many chemical carcinogens are driven by estimated upper bounds on the slopes of the dose-response functions at very low doses. The q_1* methodology commonly used to estimate these upper bounds assumes a specific parametric family of statistical risk models – the linearized multistage (LMS) family – derived from mathematical approximations that tend to over-estimate risks at low doses (Chiu et al., 1999). The LMS model is typically expressed in the form shown in Table 3, i.e.,

$$p(x) = 1 - \exp[-(q_0 + q_1x + q_2x^2 + ... + q_nx^n)]$$

with the constraints

$$q_i \geq 0, i = 1, 2, ..., n,$$

where p(x) = lifetime probability of tumor in an animal exposed to dose x. Yet, this algebraic model implies unrealistic constraints such as that the lifetime cumulative tumor hazard, -ln(1- p), must be a linear or upward-curving function of the dose, x. Its implications are significantly violated in real data sets (e.g., for 1,3-butadiene) and are not implied by the underlying exact model (Cox, 1995). (Moreover, the requirement of non-negative coefficients ignores cytotoxicity, cell proliferation and death, DNA repair, etc.) For these reasons, it is desirable to develop a more general approach to uncertainty analysis and upper-bounding of the dose-response function that better allows for model uncertainty and that relaxes some of the strong a priori constraints in the LMS model.

An alternative, non-parametric approach is the following.

- *Step 1: Select constraints* expressing knowledge or assumptions about the types of dose-response functions to be considered. For example, can it be assumed that the dose-response relation is continuous (e.g., no threshold effects), that it is smooth (has continuous derivatives), that it is upward-sloping at all points, that its global shape is sigmoid (s-shaped), i.e., convex at low doses and concave at high doses, or that it has a positive slope at very low doses? How many of these (or other) assumptions is the analyst willing to impose as a priori constraints?
- *Step 2: Randomly generate many pseudo-data sets* by Monte Carlo sampling from the Bayesian posterior distribution of the response probability in each dose group, calculated from (a) A constrained maximum-entropy prior (subject to the constraints observed in Step 1) and (b) The observed proportion of animals with tumors in each dose group. The posterior is the probability distribution for the true but unknown tumor probability in each dose group.
- *Step 3: Fit a smooth curve through each randomly generated pseudo-data set.* There are now many model-free methods available for carrying out this step (e.g., Tarter and Lock, 1993).
- *Step 4: Construct simultaneous confidence bands for the entire dose-response curve.* For example, the simultaneous 95% upper confidence band determined by the Monte-Carlo analysis is given by the lowest band such that 95% of the smoothed curves fall below it.

This method generates an entire population of curves (one for each pseudo-data set) from which to estimate the dose-response relation. Thus, it is possible to obtain very detailed information about the uncertainties in this relation. For example, the entire probability distribution of the slope of the curve at any dose level (including zero) can be estimated. Specifically, the probability that the true slope falls between any two user-specified values is approximated by the fraction of cases (in the Monte-Carlo-generated population of curves) having slopes in the specified range.

In more detail, a typical bioassay data set consists of several triples of observations:

$$D = \{(x_j, n_j, N_j), j = 1, 2, ..., M\}$$

where

D = data set from M experimental dose groups of animals.
x_j = dose administered to animals in dose group j.
n_j = number of animals in dose group j that develop tumors.
N_j = total number of animals in dose group j.

Given any such data set D and parametric model with parameter vector

$$\mathbf{q} = (q_1, q_2, ..., q_n)^T,$$

the *likelihood function* for **q** is the conditional probability of D given **q** (and assuming that the parametric model is correct):

$$L(\mathbf{q}) = \Pr(D \mid \mathbf{q}) = \Pi_j p(x_j ; \mathbf{q})^{n_j}[1 - p(x ; \mathbf{q})]^{(N_j - n_j)}.$$

Here, the notation p(x ; **q**) denotes the probability of tumor at dose level x if **q** is the parameter vector and the assumed risk model is correct. This is the same quantity as p(x), but the notation p(x ; **q**) emphasizes the dependence on **q**. It is now a matter of mathematical optimization (nonlinear programming with inequality constraints) to discover a value of **q** that maximizes the likelihood function, i.e., the MLE of **q**. It is usually numerically more convenient to find the value of **q** that maximizes the log-likelihood function:

$$LL(\mathbf{q}) = \log[L(\mathbf{q})] = \Sigma_j\{n_j p(x_j ; \mathbf{q}) + (N_j - n_j)[1 - p(x_j ; \mathbf{q})]\}.$$

The same value of **q** maximizes both L(**q**) and LL(**q**). Let this MLE value of **q** be denoted by **q*** (not to be confused with q_1*). Under regularity conditions, the asymptotic distribution of minus twice the log-likelihood ratio, -2log[L(**q**)/L(**q***)], approaches a chi-square distribution with n degrees of freedom as sample size goes to infinity (Lawless, 1981). This asymptotic result can be used to construct approximate confidence limits (UCLs) for **q** from the MLE estimate **q***. The usual regularity conditions require that the true parameter vector **q** should lie in the interior of its set of possible values (so that the constraint $\mathbf{q} \geq 0$ is not binding) and that L(**q**) is thrice differentiable with bounded third derivatives.

Instead of computing a likelihood function, the above Monte Carlo approach computes the *conditional probability density functions* for response probabilities:

$$\Pr[p(x) \mid D]$$

for each dose level, x. In other words, the true but unknown response probability at dose x is represented as the realization of a random variable whose PDF is obtained by conditioning some prior PDF on the observed data in D. In the interests of objectivity, the prior PDF for p(x) may be chosen as the *maximum-entropy PDF* satisfying any user-specified constraints. The maximum entropy methodology for PDF estimation is often motivated as the "minimum bias" and "minimum assumption" PDF satisfying the given constraints (Jessop, 1995). It introduces as little "information" as possible (in a well-defined sense) other than that expressed in the constraints. For example, if the only constraint is that $p_j = p(x_j)$ must lie between 0 and 1, then the maximum-entropy prior PDF for p_j is the uniform PDF, corresponding to the following cumulative probability density function (CDF):

$$Pr[p_j \leq t] = t \quad \text{if } 0 \leq t \leq 1, \text{ else } 0.$$

If the mean value of p_j is specified, along with upper and lower bounds, then the constrained maximum-entropy prior is either uniform (if the mean is the midpoint of the range) or exponential (Jessop, 1995). For purposes of illustration, the case in which the only required constraint is that p_j must lie between 0 and 1 will be considered further.

Starting from the constrained maximum-entropy prior PDF, which only reflects information available before looking at the data, the impact of the experimental data is incorporated by conditioning the prior PDF on it. If the p_j are treated as statistically independent for different dose groups (i.e., no a priori assumption is made that would link them, such as that the dose-response function is an increasing function of dose) then the required conditioning can be carried out for one dose group at a time via the definition of conditional probability:

$$Pr[p_j \mid D] = Pr[D \mid p_j] * Pr(p_j) / Pr(D), \quad j = 1, 2, ..., M.$$

In this formula, $Pr[D \mid p_j]$ is the likelihood function for the tumor probabilities p_j, while $Pr[p_j] = 1$ is the prior PDF for p_j (here assumed to be uniform between 0 and 1) and Pr(D) is a number such that the entire conditional PDF integrated from $p_j = 0$ to $p_j = 1$ evaluates to 1. Conditioning a uniform prior for p_j on the observed binomial data point (x_j, n_j, N_j) yields the posterior PDF

$$Pr[p_j \mid D] = \beta(n_j + 1, N_j - n_j + 1),$$

where

$$\beta(n_j + 1, N_j - n_j + 1) = \text{beta distribution with parameters } n_j + 1 \text{ and } (N_j - n_j) + 1.$$

This distribution has a variance that decreases with N_j and has a mean value of

$$E[p_j \mid D] = (n_j + 1)/(N_j + 2).$$

Monte-Carlo samples of values for p_j are drawn from this posterior distribution.

If constraints are imposed on the allowable dose-response functions – such as that the dose-response curve must be upward-curving – then the preceding calculations must be generalized to express the constraint-induced dependencies among the tumor probabilities at different doses. In many cases, this can be done efficiently during the Monte-Carlo step by generating the sample values of the p_j sequentially (from j = 1 to j = M) and conditioning the PDF of each successive p_j value on the values of the rest. For example, the joint PDF for all of the $p(x_j)$ can be factored as follows:

$$Pr[p_1, ..., p_n \mid D] = Pr[p_1 \mid D]Pr[p_2 \mid p_1, D] ... Pr[p_n \mid p_1, p_2, ..., p_{n-1}, D].$$

This factoring allows dependencies among data points to be accounted for while preserving some computational efficiency. Applied iteratively (e.g., starting with uniform marginal priors for the p_j followed by multiple cycles of conditioning and updating of the marginal PDFs until convergence occurs), it forms the basis for the computational statistics technique of Gibbs sampling. Methods for estimating the dose-response curve when only qualitative constraints are specified (such as that the dose-response function is s-shaped) are given by Schmoyer (1984) and Schell and Leysieffer (1989).

Semi-parametric models, as the name suggests, mix aspects of parametric and non-parametric modeling. The proportional hazards (PH) model for survival data is an example. In the PH formula

$$h(t \mid x, z) = e^{qx}h(t \mid z),$$

the function linking covariate values to risk is clearly parametric, namely, *exp(qx)*. However, the baseline hazard function, namely, $h(t \mid z)$, can be arbitrary and so is non-parametric. (The statistical estimation algorithms that produce point estimates and confidence regions for the parameter vector *b* treat $h(t \mid z)$ as a nuisance parameter that need not be estimated. However, if the baseline hazard function is hypothesized, perhaps based on biomathematical modeling, then the hypothesis can be tested; see Pena, 1998.) Other semi-parametric models, including the linear relative risk model in Table 3 and many multi-state transition models used in modeling disease processes (not shown in Table 3), similarly combine parametric and non-parametric components.

Survival Data Analysis

For both epidemiological data and animal bioassay data, *survival data analysis* is increasingly used by risk assessors, with proportional hazards (PH) models being among the most widely applied (e.g., Therneau and Grambsch, 2000). Classification tree methods (Zhang and Singer, 1999) and other non-parametric techniques for survival data are also becoming more widely used. Compared to survival models, models with dichotomous responses lose potentially valuable information by not fully using data on *when* responses in estimating effects of covariates on risk. Moreover, they cannot always deal well with competing risks. For example, quantal models, including the multistage and logistic regression models, are sometimes applied to experimental data in which tumor rates are reduced at high dose levels by early mortality due to toxicity. Although one can try to adjust the independent and dependent variables in such cases to correct for the effects of reduced exposure times, survival data analysis provides more powerful and natural methods for dealing with censored response data.

Statistical methods have been developed specifically for survival data analysis (Andersen et al., 1993; Klein and Moeschberger, 1997) and transition data analysis (Lancaster, 1990, Hougaard, 2001). In survival data analysis, the basic data are typically of the form $D = \{(x_i, y_i, t_i), i = 1, ..., N\}$ where x_i is a vector of covariates, including exposure history; t_i is the time at which observation of subject *i* ceased (e.g., due to "right censoring" by death or the end of the study); and y_i indicates whether a response had occurred by this time. In other study designs, the times at which subjects entered the study and the reason for withdrawal (e.g., death due to cancer or to competing risks) are included in the data. Given such data, special-purpose algorithms and methods such as nonparametric maximum likelihood estimation (NPMLE) have been developed to estimate survival, hazard, and cumulative hazard functions (Klein and Moeschberger 1997).

Suppose that a group of mice are exposed to a carcinogen (e.g., 1,3-butadiene) via inhalation and that their times of death are recorded. Aggregating the individual data, let d_t denote the number of mice that die on day *t* and let N_t denote the number of mice that survived until the start of day *t*, and that therefore were at risk on day *t*. Table 4 shows some well-known estimators for survival-related variables. Two of the most widely used are the Product-Limit (also called Kaplan-Meier) and Nelson-Aalen estimators of cumulative hazard functions. Both are step functions of time, with jumps occurring at the times of observed deaths. To estimate the corresponding hazard functions, it is common to smooth these step functions (e.g., using a parametric kernel smoother) and then take their numerical derivatives. Survival data analysis theory justifies interpreting the Nelson-Aalen estimator as a (NPMLE) estimator and develops techniques for estimating simultaneous confidence bands (with specified coverage probabilities) for the survival, cumulative hazard, and hazard functions.

Survival curves for different dose groups can be compared to test hypotheses about the relation between dose and survival times. Standard statistical software packages now include a variety of nonparametric hypothesis tests to determine whether one survival curve lies significantly to the left of another (or whether one corresponding hazard function lies above the other), indicating a stochastically decreased life duration. More generally, survival data models with covariates, including parametric and semi-parametric (e.g., proportional hazards) models as well as nonparametric estimates of survival functions, can be compared to determine whether survival is stochastically decreasing in dose, other factors held equal.

Table 4: Some Non-Parametric Survival Data Estimators

Symbol	**Name**	**Formula**
S*(t)	• Product-Limit (PL) estimator • Kaplan-Meier estimator of the survival function	S*(t) = 1 for times less than the first observed death time, else $S^*(t) = \Pi_{k \leq t}[1 - (d_k/N_k)]$ for t no grater than last obs. death time
V[S*(t)]	Greenwood's formula for variance of PL estimator	V[S*(t)] = $S^*(t)^2\Sigma_{k \leq t}d_k/[N_k(N_k - d_k)]$
H*(t)	PL estimator of cumulative hazard function	H*(t) = -ln[S*(t)]
H (t)	• Nelson-Aalen estimator of cumulative hazard function • Non-parametric maximum likelihood (NPMLE) estimator	H`(t) = 0 before first observed death time, else $\Sigma_{k \leq t}(d_k/N_k)$ for times not greater than last observed death time.
S`(t)	Nelson-Aalen estimator of survival function	S`(t) = exp[H`(t)]
V[H`(t)]	Estimated variance of Nelson-Aalen estimator	$V[H`(t)] = \Sigma_{k \leq t}(d_k/N_k^2)$

Multivariate survival models (Hougaard, 2001) allow for more transitions than just from alive to dead. As mentioned in Chapters 1 and 2, multi-state transition models represent stochastic transitions of an individual among states – and/or the probable evolution of the frequency distribution of multiple individuals among multiple states – by a matrix of state transition rates. These rates may be functions of covariates. Such models allow describe and predict how changes in transition rates (and hence in the covariates and decisions that determine them) will affect probabilities and durations of state occupancy times. However, the ways in which transition rates will change when various actions are taken may not be uniquely identifiable from observations. Assumptions must then be made – most often, about the statistical independence of different transition times – in order to predict the probable consequences of actions.

Example: Competing Risks

Some of the mice exposed to a carcinogen may die of other causes (e.g., cage door accidents, toxicity of the inhaled dose, or old age) before cancer occurs. A competing risk model represents these possibilities by allowing multiple transitions, as in the following diagram:

Dead without tumor ⇐ *Alive* ⇒ *Dead with tumor*

Under certain conditions, and depending on what can be observed (e.g., time of death, cause(s) of death, presence of tumor before death and at death, etc.), it may be possible to estimate both transition rates as a function of dose. If so, and if death rates with and without tumor for animals in dose group x at time t are $h_1(x, t)$ and $h_2(x, t)$, respectively, then the total hazard rate for mortality at time t is:

$$h(x, t) = h_1(x, t) + h_2(x, t).$$

The probability that a death that occurs at time t occurs with tumor is:

$$h_1(x, t)/h(x, t)$$

In general, cause-specific hazard functions *cannot* be uniquely identified from observed numbers of animals in each terminal state at one or more points in time. The main problem is that the effects of removing one particular cause of death (e.g., cage door accidents) on the rates of death from other causes cannot be predicted based on data alone. This is because the dependencies (if any) among the random times to die from various competing causes are not revealed by the joint distribution of survival times and causes of death (Elandt-Johnson and Johnson, 1980, p. 279). Other assumptions or data (e.g., biological measurements) are then required to identify the cause-specific hazard rates are $h_1(x, t)$ and $h_2(x, t)$ from data.

Specialized parametric, semi-parametric, and non-parametric (including NPMLE) estimators have been developed for multivariate survival models. These address the problem of estimating the transition matrix from observed transition times and/or from counts of individuals in states at different times. For example, a product-limit estimator has been developed for estimating the time-varying transition matrices of nonhomogeneous Markov processes (Andersen et al., 1993, pp 287-321). This generalizes the Kaplan-Meier estimator for a single (e.g., live-to-dead) transition. Techniques for estimating the effects of covariates on transition rates, e.g., by postulating parametric or semi-parametric models for specific transitions and then estimating their parameter values from data, have also been developed and play a growing role in analysis of health risk data (Hougaard, 2001).

Model Form-Selection and Variable-Selection Biases

Selecting *any* statistical model form to represent a dose-response relation – or, more generally, to model a multivariate exposure-covariates-response relation – raises the threat of *model-selection biases*. Three types of bias due to such selection are particularly prevalent in health risk modeling. First, the predicted risks may be very sensitive to the assumed model form (Greenland, 1989). This tends to be especially true for risks extrapolated to low dose levels outside the observed range of data. Second, confidence intervals for model parameters and predicted risks that ignore model form uncertainty tend to be too narrow (Raftery et al., 1997). This can increase the rate of false positives, i.e., finding apparently significant exposure-response relations where no true elation exists, because the artificially narrow ranges may exclude the "no effect" hypothesis even though corrected confidence intervals that reflect model uncertainty would include it. Finally, goodness-of-fit tests that are based on the same data used to select the model are typically overly optimistic, have low power to detect violations of model assumptions, and should not be used for model selection (Greenland, 1989).

Example: Selecting a Wrong Model Form Can Invalidate Conclusions

It often seems natural to hypothesize that excess illnesses in a population are approximately proportional to exposure to a causal agent, at least at typical low, dose levels. For example, the U.S. FDA proposed this proportionality postulate for quantifying health risks associated with Campylobacter in chickens (FDA-CVM, 2000). Many common dose-response models, including both the logistic regression model and the proportional hazards model, approach straight-line relations at low doses, leading to a proportional relation for excess risks. However, assuming this model form without testing it can lead to major mistakes in risk assessment. To illustrate, suppose that the following model form is assumed:

$$H = kQ \qquad \text{(Linear population risk model)}$$

where H is interpreted as expected number of excess illnesses in an exposed population and Q is interpreted as a measure of exposure. Suppose that, unbeknownst to a risk assessor, H and Q actually have no relation to each other, suggesting that the true value of k should be zero. More specifically, suppose that the measured values of H and Q are realizations of statistically independent, identically distributed random variables, each being uniformly distributed between 0 and 100. Then fitting the above model to a random sample of (Q, H) observations (e.g., by simple linear regression with a zero intercept) will give an estimated value of k of about 0.76 with a confidence interval that depends on the sample size. For large samples, a standard t-test will reveal that k is significantly greater than zero – despite the fact that H and Q are unrelated.

The problem here is the assumed model form – a straight line through the origin. All data points lie to the north-east of the origin, so the best-fitting line must slope upward to go through them. This is an artifact of the assumed model form. It reflects nothing real about risk. It illustrates the warning that: "A key assumption underlying the use of model-based estimates in epidemiology is that the structural-model form is an adequate mathematical description of the dependence of disease occurrence on exposures and covariates (that is, the model form is correctly specified). If this assumption is violated, model-based point estimators and variance estimators may be biased, standard confidence intervals may be invalid, and inferences derived from these estimators may be incorrect. In practice, the true structural-model form is usually unknown." (Maldonado and Greenland, 1996). "In the most severe instances... it is possible for a modeling result to be entirely dependent on modeling assumptions or to be an artifact of some violation of those assumptions." (Greenland, 1989, p. 342).

Fortunately, this type of model specification error is readily detected by simple diagnostic tests such as a graph of residuals, which show a strong dependence on the value of Q. Such model diagnostics, in addition to goodness-of-fit tests (e.g., Pena, 1998; Stute et al, 1998; Wilbert et al., 1997) should always be used to check whether model form assumptions are conspicuously inappropriate for a data set. This has been well recognized in principle (Greenland 1989) although too often neglected in practice (Concato et al., 1993; Bramwell and Williams, 1997).

Using a data set both to select a model form and to estimate its parameters can produce an illusion of excellent fit that disappears when the same model is applied to additional data. Although it is now relatively easy to avoid or correct for such over-fitting using modern techniques, naïve approaches that select a model form *and* its parameters to maximize fit, but without correcting for over-fitting, still appear quite frequently in applied work (Concato et al., 1993). Section 3 discusses methods for overcoming model-selection biases and for dealing with uncertainty about the correct model form.

Deciding which variables to include from among all those in a large data set is a further aspect of model selection. In any data set with many variables recorded for each individual, it is necessary to choose which variables should enter the final risk model. Many of the principles of model form selection also apply to this choice. Both model-form selection and variable selection are referred to in the statistics literature as *model-selection problems* (Greenland, 1989; 1996).

- Selecting a subset of explanatory variables to include in a multivariate risk model based on statistical criteria alone, rather than on causal knowledge, may create a spurious appearance of statistically significant associations between the dependent variable and the selected independent variables, even if there is no real relation among them. Indeed, "In the extreme case where there are many candidate predictors but no relationship between any of them and the response, standard variable selection procedures often choose some subset of variables that yields a high R^2 and a highly significant overall F value." (Raftery et al., 1997) This can be confirmed by creating an artificial data set consisting of perhaps 100 rows ("individuals") and ten columns ("variables") filled in with independent random numbers. Regressing one column against the rest using standard statistical packages with automatic variable-selection procedures (e.g., stepwise forward variable selection based on an F-test) will quite often identify several "significant" variables in the random data, illustrating the generally high rate of false positive associations (typically many times the nominal 5% significance level) created by standard variable-selection procedures. Conversely, "Variable-selection algorithms in current packaged programs, such as conventional stepwise regression, can easily lead to invalid estimates and tests of effect. ... For example, if conventional significance levels are used, it is likely that some important confounders and product terms will *not* end up in the final model" (Greenland, 1989).

- Similar challenges exist for selection of confounders to control for in multivariate risk models. For example, "Assessment of confounding by testing the statistical significance... of a potential confounding factor in a multivariate model can produce underestimates or overestimates of the true association between an exposure and an outcome. Inclusion of all covariates in a multivariate model may lead to controlling for variables that are not, in fact, confounders. This may produce underestimates or overestimates of the effect in question, as well as artificially widened confidence intervals. ... To prevent problems resulting from these misunderstandings, researchers should consider drawing causal models prior to conducting the research and use the change-in-estimate criterion, rather than a statistical test, to detect confounding." (Sonis, 1998; see also Mickey and Greenland, 1989; and Lindsey and Jones, 1998, p. 66.)

As suggested by these quotes, variable-selection biases can interact with other problems. These include artificially narrow confidence intervals due to neglected uncertainty about the correct set of variables (Lindsey and Jones, 1998); and omitted explanatory variables, i.e., failing to select model variables that truly do affect risk. Moreover, "What appear to be the best model and variable forms will be influenced by the choice of variables and product terms, and *vice-versa*" (Greenland, 1989, p. 340). Thus, model-form selection bias and variable selection biases may interact. The result is that an investigator's selection of variables to include in a multivariate statistical risk model can strongly affect conclusions about the existence and magnitudes of relations between exposure and response variables.

In general, one should *not* use statistical hypothesis tests to assess whether variables should be considered as potential confounders. Doing so inappropriately conflates sample size and magnitude of confounding effect. Instead, it is better to fit separate models with and without the suspected confounder(s) and then use the resulting change in estimated coefficients to determine which variables should be included as confounders. This is the *change-in-estimate* criterion (Sonis, 1998).

There are now many appropriate statistical methods to overcome the potential biases due to variable selection. Approaches that improve dramatically on earlier (e.g., stepwise regression) methods include the bootstrap variable selection, cross-validation, and Bayesian model-averaging (BMA) techniques discussed in Section 3, as well as formal model-selection criteria such as the Akaike Information Criterion, AIC, discussed below (Lindsey and Jones, 1998). Section 3 introduces other constructive approaches to variable selection and bias reduction.

No matter which specific technique(s) are used to correct for variable-selection bias, a guideline for sound modeling is that *all* the variables considered – both those included in and those excluded from the final model(s) used to estimate exposure effects – should be reported with the resulting effects estimates. The methods used to select the included variables and to correct for variable-selection biases should also be stated. Otherwise, it is impossible to determine how much of any reported exposure-response association is likely to be due to variable selection biases.

2.3 Estimate Risk, Confidence Limits and Model Fit

Parametric dose-response models and estimation methods have dominated the field of mathematical dose-response modeling since its inception. In a parametric model, the possible shapes of the dose-response functions are determined by the numerical values of one or more constants, i.e., the parameters. The data assembled in Step 1 of Table 1 and the model form(s) selected in Step 2 can be supplied as inputs to parameter-estimation algorithms in computer programs that estimate the parameters of specific dose-response models and calculate confidence intervals for the risks at different doses. Most such programs evaluate an estimated model by calculating global goodness-of-fit tests (although these typically have low power to reject incorrect models, as previously mentioned). Some packages also generate model-diagnostic tests and graphs that allow the user to examine whether the selected model provides a conspicuously inappropriate description of the data. If the model fit is not obviously bad, the risk estimates are reported; otherwise, a warning message is generated suggesting that a different model form should be selected. Similar statistical estimation programs are also widely available to fit proportional hazards and other semi-parametric and parametric survival models to data sets that include time-to-tumor information and indicators of when observations on individuals started and stopped.

Parameter-estimation programs typically implement numerical optimization algorithms that seek values for the parameters to maximize a measure of agreement or to minimize a measure of disagreement between the observed data points and the curve corresponding to the optimized parameter values. The most common measure of agreement is the likelihood of the data (i.e., its probability), given the parameter values and assuming that the selected parametric family of dose-response functions is correct. The parameter values that maximize the likelihood function, which expresses the probability of the data as a function of the parameter values, constitute the *maximum likelihood estimation* (MLE) of parameters.

Many statistical software packages and statistical computing environments now contain embedded optimization algorithms for finding parameter values to maximize the likelihood function (or, equivalently, the log-likelihood, to prevent underflow problems) and various approximations to it. Most of the models in Table 3 are fit to data by MLE. (If some of the data are missing, truncated, or censored, then algorithms for approximate and conditional likelihood functions may be used; see Table 6.) MLE algorithms may be quite sophisticated. They may consider uniqueness of solutions and identifiability of parameters, convergence rates, local *vs.* global maxima, and sampling statistics of the MLE obtained from data compared to the true parameter values.

The most common measure of disagreement between parametric curves and data points is the sum of the squared differences in their values. This leads to least-squares regression (with which we assume familiarity) and to its very substantial generalization, Generalized Estimating Equations (GEEs) (Shao, 1999). However, when the data points are discrete, e.g., indicating whether or not each exposed animal developed a tumor, while the model predictions are continuous, e.g., tumor probabilities, likelihood-based techniques are generally used instead of least squares.

Other optimization-based approaches to inference are also sometimes used, but less commonly. One important approach uses data to choose among members of a parametric family of models by selecting the model that minimizes an information-theoretic measure of disagreement (the Kullback-Leibler dissimilarity index) between the selected model and the joint distribution of the data (Peracchi, 2001). This approach turns out to have close mathematical ties to model-selection criteria such as the Akaike Information Criterion (AIC), which scores models based on the maximized value of the sample log-likelihood function penalized for model complexity:

$$\text{AIC} = -2(L^* - n) \qquad \text{(Akaike Information Criterion)}$$

where L^* = the maximized value of the sample log-likelihood function and n is the number of parameters. Selecting the model form that minimizes AIC is often recommended as a formal model-selection criterion for statistical risk models (e.g., Peracchi, 2001, p. 331; Lindsey and Jones, 1998).

Dose-Response Curve-Fitting Software: EPA's BMDS

Push-button programs have been created to automate as far as possible the tasks of fitting statistical dose-response models to data and of diagnosing adequacy of model fit. Some human intervention is usually still required to select the models to be fit to the data and, for some models, to choose initial estimates of parameter values as starting points for MLE estimation.

One such program, BMDS, is described as follows.

EPA's Benchmark Dose Software (BMDS)
(*Source:* http://www.epa.gov/nceawww1/bmds.htm)

"Use of benchmark dose (BMD) methods involves fitting mathematical models to dose-response data and using the different results to select a BMD that is associated with a predetermined benchmark response (BMR), such as a 10% increase in the incidence of a particular lesion... At this time, BMDS offers sixteen (16) different models for the analysis of dichotomous (quantal) data (9 models; Gamma, Logistic, Log-Logisitic, Multistage, Probit, Log-Probit, Quantal-Linear, Quantal-Quadratic, Weibull), continuous data (4 models; Linear, Polynomial, Power, Hybrid) and nested developmental toxicology data (3 models; NLogistic, NCTR, Rai & Van Ryzin). Results from these models include goodness-of-fit information, the BMD, and the estimate of the lower-bound confidence limit on the BMD (the BMDL). ... Running the models on a data set consists of four basic steps.
Step 1: Create a data set using the BMDS spreadsheet facility, or Import a data set from a text, Excel or Lotus spreadsheet file
Step 2: Select the appropriate model based on the type of data set being evaluated.
Step 3: Specify the parameters associated with the model that was selected.
Step 4: Run the Model and view the textual and graphical results."

Example: Fitting Dose-Response Models to Data

This example illustrates the use of the BMDS software to fit a multistage model to a simple hypothetical data set. The data set summarizes the number of animals in each of five exposure groups that develop a tumor. There are 100 animals in each group, as shown in Column 2 of the upper panel in Figure 1. The exposure levels, shown in Column 1, are 0, 25, 50, 75, and 100 units. The units and route of exposure are left unspecified.

Earlier statistical dose-response modeling software packages required dose unit and route of administration as inputs to the program. Typical options included parts per million (ppm) of substance in inhaled air times hours of exposure per day times number of days exposed (i.e., ppm-hours of exposure); or milligrams of substance ingested or injected per kilogram of body weight per day (mg/kg-day), times number of days of the dose regimen (mg/kg). These units were used to automatically convert dose in one species administered by a specific route to algebraically "equivalent" units of dose in other species, such as humans, possibly exposed via a different route. However, it is now widely recognized that algebraically equivalent units of exposure may have very different effects in different species, or even within the same species. Such automatic extrapolation of dose-response relations across species (and also from high exposures for less time to lower exposures for a longer duration) must be regarded as speculative and unreliable. The BMDS curve-fitting program leaves the units unspecified.

Figure 1: BMDS Curve-Fitting Software Inputs (Upper) and Outputs (Lower)

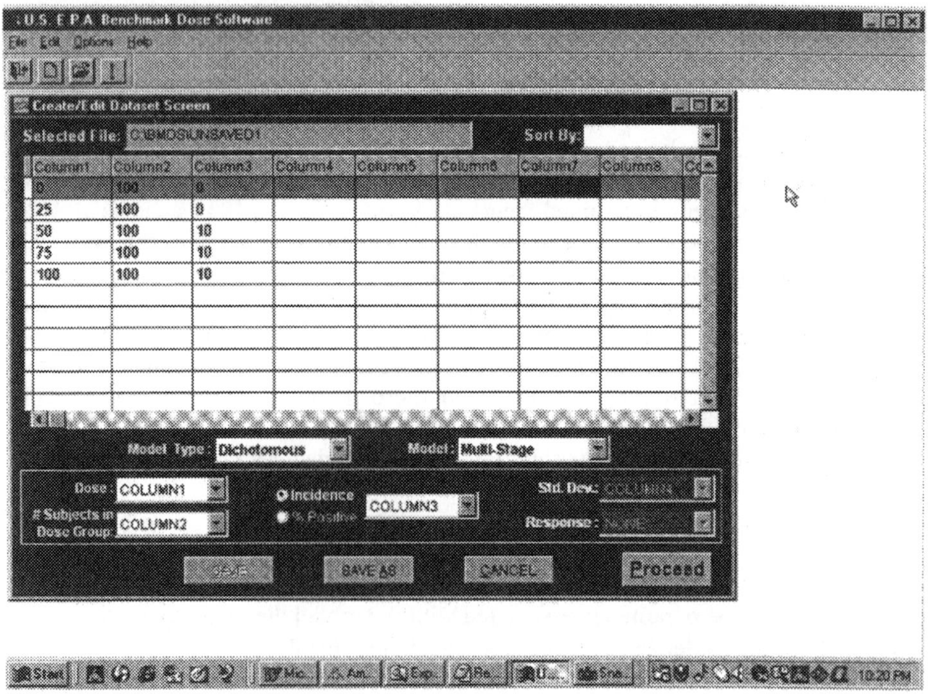

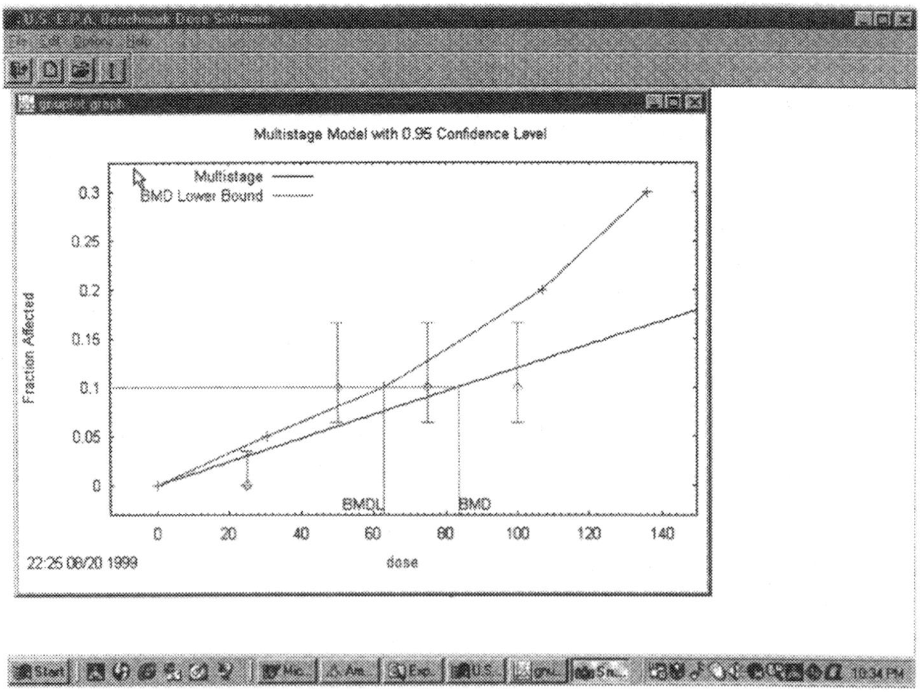

Column 3 of the input table in Figure 1 shows the number of animals that develop a tumor before death in each exposure group. (If multiple tumor types or other effects are to be considered, they may be recorded in subsequent columns and models may be fit to them separately.) For simplicity, we consider hypothetical data in which it is assumed that all exposed animals completely detoxify any dose less than 30 units. Doses of 50 units or more saturate the detoxification mechanisms and create a constant hazard rate large enough to produce detectable tumors in an average of 10% of exposed animals before terminal sacrifice.

The lower panel in Figure 1 shows sample output from the BMDS program. The multistage model does not allow for the possibility of a dose threshold below which no excess risk occurs. Instead, BMDS estimates model parameter values by MLE, typically within a set of allowed values that constrains the dose-response curve to have a non-negative slope at low doses. Unlike earlier software, BMDS allows this constraint to be toggled on or off. Whether or not it is imposed, the result for this data set is a smoothly increasing dose-response curve, represented by the lower curve – approximately a straight line – in Figure 1. This line is the MLE estimate of the dose-response relation within the model family being considered. The graphical output is complemented by tabular outputs that show the selected model form, the MLE and approximate confidence intervals for its parameters, and the estimated doses corresponding to specified levels of risk.

The MLE routine not only produces an MLE estimate of the dose-response model, but also it estimates confidence bounds on parameter values and on the lowest dose that might create a given excess risk, using theoretical formulas for large-sample confidence intervals. The theoretical conditions needed to justify these asymptotic confidence interval calculations (e.g., parameter values in the interior of their set of possible values) may not hold in practice, since parameter values are typically constrained to be non-negative, but the calculations are automated and are often used anyway. For example, the upper curve in Figure 1, i.e., the "BMD lower bound" curve, estimates the smallest dose level corresponding to each excess risk level on the y axis. In other words, it plots the 95% lower confidence limit for dose corresponding to each risk level, assuming that the selected multistage model is correct and that the asymptotic confidence interval calculations are appropriate.

For the date set and multistage model selected in Figure 1, the benchmark dose corresponding to a 10% excess risk is estimated as 83.32 units, with an estimated 95% lower confidence limit of 62.95 units. This is as close as the selected parametric family of models can come to describing the true relation, i.e., zero risk at 30 units or below, 10% risk at 50 units or above. More flexible models, such as logistic regression models with piece-wise polynomial arguments and change-points estimated from data (Pastor and Guallar, 1998) can describe such data better, but are not yet available in standard packages.

Critique of Statistical Dose-Response Curve Fitting

Statistical fitting of dose-response models to data is often criticized on the following grounds:

- Available experimental data may not adequately discriminate among models that make very different predictions for risks at low doses. In such cases, low-dose risk estimates are under-determined by the data.
- The best (e.g., MLE) member of the family of models being considered may be difficult or impossible to identify from the data (if the parameters are non-identifiable or if the estimation problem is ill-conditioned).
- Even if the best (e.g., MLE) member of family being considered can be uniquely and confidently estimated from data, the family itself may not contain any members that adequately describe the true dose-response relation. In this case, model-based risk predictions may be misleading.

A difficulty with parametric models is that none of the curves generated by varying the parameter values may correctly describe the relation between dose and response. In other words, the best-fitting or maximum likelihood member of a parametric family of curves may still provide an inadequate description of the true data-generating process. (See the discussion of Figure 1.) This problem can be severe when one or more parametric families fit the data well in the observed (usually high) experimental dose range, but do not correctly describe risks at realistic (usually much lower) dose levels. Often, alternative parametric risk models such as the multistage, quantal Weibull, and probit all describe experimental data approximately equally well, but predict low-dose risks that may differ by orders of magnitude. In such cases, the correct model form is said to be unidentifiable from the data. Then, uncertainty about the correct model family or model form can swamp uncertainties due to sampling variability and to parameter estimation errors.

Example: Identifiability of Models from Data

Problem: Suppose that a single dose group of 100 animals is subjected to a concentration of 10 ppm of a chemical in inhaled air for life and that 10 animals develop tumors. In a control group, 0/100 animals develop tumors. What risk of tumor should be extrapolated to 1 ppm?

Solution: If a one-hit model is assumed, then the MLE for the potency of the chemical can be calculated by using the data from the one dose group to estimate the model's one parameter:

$p(x) = 1 - e^{-qx}$ (One-hit model)
$0.10 = 1 - e^{-10q}$ (Fitting model to data)
$q^* = \ln(1 - 0.10)/(-10) = 0.0105$ (MLE for parameter q)
At x = 1 ppm: $p(x) = 1 - e^{-0.1005} = 0.0104$ (MLE for risk at x = 1ppm)

But if a Weibull model is postulated instead, with

$p(x) = 1 - \exp(-qx^b)$ (Weibull model)

then there are an infinite number of combinations of the two parameters (q, b) that will exactly fit the data. Hence, the correct member of this family cannot be identified from the data alone. Biological knowledge or assumptions may break this impasse. For example, if it is found that mode of action of the chemical involves breaking both strands of DNA in certain target cells within a sufficiently short time of each other, then it might be conjectured that the exponent parameter should have the value $b = 2$.

More generally, statistical theory does not by itself provide a sufficient basis for specifying or selecting a parametric model that describes the true dose-response relation. The curve-fitting paradigm requires selecting specific model forms (or families of models) instead of acknowledging that multiple statistical models are often consistent with available data and knowledge, with the true one being unknown. It provides no principled way to synthesize risk estimates from multiple plausible model forms or to quantify remaining model uncertainties.

Bayesian model-averaging, discussed in Section 3, can help to overcome some of these difficulties by incorporating model uncertainty into statistical risk estimates. Reporting the sensitivity of risk estimates to plausible variations in the data *and* in modeling assumptions is increasingly recognized as a requirement for credible risk assessment, and many practitioners now provide ranges or sets of plausible risk estimates, given the relevant data and model uncertainties.

The modeling techniques introduced in Chapter 2, such as compartmental flow models and applied probability models, can also help to reduce model uncertainties by applying biological knowledge to constrain and suggest model forms which can then be fit to experimental or epidemiological data by statistical methods.

Uncertainties about dose-response relations for particular chemicals are often gradually reduced over a period of years as more data and toxicological knowledge are accumulated. Early models are refined, or if necessary replaced, in light of new knowledge. Steps 4 and 5 in Table 1 refer to the process of testing and refining models in light of data, but formal methods for learning and refining models based on data and knowledge are still very active topics of research.

Example: Refining a Risk Model with New PBPK Knowledge and Data

Pharmacokinetic models, especially PBPK models, are often used to refine previous statistical dose-response models by adding knowledge of relevant metabolites and of relevant tissue-specific or organ-specific dose histories. The general template for such a refinement involves replacing a purely statistical dose-response model that can be diagrammed as:

$\{x\} \rightarrow \{p\}$ (Statistical model graph)

with a refined model that can diagrammed as:

$\{x\} \rightarrow \{y\} \rightarrow \{p\}$ (PBPK model graph)

In these diagrams, $\{x\}$ is an abbreviation of $\{x(t), t \geq 0\}$. It denotes an administered dose history. $\{y\}$ an internal dose history of one or more metabolites in one or more physiological compartments (e.g., tissues or organs), and $\{p\}$ a response probability history. The interpretation of the PBPK graph is that, for all times $t \geq 0$, the history of x up to time t is assumed to determine the internal dose history (of y) up to time t, and this in turn determines the probability of tumor by time t, i.e., p(t).

The PBPK model graph is a *refinement* of the statistical model graph in that it contains all the same nodes and precedence relations and also the additional node $\{y\}$. The statistical model graph can be obtained from the PBPK model graph by ignoring information about $\{y\}$, e.g., by "integrating out" internal dose information:

$$\Pr(p \mid x) = \Sigma_y \Pr(p \mid y)\Pr(y \mid x).$$

(In general, if one DAG model determines another, but not conversely, then the first may be called "finer than" or a strict "refinement of" the second. The "finer than" relation provides a partial ordering of DAG models.) Chapter 4 discusses the fact that if y can be measured, then the conditional distribution $\Pr(p \mid x, y)$ will be more informative than the conditional distribution $\Pr(p \mid x)$, in the senses that (a) the entropy of $\Pr(p \mid x, y) \leq$ the entropy of $\Pr(p \mid x)$; and (b) The expected utility of a decision that involves choosing a value of x (or a probability distributions for x values) will always be at least as high if it is made by maximizing expected utility with respect to $\Pr(p \mid x, y)$ as if it is made by maximizing expected utility with respect to $\Pr(p \mid x)$ (Laffont, 1989, Chapter 4.)

Refining a statistical model by incorporating internal doses from a PBPK model can substantially change extrapolated risks. For example, when benzene was administered by oral gavage to four dose groups of male B6C3F1 mice (50 mice per dose group), the following fractions of animals developed squamous cell carcinomas: 0 at x = 0 mg/kg/day (i.e., in the control group); 0.08 at x = 25 mg/kg/day; 0.40 at x = 50 mg/kg/day; and 0.74 at x = 100 mg/kg/day. Fitting a multistage model to this data set by MLE leads to a linear-quadratic (hence low-dose linear) model that implies excess risks at all positive dose levels, with a low-dose slope factor close to 0.08/25. However, when a PBPK model is used to estimate the amount of benzene metabolized at different dose levels, the resulting relation between administered dose x and internal dose y is well approximated by the nonlinear regression model $y = 76.4x/(80.75 + x)$ mg/kg/day of benzene metabolites generated from administered dose = x mg/kg/day of benzene. Fitting the multistage model to these y values and the observed tumor rates by MLE yields a purely cubic (low dose nonlinear) model. The multistage model fit to the (x, p) data by MLE, gives extrapolated risks at x = 12 mg/kg/day about double those from the multistage model fit to the (y, p) data by MLE. At x = 1 mg/kg/day, the MLE multistage model fit to internal dose data, i.e., to the (y, p) pairs for the four dose groups, gives estimated risks about 1% as great as those from the MLE multistage model fit to the administered dose data, i.e., to the (x, p) pairs (Cox 1996). These results appear to be robust to the selection of specific metabolites instead of total metabolites and to choice of bone marrow metabolites, hepatic metabolites, or both.

This conclusion for benzene is not unusual. Incorporating internal dose estimates based on measurements, PBPK modeling, or regression modeling into dose-response models often significantly reduces risks extrapolated below the range of the experimental data for positive dose levels (e.g., Reitz et al., 1996). Reductions in risk estimates for humans when PBPK information is incorporated are often based on the fact that humans are predicted to generate smaller concentrations of relevant metabolites in target organs and tissues than animals. However, as demonstrated in this example, using PBPK information may also lead to selection of a different model form (e.g., cubic instead of linear-quadratic), thus dramatically changing low-dose risk estimates.

2.4 Interpret Results

The final step in a statistical risk assessment is to state, discuss, and interpret its results. Here, the investigator presents conclusions about the risk implications of the data as revealed by the assessment. Part of this step consists of identifying statistically significant exposure-response relations (for hazard identification) and giving confidence intervals for model parameters and predictions, such as for the low-dose slope of the estimated dose-response relation. In addition, authors commonly suggest causal interpretations for associations and recommend risk management strategies based on the causal interpretations.

Perhaps surprisingly, strikingly different interpretations of the same data are common in applied risk assessments, especially in epidemiology. Different investigators can reach opposite conclusions starting from the same data. Table 5 gives representative examples for suspected lung carcinogens.

Explanations for such conflicting interpretations are not hard to find. One of the most common is that *different analysts make different modeling choices and assumptions*. These may be about the model forms to be considered, the variables to include, confounders to control for and methods used to control for them, treatment of measurement errors and missing data, coding of variables, modeling of unmeasured variables, and hypotheses to test. Ideally, reported conclusions should be explicitly stated as *contingent* on these modeling choices, as well as on the data analyzed. Sensitivity analyses should be used to study how sensitive the conclusions are to modeling choices (Greenland, 1996). However, in practice, uncertainties in modeling assumptions and model-based conclusions, are often not quantified or reported (e.g., Concato et al., 1993; Bramwell and Williams, 1997).

The availability of sensitivity analysis and other methods mentioned earlier makes it possible to perform and report more objective risk assessments. These show the potential for conclusions to change as assumptions are varied within reasonable ranges. Doing so would reduce the types of inconsistencies illustrated in Table 5.

Table 5: Contrasting conclusions for suspected lung carcinogens

Agent	Positive conclusions	Negative conclusions
Benzene	"A large cohort of 74,828 benzene-exposed and 35,805 non-exposed workers... in China was followed to determine mortality from all causes. ... **Deaths due to ... lung cancer** (ptrend = 0.01) **increased with increasing cumulative exposure to benzene**." (Hayes et al., 1996)	"Workers employed in a wide variety of industries were included in the Chinese benzene study, and were exposed to not only benzene but also a wide range of other industrial chemicals. **To attribute any or all health effects observed in the exposed cohort to benzene without examining other concomitant exposures is not appropriate.**" (Wong, 1999)
Diesel exhaust	"**A relative risk of 1.45 (95% CI = 1.11, 1.89) for lung cancer was obtained** in the group of workers 40 to 44 yr of age in 1959, the group with the longest possible duration of diesel exposure." (Garshick et al., 1988)	"A... dose-response trend in lung cancer mortality with increasing duration of exposure found by [Garshick et al.] was not present when age was controlled more carefully and years of exposure quantified more accurately. Instead, **a negative dose-response trend for lung cancer was seen** ... based on either duration of exposure or quantitative measures of cumulative exposure." (Crump, 1999)
ETS	"**Exposure to environmental tobacco smoke (ETS) is an established cause of lung cancer**" (Merletti et al., 1998)	"**It seems impossible to conclude with any certainty that ETS causes lung cancer.**" (Lee, 1998)
Formald e-hyde	"The analysis... by Blair et al. in 1986 failed to allow properly for the Healthy Worker Effect and to evaluate time integrated exposure and length of exposure simultaneously. In our reanalysis of the same data **we find a risk for lung cancer, increasing with increasing cumulative exposure to formaldehyde**." (Sterling and Weinkam, 1989)	"A recent meta-analysis by Blair and his co-workers stimulated the undertaking of a reanalysis of the epidemiologic evidence of an association between formaldehyde and respiratory cancer. ... In close agreement with the original analysis, the **aggregated evidence did not indicate an excess risk or an exposure-response gradient for lung cancer**." (Partanen, 1993)
Radon	"**Residential exposure to radon is an important cause of lung cancer** in the general population." (Pershagen, 1994)	"**Indoor radon exposure does not appear to be an important cause of lung cancer**." (Auvinen, 1998)
Vinyl chloride	"Vinyl chloride, a confirmed carcinogen, has been associated with, among other malignant tumors, a **significant increase in the incidence of lung cancer.**" (Lilis, 1981)	Nine retrospective mortality studies of workers exposed to vinyl chloride were reviewed... **For lung cancer, the data were not consistent with an etiologic role** in that two studies with very high power yielded negative results. (Beaumont and Breslow, 1981)

Multiple Comparisons Biases

A second reason for conflicting conclusions is the very widespread occurrence of *false positives* in epidemiological research. Specifically, conclusions driven by *multiple testing bias,* as well as misinterpretations of small p-values as necessarily indicating statistically (or clinically) significant associations, are surprisingly common in published epidemiological risk assessments (Beck-Borholdt and Dubben, 1994; Ludbrook, 1998; Ottenbacher, 1998). This creates an excess (greater-than-nominal) rate of false positives, i.e., of random patterns of exposure-response associations incorrectly identified as statistically significant. These associations cannot be reproduced reliably from study to study since they do not represent true cause-and-effect relations. Moreover, to the extent that conclusions are based on false positives arising from multiple testing, different investigators may report different findings even from the same data set because they use it to test different sets of hypotheses.

Multiple testing bias occurs when an investigator tests the same hypothesis – typically, that exposure to a substance is positively associated with risk of an adverse health response – in multiple subsets of subjects independently, using standard tests and significance levels. Related biases arise when multiple hypotheses are tested using the same data set. If multiple sub-populations are examined for possible exposure-response associations using multiple health response endpoints and exposures, then multiple testing bias will tend to create false positives unless it is corrected in the analysis and interpretation of results. A similar logic holds when different exposure metrics, models, response definitions, etc. are tried and only positive results are reported.

Example: Multiple Testing Bias Produces Excess False Positives

The logic of the multiple testing bias is simple. Standard tests and estimates are usually based on a 5% significance level, i.e., a 5% cap on the probability of "Type 1 error", assuming that the selected statistical model is correct. A Type 1 error, also called a *false positive,* is defined as an error in which the "null hypothesis" of no relation between exposure and response is incorrectly rejected in favor of the "alternative hypothesis" that exposure is associated with response. Now, suppose that a statistical test with a 5% significance level is applied to many subsets of subjects in a study. Then, on average, up to 5% of the tests might be expected to yield false positives. (If model selection or other modeling biases operate, then the true rate of false positives may differ from this nominal rate based on the assumption that the model is correct.) So, for every 20 subgroups examined, one might expect one false positive. For example, in a study with six age groups, four exposure groups (e.g., home *vs.* out-of-home exposure and heavy *vs.* light exposure), three diagnostic categories (e.g., for types of lung cancer), and two sexes of subjects, there are 144 subgroups. Thus, about 7 false positives might be expected if a test with a

5% probability of false positives (the "p-value" of the test) is applied independently to each group.

If the hypothesis being tested is that there is a positive association between exposure and risk, one might expect it to be confirmed by chance alone in several different groups. Varying the boundaries of the age, exposure, and response categories considered – or trying out multiple exposure metrics, statistical model-forms or subsets of variables – as part of a search for sub-groups with significantly elevated risks, exacerbates the multiple testing problem.

Multiple testing bias is well recognized in epidemiological methodology and has often been warned against. However, because it is so important in applied risk assessment, it is worth surveying some of these past warnings. The following excerpts from the literature are representative.

Recognition Of Multiple Testing Bias In The Literature

- "When examining subgroups of data, care should be taken not to turn it into a 'fishing expedition' in which significant subgroups are highlighted at the expense of the totality of the evidence. Subgroup analyses exacerbate the multiple comparisons problem." (Shore et al., 1992)
- "Epidemiologic and public health researchers frequently include several dependent variables, repeated assessments, or subgroup analyses.... These factors result in multiple tests of statistical significance and may produce Type 1 experimental errors." (Ottenbacher, 1998)
- "Multivariate analyses often impress with extremely low p-values, but frequently these results turn out to be non-significant when the influence of multiple testing is considered. As a consequence the actually relevant results might be overlooked, because of the large number of spurious results." (Beck-Bornholdt and Dubben, 1994)
- "When many statistical tests are performed simultaneously, the overall chance of a type I error (incorrect rejection of a true null hypothesis) can substantially exceed the nominal error rate used in each individual test." (Parker, 1998)
- "The inflationary effect of multiple comparisons has been repeatedly pointed out to biomedical investigators in the journals they may be expected to read. Yet, when statistical reviewers of manuscripts criticize authors for not having taken this into account in analysing their experimental results, the authors often react with surprise or, even, disbelief." (Ludbrook, 1998, p. 1033)

A problem very similar to multiple testing bias is *multiple comparisons bias,* in which an investigator examines many possible effects instead of many possible subsets of the target population – but without correctly adjusting the p-values used for each individual test. Multiple comparisons bias also tends to increase the rate of false positives. An indicator that multiple comparisons bias is present is that different investigators reporting significant exposure-response associations find the associations in distinct response categories.

General guidelines to help practitioners avoid multiple testing bias include the following (Feinstein, 1988; Shore et al., 1982):

a) Specify the research hypotheses to be tested and the coding methods to be used in recording the data in advance of data collection. Research hypotheses may be based on the results of a separate earlier exploratory effort, sometimes called a "hypothesis-generating" study, but should be tested on new data with procedures specified before collecting the data. If apparent associations detected in an initial study are only due to the way in which category boundaries are selected, for example (another form of multiple-testing bias) then following this guideline makes it unlikely that the same category boundaries would yield the same apparent associations in a follow-up study.
b) Avoid data-dredging and multiple testing during the data analysis. Test *only* the specific hypotheses identified in advance, and resist the temptation to search for other patterns. (Note that this recommendation applies to *hypothesis-testing* studies that might be used to support a decision or to provide evidence for or against the hypothesis of a causal relation between exposure and response. It does not apply to exploratory and *hypothesis-generating* data analyses.)

It is often desirable to use one data set to test multiple hypotheses, and the threat of multiple testing bias should not deter such investigations. But, it does necessitate making some statistical adjustments (e.g., in calculating p-values for hypothesis-testing results) before the statistical significance of results can be correctly calculated and interpreted.

If multiple tests are performed during the data analysis, then appropriate technical methods are available to estimate and remove the effects of multiple testing and multiple comparisons biases. These methods make it practical to follow the methodological recommendation that "Methods to determine and correct Type 1 errors should be reported in epidemiological and public health research investigations that include multiple statistical tests" (Ottenbacher, 1998).

Effective approaches for eliminating the effects of such biases include the following:

Methods to Avoid Multiple-Testing Bias

- *Estimate the inflation in probability of a Type 1 error* due to multiple testing bias. Then, correct reported results by this factor (Ottenbacher, 1998)

- *Adjust reported significance levels (p-values) of individual tests* to obtain the desired nominal significance level for the whole study. Many appropriate technical methods for making such adjustments are now available (e.g., Aickin and Gensler, 1996; Blair et al., 1996; Ludbrook, 1998; Sarkar and Chang, 1992; Troendle, 1996; Westfall, 1997; Zhang et al., 1997). *Step-down* methods avoid the need to increase false negatives as false positives are reduced.

- *Apply empirical-Bayes adjustments* (Greenland and Robins, 1991). These estimate the potential extent of multiple testing in a population and apply it to correct the results of individual tests.

- *Use a mixture of distributions* to test whether the observed p-value provides evidence of a false null hypothesis (Parker and Rotherberg, 1988).

- *If possible, test a single global hypothesis (e.g., "no association") instead of making multiple pairwise comparisons* (Ludbrook, 1998).

Advances in the past 10 years often make it unnecessary to increase probabilities of Type 2 errors (failing to find an association that is there) in order to reduce Type 1 error probabilities (false positives). One key idea is to use Monte-Carlo simulation to solve for the needed adjustments exactly (Blair, 1996; Westfall, 1997) instead of using the approximate probability bounds (known as Bonferroni or Sidak bounds) that used to be the main practical adjustment method. Another key idea is to use "step-wise sequentially rejective" procedures (reviewed in Ludbrook, 1998). These rank-order multiple hypotheses based on their uncorrected p-values and then correct each p-value in turn, using simple formulas such as $p^* = 1 - (1 - p)^m$ or $p^* = mp$, where p^* is the corrected p-value, p is the original (nominal) p-value, and m = number of hypotheses with uncorrected p-values at least as large as the one being corrected. (These simple correction formulas are derived from the Sidak and Bonferroni inequalities, respectively.)

In light of the simplicity and power of these modern methods, the burden of quantifying the likely effects of multiple testing bias – and of showing that it does not provide a plausible explanation for reported associations, if this is the case – should be borne by the data analyst. If such tests are not performed or are not reported, then the reported associations should be interpreted and identified as exploratory only, meaning that they must be confirmed by further study before being interpreted as providing evidence useful for decision-making. This has been also been described by saying that studies that report "significant" associations based on p-values not corrected for multiple testing should be viewed as hypothesis-generating rather than hypothesis-testing (Beck-Bornholdt and Dubben, 1994).

Reporting and Interpretation Biases

There are obvious logical and rhetorical errors that should be avoided in interpreting and reporting the results of statistical risk assessments. These include:

- *Interpreting statistical associations as causal relations* without testing for or refuting other possible interpretations (e.g., multiple testing bias or modeling biases due to model selection, variable coding, selection and information biases, confounding, and so forth). Many statistical risk assessments draw conclusions and recommend risk management actions based only on statistical associations, without discussing plausible non-causal interpretations. Chapter 4 discusses tests for causation that can help to separate true cause-and-effect relations from non-causal ones.
- *Stating conclusions that do not follow from the data,* or whose negations are just as strongly supported by the data.
- *Over-generalizing* from the data and analysis. For example, an investigator might test for effects of one or a few confounders, find that they do not explain away an observed exposure-response association, and then state that confounding in general does not provide a plausible explanation for the observed association – even though other possible confounders have not been addressed.
- *Hidden assumptions.* Conclusions are typically contingent on model assumptions as well as on the data collected. This contingency should be acknowledged when conclusions are stated. More constructively, conclusions should be presented in the context of realistic sensitivity analyses and of presentations of uncertainties (e.g., confidence bands) that include model uncertainties.
- *Gratuitous interpretations.* Authors often offer suggested interpretations for results that seem anomalous with respect to pre-existing expectations or previous papers. If these interpretations are purely speculative, without foundation in the data presented, then they may not help to clarify the truth.

Example: Interpreting Dietary Confounding for ETS

Fontham et al. (1994) considered whether poor diet might confound the relation between ETS and lung cancer risk by estimating "dietary cholesterol intake and an index of dietary antioxidant consumption based on weekly consumption of fruits and vegetables and supplemental vitamin use". From this limited basis, they drew the logically impossible generalization that "These observations indicate that the strong association in this study between adult ETS exposure and lung cancer cannot be attributed to *any* likely confounder" (emphasis added). However, the study did not consider other likely confounders for ETS and lung cancer, such as consumption of fried meats, alcohol, and beer; or socioeconomic variables associated with both lung cancer risk and ETS exposure, such as income, and blue-collar *vs.* white-collar employment. Its stated conclusion is therefore an overgeneralization.

Example: Interpreting Positive and Negative Associations for ETS

A large study of human health risks from environmental tobacco smoke (ETS) by Boffetta et al. (1998) begins its discussion section as follows. "The results of our study of the risk of lung cancer from ETS in several European countries showed a reduced risk of exposure during childhood and a measurable effect of exposure to ETS from the spouse and at the workplace." This is based on a statistically significant reduction in lung cancer risk among adults ever exposed to ETS in childhood (OR = 0.78, 95% CI = 0.64-0.96) and non-significant increases in lung cancer risk among adults ever exposed to spousal or workplace ETS (OR = 1.16, 95% CI = 0.93-1.44 for spousal ETS; OR = 1.17, 95% CI = 0.94-1.45 for workplace ETS. None of these estimates is adjusted for model uncertainty.) The authors interpret these non-significant associations as “a measurable effect” but dismiss the result for childhood exposure to ETS as follows: "In the light of the inconsistent findings of other studies, our results on childhood ETS exposure can be plausibly interpreted as sampling fluctuation around a relative risk of 1 (no effect) and do not allow us to conclude that ETS exposure during childhood is protective against lung cancer." This risk analysis suffers from uncontrolled multiple testing bias, so sampling variability could indeed explain some false positive associations (including a perhaps spurious negative association between childhood exposure to ETS and subsequent lung cancer risks.) However, offering this interpretation for negative associations but not positive ones seems gratuitous.

Similarly, Brownson et al. (1992) summarize their key findings from a case-control study of ETS as follows. "No increased risk of lung cancer was associated with childhood passive smoking exposure. Adulthood analyses showed an increased lung cancer risk from lifetime nonsmokers with exposure of 40 pack-years from all household members (odd ratio OR = 1.3; 95% confidence interval CI = 1.0, 1.8) or from spouses only (OR = 1.3; 95% CI = 1.0, 1.7)." Based on these results, they conclude that "Ours and other recent studies suggest a small but consistent increased risk of lung cancer from passive smoking."

This summary omits the finding of a statistically significant *negative* association between childhood ETS exposure and lung cancer (OR = 0.7, 95% CI = 0.5, 0.9 for "ever" *vs.* "never" exposed in childhood.) Among 24 odds ratios reported for childhood ETS exposure in different exposure categories (*ibid.,* Table 1, p. 1527), 6 were statistically significantly less than 1, 10 others were less than 1, though not significantly so, and none was greater than one. Collectively, these results constitute evidence of a negative association. Brownson et al. (p. 1526) describe them as follows. "There was little evidence of increased lung cancer risk associated with passive smoke exposure in childhood." This characterization interprets the results as showing "lack of association" when in fact they suggest a negative association.

For adult exposure to ETS, Brownson et al. begin their "Dicussion" section with a causal interpretation, as follows. "Our study suggests that exposure to high levels of environmental tobacco smoke in adulthood increases the risk of lung cancer in nonsmokers. Exposure of more than 40 pack-years duration increased the risk of lung cancer among nonsmokers by approximately 30%. ...Our findings are similar to those of another large study of lung cancer in nonsmoking women [by Fontham *et al.*] that identified an OR of approximately 1.3 due to exposure to greater than 40 pack-years of spousal smoking."

The findings behind these conclusions tell a different story. If only confidence intervals that fall entirely to one side of 1 (i.e., nominally statistically significant results) are considered, then there are only two "significant" results. Both show odds ratios significantly *less* than 1. (They occur among all subjects exposed to 0-50 and 50-175 pack-years x hours/day, respectively.) If confidence intervals that include 1.0 as one endpoint are counted, then the number of "significant" results increases to 13. Of these, 8 are negative associations between ETS exposure and lung cancer risk, while 5 are positive associations. One of these is the "30% excess risk number" cited by the authors in summarizing and interpreting their results. The highest exposure category has a barely significantly elevated risk, according to the author's calculations (uncorrected for model uncertainty). But the use of discrete categories to code a continuous exposure variable (pack-years), coupled with a significant negative association between ETS and lung cancer in the next lower exposure category, raises the possibility of multiple testing bias from selection of category boundaries. This possibility was not discussed.

Brownson et al. conclude "In summary, our study and others conducted during the past decade suggest a small but consistent elevation in risk of lung cancer in nonsmokers due to passive smoking. The proliferation of federal, state, and local regulations that restrict smoking in public places and work sites is well founded." This policy conclusion is offered despite the fact that, according to the study's own Results section, "In general, there was no elevated lung cancer risk associated with passive smoke exposure in the workplace." The findings in this study show substantially more negative associations than positive ones between ETS exposure and lung cancer. Overall, the data could equally well have been used to support a conclusion that ETS poses zero or negative risk of lung cancer, especially since the only two clearly statistically significant results indicate negative associations. This is not obvious from the stated conclusions and discussion. In fact the Brownson et al. paper has often been cited as supporting the hypothesis that ETS exposure increases risk of lung cancer, even though all of its clearly statistically significant results support the opposite conclusion.

Chapter 4 considers methods for formally testing causal interpretations. They can reduce the potential to unintentionally place unjustified interpretations on data.

3. PROGRESS IN STATISTICAL RISK MODELING

Advances in statistics and statistical computing over the past twenty years have led to new algorithms and methods that can dramatically improve the practice of statistical risk assessment (e.g., van Houwelingen, 1997). This section summarizes key ideas that are revolutionizing biostatistics and statistical risk modeling. They overcome many of the limitations of the traditional parametric statistical risk-modeling framework in Table 1.

Table 6 lists some of the most important current computational statistical methods being used in biostatistics and risk assessment. These algorithms provide practical methods for statistical inference from incomplete and error-prone observations, including the types of data commonly encountered in health risk assessments based on epidemiological and clinical data. Several key themes cut across them. For example:

- *Cross-validation, bootstrap estimation, multiple imputation,* and other *re-sampling* techniques all create *multiple sampled data sets* by repeatedly sampling from the original data or from its (perhaps smoothed) empirical distribution function. The multiple sampled data sets can be used for variable-selection (in what fraction of them does each explanatory variable appear to be significant?) model-selection (which models fit each sampled data set best, and what fraction of the sampled sets thus support each model?), parameter estimation, and uncertainty analysis (what is the frequency distribution of parameter estimates across sampled data sets?). Using multiple samples rather than estimated parameters to quantify probability distributions for model predictions and performance statistics allows one to avoid making parametric modeling assumptions.
- *Gibbs Sampling* and other *Markov Chain Monte Carlo (MCMC) algorithms* provide for efficient iterative numerical estimation of conditional and Bayesian posterior multivariate CDFs by approximating hard-to-calculate joint distributions as limits of sequences of samples from easily calculated conditional distributions.
- *Data Augmentation* and *EM algorithms* make use of a self-consistency principle to estimate both missing data and model parameters. These algorithms are iterative. At any iteration, the current estimated model determines the likelihood of missing data, while the complete data (approximated as the union of the observed data and the model-estimated values of the unobserved data) provides the information needed to estimate model parameters. Iterating model-estimation (given the estimated data) and data-estimation (given the estimated model) can create a mutually consistent estimate of the model and the missing data.

Table 6: Key Computational Techniques for Statistical Risk Assessment

Technique	**Key Ideas**
Bootstrap (Politis et al., 1999)	Sample repeatedly from the empirical distribution function of the data (or a smoothed version it). Use the multiple sampled values to estimate the distributions of quantities of interest.
Bootstrap variable selection	Apply variable-selection techniques to many random samples from the data. Then, keep the variables that are most frequently selected in those samples. (Sauerbrei and Schumacher, 1992)
Cross-validation for model evaluation (Harrell, 1996)	Delete each data value in turn. Use the model and remaining data to predict the deleted values. Repeat for all data points (or for many random subsets). Use the distribution of predicted *vs.* true values to estimate the model's predictive accuracy.
Data augmentation (Schafer, 1997)	1. Given a current guess at parameter values, q_t, sample missing data values from: $Pr(D_{missing} \mid D, q_t)$. (D = observed data.) 2. Draw a new sample value of q from its complete-data posterior, $Pr(q \mid D, D_{missing})$, using the sampled values of $D_{missing}$ from Step 1. Call this new estimate q_{t+1}. 3. Iterate Steps 1 and 2.
EM algorithm (Expectation-Maximization algorithm)	0. Guess at parameter values and unobserved data values. 1. Then use complete data (= observed data + guessed-at values for missing data) to re-estimate the parameters. 2. Use the new parameter values from Step 1 to re-estimate the missing data. 3. Iterate Steps 1 and 2.
Gibbs sampling (Gamerman, 1997)	Given a multivariate joint distribution $Pr(x_1, x_2, \ldots, x_n)$, sample n-vectors from it as follows. Repeatedly cycle through all n variables, sampling the value of each one from its conditional distribution given the most recently sampled values of the rest. Repeating many times generates a sequence of n-vectors that, under mild regularity conditions, form a Markov chain converging to $Pr(x_1, x_2, \ldots, x_n)$ as its stationary distribution.
MCMC (Markov Chain Monte Carlo) (Carlin and Louis, 2000; Shafer, 1997)	A generalization of Gibbs sampling for sampling from a multivariate (often, high-dimensional) joint probability distribution by sequentially sampling its components from easily-sampled distributions. The components are sampled (typically using a transition kernel and an acceptance rule) so that their sampled values create a Markov chain whose steady-state distribution is the joint probability distribution of interest.
MLE (Maximum likelihood estimation).	Given a parametric model $y = f(x ; q)$ with parameters q, and a data set $D = \{(x_i, y_i), i = 1, \ldots, N\}\}$, choose q to maximize the likelihood of the data: $L(q) = \Pi_i Pr(y_i \mid x_i, q)$.
Resampling methods (Politis et al., 1999)	Methods that generalize the Bootstrap by creating many random subsets of a data set and basing estimates or tests of hypotheses on sample distributions across these "replicates".
Smoothing algorithms	Pass a smooth curve through data points, e.g., via splines, kernel smoothers, wavelets and series expansions (Eframovich, 1999).

These basic ideas and techniques are often used together. For example, cross-validation may be used to optimize the parameters of smoothing procedures in fitting smooth non-parametric curves to scatter plots (Cummins et al., 2000). Conceptually, for each value of the parameters, (e.g., the penalty weight for "roughness" of the fitted curve), the performance of the fitted curve can be evaluated in each of the cross-validation samples. Then, the parameter values that lead to the best estimated performance can be selected. MCMC can be used to compare and choose between two alternative nested families of parametric models based on the posterior distribution of the discrepancy (minimized Kullback-Leibler distance) between them (Goutis and Robert, 1998). Gibbs sampling can be used to estimate the Bayes' factors in Bayesian model-averaging (BMA) analyses of model uncertainty, discussed below (Chib and Jeliazkov, 2001). In general, the key ideas of smoothing, resampling from data, iteration, and sampling from estimated distributions as a computational technique can be combined in many ways to constructively address statistical challenges.

Two of the most important advances for statistical risk assessment are techniques for including or avoiding *model uncertainty;* and methods for dealing with *missing data,* including unobserved variables and measurement errors, which can be interpreted as making the corresponding true variables unobserved. The following sections address these topics.

3.1 Dealing with Model Uncertainty and Variable Selection

This section summarizes several constructive methods for quantifying and dealing with uncertainties about risk models. For ease of reference, the following list summarizes the main approaches to dealing with model uncertainty in statistical risk modeling.

Methods to Avoid and Mitigate Model-Selection Bias

1. Present **formal model diagnostics** (Greenland, 1989)) as well as results of model validation and goodness-of-fit tests (e.g., Qin and Zhang, 1997; Wilbert et al., 1997). These tests and diagnostics address whether models (typically composed of model forms, variables, and parameter values) are obviously inconsistent with the data being modeled.
2. Use the data to **select a specific model form** from a broad range of alternative model forms (Maldonado and Greenland, 1996)
   Example: Use data to estimate wavelet coefficients (Antoniadis et al., 1997)
3. Perform and report **sensitivity analyses** for model-form uncertainty.
   "Attempt to fit a complete family of models in order to see what conclusions seem to follow regardless of the model one selects. ... At a minimum, one should validate modeling results by seeing whether similar results [hold for] simple stratified analysis." (Greenland, 1989)

4. Use **formal model-selection criteria** (e.g., the Akaike Information Criterion, AIC, or the Bayesian Information Criterion, BIC) to select a model form from a broad class of models (e.g., generalized linear models) (Lindsey and Jones, 1998; Goutis and Roberts, 1998. Chapter 4 discusses the BIC further in the context of selecting causal graph models for use in Bayesian Model-Averaging.)
5. Use **model cross-validation** to select a least-biased model.
6. Use **nonparametric regression** procedures (classification trees, neural nets, splines, robust binary nonparametric regression, etc.) to estimate unknown functions linking independent variables to response probability. (Denison et al., 1998;Ye, 1998) to avoid selecting a parametric model.
7. Use **Bayesian model-averaging** (Raftery et al., 1997; Carlin and Louis, 2000) for logistic regression, linear, proportional hazards, and other models to avoid assuming any single model.
8. **Optimally combine predictions** from multiple regression models (possibly non-parametric as well as parametric) using adaptive weighting schemes (Yang, 2001).

They address the following two questions:

a) How can one *quantify uncertainty in model predictions* while taking into account uncertainty about correct model forms? (Recall that traditional statistical risk modeling quantifies confidence intervals for model parameters and risk predictions by assuming that the model form is correct – ignoring model uncertainty – and then estimating sampling error based on this assumption. Something else is needed to capture the uncertainty due to the assumption of a model form.)
b) How can one *reduce or avoid biases* in risk estimates, confidence intervals, and model diagnostic tests due to the selection of specific model forms and/or estimation of model confidence intervals from the same data used to estimate model parameters?

Two related principles for estimating and correcting potential biases, including model-selection biases, are bootstrapping and model cross-validation. *Bootstrap resampling,* as it is applied to model-form selection (Hjorth, 1994; Stute et al., 1998), draws many random samples from the empirical cumulative distribution function of the data (or, a smoothed version of it, in some implementations). For each random sample, a large set of models (e.g., all polynomials with a specified maximum degree) are fit to the data. Model parameters are estimated and risk estimates and model diagnostics are quantified for each sample. The frequency distributions of the model parameters, effects estimates, and model diagnostic tests based on these many resampled data sets are used to estimate the biases expected in these quantities for the model fit to the complete data set.

Model cross-validation (Harrell et al., 1996) can help select model forms as well as model variables. To evaluate the predictive performance of a proposed model, cross-validation selects many subsets of the data (which may be interpreted as "training" sets), e.g., all data points but one, with each data point in turn left out. The model is fit to each training set. Its predictive accuracy, i.e., how well the fitted model predicts the left-out values, is assessed across all training sets. The frequency distribution of results across subsets indicates the probable performance of the model. Cross-validation can help overcome model-selection biases that lead to overly narrow confidence intervals (and therefore perhaps to excess false positives) when model uncertainty is ignored and the same data are used both to estimate model parameter values and assess goodness of fit and average prediction error. These ideas can be extended to estimate and correct model-selection biases in model-based risk estimates and confidence intervals.

When the correct model form is unknown or uncertain, non-parametric techniques may be used to avoid artificially restricting models to members of a parametric family. Examples include classification trees (Ye, 1998) and algorithms for fitting smooth curves or surfaces directly to data. The *nonparametric bootstrap* avoids models altogether by creating many random subsets of the original data set and estimating quantities of interest from their sample values in these subsets. However, this strategy does not work when the quantities of interest lies outside the range of data.

Bayesian model-averaging (BMA) acknowledges that more than one model may be plausible, given available data and knowledge. It therefore uses predictions from multiple plausible models. Model-specific predictions are weighted by *Bayes factors* that reflect their relative plausibilities given the data. These factors are ratios of marginal likelihoods or of posterior odds to prior odds for different models. MCMC algorithms can be used to calculate them even in complex situations for logistic, multivariate probit, Poisson regression, and other models (Chib and Jeliazkov, 2001). Considering results from multiple plausible models – or optimally combining them via weighting schemes (Yang, 2001) – can give more realistic estimates of uncertainty in model-based results than confidence intervals from any one model (Raftery et al., 1997). Since model uncertainty often dominates sampling variability and parameter uncertainty in risk assessments, such techniques for dealing with model uncertainty are especially relevant for risk assessment.

Many of these same ideas apply to the problem of selecting the variables to include in a multivariate risk model from a large set of variables. However, detailed algorithms for variable-selection differ from those from model-form selection. The main techniques are summarized in the following box for ease of reference.

Variable-Selection Methods that Avoid Variable-Selection Biases

- **Bootstrap resampling** (from empirical CDF): Include variables in the model if and only if they are selected in a large fraction of the bootstrap replications (Sauerbrei and Schumacher, 1998). [Similar ideas are used for final variable selection via classification trees, which can otherwise be quite unstable, under the name "bagging" (Breiman, 1996).]
- **Bayesian Model-Averaging.** Use the data to weight risk predictions from different possible subsets of variables. (This may require MCMC methods) (Brown et al, 1998; Raftery et al., 1997)
- **Model cross-validation for variable selection** (Ronchetti et al., 1997). Identify which subset of variables best predicts the values of all known data points. Predictive accuracy is averaged over all known data points (treating other data points as a "training set".)
- **Formal variable-selection methods (e.g., AIC, BIC) and sensitivity analyses.** See Greenland, 1989 and 1996; Maldonado and Greenland, 1993 for basic sensitivity analyses for selecting confounders.
- **Bayesian variable selection** (e.g., Ibrahim and Chen, 1997)
- **Model comparisons** based on generalized degrees of freedom can help to quantify and correct for the effects of variable selection. (Ye, 1998)
- **Modified Akaike Information Criterion (AIC)**. Candidate models corresponds to different subsets of independent variables (Ye, 1998).

These techniques can improve selection of variables and reduce false positives compared to naïve (e.g., standard stepwise regression) approaches. However, a fully satisfactory approach to variable-selection often requires causal modeling of the relations among variables. Chapter 4 discusses causal graph models and causally-motivated selection of variables.

3.2 Dealing with Missing Data: New Algorithms and Ideas

Often, one or more explanatory variables are measured with error, are not observed at all, or contain missing data values. Then the traditional framework must be modified. Modern computational statistics provides techniques for estimating parameters and model functional forms from incomplete, possibly incorrect (noisy) data values.

Several statistical algorithms now make use of a *self-consistency principle* to iteratively refine models and model-based risk estimates. An initial model and set of observed data may be used to guess at values of unobserved quantities, e.g., by choosing their estimated values to maximize an approximate likelihood function constructed from the initial model and the observed data. The resulting estimates of the unobserved values, together with the observed data, are then used to refine the model by re-estimating its parameters to increase consistency with the observed and guessed-at data values. Next, the refined model is used to re-estimate the unobserved data values. This cycle is repeated until a convergence criterion is met. It produces a sequence of estimates of both the unobserved data values and the model parameters. This is the basic idea behind the famous *Expectation-Maximization (EM) algorithm* for estimating missing data values together with parameters of interest. Table 7 outlines the algorithm. Its mathematical theory establishes its convergence properties and gives conditions under which the final estimates constitute maximum-likelihood estimates of the model parameters (Shafer, 1997).

Other computationally intensive methods for dealing with missing data are even more ambitious. For example, the *Data Augmentation algorithm* improves on the EM algorithm by estimating the entire conditional probability distributions for unobserved data values in the presence of missing data. It does so by replacing the E step of the EM algorithms with a sampling step, thus combining the self-consistency idea with the idea of Gibbs sampling (Table 6.) A key idea in most current approaches to handling missing data is to use *multiple imputed values* for each missing value, rather than using any single value (Greenland and Finkle, 1995). Replacing missing values with point estimates, no matter how they are constructed, tends under-estimate uncertainty about parameter estimates and risk estimates. In most settings, imputing a single value for each missing value also creates some biases in the estimated quantities. Imputing multiple values, executing multiple analyses using them, and presenting the frequency distribution of results can help to address the bias problem as well as more realistically depicting uncertainties about model-derived risk estimates.

Specific algorithms and techniques have been developed for missing data on confounders and covariates when the pattern of missingness is uninformative about the omitted values. Examples include generalizations of Joint Estimating Equations (JEE) for linear and logistic regression (Zhao et al., 1996) and techniques analogous to maximum likelihood (e.g., maximum psuedo-likelihood for the proportional hazards model, see Paik and Tsai, 1997). These maximize approximations to or substitutes for the likelihood function that can be estimated from the incomplete data. Even without these techniques, simple sensitivity analyses or more sophisticated regression-based sensitivity analyses and bias-correction adjustments can be used to bound and reduce the impacts of missing data (Lin et al., 1998).

Table 7: Outline of EM Algorithm for Missing Data

INPUTS

1. A *data set* $D = \{(x_i, y_i), i = 1, ..., N\}$
   - x_i = vector of explanatory variables for case i.
   - y_i = vector of responses for case i.
   - Covariates, previously denoted by z_i, are here subsumed into x_i.
   - Some components of x_i and y_i may be missing.

2. A *parametric statistical model,* $Pr(y \mid x ; q)$ = probability of response y in a case with explanatory variables x. This model is often represented via a *likelihood function* for the parameters given the complete data:

$$L(q) = Pr(D^* \mid q).$$

Here, D* is a complete data set obtained from the observed data set D by filling in all missing values. The probability distribution for the complete data set D* given the observed data D and assumed parameter values q can be factored as:

$$L(q) = Pr(D^* \mid q) = Pr(D \mid q)Pr(D_{missing} \mid D, q),$$

where $D_{missing}$ = the missing data values.

PROCEDURE

1. Start with an initial parameter estimate, q_0
2. Given a current parameter estimate q_t, improve it as follows:
   E step: Compute $J(q, q_t) = E[\log L(q) \mid q_t]$
   M step: Find q_{t+1} as the value of q maximizing $J(q, q_t)$.

The E step is accomplished by averaging over all possible completions of the data set, weighted by $Pr(D_{missing} \mid D, q_t)$.

3. Iterate until $q_{t+1} = q_t$ to within an acceptably small tolerance, ε.

OUTPUTS

A sequence of parameter estimates for which the likelihood of the observed data, D, is non-decreasing.

References: Schafer, 1997; Titterington, Smith and Makov, 1985.

3.3 Mixture Distribution Models of Population Risks: Dealing with Unobserved Variables and Variability

This section examines statistical risk models in which the risk analyst is unable to observe one or more of the explanatory variables that determine individual responses to exposures. These "hidden" variables have a perhaps unknown frequency distribution in the population. Examples include individual biochemistry parameters and covariates such as enzyme activity levels or genetic predispositions that vary among individuals and that affect their responses to exposures. They also include exposures to risk factors that affect the health response of interest, such as aspects of diet or lifestyle. The challenge for statistical risk assessment in this context is to isolate and quantify the effects of the exposure of interest on the frequency distribution of individual risks (i.e., response probabilities or response rates) in the population. Typically, this requires drawing inferences about the hidden variables, too.

Finite mixture distribution modeling

Finite mixture modeling, also called discrete mixture distribution modeling, provides a powerful approach for estimating the statistical effects of exposures on population risks when individual risks also depend on unobserved quantities. The conceptual framework is simple. It assumes that the relevant unobserved quantities lead to inter-individual heterogeneity in observed exposure-response relations. (Otherwise, mixture modeling is not needed.) This heterogeneity is approximated by partitioning the individuals into a finite number of groups or *types*. Individuals belonging to the same type are assumed to be homogeneous, i.e., to have the same exposure-response relations, at least to a useful approximation.

Mathematically, this model implies that the probability that a randomly selected individual with exposure x will exhibit response y can be represented by the decomposition

$$\Pr(\text{response} = y \mid \text{exposure} = x) = \Sigma_g \Pr(y \mid x, \text{type} = g)\Pr(\text{type} = g).$$

(Here, both x and y can be multivariate indicators of exposure and responses, respectively.) The importance of this representation is that *it is not necessary to know what the types, g, represent in order to quantify the model.* The groups indexed by g may correspond to phenotypic variations in the population, sub-populations with different sensitivities, groups with different dietary habits and lifestyles, or anything else that differs among individuals and that affects the relation between exposure and response.

Knowing what they represent is not required in order to accomplish the statistical tasks of estimating the two pieces of the representation, i.e.,

- The weights Pr(type = g). These are nonnegative weights summing to 1. Together, they constitute the *mixing distribution*.
- The type-specific exposure-response relations Pr(y | x, type = g). These constitute the *components* of the mixture.

Given a data set of (x, y) pairs for many individuals, possibly with missing data on some components of x and y for some individuals, statistical algorithms can estimate the model under certain assumptions, e.g., that all of the components belong to the same parametric family. The EM algorithm is widely used for this purpose, since the types g may be interpreted as missing (unobserved) data for each individual (Titterington et al., 1985).

In general, more than one mixture distribution model may fit the data equally well. This *non-identifiability* problem, discussed further below, can be resolved by restricting the components of the mixture to specific parametric families. *Gaussian mixtures,* for example, in which all components are assumed to be normally distributed, have led to efficient algorithms for accurately approximating most continuous distributions when y is a continuous variable. The number of needed components (usually only a few), the mean and variance of each component, and the weights of the components, i.e., the mixing distribution, usually can be quickly calculated from data. Such approximations can be useful for computational purposes even if there is no heterogeneity in the population.

Mixture distributions for population survival data

A population risk consists of a set of individual risks, one for each member of the population. If each individual's risk is described by an individual risk model:

$h(t \mid x, z)$ (individual risk model)

and if the covariate vector z has a frequency distribution in the population described by a cumulative joint frequency distribution:

$G(z)$ (population CDF of z)

then the pair

(h, G) (mixture distribution model)

is a mixture distribution model with G as the CDF of its mixing distribution. It is a *continuous mixture distribution model* (rather than a finite one) if G is a continuous CDF and if z indexes a continuum of types. Here, $G(s)$ is the fraction of the population for which $z \leq s$. When there is no need to emphasize the arguments of these functions, i.e., x and z for h and z for G, they may be denoted simply by h and G.

To clarify the relations between individual and population risk concepts, consider individual risk models in which $h(t \mid x, z)$ represents the individual hazard function for occurrence of an irreversible binary event, such as death, that occurs at most once for any individual. Assume that individual responses are statistically independent, and let G now denote the joint distribution function for all the unknown input quantities, namely, z if z is unknown but x is known; or (x, z) if both are unknown.

A population hazard function can easily be calculated from (h, G). The population hazard rate at any time $t \geq 0$ is defined as the fractional rate per unit time at which individuals who have not yet undergone the transition of interest (e.g., death, development of a tumor, etc.) do so at time t. In other words, it is the expected number of transitions per unit time at time t for a randomly selected survivor, i.e., for an individual who has not yet undergone the transition. It is calculated from the following two components:

a) The conditional probability density of z (or of (x, z), if both inputs are uncertain) for a survivor, i.e., an individual who has not yet undergone the transition. Applied probability (or Monte-Carlo simulation, for analytically intractable problems) can calculate this PDF as a function of elapsed time, given the model (h, G).
b) The conditional expected transition rate per unit time, given the values of the input quantities. This is just the individual hazard function, h.

While the analytic form of the population hazard function can be worked out for some specific (h, G) pairs, Monte-Carlo simulation is needed in general. The simplest form is as follows: (a) Repeatedly sample inputs from G; (b) Simulate individual transitions using individual hazard functions conditioned on inputs drawn from G; and (c) Collect the results and tabulate the fraction of survivors who undergo transitions in each time period after surviving any amount of elapsed time, t.

Example: Individual and Population Hazards for Weibull Mixtures

Suppose that each individual hazard function is constant. Then, in general, the population hazard function decreases over time, unless all individuals have identical hazard functions (Lawless, 1982, p. 45). This is because individuals with relatively high hazard rates will tend to be lost (i.e., make the transition) first. Subsequent transitions among survivors (those who have not yet made the transition) tend to take place less quickly as individuals with the smallest hazard rates become an increasing proportion of the surviving population. If the population CDF for the individual hazard rates is known, then it is sometimes possible to solve for the population hazard function analytically. Table 8 presents several examples.

Table 8: Individual and Population Risk Models for Hazard Functions

***h(t | z)* = individual hazard model**	***G = mixing distribution***	***$h_{pop}(t)$ = population hazard model***	***Reference***
h(t | z) = *z*, with *z* a scalar. The hazard function is constant for each individual, but different individuals have different values.	*z* has a gamma distribution in the population. The gamma distribution has two parameters, *a* and *k*.	Generalized Pareto $h_{pop}(t) = ka/(1 + at)$	Lawless, 1982
h(t | z) = *zh(t)*, *z* is a scalar. *z* is called the *frailty* for an individual.	Frailty parameter *z* initially has an inverse Gaussian distribution in the population at time t = 0.	*z* still has an inverse Gaussian distribution among survivors at any later time *t* > 0.	Johnson *et al.*, 1994, p. 291
Frailty model: *h(t | z)* = *zh*, *z* is a scalar. *h* is a constant with the same value for all members of the population, but "frailty" *z* differs among individuals.	Frailty parameter *z* has a gamma distribution in the population.	Number of accidents in the population in [0, t) has a negative binomial distribution.	Prakasa Rao, 1992, p. 178
Contagion model: h(t | n, k) = (k + n)u/(1 + ut) for an individual who has had n accidents in (0, t)	All individuals have the same parameters, k and u, but n is stochastic.	Number of accidents in the population in [0, t) has a negative binomial distribution.	Prakasa Rao, 1992, p. 178
Weibull hazard model: $h(t \mid a, b) = (a/b)(t/b)^{a-1}$	A fraction p of the population has parameter vector (a_1, b_1). The rest have a different parameter vector, (a_2, b_2).	Eight different shapes of the population hazard function are possible. None is bathtub-shaped.	Jiang and Murthy, 1998.

For example, if the population CDF for constant individual hazard rates has a gamma distribution, then the population hazard function has the following form:

$$h_{pop}(t) = ka/(1 + at),$$

in which k and a are parameters of the gamma distribution (Lawless, 1982, p. 49). It is a decreasing function of elapsed time, t. Many such analytic results are available in the literature on reliability engineering. Table 8 summarizes some of them.

As another example, consider a population consisting of a mixture of only two types of individuals, with all members of the same type having identical Weibull individual hazard functions. These have the form:

$$h(t \mid a, b) = (a/b)(t/b)^{a-1} \qquad \text{(two-parameter Weibull hazard function)}$$

where a and b are the shape and scale parameters of the Weibull distribution for the random time until transition (e.g., failure of a system). Eight qualitatively different shapes for the population hazard function are possible (Jiang and Murthy, 1998). Some of the shapes are complex, e.g., first increasing, then decreasing, then increasing again, and finally gradually decreasing and leveling off to approach a horizontal asymptote. However, if all individuals in a population have Weibull individual hazard functions, then the population hazard function can never have the familiar bathtub shape (first decreasing, then leveling off, and finally increasing), even if arbitrarily many different types of individual Weibull hazard functions are allowed in the mixture.

Example: Tolerance Threshold Models

The idea of a mixture distribution can be extended to various types of probability and stochastic process models for individual risks by assigning the parameters of the individual processes a joint frequency distribution in a population. A well-known example involves *tolerance threshold* dose-response models. In such a model, an individual develops an adverse health response if and only if she has an exposure that exceeds her tolerance threshold. Thus, the individual response probability model is:

p = 1 if x > tolerance threshold (tolerance threshold risk model)
p = 0 otherwise

If the frequency distribution of tolerance thresholds in the exposed population follows a standard normal distribution, then the probability that a randomly selected individual will respond to dose level x is said to be described by a *probit* dose-response model. In this case, the mixing distribution G is just Φ, the CDF for the standard normal distribution.

For a dynamic version of the tolerance threshold concept, Lenk and Rao (1995) give formulas for the population hazard functions in cumulative damage models where both the average damage received per unit time and the threshold for response differ across individuals. Prakasa Rao (1992) also considers random shock models in which all individuals receive the same exposure intensity of shocks per unit time but have different response thresholds. It is easy to calculate the population probability distribution of the number of responses generated in any interval of time from knowledge of the exposure intensity and the frequency distribution of thresholds. However, the inverse problem of reconstructing the frequency distribution of individual response thresholds and the exposure intensity from observations of individual responses has no unique solution.

Example: Heterogeneity in Poisson Regression Models

Suppose that accidents occur in households according to Poisson processes with rates that depend on both observed and unobserved covariates. Unobserved heterogeneity in the Poisson rates necessarily leads to over-dispersion (i.e., excess variance) in frequency counts of accidents per unit time, and also to an excess of zeros and heavy-tailed distributions, relative to a Poisson distribution with the same mean (Cameron and Reivedi, 1998). Indeed, tests for unobserved sources of heterogeneity can be based on these departures from pure Poisson statistics.

Identifiability in mixture distribution models

The same population risk sometimes can be generated by very different individual risks. For example, Table 8 describes a contagion process for individual risks in which each individual's hazard function for the occurrence of the next event increases with the number of events already experienced. (This is an example of a stochastic hazard function in which the filtration for each individual is her event history.) This model generates exactly the same probability distribution for number of accidents in the population over any time interval as a frailty model in which intrinsic susceptibilities or frailties differ among individuals. Yet, in the contagion model, all individuals are initially identical, whereas in the frailty model, they are different. Because both processes generate identical probability distributions for observed accidents in the population over any time interval, no amount of data can show which is correct. As discussed in Chapter 4, it may therefore be impossible to determine from available data whether occurrence of one event makes subsequent occurrence of another event more likely, i.e., whether the first is a probabilistic cause of the second.

This issue of *non-identifiability* issue arises in many stochastic models (Prakasa Rao 1992). It implies that population data alone are not always sufficient to determine the frequency distribution of individual risk parameters or to determine what causes what. Therefore, it may be necessary to augment population data with biological or other data (see Chapters 5 and 6) to determine the effects of different decision options on the frequency distribution of individual risks.

Necessary and sufficient conditions have been developed under which the frequency distribution of individual risks in a population can be identified from population data. These conditions involve linear independence of conditional probability density functions (Prakasa Rao, 1992). They imply many useful corollaries, such as that a mixture of a finite number of Weibull hazard functions *can* be identified from data on observed failure times in a population. In other words, from these population data, the number of different types of Weibull hazard functions (i.e., the number of components in the mixture distribution), their relative frequencies in the population, and the values of the parameters for each type, can all be estimated.

Non-identifiability of individual risks from population data can have important implications for risk management. For example, a recent finding in the study of criminal behavior is that individual careers tend to follow contagion processes, with number of crimes to date tending to increase the expected rate of future crimes (Bushway *et al.*, 1999). This model is supported by data over an alternative frailty model in which differences in individual arrest rates are attributed to intrinsic differences in propensity to commit crimes. Risk management tactics for reducing crime rates typically differ in acquired-behavior models compared to intrinsic-rate models. Therefore, population data that cannot distinguish between these different models of individual behavior are less useful for supporting intervention decisions than data that identify the individual risk models. Either type of data can be equally useful in predicting population-level data on crime rates in the absence of interventions.

3.4 Summary of Advances in Statistical Risk Modeling

Table 9 recapitulates key challenges for statistical risk modeling and lists some of the most promising approaches for overcoming them. These approaches and relatively sophisticated numerical algorithms to support them are increasingly available through statistical software packages and programs in repositories such as StatLib. An understanding of the capabilities, limitations, and basic principles of these methods should be part of each risk analyst's working tool kit for performing and evaluating statistical risk assessments.

Many other important techniques and ideas in contemporary statistics can be enormously useful to risk assessors. For example, analysis of spatial data to detect and quantify patterns of disease risk around a point source is becoming a well-developed topic in applied statistics (e.g., Wakefield and Morris, 2001). Spatial and spatio-temporal analysis of environmental data generalize many ideas from survival data analysis by allowing multiple coordinates, rather than time alone, as arguments for hazard functions. Quantifying the variance in multiple imputation schemes for missing data (e.g., nearest-neighbor imputation with occasional re-imputation to reduce bias) is an active research topic in analysis of survey data with non-responses, as well as being relevant to spatial data analysis (e.g., Chen and Shao, 2001). Countless other topics and research areas that can extend the generality or improve the practical application of the techniques in Table 9, as documented in the current statistics research literature.

Table 9: Summary of Some Problems and Solutions in Statistical Modeling

Problem	Solution Techniques
Model form selection	• AIC and Kullback-Leibler information metric • Non-parametric methods • Bayesian model-averaging (BMA) • Model cross-validation • Sensitivity analysis for model-forms
Variable selection for exposures, responses, confounders, covariates	BMA, cross-validation, bootstrap resampling, classification trees with "bagging" Breiman, 1996. This is a form of resampling that reduces sensitivity of selected variables to random variations in data.)
Omitted explanatory variables	Test for latent variables. Use discrete mixture distribution models to estimate effects of unobserved variables.
Variable coding	• Automatic quantization algorithms, classification trees, MCMC-based adaptive random partitioning to set boundaries • Do not discretize continuous variables
Missing data (missing at random. See Table 2 for selection biases.)	Multiple imputations, sensitivity analysis, EM algorithm, Data Augmentation, Bayesian modeling,
Measurement errors in explanatory variables	Sensitivity analyses, bias correction algorithms, simulation-based corrections (SIMEX), EM algorithm, Data Augmentation, Bayesian modeling.
Exposure metric selection	Use multivariate exposure description rather than any single metric.
Multiple testing and multiple comparisons	Adjust p-values of individual tests

The techniques in Table 9 are especially useful for dealing with the problems most commonly encountered in applied statistical risk assessment and provide a useful foundation for more in-depth study. Rather than surveying extensions, refinements, and applications of these methods, or discussing additional important topics such as spatial statistics for risk assessment, the following section presents a case study to illustrate how the basic ideas already covered can be used to help overcome data analysis challenges commonly encountered in statistical risk assessment modeling.

4. A STATISTICAL CASE STUDY: SOIL SAMPLING

The case study in this section suggests how modern statistical ideas, especially multiple imputation of missing data values and cross-validation for model selection, can be applied to solve a statistical risk management decision problem. It also introduces principles from statistical decision theory to help formulate and clarify the problem and its solution.

Suppose that a residential neighborhood may have been contaminated by a nearby abandoned hazardous waste site. The suspected contamination consists of elevated soil concentrations of chemicals that are also found in the absence of site-related contamination. Figure 2 shows a typical example of sampling locations and estimated soil concentration contours for polycyclic aromatic hydrocarbons (PAHs) in residential surface soils around an abandoned hazardous waste site located at the origin. Estimated average concentrations diminish with increasing distance from the site. Given such data and a desired maximum acceptable surface soil concentration level, how should a responsible risk manager decide which additional residential properties to sample, if any, and which ones to clean?

Alternative Approaches to Statistical Risk Management

Table 10 defines the decision problems addressed. Table 11 summarizes a traditional solution approach that we call the *bright-line* sampling strategy. Its parameters (S, k, and T), are chosen to give high statistical confidence that a *trigger condition* will be satisfied (so remediation will be undertaken) unless a desired *target condition* is met. A target condition often used is that mean concentrations at uncleaned locations should not exceed those at "background" locations not detectably affected by the site. Other common target conditions require 99% confidence that at least 95% of measurements at a location will fall below a background-based Maximum Contaminant Level (MCL) or health-based Alternative Concentration Limit (ACL).

Figure 2: Estimated P(a)H Soil Concentration Contours based on Sampled Data at Multiple Locations (500 - 2500 feet from Center)

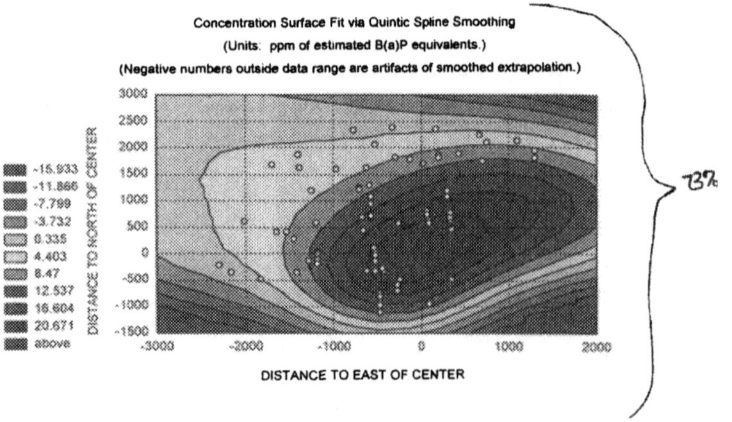

Table 10: Elements of Spatial Sampling and Remediation Decision Problem

1. A *spatial distribution* of soil contamination, initially unknown.
2. A list of feasible *sampling locations* at which samples may be collected.
3. An *initial data set* of sampled concentrations at various locations.
4. A set of feasible sampling *decision rules.* A decision rule specifies what to do next (e.g., where to sample next and whether to stop sampling), given the data observed so far.
5. A *terminal action decision rule,* specifying which locations to clean when sampling halts.
6. A *cost function* specifying the dollar cost of any sequence of sampling and terminal actions.
7. A set of *outcome evaluation criteria.* Possible criteria include:
- Total amount of contaminant removed from residential soils
- Total amount of contaminant remaining in residential soils
- Amount of contamination *due to the waste site* that has been removed;
- Amount and distribution of remaining contamination *from the waste site*
- Number of properties cleaned; contamination removed/remaining at each
- Dollars spent (sum of sampling costs and remediation costs).
8. A *criterion or objective function for choosing among probability distributions for outcomes,* e.g., a utility function (Chapter 5).

Table 11: Traditional Bright-Line Sampling Strategy

(a) Draw a fixed number, S, of samples from each location (e.g., each residential property);
(b) Compute sample statistics (e.g., the arithmetic or geometric mean and standard deviation) from the sampled values at each location; and
(b) Clean a location if and only if its statistics satisfy a *trigger condition,* e.g., if and only if the sample mean minus k sample deviations exceeds *an action threshold,* T. (T is the "bright line" above which action is required.)

Confidence limit and tolerance limit formulas for calculating trigger conditions have been provided by EPA for soil and ground water contamination (e.g., 40 CFR Part 264 for RCRA Subtitle C hazardous waste sites). Gibbons (1994) concludes that these recommendations are prone to false positives and recommends methods for computing nonparametric tolerance intervals when the concentration frequency distribution being sampled is unknown. However, such methods treat concentrations at different locations as statistically independent. As noted by Gibbons, in reality, "Both temporal and spatial correlations violate the independence assumptions (p. 33). ...[When] comparisons are needed to control for inherent spatial variability, nonparametric procedures are often simply not available (p. 76)." The following discussion fills this gap by introducing a class of nonparametric sampling procedures that exploit spatial correlations to improve sampling and remediation decisions.

Table 12: Outline of a Bayesian Decision Analysis Strategy

(a) Assume that the spatial distribution of contamination is described by a parametric model, say $\rho(x, y\ ;\ \mathbf{q})$, where $\mathbf{q}$ is a vector of parameters. Assume a prior probability density function for $\mathbf{q}$, say, $f_q(\mathbf{q})$.
(b) After sampling values for soil concentrations at different locations, update beliefs about $\mathbf{q}$ by conditioning on the observed values using Bayes' rule. (This requires knowing or assuming a likelihood function relating $\mathbf{q}$ to the observed values.)
(c) Choose the terminal decision, a*, to maximize the expected value of a utility function u(c, **b**), where the expectation is taken with respect to the posterior probability distribution for $\mathbf{q}$.
(d) Recursively choose the sampling locations and stopping rule via backward dynamic programming to maximize expected utility.

An alternative to the bright-line approach is a *Bayesian decision analysis framework* (see Chapter 5) in which sampling and clean-up decisions are based on the estimated distribution of contamination, conditioned on all data collected so far. The procedures are typically sequential, reducing the average sample size needed to make a confident decision (DeGroot, 1970). Table 12 outlines the Bayesian strategy.

The Bayesian decision analysis framework is, arguably, a uniquely coherent and powerful guide to "rational" individual decision-making. It may incorporate much sophistication in quantifying prior beliefs about complex quantities, such as spatial distributions, and in the dynamic programming algorithms used to decide where to sample next, how many samples to take, and when to stop (DeGroot, 1970).

However, the Bayesian framework suffers from practical limitations, especially when multiple stakeholders such as PRPs, the community, and regulators are involved. These include the following:

- *How to convince others about models?* Even if one decision-maker has clear beliefs about the model describing the spatial distributions of contamination, it may be difficult to convince anyone else that it is true or useful. Ripley (1981) discusses the limitations of many parametric models for spatial distributions, suggesting that doubts and disagreements can easily arise.

- *Unknown and ill-defined priors for spatial density functions.* It may not be easy for even one decision-maker to form or justify a coherent probabilistic model for the prior distribution of the spatial distribution of contaminants (e.g., the joint prior distribution of the model parameters).
- *Hard-to-calculate decision rules.* Dynamic programming is often computationally impractical unless the class of decision rules considered can be restricted.
- *Unknown or ambiguous utility function.* The utility function which roots the dynamic programming recursion with an evaluation for the terminal act may be unknown or disputed. For example, different stakeholders (the PRP, the residential property owners, and regulators) may have different degrees of risk aversion for uncertain health benefits, or may weight the rights and interests and assess the duties and responsibilities of the different stakeholders differently. They may therefore disagree about what should be done. More generally, Bayesian decision theory is not designed to show why the public should accept its prescriptions when these depend sensitively on subjective preferences and risk attitudes.

These challenges must be overcome by practical sampling and decision-making approaches. The fact that different people have different preferences and risk attitudes may perhaps best be addressed by bypassing subjective utility-theoretic calculations in favor of simpler criteria that most stake-holders accept as fair. The other challenges can be addressed by technical devices such as data analysis methods that do not require parametric modeling or prior beliefs These are developed next.

Optimized Adaptive Spatial Sampling

This section introduces a new heuristic for identifying approximately optimal adaptive sampling decision rules. It is based on the following steps:

- Bound the area(s) of greatest concern by analysis of spatial sample data.
- *Restrict the class of decision rules considered.* This step seeks to identify a set of high-performance decision rules that can be described by a small number of decision parameters. We will focus on *adaptive cluster sampling rules based on order statistics* (Thompson and Seber, 1996, Chapter 6). Such rules address the fact that the spatial distribution of soil contamination concentrations is unknown, while helping to identify locations with the highest concentrations. They can be indexed by a small set of decision parameters that define specific adaptive sampling procedures.

- *Evaluate decision rules in the restricted set using simulation-optimization.* Estimate the expected performance of each decision rule by simulating its application starting from the current data set. Assign multiple imputed values to unknown quantities encountered during the simulation (e.g., contaminant concentrations at locations that have not yet been sampled) by repeatedly sampling the available data from nearby locations. The performance of the decision rule is averaged over the multiple complete data sets, formed from the available data and the imputed values of currently missing data, to estimate its expected performance value. The neighborhood sampled from to impute missing values is selected to minimize the average error of the imputed values, using the currently available data points as a training set. We call this entire process *simulation-optimization,* i.e., simulation with decision parameters optimized with respect to performance criteria estimated from multiple imputations of missing data. It is motivated in part by computational Bayesian methods for missing data (Tanner, 1996; O Ruanaidh and Fitzgerald, 1996; Schafer, 1997).

Simulation-optimization yields the estimated expected value of the decision rule corresponding to each choice of decision parameter values. Searching a grid of parameter values yields an approximation to the optimal choice of the sampling decision rule. Expected values of multiple criteria emphasizing different goals or making different value trade-offs can be estimated simultaneously and used to examine the sensitivity of the optimal decision rule to changes in the evaluation criterion. These ideas are developed next..

Bounding The Area of Greatest Concern

Non-parametric methods for establishing the distances and directions where site-related contamination is most likely include the following.

1. *Gradient methods:* Create several "cells" of neighboring sample points (e.g., via a Voronoi tessellation of sampled locations (Ripley 81)). Estimate the direction of steepest increase in concentration for each cell, e.g., by fitting a plane to its sampled concentration values. Then likely source(s) of contamination can be identified as sinks in the estimated gradient field, while the boundary of site-related contamination is roughly indicated by where the gradient directions become randomly distributed or no longer biased toward the site. Steepest ascent methods such as stochastic approximation (Kushner and Yin, 1997) or Evolutionary Operations (Box and Draper, 1969) can also be used to seek the location of the maximum value of the contamination density function based on sequentially sampled values.

2. *Nonparametric regression.* If sampled values are adequately spaced, then the spatial distribution of contamination may be approximated using nonparametric surface-fitting technique such as loess, splines, or kernel regression, as in Figure 2 (Venables and Ripley, 1994). Geostatistical techniques such as kriging may also be used to estimate iso-concentration contours (Isaaks and Srivastava, 1989; Venables and Ripley, 1994). Source(s) of contamination should be located approximately at the centers of concentric contours. Such methods may give unrealistic values outside the spatial range of the data, however (e.g., the negative numbers in Figure 2.)
3. *Nonparametric regression estimates of boundaries.* In each octant (or smaller angular sector, if there are enough data points) fit a nonparametric smooth curve to the relation between distance from the site and concentrations at sampled locations, e.g., using nonparametric regression. The distance at which this curve levels out (if any) indicates an approximate practical boundary for site-related contamination in each direction. An alternative that corrects for multiple hypothesis testing is to apply classification tree algorithms to divide the sampled data points into relatively homogeneous sectors (Biggs 1991). Although such algorithms are not inherently optimized for spatial data, they can be applied directly to sample data (eastings, northings, and sample concentrations) to obtain an empirical histogram. If sample locations are expressed in polar coordinates, recursive partitioning will create a histogram from annular segments (defined by inner and outer radii and subtending angular directions).
4. *Nonparametric changepoint detection and image segmentation techniques.* A more sophisticated algorithm approximates the spatial density function with a histogram formed from polygonal prisms obtained by iterative improvement of randomly generated Voronoi tessellations (Green, 1995) using Markov Chain Monte Carlo methods (Gamerman, 1997).

For the case study described later, these approaches gave similar estimated boundaries for the contaminated area. However, no statistical approach can preclude the possibility that *some* site-related contamination extends beyond the statistically identified boundaries. Rather, boundary-setting helps to focus attention and resources on the most contaminated areas.

Adaptive Sampling: A Restricted Class of Decision Rules

Decision rules based on adaptive cluster sampling (Thompson and Seber, 1996) use four decision parameters to determine where to sample next and when to stop based on the data obtained. They are defined as follows:

- *N = initial number of random samples* to be drawn from the area of concern. N is the number of *seeds* for the algorithm. They may be selected by listing all residential properties in the area of concern and drawing N of them at random. Where to sample within a selected property is determined by a standard field protocol (e.g., a randomly selected point with no surface discoloration, evidence of oil spills, or other obvious non-site-related sources of contamination.) These N samples are in addition to those that have already been obtained, if any, when adaptive sampling begins.
- *S = number of soil samples* to take from each newly sampled property.
- *K = maximum allowed length of a "search list".* This is a list of properties with the highest average sample concentrations seen so far.
- *T = action threshold.* A property is cleaned if and only if the average of its S sample values exceeds T.

The adaptive sampling procedure for parameter vector (N, S, K, T) is as follows. Call a property *fathomed* if S samples have been taken from it and from each of its immediate neighbors. (The "immediate neighbors" of a property may be defined in various ways without changing the algorithm. A common choice is to define them as the four properties that are closest to it from the north, south, east, and west directions.) Initially, place the K properties with the highest average sampled values seen so far on the search list. Fathom each property on the list (by sampling), and remove it from the list when it has been fathomed. Whenever a property is found that has an average sample value greater than the K + 1st initial value (i.e., that is higher than any of the values for properties not included on the initial search list), place it on the search list. Continue until the search list becomes empty (i.e., all properties previously placed on it have been fathomed and no new, unfathomed properties are added to it.) This ends the adaptive sampling phase. Then, clean all and only those properties for which the sample average concentrations meet or exceed the action threshold, T. Table 13 summarizes the procedure.

Table 13: Worst-First Adaptive Sampling With Parameters N, K, S, T

1. Randomly sample N previously unsampled locations in the area of concern.
2. Place on a search list the K properties with the highest sample average concentrations observed so far.
3. Fathom each property on the search list by taking S samples from it and from each of its previously unsampled neighbors. Remove each property from the list when it has been fathomed.

4. Place onto the search list any property with a sufficiently high sample average concentration, where "sufficiently high" means higher than any of the initial sample values not included on the initial list in step 2.
5. Continue steps 3 and 4 until the search list is empty.
6. Clean those properties with sample average concentrations of at least T.

This procedure may be called *worst-first adaptive search* for seeking properties with high soil contamination levels. As shown by Thompson and Seber (1996, p. 163), such sampling allows unbiased estimates of the average soil concentration remaining in unsampled locations at any stage. The terminal decision rule specified by T is motivated by the Neyman-Pearson Lemma (DeGroot, 1971, p.146).

Simulation-Optimization of Decision Parameters

To find an approximately optimal adaptive sampling strategy, the strategies considered are restricted to worst-first adaptive search rules parameterized by the decision parameter vector $\mathbf{p} = (N, K, S, T)$. Seeking an optimal strategy in this class is straightforward, even using an exhaustive grid search over all values of the integer variables N, K, and S of practical interest, e.g., with N, K, and S each ranging from 0 to 20. (In the case study example below, performance is optimized well within the interior of this region.) For each combination of values for the integer variables, a line search is conducted to optimize the value of T (M. Box, 1969).

Given a proposed parameter vector, say $\mathbf{p}_0 = (N_0, K_0, S_0, T_0)$, its expected performance on the selected criteria (e.g., utility, Type 1 and Type 2 error rates, amounts of contamination removed and remaining, etc.) is estimated by simulation. During the simulation runs, missing data values are imputed from available data multiple times and the average performance of $\mathbf{p}_0$ is calculated by averaging over these multiple imputations. In more detail, $\mathbf{p}_0$ is evaluated as follows.

1. IMPUTE CONCENTRATION DISTRIBUTIONS: For each location, impute a number, C, of values for its estimated concentration distribution by drawing C times at random with replacement from among the values of the M nearest neighboring locations for which sample data are available (possibly including the location itself). (Alternatively, values can be imputed from all sampled values within R feet of the location, if there are enough of them.) More elaborate methods of imputation, such as distance-weighted sampling probabilities or kriging among nearby values, could also be used in this step (Ripley, 1981), but sampling from nearby data values is simple and reflects the local distribution of concentrations. The parameters C and M (or R) are *meta-parameters*, and are set via another round of optimization (Step 5).

2. SIMULATE ADAPTIVE SAMPLING. Simulate the procedure corresponding to $\mathbf{p}_0$. Thus, select N_0 new locations at random from among those not yet sampled. Take S_0 samples from each new location (drawing S_0 times at random with replacement from the C available values where imputed data values must be used). Form a search list of length K_0, recursively fathom its members, and identify locations to clean, i.e., whose average sample concentrations exceed T_0.
3. EVALUATE PERFORMANCE. Evaluate how well the procedure $\mathbf{p}_0$ has done at each location, using the location's imputed concentration distribution to evaluate the decision made there (skip or clean). (For example, the value of the true average concentration would be imputed by taking the average of the C imputed values.)
4. ITERATE. Repeat Steps 1-3 a total of I times, where I = number of iterations. (I = 20 is a typical value in our applications.) Average the performance of $\mathbf{p}_0$ over the I iterations. Thus, though Step 3 treats imputed values as if they were correct within each iteration, evaluating $\mathbf{p}_0$ for multiple imputations (as well as for multiple random seeds) builds up a frequency distribution of evaluations. The mean of this distribution is an estimate of the expected value of $\mathbf{p}_0$ given the available data used for imputations (Schafer, 1997).
5. OPTIMIZE META-PARAMETERS. Steps 1-4 involve the meta-parameters C, M (or R), and I. The final step is to set the values of these meta-parameters. The value of I is set by satisficing, with 20 or 30 iterations generally proving adequate. The criterion for adequacy is that increasing the number of iterations further (e.g., doubling or quadrupling it) should make no significant difference in the expected values estimated. The values of C and M, by contrast, are set by a separate round of optimization based on cross-validation (Gamerman, 1997). Values of C and M are selected to minimize the average error in imputed means *vs.* empirical means at locations where sample data are available. (In the case studies reported below, we used C = 10 and M = 5.)

Applying the evaluation loop 1-4 to each trial value of **p** identifies the procedures giving the best estimated average performance with respect to any performance criterion desired (e.g., error rates or expected utilities) that can be calculated from the imputed values in Step 3. Assessing multiple criteria increases the computational burden very little: it is only necessary to store a performance vector for each **p** value instead of a single value during each iteration. The stored performance data then allow correlations and trade-offs among criteria to be displayed.

In decision analysis, "optimal" decisions are typically made based only on the *probabilities* of consequences induced by the decisions (DeGroot, 1970). The actual consequences of alternative risk management acts are seldom known when a decision among them must be made. In the soil remediation context, the actual consequences may remain unknown, as the amounts of contaminant removed and remaining are usually never learned. To the extent that multiple imputations produce a probability distribution for consequences that is close to the true one, however, simulation-optimization provides the information needed to make optimal (e.g., expected utility-maximizing) statistical decisions, given the available data.

A Case Study Example

In 1995, a regional Environmental Protection Agency had reviewed the surface soil sample data underlying Figure 2 with two potentially responsible parties (PRPs). All parties had agreed on the following policy goals:

- *Health-protection/efficiency:* Remove as much as possible of the residential surface soil contamination created by the site, for any level of resources applied. (This motivated the "worst-first" approach.)
- *Fairness:* Avoid having the PRPs clean soil contamination not due to their site. Since the hypothesized mechanism of contamination was airborne deposition, the exact extent of site-related contamination in different directions was uncertain.
- *Clarity:* Be able to explain and justify to members of the community which locations were selected for sampling and/or remediation, and why.

Both the EPA Project Manager and the PRPs recognized that these goals partially conflict. The statistical methodology developed to pursue them also had to deal with the following realities. First, *it is impossible to determine whether (or by how much) any specific individual residential property has been contaminated by the site.* For example, some background locations have higher sample soil concentrations than some of the locations closer to the site. Second, soil concentrations are so variable, even on the same residential property, that *it is impractical to determine which locations have the highest true mean soil concentrations*. Finally, the spatial distribution of soil contamination concentrations is unknown.

To reach a pragmatic solution, the EPA Project Manager, in agreement with the PRPs, made the following simplifying assumptions:

1. The objective is to *minimize the expected number of decision errors*, defined as the sum of Type 1 and Type 2 errors ("false cleans" and "false skips" of residential locations, respectively). This objective considers only the number of errors made and not their sizes. Yet, it was considered simple and clear enough to be understood by the community (unlike utility-based calculations) and likely to produce reasonably fair, efficient, and politically acceptable outcomes. This objective function was later generalized to a *weighted sum* of Type 1 and Type 2 errors.

2. *No location should be cleaned that has not first been sampled.* Remediation is sufficiently disruptive for home-owners so that it should not be undertaken without strong evidence of elevated PAHs. In addition, sampling a location is orders of magnitude less expensive than cleaning it, making it worthwhile to sample extensively before allocating remediation resources.

To define the minimum expected-errors criterion, a risk-based health protective concentration level for benzo(a)pyrene [B(a)P] in residential soils was first estimated via probabilistic risk assessment, taking into account likely soil ingestion, inhalation, and dermal contact scenarios and bioavailability factors. This set an action threshold such that it was deemed to be a "correct" decision to clean a residential location if and only if the average concentration of PAHs there (expressed as B(a)P equivalents, using conversion tables provided by the EPA) exceeded the action threshold.

Identifying the Area of Greatest Concern

Table 14 shows the results of applying a classification tree algorithm (Biggs *et al.*, 1991) to the initial sample data underlying Figure 2 to help bound the area of greatest concern and detect any potential "hot spots". It automatically partitioned the available soil concentration data (B(a)P equivalent concentrations measured at different polar coordinates) into five distinct "sectors", represented by the five rows in Table 14. Three of these sectors lie in the north-northeast octant, at different distances from the center of the site. The algorithm creates sectors with statistically significantly different frequency distributions of sampled soil concentrations, such that no sub-sectors (identifiable by subdividing the angle or distance coordinates) are statistically significantly different. The greatest mean concentration occurs in Sector 1, which lies immediately to the north-northeast of the site, from the fence line out to 765 feet. Seven distinct residential locations had been sampled in this sector, some more than once, suggesting that a confident decision about the true mean concentration at a residential property will require many samples.

Table 14: Sample Soil Concentration Data Partitioned Into Sectors

Sector	Angle (degrees)	Distance Range	Mean Conc. (B(a)P Equiv)	Std Deviation	N
1	48 - 89	< 765 feet	36.6	15.0	10
2	48 - 89	765 - 1193	16.8	6.5	12
3	48 - 89	> 1193	4.9	5.1	21
4	89 - 185	All (< 2500)	4.7	5.8	56
5	> 185	All (< 2500)	10.3	10.4	26

Analysis of the correlation between concentrations at different locations, expressed as a function of the distance and direction between them, revealed that concentrations are positively correlated at distances on the order of 100 feet throughout the sample. Concentrations at the nearest sampled neighbor of a location (typically on the order of 100 feet away) is a better predictor of its concentration than are concentrations at more distant locations. This observation motivated the need for an approach that uses observed high concentrations to guide the search for other locations with high concentrations.

Implementation Experience and Results

The PRPs proposed the adaptive spatial sampling approach and simulation-optimization method to the regional EPA team in 1996. Several rounds of discussion showed that the approach seemed suitable to address the needs to learn from data, allocate sampling and remediation resources effectively, and maximize the probability of finding high soil concentrations within the area of concern. Refinements and extensions made during these discussions included using traditional statistical methods such as linear regression to confirm results of the spatial data analysis and the recursive partitioning algorithm (Table 7).

During 1997, it became clear that the relative value weights for different policy goals might change, depending in part on the effects of local environmental activists in shaping community perceptions and preferences. In particular, the relative priorities to be given to Type 1 and Type 2 errors were identified as uncertain and labile. Therefore, the EPA Project Manager identified a need to be able to minimize any weighted sum of Type 1 and Type 2 errors with arbitrary non-negative relative value weights (summing to 1). In addition, the PRPs wished to use a decision method that could also be applied to make normatively "rational" decisions, i.e., to maximize the expected value of a single- or multi-attribute utility function (see Chapters 5 and 6). Both the regional EPA team and the PRPs judged worst-first adaptive spatial sampling with optimized decision parameters to be the best method proposed to date for balancing the competing policy goals and making effective use of existing data in sampling and remediation decisions. The EPA Project Manager approved the method for use in 1997 and it was implemented in 1998.

Figure 3 shows the expected Type 1 and Type 2 error rates, assessed via simulation-optimization, for many different parameter vectors **p** = (N, K, S, T). Table 14 shows example outputs from the 20 vectors having the lowest total error rates. (The strategies used in these illustrations are a subset of over 5000 evaluated by grid search.)

Figure 3: Error Trade-Offs Among Adaptive Sampling Strategies

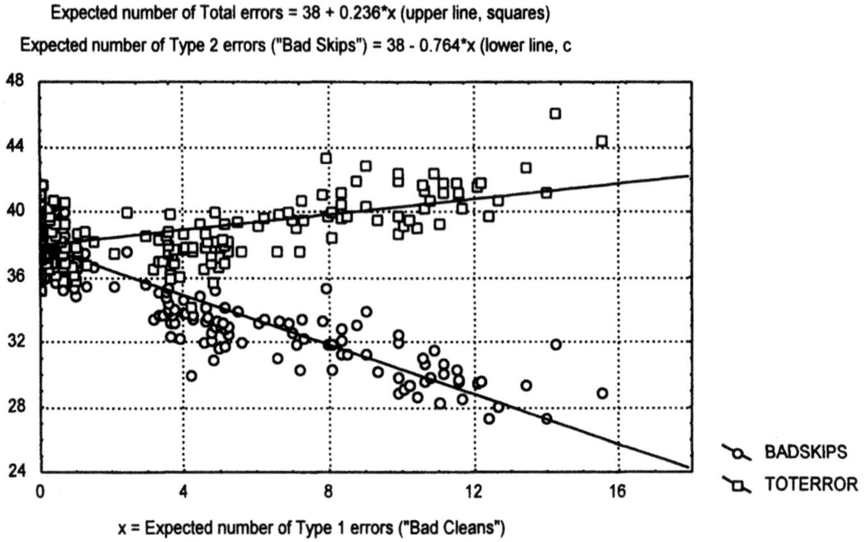

Table 15: Simulation-Optimization Outputs: Evaluations of 20 Adaptive Strategies

N	S	K	T	Error Rank	Cost Rank	Type 1* (Bad Cleans)	Type 2* (Bad Skips)	Total Errors	Expected Total Cost*
5	16	1	3.3	**1**	**142**	**4**	**30**	34	966571
20	12	3	3.9	**2**	**97**	**0**	**35**	35	685,714
20	8	3	3.3	3	141	5	31	36	951333
0	8	3	3.6	**4**	**43**	1	35	36	410,190
0	12	3	3.6	5	63	**0**	**35**	36	512,095
5	20	1	3.3	6	161	4	32	36	1,066,667
0	16	1	3.9	7	56	0	36	36	474667
20	4	3	3.9	**8**	**18**	0	36	36	245,048
5	30	1	3.3	9	200	4	32	36	1414286
0	30	3	3.9	10	159	0	36	36	1060000
5	30	3	4.2	11	180	0	36	36	1232857
20	12	3	4.2	12	93	0	36	36	669143
10	16	3	3.6	13	114	0	36	36	789429
20	8	1	3.6	14	61	0	35	36	496286
10	20	3	3.3	15	201	5	32	37	1414762
10	8	3	3.6	16	49	0	36	37	439238
0	8	1	3.9	**17**	**17**	0	37	37	236,190
20	4	1	4.2	**18**	**12**	0	37	37	198,286
0	30	3	3.6	19	167	0	36	37	1123810
10	20	1	3.3	20	175	**3**	**33**	37	1,174,286

* Expected values, rounded to nearest integer, assessed by simulation-optimization. Bold numbers indicate strategies undominated with respect to the attributes in their columns.

Each strategy takes several CPU-seconds to evaluate via simulation-optimization. The two superimposed scatter plots in Figure 3 show the estimated total error rate (upper scatter plot, square points) and Type 2 error rate (lower scatter plot, circles) for each strategy on the vertical axis, while the horizontal axis shows the corresponding Type 1 error rates.

The regression lines indicate that:

(a) Strategies with the lowest Type 1 error rates tend to have higher Type 2 error rates (as expected); and

(b) Strategies that spend resources to reduce Type 2 errors tend to create more than one Type 1 error for each Type 2 error removed. Thus, the total error line slopes up while the Type 2 error line slopes down.

Intuitively, cleaning all locations would eliminate all Type 2 errors, but at the cost of introducing a larger number of Type 1 errors. If Type 1 and Type 2 errors are weighted equally, then the optimal strategy will lie toward the left in Figure 3. Simulation-optimization method allows such trade-offs to be quantified. It also allows the sensitivity of the "best" strategy to the relative value weights placed on different goals or criteria to be assessed

In the absence of known, stable relative value weights for Type 1 and Type 2 errors, it is worthwhile to identify *undominated strategies*. A strategy is undominated with respect to any set of criteria if no other strategy is better on all of them. For example, strategies 8, 17, and 18 in Table 1 are undominated with respect to the criteria of "error rank" and "cost rank". Here, the "error rank" column is used to number the strategies. This column shows the ordinal ranking of each strategy based on total error rate. (Bold numbers in Table 6 indicate the strategies that are undominated with respect to the attributes in the corresponding columns.) The Type 1 and Type 2 error columns and the cost rank column show information for these criteria. "Cost" in Table 6 is defined as the sum of sampling plus remediation costs, using estimated unit costs of sampling and cleaning each location provided by the PRPs. Sensitivity analyses showed that the undominated strategies are robust to variations in these cost estimates.

Although expected Type 1 and Type 2 error rates were primary criteria for decision-making, some attention was also paid to cost-effectiveness. In part, this recognized that any remediation activity would remove some contamination that was not site-related. (Indeed, the site may not be the source of the elevated soil contamination levels. For example, Table 14 shows no apparent distance-related decline in soil concentrations around the site except in the north-northeast octant. This contrasts with the impression given by the contours in Figure 2, which reflect the assumption of a *smooth* relation.) In light of these considerations, undominated strategies were sought that ranked in the top 10% of all strategies on *both* total error rate and cost. Strategies 4, 8, 17, and 18 in Table 14 satisfy these requirements.

The adaptive sampling procedure corresponding to strategy 17 (N = 0, S = 8, K = 1, T = 3.9) was implemented in 1998. As anticipated based on previous multiple imputations, none of the four neighbors of the most-contaminated site (K = 1) identified in the initial sample had higher average concentrations than those already observed, suggesting that the locations with the highest concentrations may have already been identified. By contrast, an originally discussed "bright line" sampling plan would have required sampling residential properties out to 2,000 feet in all directions, yet probably would have produced little additional information for improving the final decisions on where to skip and where to clean. Both the PRPs and the regional EPA felt that worst-first adaptive sampling had helped to focus the search and greatly reduce the time and effort needed to obtain information useful for resource-allocation decisions.

Discussion, Extensions, and Conclusions

Simulation-optimization can be applied whenever the set of decision rules considered is described by a vector of decision parameters and the expected performance of each rule can be accurately simulated. Unlike recent model-based spatial data analysis techniques (Byers and Raftery 1998; Dasgupta and Raftery 1998), it does not require specific assumptions about the spatial distribution of contamination. As sampling progresses, new data are used to obtain increasingly refined estimates of this distribution.

For example, suppose that a project manager decides that adaptive sampling is impractical, e.g., because it requires too many intrusive visits to the community, and instead elects to use a two-stage sampling procedure, in which an initial sample is used to screen each residential property for further investigation, and a subset of the screened properties is then investigated further to determine whether remediation is needed. No location may receive more than two sampling visits. The appropriate class of decision rules in this case might be indexed by a vector of decision parameters of the following form:

$$\mathbf{p} = (S_1, T_1, S_2, T_2),$$

where S_1 samples are first taken from each location, an additional S_2 samples are then taken from all and only those locations for which the arithmetic average of the first S_1 samples exceeds the screening threshold T_1, and then those locations are cleaned for which the average of the $(S_1 + S_2)$ samples exceeds the action threshold T_2. Although such two-stage sampling (or its n-stage generalization) are very different from the adaptive sampling rules considered in this section, the simulation-optimization heuristic can be applied to optimize its decision parameters.

Much could be done to refine the simulation-optimization method presented in this case study, especially by studying the bias and variance associated with different multiple imputation schemes. However, it seems clear that the basic techniques of multiple imputation for missing data values, together with cross-validation estimates of performance for different decision rules, provide a practical basis for solving decision problems even when much of the data required for completely informed decision-making is not available. This is a major contribution from recent statistical methods.

In summary, the mathematical forms of decision rules that can be implemented are often determined by pragmatic constraints rather than by purely statistical considerations. Once the form has been decided, however, simulation-optimization using key ideas from modern statistical risk assessment, including multiple imputation of missing values and cross-validation estimation of the probability distribution of objective functions, can be used to identify the best specific procedure, provided that data are available to support the required imputations. Thus, the simulation-optimization framework can potentially be applied to improve statistical decision-making in a wide variety of practical applications.

Chapter 4

Causality

1. INTRODUCTION

The most important question for applied health risk assessment is: *What change in health risks would be caused by a given change in exposures?* If this question can be answered reliably for different proposed changes in exposures, then the answers provide a basis for choosing among different risk management actions that affect or constrain exposures. More generally, risk assessment deals with the causal link between actions and their probable consequences. Therefore, techniques for modeling such causal relations are crucial to successful risk assessment.

This chapter describes techniques for causal modeling. It emphasizes the construction and use of causal graphs to support risk modeling, decision-making, and uncertainty analysis, but also pays some attention to more traditional statistical and epidemiological approaches to causal modeling. The major points of the chapter are as follows:

a) Causal relations are not statistical relations. In extreme cases, the statistical (i.e., inferential) relation between exposure and response may be the opposite of the causal relation between them. For example, there may be a statistically significant positive association between exposure and response even if there is a negative causal relation between them, i.e., even if greater exposures cause reduced probabilities of response.
b) Some traditional epidemiological criteria and guidelines for judging or inferring causality are essentially useless. They provide neither

necessary nor sufficient conditions – nor even useful evidence, in many cases – for inferring causation or its absence.

c) One approach to establishing causation is to refute competing (non-causal) explanations for an observed statistical association between exposure and response variables. Chance, biases, and confounding are the three main categories of alternative explanations that must usually be refuted. As discussed in Chapter 3, biases may arise from sampling, information collection, or modeling choices. Statistical methods and algorithms can be used to test whether chance, biases, or confounding might plausibly account for observed associations. If not, the hypothesis that the relation is causal gains strength.

d) The relation between exposure and response variables (and other mediating variables, if any) can be described quantitatively by statistical methods, including curve-fitting, once the qualitative causal relations among the variables have been established by hypothesis-testing or other theory-constructing methods. These qualitative relations can be summarized in a path diagram or causal graph that clearly distinguishes between direct and indirect causal paths linking variables.

e) It is not always possible to learn a unique causal graph directly from epidemiological data. Other data, e.g., dealing with biological mechanisms of disease causation, may also be needed.

f) An alternative approach to modeling causality, widely used throughout applied mathematics and engineering, is to create a mathematical or simulation model of the processes by which changes in inputs (including decision variables controlled by a decision-maker) propagate to cause changes in outputs (e.g., health risks and their precursors). Such models need not be completely detailed in order to be descriptively accurate and sufficient to support risk management decision-making.

g) The qualitative structures of simulation or mathematical models, as well as of multivariate statistical causal models, can be represented by causal graphs. These directed acyclic graphs (DAGs) and closely related DAG constructs (such as Bayesian belief networks whose structure and/or probability relations at nodes represent expertise; and influence diagrams, which are DAGs with decision and value nodes) imply testable qualitative conditional independence relations. Fully quantified causal graphs (i.e., specifying conditional probability distributions for the value of each variable given the values of its parents) also imply testable quantitative relations, namely, the composition relations referred to in Chapter 2. These relations can be used both to help validate causal graph models and to automatically discover or "learn" them from data.

h) Once a causal graph model (or influence diagram model, etc.) relating actions to their probable consequences has been specified, it can be used to optimize decision inputs, draw inferences from observations, and allocate blame for an outcome among multiple causes. Algorithms are available for these purposes. Thus, causal graph models provide an effective framework for representing and using knowledge of causal relations to support improved risk-management decision-making.

Philosophers and statisticians often regard causality as a mysterious, perhaps somewhat ill-defined, concept. Yet, engineers routinely create and use successful models of deterministic and probabilistic cause-and-effect relations to support design, optimization and risk management decisions. This chapter seeks to clarify the concept of causation while retaining a practical emphasis on how to develop and use causal models.

2. STATISTICAL VS. CAUSAL RISK MODELING

Both prospective and retrospective risk assessments focus on the causal relation between actions and their probable consequences. This may be quite different from the statistical association between them, as the following examples show.

Example: Causal vs. Statistical Regression Relations

Suppose that the true relations in a population among (a) Risk of an adverse health effect, (b) Age, and (c) Cumulative exposure to a chemical are as follows:

$$\text{E(EFFECT} \mid \text{AGE, EXPOSURE)} = \text{AGE} - \text{EXPOSURE} \qquad (1)$$

$$\text{EXPOSURE} = (1/3)*\text{AGE} \qquad (2)$$

The corresponding causal graph (DAG) model, augmented with signs and coefficients to show the directions and sizes of direct effects, is:

$$\begin{array}{ccc} & \text{E(EFFECT)} & \\ \uparrow^{-1} & & \uparrow^{+1} \\ \text{EXPOSURE} & \underset{+1/3}{\leftarrow} & \text{AGE} \end{array}$$

Here, E(y | x) denotes the conditional expected value of quantity y, given the observed value of x. Note that equation (2) suggests a sensible causal interpretation, in which causality flows from the right side of the equation to the left i.e., exposure increases with age because longer time exposed causes greater cumulative exposure. The algebraically equivalent equation:

$$\text{AGE} = 3*\text{EXPOSURE}$$

lacks such a causal interpretation, since exposure does not cause age. This asymmetry suggests that there is more to the intuitive concept of causation than is captured by isolated mathematical equations alone. The theory of *causal ordering* of equations (Iwasaki and Simon, 1994) explains this phenomenon in terms of which subsets of variables determine the values of each variable in a system of equations. In the above system, age determines exposure, and age and exposure together determine the expected effect. The DAG diagram makes this ordering clear.

A risk assessor who does not know the true relations (1) and (2) may find that the following regression model perfectly describes exposure and effect data:

$$E(\text{EFFECT} \mid \text{EXPOSURE}) = 2^{*}\text{EXPOSURE}. \quad (3)$$

(Note that substituting (1) and (2) into E(EFFECT | EXPOSURE) and applying laws for expected values gives E(EFFECT | EXPOSURE) = E(AGE – EXPOSURE | EXPOSURE) = E(AGE | EXPOSURE) – EXPOSURE = 3* EXPOSURE – EXPOSURE = 2* EXPOSURE.) Thus, linear regression of EFFECT against EXPOSURE will indicate that they are significantly positively correlated ($R^2 = 1$ in this example) with risk of effect proportional to exposure.

However, *this statistical relation is useless for predicting the effects of interventions.* For, equation (1) shows that increasing EXPOSURE will *reduce* the expected value of EFFECT, even though equation (3) correctly shows that the expected value of EFFECT is greater among people with higher observed values of EXPOSURE. This illustrates the key difference between causal relations and statistical inference relations. For purposes of *inference,* the conclusion that risk is proportional to exposure is valid. If an individual is observed to have a certain exposure, then the expected value of the EFFECT variable is 2*EXPOSURE. This may be useful, for example, for inferring the current risks in the population from cross-sectional data on the current frequency distribution of exposures. But it does not help in predicting the effects of risk management interventions. The statistical regression model (3) says nothing about how risk will *change* if exposure is changed. Indeed, if risk for each individual increases in direct proportion to increases in age and decreases in direct proportion to increases in exposure (at any age), then increasing exposure will reduce risk – the opposite of the conclusion suggested by the statistical relation (3).

Example: Simpson's Paradox

Cohort data are often summarized in a 2 x 2 table as follows:

	Exposed	Unexposed	Sample sizes:
Cases (Diseased)	a	c	$n_1 = a + c$
Controls (Not diseased)	b	d	$n_0 = b + d$

For such tables, the most common measures of the association between exposure and disease are the estimated relative risk (RR) and the estimated odds ratio (OR). They are defined as follows:

- Estimated RR = $[a/(a + b)] \div [c/(c + d)] = [a(c + d)]/[c(a + b)]$
- Estimated OR = $(ad)/(bc) \cong$ RR if a and c are small compared to b and d.
- Approximate 95% confidence interval for OR = $\exp\{\ln[(ad)/(bc)] \pm 1.96[(1/a) + (1/b) + (1/c) + (1/d)]^{0.5}\}$

In addition, the *population attributable risk* (PAR) is calculated as:

- Estimated PAR = estimated fraction of cases that are due to exposure
  = (number of cases – expected no. of cases if no exposure)/(number of cases)
  = 1 – (expected number of cases if there were no exposure)/(number of cases)
  = $1 - (c/d)(n_0/n_1)$. (Le, 1998, p. 39).

Although these statistical measures of association and attributable risk are widely used, they can be misleading if they are interpreted causally without justification. To illustrate, Table A shows a 2 x 2 table for exposure and illness that demonstrates a clear, statistically significant, positive association between them. The estimated relative risk has a sample value (which is also the maximum likelihood estimate) of $RR = (505/1100)/(110/1011) = 4.2$. This indicates a strong positive association between exposure and subsequent illness incidence.

Table A: 2 x 2 Table Showing a Positive Exposure-Illness Association

	Exposed	Unexposed	
Illness	505	110	RR = 4.2
No Illness	595	901	(Aggregate)
	1100	1011	
Illness Incidence Rate:	46%	11%	

Table B: Occupational Sub-Table Showing Negative Association

	Exposed	Unexposed	
Illness	500	10	RR = 0.55
No Illness	500	1	("Occupational Subgroup")
	1000	11	
Illness Incidence Rate:	50%	91%	

Table C: General Population Sub-Table Showing a Negative Association

	Exposed	Unexposed	
Illness	5	100	RR = 0.5
No Illness	95	900	("General Public")
	100	1000	
Illness Incidence Rate:	5%	10%	

Now, suppose that the population consists of two subgroups: an "occupational" component shown in Table B, and a "general public" component shown in Table C. (In reality, there could be dozens or hundreds of subgroups, based on race, family and medical history, diet, geographic location, and so forth. But two groups suffice to illustrate the point.) Tables B and C present a disaggregation of the original data set, i.e., they sum cell-by-cell to give the aggregate data in Table A. The relative risk rates for these two subgroups are only 0.55 and 0.5, however. Both are much less than 1, indicating a *negative* association between exposure and illness Thus, *the positive association observed at the group level is reversed at the subgroup level.* The causally relevant associations are those that hold for individuals (negative in this example, if all individuals in the same cells of Tables B or C have the same risks), rather than the aggregate ones that hold for populations (positive in this case.)

The apparent paradox that exposure is significantly positively associated with risk of illness in the population even though exposure reduces illness risk for each individual can be explained with the help of the following graph:

Exposure
+↑ ↓ -
Occupation →(+) Risk = Probability of illness

Here, membership in the occupational subgroup is strongly positively associated with both risk and exposure, creating a positive association between them in the population. If it is observed that a randomly selected individual is exposed, the effect of the *inference* that he probably belongs to the occupational subgroup (and therefore has a high risk) overwhelms the effect of the *causal* relation, that exposure reduces risk. Thus, observing that someone is exposed makes it more likely that illness will be observed.

In algebraic notation,

$$\Pr(\text{illness} \mid \text{exposed}) = \Sigma_{\text{group}}\Pr(\text{illness} \mid \text{group, exposed})\Pr(\text{group} \mid \text{exposed}).$$

The essence of Simpson's paradox is that the conditional probability of membership in the occupational group given exposure (namely, the term Pr(group | exposed)) is the main determinant of the probability of illness given exposure. The causal exposure-response relation is not apparent in this conditional probability relation. As illustrated in this example, the two can be quite different.

Example: Heterogenity vs. Causation in Accident Proneness

Suppose that an insurance company observes that the event of a policy-holder having an accident makes subsequent accidents more likely, i.e., the mean time between accidents decreases with the number of accidents that have already occurred. Then *should the first accident in a sequence be interpreted as a probabilistic cause (or at least as a contributing cause) of the next accident?* Not necessarily. The mean time until the next accident may be reduced by an accident even if each individual has a constant hazard rate (and hence constant mean time between accidents) if different individuals have different (but constant) accident rates. The reason is that observing an accident makes it more likely that the policy-holder who suffers it has a high accident rate rather than a low accident rate, even though the rate may not be affected by the accident. In other words, *inference* about the rate changes in light of the fact that an accident has occurred, even if the rate itself does not. [Another possibility is that occurrence of an accident signals that the policy-holder is probably experiencing a period of high accident intensity (e.g., an icy winter) that may also cause subsequent accidents – and yet, the first accident is a symptom, not a cause, of this dangerous spell and of increased future accident rates.]

Some philosophers of causation have proposed to define one event A to be a *probabilistic cause* of a subsequent event B if occurrence of A makes occurrence of B more likely. As suggested by these different possible interpretations of serial accident data, causation in the sense that we will mean it – that preventing A would reduce the probability or occurrence rate of B – does not necessarily follow from the statistical inferential relation between them, i.e., that A makes B more probable.

Example: Spurious Regression Between Trends

If two quantities are significantly correlated over time even after removing any trends, then some risk managers may interpret their correlation as evidence for a possible causal relation between them. For example, suppose that Y and X are two variables measured in several different time periods (e.g., X_t might be the amount of antibiotic-resistant Campylobacter found in samples of uncooked chicken in year t and Y_t might be the number of cases of resistant campylobacteriosis reported by the Centers for Disease Control in year t.) If the linear regression model

$$E(Y \mid X) = \beta_0 + \beta_1 X$$

produces a value of β_1 statistically significantly greater than zero (using a standard t-statistic) and a high correlation coefficient (R^2) for the correlation of Y and X, then this might be interpreted as evidence for a causal relation between X and Y – or at least as evidence that both X and Y are affected by one or more common causes.

Such an interpretation seems plausible at first, since if X does cause Y via the linear regression model as shown, then their values will indeed be correlated. However, further analysis shows that such temporal association is *not* necessarily evidence of a causal (non-random) relation. For example, suppose that X and Y follow statistically independent random walks, i.e.,

$$X_t = X_{t-1} + e_t$$
$$Y_t = Y_{t-1} + n_t$$

where e_t and n_t are independent, zero-mean, white noise series of random variables. Then high R^2 and apparently significant t values typically occur (e.g., in 87% of simulations performed by Granger and Newbold, 1974, reviewed by Granger, 1990) when Y_t values are regressed against X_t values. Such "spurious regression" results because the assumptions of standard classical test statistics such as the t-test statistic are violated by the error term for the regression of Y_t against X_t (Granger, 1990; Phillips, 1998). It is likely that two random walks will each exhibit enough random drift over a period of observation so that their values are correlated, even though neither affects the other. More generally, multiple regression of one random walk against one or more others is likely to produce apparently significant regression coefficients even though, by construction, none of the variables affects any of the others. Thus, while a true causal relation between X and Y is expected to produce correlations between their values over time, the converse is not true, i.e., correlations between X and Y over time does not necessarily indicate a true causal (as opposed to spurious) relation between them. To constitute useful evidence for causation, a statistical test must show something that is likely if and *only if* causation is present. Correlated values do not meet this requirement.

Such examples show that, no matter how well statistical models describe data and predict values of some variables (e.g., risk) from values of other variables (e.g., exposure), they may not be useful for predicting impacts of interventions. Yet, effects of interventions are usually the chief concern of risk managers. Therefore, it is necessary to use causal risk models, rather than only models of statistical associations, to predict the effects on health risks of alternative interventions or decisions.

3. CRITERIA FOR CAUSATION

Causation has a simple practical meaning in applied risk analysis. If exposure to a substance causes an adverse health response, then preventing exposure should reduce the risk of response. Thoroughly understanding this idea requires a detailed discussion of risk and can benefit from careful consideration of philosophical subtleties such as the logic of counterfactuals (Galles and Pearl, 1997; Rubin, 1991; Schaffner, 1993), distinctions among potential, predictive, and retrodictive causal assertions (Kramer and Lane, 1992), relations among causation, explanation, and inductive and deductive inference (Vineis, 1991; Nayak and Joskowicz, 1996) and necessary, enabling and probabilistic causation (Renton, 1994; Susser and Gertrude, 1991). However, the *manipulability* account of causation (Schaffner, 1993), based on the intuitive idea that preventing exposure reduces risk caused by the exposure, suffices for many practical purposes.

Several criteria for judging the plausibility of a hypothesized causal relation have been proposed. They are reviewed next, starting with traditional epidemiological criteria and progressing to alternatives.

3.1 Traditional Epidemiological Criteria for Causation

Epidemiologists have long used a set of qualitative criteria to help judge whether an identified exposure-response association is likely to be causal.

Originally formulated by Sir Bradford Hill, these criteria have been extended and refined by many subsequent authorities, including the 1964 Surgeon General's Report (Weed and Gorelic, 1996; Susser and Gertrude, 1991).

Table 1 lists versions of the criteria that are now in widespread use. Many epidemiologists consider them useful aids for assessing the plausibility of a hypothesized causal relation from epidemiological data, even while acknowledging that they are neither necessary nor sufficient for establishing causality (e.g., Susser and Gertrude, 1991).

Table 1: Traditional epidemiological criteria for judging causality

1. *Strength of association* between exposure and response. A relative risk of 2 or more is often taken as suggestive evidence that an association is not due only to unrecognized confounders (e.g., Lin et al., 1998, p. 958).
2. *Consistency of association* between exposure and response. Consistency means that an association can be repeatedly observed by different investigators under different conditions.
3. *Temporality* (i.e., exposures precede responses, exposure-response lag is similar among different studies, changes in exposures are followed by changes in response rates, etc.)
4. *Biological gradient* (i.e., there appears to be a dose-response or exposure-response relation, with greater exposures or doses being associated with greater risk levels.)
5. *Biological plausibility* (i.e., plausible causal mechanisms relating exposures to response probabilities can be suggested.)
6. *Specificity* of the exposure-response relation, i.e., is exposure necessary and sufficient for the response? Does exposure cause a specific effect (e.g., a cancer of a particular type or at a particular anatomic site?)
7. "*Coherence* in all its forms (e.g., theoretical, factual, biologic, and statistical)" (Susser and Gertrude, 1991).

In practice, these criteria have often been used inconsistently in published articles (Weed and Gorelic, 1996; Weed, 1997). They may not be useful for diseases with complex etiologies unless they are reinforced with detailed understanding of biological causal mechanisms (e.g., Gori, 1980). Objections to the individual criteria include the following:

- *Strength of association* does not rule out confounding or biases as possible non-causal explanations. Modeling biases (e.g., from multiple testing or spurious regression) and sample selection biases can be strong, producing relative risks greater than 2.

  Conversely, absence of association between exposure and response may not indicate lack of a causal relation between them. Even a strong deterministic relation between two variables may produce no statistical association if the relation is nonlinear. For example, if $y = x^2$, where x is uniformly distributed between −1 and 1, then there is zero correlation between x and y.

- *Consistency of association* between exposure and response variables, like strength of association, can result from biases and confounding as well as true cause-and-effect relations. Conversely, if signs of associations are opposite in different contexts (e.g., in different sub-trees of a classification tree) then there may be no apparent overall consistency despite strong context-dependent effects. Consistency may also fail to be apparent if exposure and response are not linearly (or monotonically) related. For example, the following tree presents data from a survey of 3,421 individuals (2104 cases and 1317 controls) collected by the Centers for Disease Control (CDC) from several states for a 1998-1999 Campylobacter Case Control Study. Campylobacteriosis risk is indicated by the case-control variable CACO, defined as CACO = 1 for cases, CACO = 0 for controls. Splitting CACO against number of fast food commercial meals consumed in the past week (the FASTFD variable), shows that risk increases with number of fast food restaurant meals whenever the number of meals is positive. However, people with zero restaurant visits (the [0, 1) branch) have significantly higher risks than people with one or two few restaurant visits (the [1, 3) branch). This type of nonlinear (u-shaped) relation can cause correlation-based analyses and regression models to overlook relations that are consistent but not monotonic.

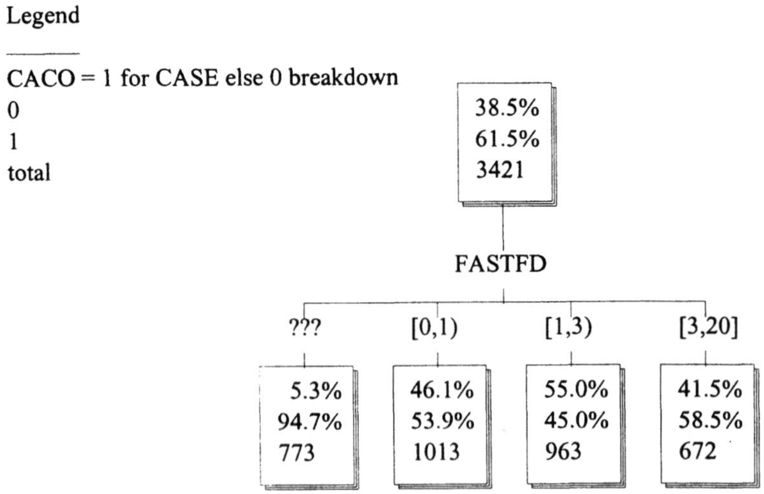

- *Temporality* is often interpreted as meaning only that responses usually follow exposures. For diseases such as lung cancer that occur primarily in elderly and retired populations, occurrences will tend to follow any occupational exposures, even if there is no causal relation between them. Thus, "temporality" may not be useful for judging causation. If consistent time lags (possibly with a distribution of values) are observed between exposures and changes in the hazard rates for adverse responses, then the evidence for causation can be much stronger. Survival data analysis and causal analysis of multiple time series (Table 3) can add force to the temporality criterion.

Example: Analysis of Temporality for Resistant Campylobacter

Rates of fluoroquinolone-resistance in *Campylobacter* (CP) isolates from human patients rose rapidly in many countries after 1987, as shown in a plot of resistance rates over time. A widely cited explanation for these increasing trends in resistance is introduction of fluoroquinolones (FQs, especially, enrofloxacin) to curb mortality in infected chicken flocks. This causal attribution is based largely on the fact that, in multiple countries, introduction of enrofloxacin was followed by significant increases in resistance rates among human patients (e.g., Engberg, 2001).

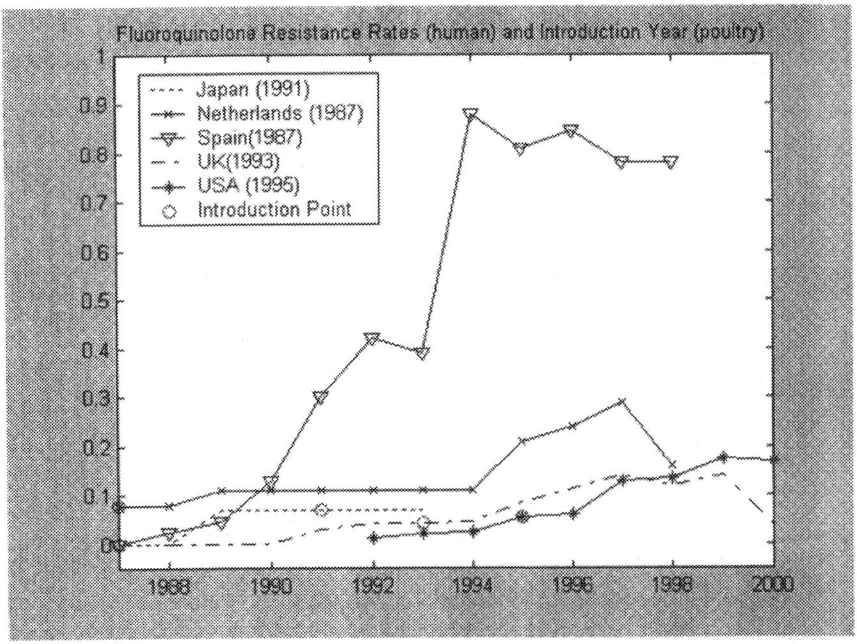

Source: Cox LA, Popken DA, 2001. *The causes of campylobacteriosis.* Technical Report to the Animal Health Institute. Cox Associates, Inc., Denver, CO

As suggested by the figure, however, this *ex post* argument does not fully fit the available data. Although resistance rates in humans did rise following introduction of FQs into veterinary use for sick chicken flocks, they also rose *before* such introductions, undermining any simple direct causal interpretation. (Travel among countries does not appear to explain this result where data are available to check.) There is no consistent lag between introduction in chickens and increase in human resistance. Moreover, well after introduction, resistance rates in humans have leveled off or declined in several countries despite continued use in animals, casting doubt on veterinary use of FQs as the main driver of human resistance rates.

To further test the hypothesis that fluoroquinolone use in chickens is a significant cause of resistance in humans, it is desirable to compare levels of use in chickens, other farm applications, human medicine, and in domestic pets such as cats and dogs, to levels of resistance in humans. Without such detailed comparisons and analyses of multiple time series, the criterion of "temporality" is too ambiguous to provide help in establishing or refuting the hypothesis.

- *Biological gradient* suffers from the same limitations as strength and consistency of associations. For example, suppose that greater cumulative exposures to a chemical are consistently associated with increased risk of lung cancer, with their observed relation exhibiting a well-defined gradient. If this relation were explained away by the fact that both cumulative exposure and lung cancer risk increase with an uncontrolled confounder such as age, then the apparent gradient would not provide evidence of a possible causal relation.
- *Biological plausibility* is not well-defined. For example, the less one knows about relevant toxicology, the more biologically plausible an observed or hypothesized association may appear to be. Plaintiffs and their experts often consider unproved causal relations between exposures and adverse health conditions to be biologically plausible while defendants and their experts consider them implausible.
- *Specificity* may not hold even if there is a cause-and-effect relation. For example, in animal experiments, some chemical carcinogens increase tumor rates at several anatomic sites and reduce them at others. While occasionally useful (e.g., for asbestos exposure and mesothelioma), specificity is too stringent a criterion to be useful for most health outcomes (e.g., lung cancer or early mortality) of interest in current public health debates.

  Conversely, ambiguous descriptions may create an illusion of specificity even if none exists. For example, many epidemiological studies of the relation between environmental tobacco smoke (ETS) exposure and lung cancer report a significant positive association, suggesting that ETS may specifically cause lung cancer, as one might expect. Yet, Fontham et al. (1994) identified significant increases in risk of lung *adenocarcinomas,* but not other histopathological types, associated with ETS exposure; while Stockwell et al.(1992) found that *non-adenocarcinomas* were most associated with ETS; Garfinkel et al. (1985) and Pershagen et al. (1987) found that ETS exposure is significantly associated with *squamous cell carcinomas* but not adenocarcinomas; and Lam (1987) found that ETS is associated with elevated risks of both *small cell carcinomas* and adenocarcinomas, but not squamous cell carcinomas. Such inconsistencies suggest that modeling biases (e.g., multiple testing bias) may be inflating the rate of false positives, undermining the interpretation of a cause-and-effect relation between ETS and lung cancer. Yet, all the individual findings have been cited as supporting a specific (and therefore probably causal) relation between ETS exposure and "lung cancer".

- *Coherence*, like biological plausibility, is often not clearly defined. What seems coherent to one reviewer may seem incoherent to another with different training or experience or with more or les willingness to speculate in the absence of compelling data or biological knowledge. Moreover, a body of theoretical hypotheses and observations may be mutually coherent without being true, undermining coherence as an indicator of true causality.

In summary, the traditional epidemiological criteria in Table 1 are not always helpful in distinguishing between exposure-effect relations that are causal and those that result from biases or confounding. While they will typically hold if exposure does cause increased risk of effects, they may also hold if biases or confounding create a non-causal relation between exposure and effects. Interpreting compatibility of data with these criteria as evidence of causality runs the risk of confirmation bias, i.e., the phenomenon of investigators finding whatever they expect to find (Bar Hillel et al., 1998).

3.2 Proposed Criteria for Inferring Probable Causation

This section considers alternatives to traditional epidemiological criteria for establishing whether an observed exposure-response relation is likely to be causal. Table 2 outlines an alternative approach, synthesized from ideas developed primarily outside the mainstream of epidemiology. Table 2 outlines some of these ideas, which offer constructive tests for causality.

It is useful to divide assessment of the plausibility of a causal relation between exposure and response into three phases, as follows:

Phase 1: Evidential reasoning. Assess whether available evidence is *more consistent* with the hypothesis of causation than with other (non-causal) explanations. If not, then it cannot be used to justify an inference that observed associations are causal. The consistency of evidence with different hypotheses is often expressed via likelihoods or conditional probabilities.

Phase 2: Theory-testing. Assess whether the available evidence is consistent not only with the specific causal hypothesis of interest (e.g., "Exposure to substance X causes lung cancer") but also with a larger causal theory to which the hypothesis belongs (e.g., "X creates conditions, such as mutations or proliferation in lung cells, that in turn increase the risk of lung cancer.") In other word, test whether available data support the *testable implications* of the theory that justifies the hypothesis, as well as supporting the hypothesis itself.

Phase 3: Law-based explanation. *Explain* the hypothesized causal relation by showing how it can be deduced from the operation of known general causal laws or mechanisms. For complex systems, such an explanation typically involves composing instances of general laws whose variables can be instantiated to describe the specific situation being explained (e.g., Nayak and Joskowicz, 1996).

Statistical methods can assist only in the first two of these phases (Nurminen, 1997; Robins and Wasserman, 1998.) Moreover, if the third phase can be completed, then causality can be inferred or deduced without the first two. Therefore, causal inference is sometimes characterized as "not statistical" (e.g., Nurminen, 1997), although "not only statistical" might be a more accurate description.

Table 2 proposes a more detailed sequence of steps for supporting causal inferences about a hypothesized exposure-response relation. Steps 1-3 correspond to evidential reasoning (phase 1), step 4 to theory-testing using data (phase 2), and steps 5 and 6 to explanation based on composition of known causal laws and relations (phase 3). The more of these steps can be completed, the stronger is the inference that there is a causal relation between exposure and risk.

Completing phases 1 and 2, evidential reasoning and hypothesis testing, establishes exposure as a *possible cause* of the response, meaning that this hypothesis is consistent with data. Even if the theoretical or empirical knowledge needed for phase 3 (explanation) is not available, a causal hypothesis that survives phases 1 and 2 may be considered supported by the data (Susser and Gertrude, 1991). Campbell and Stanley (1963) emphasized the importance of phase 1 for valid interpretation of causal relations in observational (non-experimental) settings in the social sciences and provided a systematic approach for enumerating competing (non-causal) hypotheses and using observational data to test them rigorously. Today, the resulting "quasi-experimental" designs are commonly used to infer the likely causal impacts of health risk management interventions such as smoking-cessation programs (e.g., Conrad et al., 1996; Lamkin et al., 1998).

Table 2: Steps to establish a possible causal exposure-risk relation

1. *Identify an association.* Demonstrate that there is a consistent, non-random positive association between exposure histories or events and their hypothesized consequences.
2. *Eliminate confounding* as a possible cause of the association. Show that it is not due to or explained by other (non-exposure) causes such as differences in lifestyle factors, age, or exposures to other confounders. (Nurminen, 1983; Lin et al., 1998)
3. *Eliminate biases from sampling, information, and modeling as possible causes.* Show that the association is not explained by biases in who was selected (as study subjects or as controls) or in how information about them was collected and analyzed. (Choi and Noseworthy, 1992)
4. *Test and confirm hypothesized causal ordering and conditional independence* relations among observed values of variables. For example, show that the response is not conditionally independent of its hypothesized direct causal predecessors (e.g., exposure), but that it is conditionally independent of more remote causal predecessors given the direct predecessors. (See http://www.sis.pitt.edu/~dsl/UAI/uai97.html.)
5. *Confirm efficacy of interventions.* Confirm that changes in the levels of direct causal predecessors (e.g., exposures) are followed by the predicted changes in the levels of the variables they affect. This may often be done from time series observations, even if direct experimental manipulation is impossible (Granger, 1980).
6. *Identify and elucidate causal mechanism(s).* Explain how changes propagate via one or more causal paths to produce effects (Nurminen, 1997). A "causal path" is a sequence of steps in which completion of the earlier steps creates conditions that trigger or increase occurrence rates of subsequent steps. Such steps may be identified from experimental data and/or by applying generally accepted laws (Renton, 1994).

Most statistical methods used in epidemiological analysis focus on steps 1 to 3, i.e., identifying associations and eliminating potential biases and confounders as likely explanations (see Chapter 3). These steps can often be carried out using data from quasi-experimental designs for observational studies (Campbell and Stanley, 1963) without requiring direct manipulations and experimental verification of predictions. Quasi-experimental designs and methods of causal analysis and intervention analysis developed in time series econometrics are also useful for carrying out step 5 even when the opportunity to perform controlled experiments is not available.

Example: Historical Interventions and Campylobacter Rates

In the first half of 1999, contamination of feeds with dioxin and PCBs in Belgium led to a scare and rapid decline in consumption of chicken. This was followed by a decline in campylobacteriosis cases, consistent with a causal relation between chicken consumption and campylobacteriosis risk.

On the other hand, Canada banned the veterinary use of fluoroquinolones (FQs), which had been used in turkey egg dips, in 1997. The rates of FQ-resistance in human Camplobacter isolates during the years before and after the ban were as follows (Gaudreau et al., 1998; Farber et al., 2001):

1995: 0.13, 1996: 0.13, 1997: 0.124, 1998: 0.14, 1999: 0.14.

The increase in resistance rates following withdrawal suggests that veterinary use of FQs was probably not the main driver of resistance rates in these years.

Step 4 requires testing the implications of a causal theory. A *theory* specifies constraints on the possible joint values of observable variables. If values are observed for some variables, then the theory predicts or constrains the values of other variables. A *causal theory* specifies how changing the values of some variables will cause values of other variables to change.

Example: A Simple Causal Theory for Impacts of Smoking Prevention

The following DAG summarizes the logical structure of a simplified causal theory about the relation between observed smoking and mortality data. It contains an unobserved confounder, labeled *Unhealthy lifestyle,* shown in parentheses to indicate that it is an unobserved ("latent") variable.

```
        (Unhealthy lifestyle)
         ↓        ↓         ↓
Smoking  (Exercise)  Poor diet
         ↓        ↓         ↓
| Age-specific mortality risk |
```

This latent variable has *Smoking* and *Poor diet* as its observable correlates and *Exercise* as an unobserved correlate. The intended interpretation of the graph is that the probability distribution for each variable is determined by the values of the variables that point into it. According to this theory, preventing smoking (i.e., setting *Smoking* to 0) will have a direct impact on age-specific mortality risk. However, not all of the risk statistically associated with smoking compared to non-smoking in the population will necessarily disappear if smoking is prevented. If would-be smokers continue to have other risk factors (such as poor diet and exercise factors) that contribute to early mortality even in the absence of smoking, then setting Smoking to 0 but leaving *Exercise* and *Poor diet* unchanged will remove only part of the risk that is statistically associated with Smoking. How much of the total mortality risk in a population is "due" to smoking, in the sense that (a) it will disappear if smoking is prevented; or (b) smoking is to blame for it, is the key question addressed by *attributable risk* calculations.

Table 3 lists several approaches for testing whether causal theories are consistent with various types of data. Key ideas include examining time series to see how well one (or more) can predict another; ordering variables based on which ones determine the values of which others; identifying conditional independence relations among variables; and studying how correlations among variables may be explained by the paths joining them in a "path analysis" graph, assuming that the values of variables are linearly related to each other. Causal modeling in econometrics and artificial intelligence has been based largely on which sets of variables determine which others (Iwasaki and Simon, 1994; Sims 1990) and on conditional independence relations (Pearl, 2000.) Each approach has led to statistical tests and models that may help carry out the steps in Table 2. They are especially useful for steps 4 and 5, i.e., verifying (or learning) hypothesized causal structures from data and testing whether predicted effects follow changes in causal predecessors.

These methods are starting to be applied to epidemiological problems (Robins and Wasserman, 1998; Richardson and Gilks, 1993; Richardson and Leblond, 1997) and medical risk analysis (Owens et al., 1997). They can help to extend and support traditional epidemiological criteria for causation, e.g., by providing more specific definitions of "temporality" based on time series analysis and of "coherence" based on causal graph models. They also support the steps of the alternative approach outlined in Table 2.

Example: Using Epidemiological Data to Test Alternative DAG Models

In epidemiological risk assessments, it is common to find several variables associated with each other in ways that suggest several alternative possible causal explanations. For example, consumption of beer and alcohol, fatty foods, and tobacco products may be statistically associated with each other, with other aspects of lifestyle (e.g., poor exercise habits) and with occupational factors (e.g., blue-collar vs. white-collar work), as well as with adverse health outcomes such as lung cancer or ischemic heart disease. Then it is natural to ask which of these factors actually help to cause the adverse health responses, instead of merely being associated with them.

The following three causal graph models are all consistent with mutual pairwise positive associations among a *Lifestyle* variable, an *Exposure* variable, and an adverse health *Response* variable.

Causal model 1: *Exposure* → *Response* ← *Lifestyle*
Causal model 2: *Exposure* ← *Lifestyle* → *Response*
Causal model 3: *Lifestyle* → *Exposure* → *Response*

Section 4 discusses the exact meanings and intended interpretations of such graphs. In models 1 and 3, exposure causes a risk of adverse health effect (model 3) or contributes to causing it (model 1), whereas in model 2, *Lifestyle* is a confounder that *explains away* the positive association between exposure and response (Pearl, 2000). To discover which of these competing models is correct based on data, it suffices to test their implied conditional independence relations. (For simplicity, we assume that exactly one of the three models is correct, but that data must be used to discover which one.) If *Response* is conditionally independent of *Exposure* given *Lifestyle*, abbreviated as (*Response* ⊥ *Exposure* | *Lifestyle*), then model 2 is correct and models 1 and 3 are not. If (*Response* ⊥ *Lifestyle* | *Exposure*), then model 3 is correct. If neither of these two conditions holds, then model 1 is correct. (Recall that, by definition, $(A \perp B \mid C)$ if and only if $\Pr(A \mid B, C) = \Pr(A \mid C)$, as in the diagram $B \rightarrow C \rightarrow A$ or in $B \leftarrow C \rightarrow A$.) This relation can be tested by determining whether the association between A and B vanishes at each level of C or, more generally, after conditioning or "splitting" on C, in classification tree terminology. Thus, classification tree analysis or other statistical tests for conditional independence provide a means to eliminate models that are inconsistent with data.

3.3 Bayesian Evidential Reasoning and Refutationism

Steps 2 and 3 in Table 2 consist of eliminating or refuting biases and confounding as plausible explanations for observed exposure-response associations. This section reviews the logic and methodology of the "refutationist" approach for establishing plausible causation. Sound evidential reasoning about probable causation can be guided and constrained by the logic of Bayes' rule. Even a simple abstract analysis suffices to make the main points.

Accordingly, let P denote a causal proposition of interest, e.g.,

P = "exposure to antibiotic X increases risk of adverse response Y".

Let

- Q = body of evidence offered in support of P. Q is typically a set of statements, e.g., about compatibility of P with the criteria in Table 3.
- Pr(P) = prior probability of P before examining the evidence in Q. (This probability need not be known to make the logical points below.)
- Pr(Q | P) = conditional probability of observing evidence Q given that P is true.
- Pr(Q | not-P) = conditional probability of observing evidence Q given that P is not true. (For simplicity, P is modeled here as a Boolean proposition that is either true or false, without attempting to consider important refinements such as by how much exposure increases risk.)

Table 3: Alternative approaches to testing causal models

DATA	HYPOTHESIS	TEST	PRINCIPLE
Two time series	**Temporal Causation:** The association between two time series is causal.	Granger-type tests (Granger, 1980; Sims, 1990; Boudjellaba, 1992; Hosoya, 1997)	The cause occurs before the effect and contains unique information about it.
Multiple variable, multiple periods	**Exogeneity**: A variable is determined from outside a system of equations (i.e., from outside a model)	Tests for exogeneity (Geweke, 1984; Ericsson and Irons, 1994)	Future values of exogenous variables do not help to predict past values of endogenous ones.
Multiple variables	**Conditional Independence:** One set of variables (e.g., health effects) is conditionally independent of a set of proposed causes, given the values of intervening variables. http://www.stat.washington.edu/graphical97/titles.html	Directed graph tests, tests for d-separation (Jensen, 1996; Pearl, 1996; Shafer, 1996; Yao and Tritchler, 1996). http://singapore.cs.ucla.edu/frl_papers.html http://weber.u.washington.edu/~soc582/henryk8.html	If X causes Y and Y causes Z, i.e., $X \rightarrow Y \rightarrow Z$, then the positive association between X and Z should disappear when conditioned on the level of Y.
Correlations among multiple variables	**Structural Equation Models:** A system of linear equations is consistent with a postulated causal structure, represented by a path diagram (Bentler and Stein, 1992).	Path analysis (for linear models) (Bentler and Stein, 1992, Heise, 1975; Kenny, 1979; Yao and Tritchler, 1996; Cohen et al., 1996) http://nw3.nai.net/~dakenny/sem.htm	Linear effects are transmitted along directed arrows in a causal graph from some variables to others. Effects are multiplied along paths and summed over paths.
Multi-equation model relating values of variables	**Causal Ordering:** One variable precedes another in the causal graph showing what is determined from what. (Iwasaki 94)	Simon-type algorithms for partial causal ordering of model variables (Iwasaki and Simon, 1994; Simon, 1977) http://net.cs.utexas.edu/users/clancy/dissertation/node25.html	Some subsets of variables suffice to determine their own values and the values of other variables. Thus, a system of equations creates a causal partial ordering among its variables.

Bayes' rule quantifies the posterior probability of P *given* the evidence Q as follows (using "|" to mean "given that" or "conditioned on"):

$$\Pr(P \mid Q) = \Pr(Q \mid P)\Pr(P) / [\Pr(Q \mid P)\Pr(P) + \Pr(Q \mid \text{not-}P)(1 - \Pr(P))].$$

For example, if Pr(Q | P) = Pr(Q | not-P), then necessarily Pr(P | Q) = Pr(P), i.e., Q contains no useful information about whether P is true. Thus, correct probabilistic reasoning about the impact of evidence Q on the probability that hypothesis P is true depends on both the likelihood of Q given P and also on the likelihood of Q holds given that P is *not* true, Pr(Q | not-P). Other things being equal, the support for P given by Q decreases as the probability of non-causal explanations for Q increases. (That is, Pr(P | Q) decreases as Pr(Q | not-P) increases.) Therefore, causal inference based on evidential reasoning (compatible with Bayes' rule) requires refuting or rendering implausible non-causal explanations for the evidence. Otherwise, Q (e.g., the usual epidemiological criteria) may convey little or no information about the plausibility of a causal hypothesis such as that exposure causes increased risk of response.

The importance of refuting non-causal explanations has also been derived, without appeal to Bayes' rule or evidential reasoning, within the well-developed framework of regression-based causal analysis and causal inference for multivariate linear models in the social sciences. For example, Kenny (1979, pp 50-51) states that:

> "Regression coefficients can be interpreted as causal coefficients if certain assumptions are met. These assumptions are the same as those of multiple regression [i.e., statistically independent, normally distributed sampling errors across individuals, variance of errors is not a function of the predictor variables, and error terms are uncorrelated with the exogenous predictor variables]. The assumption of *uncorrelated errors* implies that:
> a) The endogenous variable must not cause any exogenous variable: that is, there is no reverse causation.
> b) The exogeneous variables must be measured without error and with perfect validity.
> c) None of the unmeasured causes must cause any of the exogenous variables: that is, there are no common causes, or third variables.
>
> Obviously, lack of reverse causation, perfect measurement, and lack of common causes are rather stringent assumptions."

The monograph then examines methods to refute reverse causation, measurement errors, and common causes as potential non-causal explanations for observed associations between the explanatory ("exogenous") and dependent ("endogenous") variables.

Many epidemiologists have recognized the importance for valid causal inference of refuting competing explanations for observed exposure-response associations (Maclure, 1991). The resulting *refutationist agenda* consists of two main parts:

- Systematically enumerate possible alternative non-causal explanations for an identified exposure-response association. These alternatives usually include potential *biases* and *confounders*, and various taxonomies have been proposed to systematically describe them (Maclure, 1991; Choi and Noseworthy, 1992). Table 4 lists biases and confounders that can potentially explain away exposure-response associations unless they can be refuted by analysis or design. Until they are refuted, they must be considered threats to the validity of causal inferences based on the data (Campbell and Stanley, 1963).

- Develop statistical tests and data analysis methods to refute and/or bound the importance of each potential non-causal explanation (Choi and Noseworthy, 1992; Greenland, 1996; Nurminen, 1983). For convenience, Tables 4 and 5 list the main threats and modeling choices that can create non-causal associations and recommended statistical techniques for refuting them as explanations.

Although this is a busy agenda, it has largely been accomplished by the efforts of many epidemiologists and statisticians. Chapter 3 discusses potential confounders, data collection biases (mainly selection biases and information biases) and data analysis biases (mainly modeling biases, interpretation biases, and reporting biases) identified in the epidemiological and statistical literatures. It also recommends statistical study design and data analysis methods (See Table 5) for refuting these confounders and biases as potential explanations for exposure-response associations, correcting for their effects, or showing them to be unimportant if that is so.

The need to refute non-causal explanations in analyses of causation seems logically undeniable. And the viability of eliminating competing non-causal hypotheses as a strategy for establishing causation from observational study data has been well demonstrated in the social statistics (e.g., Campbell and Stanley, 1963). Yet, the refutationist agenda can be perceived as unhelpful when it is confined to merely identifying logical possibilities without also addressing their plausibility and the likely magnitudes of their impacts on risk estimates. For example, Savitz et al. (1990) state that "Biases that challenge a causal interpretation can always be hypothesized. ...It is essential to go beyond enumerating scenarios of bias by clearly distinguishing the improbable from the probable and the important from the unimportant." They argue that those who do not like a causal interpretation of epidemiological data might readily construct an impenetrable thicket of speculative hypothetical potential biases and confounders that can not all be refuted with available resources. This strategy could prevent any conclusions about causation from being drawn when common sense and sound policy would be better served by accepting that causation is plausible, even if it is not practical to refute all conceivable alternative explanations.

On the other hand, to accept an association as causal when equally plausible alternative non-causal explanations are available runs the reverse risk. Interpreting a statistical association as causal without rigorously examining and excluding competing hypotheses may make it too easy to launch expensive control actions that would be effective if the association were causal, but that will not produce the anticipated benefits in practice.

Table 4: Potential threats to valid causal inference in epidemiological studies

A. Threats from analysis and modeling choices

1. Multiple hypothesis testing/multiple comparisons bias.
2. Misspecified statistical models
   - Ignored estimation and classification errors in independent variables.
   - Omitted heterogeneity in individual response probabilities/parameters
   - Omitted explanatory variables (including omitted confounders)
   - Wrong model form
3. Biases due to data aggregation (e.g., Simpson's Paradox) (Saari, 1987) and to quantization of measurements.

B. Threats due to study design limitations

4. *Non-randomly missing and censored data* in which the censoring pattern is "informative" (associated with the dependent and/or independent variables)
   Example: If cause-of-death information is more complete among exposed than among unexposed workers, then relative over-counting of lung cancers may occur in the exposed group.
5. *Sampling and selection artifacts* that make sample observations non-representative of the population for which inferences are to be drawn.
   Examples:
   - Artifacts in which the experience of being surveyed changes reported job classifications, exposure histories, or lung cancer incidence.
   - Artifacts in which occurrence of the response (e.g., lung cancer) changes reporting behavior (e.g., reporting completeness) for deaths or exposures among lung cancer patients and their families.

   Other sampling and selection artifacts in pre-post designs include *history* (changes in exposure coincide with other events that effect response rates, or *vice versa), maturation, testing, instrumentation, regression to the mean, selection,* and *differential mortality* artifacts (Campbell and Stanley, 1963).

C. Threats due to patterns in the data

6. Multicollinearity of variables makes effects of individual variables impossible to isolate (e.g., because only their sums affect observations)
7. Interactions among independent variables (may obscure main effects)

Table 5: Some potential statistical problems and suggested solutions

Potential Problem	**Suggested Solutions**
Sample selection biases	Randomly sample *all* cohort members if possible
Data set selection biases	Use causal graph models to integrate diverse data sets
Variable selection bias (including covariates)	Bootstrap variable selection, Bayesian model averaging (BMA), cross-validation for variable selection. Include potential confounders in an explicit DAG model
Variable coding bias	Automated variable-coding methods (e.g., classification trees). Don't code/discretize continuous variables.
Choice of exposure metric or dose metric	Use multiple variables (e.g., concentration and time). (Don't combine.)
Choice of response metric	Define responses as conditional survival functions and/or transition rates among observed health states
Model form selection bias and uncertainty about the correct model	Non-parametric models (e.g., smoothers, wavelets); Bayesian Model-Averaging. Report model diagnostics and sensitivity analyses of results to model forms
Missing data	Data augmentation, EM algorithm, MCMC
Measurement errors in explanatory variables	Bayesian measurement error models, data augmentation and other missing-data techniques
Heterogeneity	Latent variable and mixture distribution models, frailty models of inter-individual variability
Bises in interpreting and reporting results	Report results (e.g., full posterior PDFs) *conditioned* on choices of data, models, assumptions, and statistical methods. Show sensitivities of results to these choices.

A partial solution to this dilemma is to focus on those non-causal explanations that appear to be *likely* and *important* (Savitz et al., 1990) i.e., those (if any) that might plausibly explain most or all of the observed exposure-response associations. Appropriate data analysis methods can often reveal which potential biases and confounders are most likely to provide non-causal explanations in specific studies. They can also help to eliminate logically conceivable biases that do not, in practice, play a large role. Most importantly, they can help to eliminate the most likely and important non-causal explanations when those do not, in fact, apply. Chapter 3 discusses statistical design and test principles for eliminating non-causal explanations. Evidence that makes non-causal explanations unlikely makes causal explanations more likely, even if the evidence is not definitive. For convenience, Table 5 recapitulates some of the most important potential problems and suggested ways to avoid or solve them.

In summary, the refutationist approach suggests a key criterion for establishing causation for an observed exposure-response association: *have competing non-causal explanations been eliminated?* If so, then the hypothesis of causation is supported by the data used to refute them.

4. TESTING CAUSAL GRAPH MODELS WITH DATA

Eliminating competing explanations provides an indirect approach to establishing causation. A more direct approach is to use multivariate data to construct and test causal graph models of risk. This section examines techniques for using data to form and test causal theories. The following sections build on this background and extend it to consider how to automatically learn causal theories from data.

Suppose that a health risk data set, D, consists of measured, estimated, or observed values (possibly including some missing or incorrect ones) for each of *m* variables for each of *n* individuals. The recorded data values typically summarize estimated individual exposure histories, health responses and values of covariates representing attributes, properties, events, and conditions describing individuals and their histories. From this *n x m* data matrix, it is desired to construct a *causal graph* (Jensen, 1996), i.e., a directed acyclic graph (DAG) showing dependence relations (represented by arrows, also called *arcs*) and conditional independence relations (represented by absence of arrows) among the variables.

A causal graph is a DAG model that represents dependence and conditional independence relations among variables for individuals in a data set, and perhaps for a larger population from which they were sampled. Such graphs address causation by showing which variables can affect which others and, conversely, which variables are independent of which others. They can also express probabilistic relations among variables and then use them to infer probable values of output variables, e.g., health responses, from values of input variables, e.g., exposure scenarios.

Case Study: A Causal Graph Model for Campylobacetriosis

The following table is a small excerpt developed from the CDC case-control data on risk factors for campylobacteriosis. It shows $m = 5$ variables (out of over 120 in the complete data set) for $n = 6$ individuals.

CACO	SEX	SLOWFD	FASTFD	FARM
1	M	0	0	0
0	M	2	0	1
1	M	2	0	0
0	M	0	1	0
1	F	1	1	0
1	F	0	0	0

The variables are: case-control status (CACO = 1 for cases, 0 for controls), SEX of the respondent, total non fast food restaurant meals consumed in the prior 7 days

(SLOWFD), total fast food restaurant meals consumed in the prior 7 days (FASTFD), living on or visiting a farm (FARM). From such a data table (but for all individuals and variables), we wish to learn *what are the causes of increased risk of campylobacteriosis?* Ideally, an algorithm would extract a causal graph relating the variables directly from the data. This section focuses on the easier problem of using the data to manually construct hypothesized causal graphs. The techniques introduced can also be used, as shown, to test the consistency of a hypothesized causal graph structure such as the above one with a data set.

Several methods can help to rapidly identify the factors that are most strongly associated with increased risk of campylobacter. For example, Tree A shows that travel outside the USA (TRAVEL), drinking raw milk (RAWMILK) or water (DRINK), living on (LIVFARM) or visiting (VISFARM) a farm, and contact with a puppy (ANI08) or chicken (ANI01) are significantly associated with risk (i.e., case status, CACO), even though some (e.g., RAWMILK) affect only a small sub-population. The 1,937 people who answered "No" (0) to these seven questions (i.e., did you travel outside the USA, drink raw milk or water, live on or visit a farm, or have contact with a puppy or chicken in the 7 days prior) are analyzed further below.

Similarly, multiple linear or logistic regression with stepwise variable selection can quickly screen for risk factors, provided that one is willing to risk overlooking non-linear (e.g., u-shaped or n-shaped) relations and to accept the excess false positives for main effects and potential false negatives for interactions typical of such procedures (Raftery et al., 1997; Greenland, 1989). For example, Table 6 lists variables identified by multiple linear regression with forward stepwise variable selection and mean substitution for missing data values.

Table 6: Selected variables identified by forward stepwise variable selection

VARIABLE	MEANING	t(3386)	p-level
Intercpt		4.83	.0000
TRAVEL	Travel outside USA	6.30	.0000
ANI08	Contact with puppy	4.54	.0000
CHK09D	Chicken wings cooked at a commercial establishment	3.12	.0018
CHK04D	Rotisserie chicken cooked at a commercial establishment	3.17	.0015
ANI01	Contact with chicken	3.02	.0026
DRINK	Drink raw water	2.94	.0033
VISFARM	Visit farm	2.71	.0068
ANI07	Contact with calf	2.51	.0121
CHK08PIN	Microwaved chicken pink inside	2.50	.0126
CHK06PIN	Fried chicken pink inside	2.37	.0180
RAWMILK	Drink raw milk	1.84	.0657
ANI11	Contact with horse	-2.15	.0320
CHK03A	Broiled chicken cooked at home	-2.25	.0242
OCHK1A	Other chicken cooked in home	-2.94	.0033
TOUCHICK	Touch raw chicken	-2.96	.0031
CHK05A	Outdoor grilled chicken cooked at home	-3.13	.0018
CHK06A	Fried chicken cooked in home	-3.78	.0002
CHK10A	Chicken fingers/nuggets/patties in home	-5.22	.0000
CHK07A	Stir fry chicken cooked in home	-5.42	.0000

Tree A: Some potential risk factors for campylobacteriosis

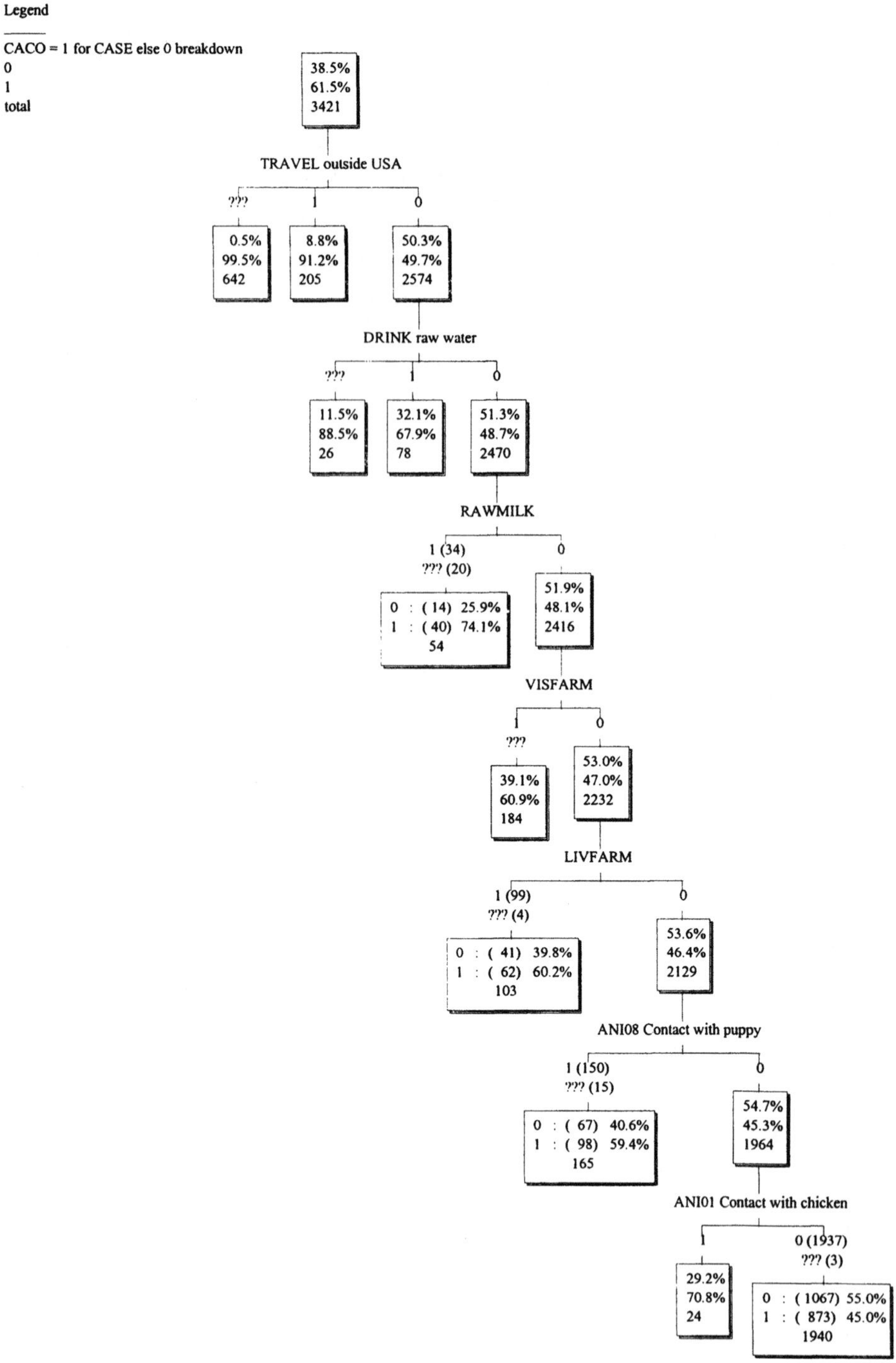

Similar results follow from logistic regression with stepwise or BMA variable selection. Variables in bold were also identified in Tree A.

It is noticeable that risk is positively associated with chicken prepared at a commercial establishment but negatively associated with chicken cooked at home (in any style, e.g., fried, broiled, nuggets, etc. – CHK variables ending in "A" are cooked at home while those ending in "D" are cooked at commercial establishments) This qualitative pattern might suggest the tentative hypothesis that chicken meals (or other meals) prepared at home reduce risk *because* they reduce the frequency of eating at commercial establishments. This hypothesis can be diagrammed in DAG form as:

Home-cooked meals $\xrightarrow{-}$ Restaurant meals $\xrightarrow{+}$ Restaurant chicken $\xrightarrow{+}$ Risk

Model 1 shows an instance of this schema for the relation between fried chicken eaten at home and reduction in risk. To determine whether this hypothesis might be correct, i.e., whether it is consistent with the data, we can test its implications.

Model 1: Eat chicken fried at home (CHK06A) → Home-cooked meals

↓ -

Risk ← commercially cooked chicken eaten ← Meals eaten out of home

For example, it suggests that the number of non fast-food restaurant meals in the past week, SLOWFD, should be negatively associated with the number of home-cooked fried chicken meals eaten in the past week, CHK06A. The following tree for SLOWFD shows that this is indeed the case:

Legend

SLOWFD # of restaurant meals breakdown

???
0
1
2
3
4 or more
total

???	:	(83)	4.3%
0	:	(752)	38.8%
1	:	(463)	23.9%
2	:	(236)	12.2%
3	:	(131)	6.8%
4 or more	:	(272)	14.0%
		1937	

CHK06A fried chicken at home

??? (1565)

???	:	(70)	4.5%
0	:	(581)	37.1%
1	:	(386)	24.7%
2	:	(196)	12.5%
3	:	(107)	6.8%
4 or more	:	(225)	14.4%
		1565	

0 (163)

???	:	(7)	4.3%
0	:	(47)	28.8%
1	:	(36)	22.1%
2	:	(19)	11.7%
3	:	(21)	12.9%
4 or more	:	(33)	20.2%
		163	

1 (209)

???	:	(6)	2.9%
0	:	(124)	59.3%
1	:	(41)	19.6%
2	:	(21)	10.0%
3	:	(3)	1.4%
4 or more	:	(14)	6.7%
		209	

Similarly, a classification trees can be grown for each node (variable) in Model 1 in turn to check whether its hypothesized parent(s) appear in its tree and whether more remote ancestors (a) Appear as potential top splits; but (b) Drop off the list of potential splits after conditioning (i.e., splitting) on its parents.

To further test and refine model 1, Tree B was grown for the 1,937 respondents who replied "No" to the questions corresponding to the seven risk factors in Tree A.

Tree B: Further risk factors for campylobacteriosis

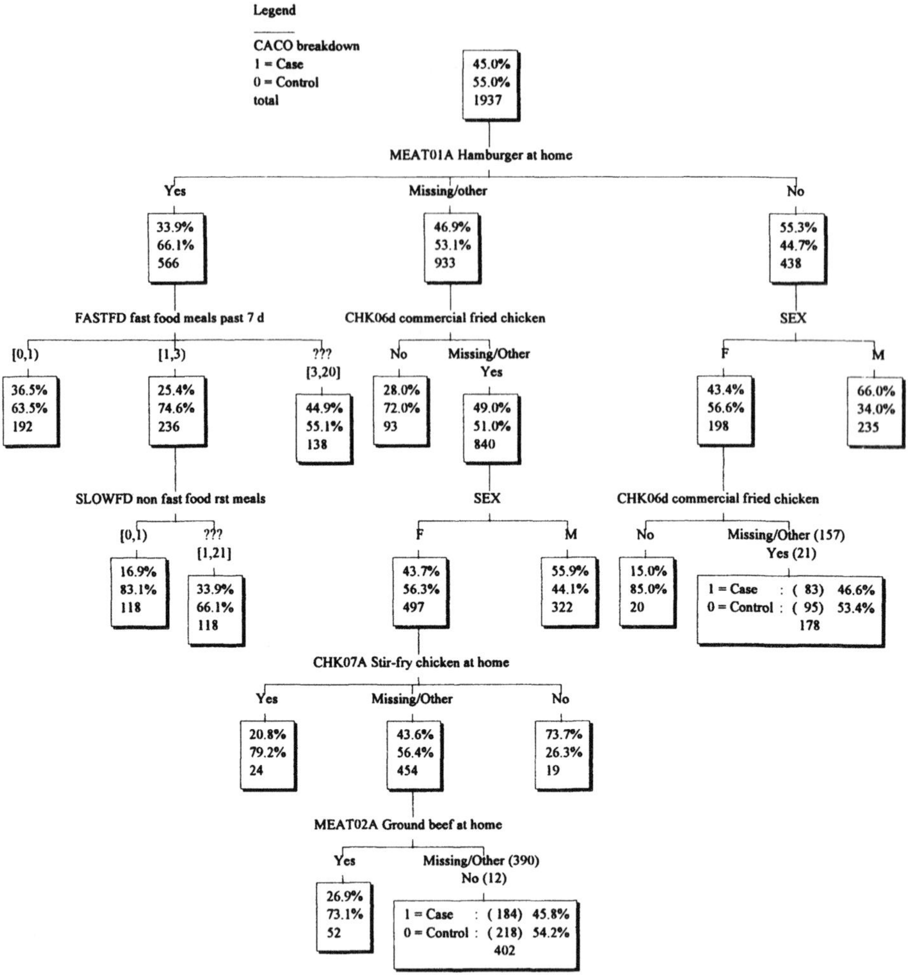

The most important additional risk factors in Tree B appear to be related to consumption of chicken, hamburger, and fast food (FASTFD = number of fast-food meals in week prior), with non-fast food restaurant meals (SLOWFD) and subject's sex also appearing. Chicken cooked at home is associated with lower risk.

A single data set can be described by many trees, obtained by changing the order in which candidate splits at nodes are selected for expansion. However, if there is adequate variability in the sample data, then *all* parents of a node should appear in *every* tree for that node, and no more remote ancestors should have this property. In particular, splitting a variable such as CACO first on all of its parents (in any order) should make all of its more remote ancestors drop out as candidate splits at the leaves of the parent-only sub-tree, since the node's value is conditionally independent of them, given the values of its parents. Applying this principle to CACO and summarizing the results by introducing the aggregate categories "Commercial chicken" and "Home-cooked chicken" leads to Model 2. (TOUCHIK has been subsumed into "Eat home-cooked chicken" for convenience.) It can easily be checked that Tree B conforms to this DAG model.

Model 2: Pink chicken, pink pork chops

↓+

Eat commercially cooked chicken →(+) [Risk of CP (CACO)] ← SEX

+↑ (-) ↑-

{FASTFD, SLOWFD} ← Eat home-cooked chicken

In this graph, signs have been placed next to arrows where there is a consistent direction (Positive or negative) of association between variables in all subtrees examined. Thus, a plus sign on an arrow from A to B means that B is stochastically increasing in A, regardless of the values of other variables that may affect A.

This approach to constructing causal graphs can be applied recursively to seek the parents of each node in the graph. For example, a respondent's age in years (AGE_YR) and SEX can be added as parents of FASTFD. Moreover, rather than substituting for each other as might be expected, FASTFD and SLOWFD turn out to be positively associated with each other, suggesting that both are indicators of a latent variable that might be interpreted as "Propensity to dine out." Similarly, "FASTFD" and "SLOWFD" turn out to be parents for Pink Chicken (specifically, CHK01PIN, "Oven roasted or baked chicken, pink inside", providing yet another path by which restaurant dining (fast food or not) may increase risk of campylobacteriosis.

This example illustrates how classification trees can be used to identify conditional independence relations and to help create causal graphs of the relations among variables. The main practical conclusions from the analysis are that (a) Restaurant dining appears to be a risk factor for campylobacteriosis in this data set; and (b) Home-cooked chicken (of several types) appears to be a protective factor, even after controlling for its indirect effects due to decreased restaurant dining.

4.1 Causal Graphs and Knowledge Representation

That conclusions about the probable health effects of exposures, activities or risk management decisions should be derived from explicit, well-documented, publicly disclosed models is increasingly regarded as a hallmark of useful risk assessment. The growing use of DAG models by decision and risk analysis practitioners supports this principle. A new generation of software tools for decision and risk analysis promotes the use of DAG models and interactive graphics displays to clarify the analytic basis, assumptions, and uncertainties involved in risk management decision-making (Buede, 1998). Causal graphs and related DAG models (including Bayesian belief nets, influence diagrams, and graph models from multivariate statistics, such as log-linear model graphs) have been used in medical expert systems to represent knowledge and beliefs about probabilistic relations among variables. Now they are being applied in epidemiology, biostatistics and risk analysis to assess multivariate exposure-covariate-risk relations.

Table 7 summarizes terminology and basic ideas for causal graphs and closely related DAG models, especially influence diagrams. An *influence diagram* is a DAG model that may include decision nodes or *choice nodes,* representing decision variables; *chance nodes* representing random variables, a *value node* representing the value or utility of outcome variables, and *deterministic nodes* representing deterministic functions of its inputs. In an influence diagram, arcs pointing into a decision node (i.e., choice node) indicate information available when the decision must be made. The value node indicate the values assigned to outcomes.

Figure 1 shows the top-level view of an influence diagram prepared using the commercial software package *Analytica*™ (originally developed at Carnegie-Mellon University and now sold by Lumina Decision Systems, Inc. of Palo Alto.) Double-clicking on a node will display its definition (e.g., a formula), title and description, type (choice, chance, function, or value, also indicated by its shape in the diagram), inputs, and outputs. The formula or probability table, distribution, etc. used to determine the node's value from the values of its inputs will also be displayed. Node types and defining formulas may be redefined by simply editing them in the model. For purposes of sensitivity analysis, an input such as the "Emissions Reduction" decision variable in Figure 1 may be defined as a list of different values, each of which will then be automatically propagated through the model.

Once the model has been created, results for any node can be obtained by simply clicking on it and then choosing "Result" from the menu bar. For example, in Figure 1, the "Total Cost" node is the output of greatest interest. Double clicking on it shows that its definition is given by a formula:

Table 7: Basic ideas of causal graphs and related DAG models

Terminology

Directed Acyclic Graph (DAG) models include

- Causal graphs (nodes represent variables in a data set)
- Bayesian Belief Networks (may represent subjective beliefs)
- Influence Diagrams (can include decision variables and a value node)

Representing Causal Knowledge by DAGs

Example: Exposure → Internal dose → Response

- Nodes represent model *variables* (e.g., random variables, decision variables, and deterministic functions of other variables)
- Absent arrows represent *conditional independence* relations. Arrows represent conditional probability or function relations among variables.
- The *conditional PDF* for a node's value, given the values of its parents, is stored at the node. Marginal PDFs are stored at input nodes. Expected values are calculated at output (value/utility) nodes in influence diagrams.

Model Validation Tests

- Check the internal consistency of a DAG model by testing its implied *composition relations.*
- Check consistency with data by testing the DAG model's implied *conditional independence relations.*

Using DAG Models for Inference, Prediction, and Control

- Graph algorithms for inference (e.g., Arc Reversal) implement Bayes' Rule for networks of variables. These map observed values of some variables to posterior PDFs for all variables.
- DAGs allow *different sources* of knowledge, evidence, and data to be integrated into a single testable predictive model.
- Predictions are made by propagating input values or distributions through the DAG model, e.g., using Monte Carlo simulation.
- Influence Diagram algorithms identify the values for controllable inputs (i.e., decision variables) to maximize the expected value of an output variable (e.g., expected utility of consequences).

Definition: (Control_cost + (Value_of_a_life*Excess_deaths))

while its text description is as follows:

Description: Total cost to society, calculated as the difference between the value of the lives saved, and the cost of the Emissions Reduction required to save those lives.

Figure 1: An Influence Diagram Drawn in *Analytica*™

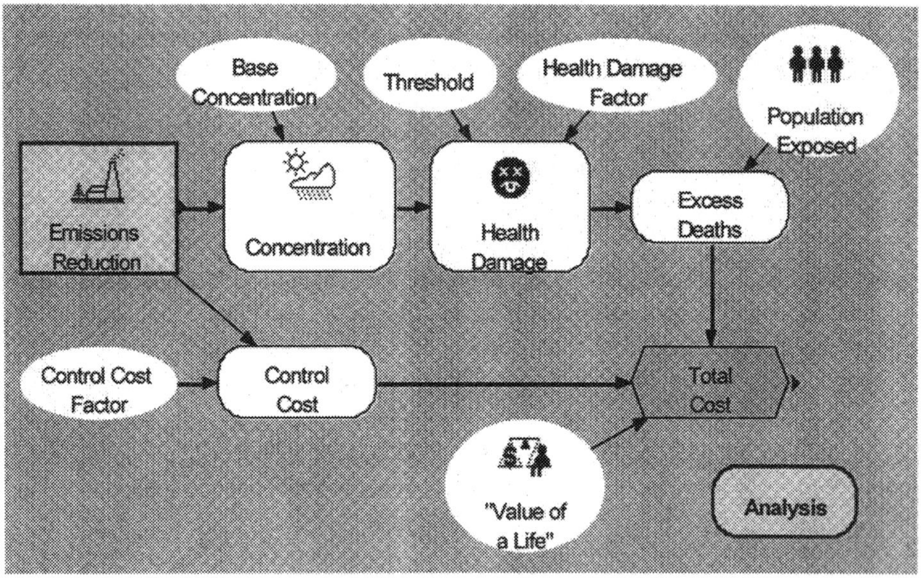

(The same dialog box also lists the inputs and outputs for this node, shows that its units are $/yr., and specifies "Total Cost" as its title and "Objective" as its type.) To view results, one may simply click on the "Total Cost" value node and then select "Mean Value" as the type of "Result" to display. Doing so promptly yields the graph in Figure 2. This shows that choosing an Emissions Reduction factor of about 0.7 minimizes the expected value of "Total Cost". To view uncertainty about this conclusion, one selects a different "Result" to display, such as Cumulative Probability, Probability Density, or Probability Bands. Choosing Probability Bands in this example leads to Figure 3, in which the different curves represent upper and lower percentiles (shown in the key) from the probability distributions for "Total Cost" conditioned on the different values (i.e., decisions) for the "Emissions Reduction" decision node. Execution time is on the order of one second.

Figure 2: Sample output: Mean total cost *vs.* emissions reduction

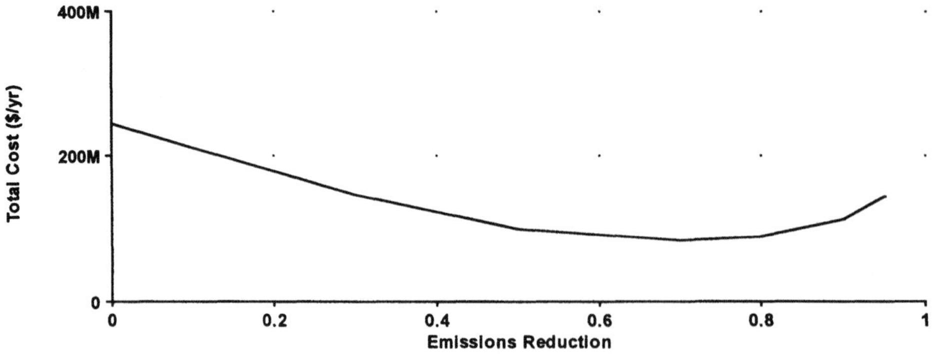

Figure 3: Probability bands for total cost *vs.* emissions reduction

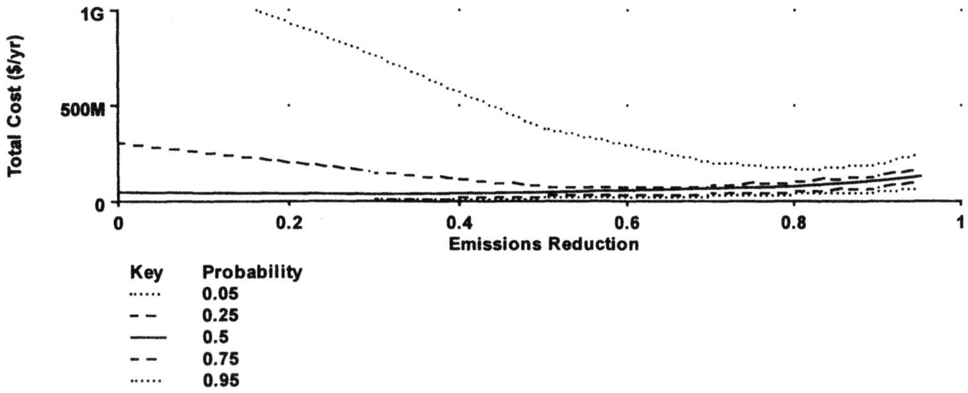

A human modeler drew the DAG model in Figure 1 and supplied the formulas and conditional probability tables at its nodes. It therefore reflects subjective judgments – albeit perhaps ones well-grounded in common sense or expertise – about how the world works and what causes what. Many expert systems and expert decision support systems use influence diagrams and related DAG models to represent expert knowledge and beliefs about the relations among variables. Indeed, knowledge from multiple experts can be combined by having them first represent and share their mental models as explicit causal graphs or influence diagrams and then work on reconciling the graphs (and the conditional and marginal distributions stored at their nodes), rather than directly seeking to combine or reconcile numerical probability judgments that can be derived from the graph.

Techniques for eliciting, representing, and combining subjective judgments and expertise have also been examined by various regulatory agencies and applied research groups as potentially important, relatively low-cost approaches to risk assessment. However, a less subjective, more data-driven approach is often desired. Most health risk decisions made in public forums still emphasize data-based processes in which analyses of shared data and their implications are followed by deliberation about what control actions to take. The problem of how to develop objective models of exposure-risk and action-exposure causal relations from data is crucial for informing such data-driven analysis-deliberation approaches to risk analysis.

After reviewing various interpretations of causal graphs and DAG models and how they are used to represent knowledge of probabilistic and causal relations among variables, the following sections examine techniques for constructing them automatically from data and for using them to draw probabilistic inferences and to make causal predictions to support decisions.

4.2 Meaning of Causal Graphs

The intended interpretation of a causal graph is that the probability distribution for the value of each variable is determined by the values of the variables that point into it. Each variable is conditionally independent of all of its more remote ancestors in the graph, given the values of its immediate parents. (Thus, the graph $x \to y \to p$ implies that p is conditionally independent of x given y.) Input variables (i.e., those variables, such as x, that have no inward-pointing arrows) may represent random variables, constants, or decision variables.

Associated with each derived variable (i.e., node with inward-pointing arrows) in the graph is a conditional probability distribution for its value, given the values of the variables pointing into it. In many commercial causal graph and influence diagram software packages, including *Analytica*™, these conditional distributions are stored at the nodes in the graph, i.e., double-clicking a chance node will cause the table or parameters of its conditional distribution to be displayed. In the special case of deterministic nodes, a formula is stored at the node and the value of the node's variable is uniquely determined via the formula from the values of the nodes that point into it (its "parents" in the graph.)

Unconditional marginal distributions are stored at input nodes. They may be degenerate (i.e., place probability 1 on a specific value) for inputs representing known constants or decision variables. *Any* joint PDF may be expressed (or "factored") as a product of marginal and conditional PDFs for input and derived variables, respectively (Schachter, 1990). For example, $\Pr(x, y, p) = \Pr(x)\Pr(y \mid x)\Pr(p \mid y)$ is the factoring corresponding to the model graph $x \to y \to p$. Such factoring exploits the natural sparseness of most DAG model structures, indicating that not all variables directly affect each other, to develop a computationally efficient representation of joint probability distributions. If each of m variables has L possible levels, then the potential need to assess probabilities for L^m members of the sample space – a daunting task for all but the smallest model – can be replaced by the much simpler one of quantifying the marginal and conditional probability distributions at each node.

In general, let there be m variables in a model, arbitrarily numbered and denoted $x_1, x_2, \ldots, x_m$. Then the joint probability distribution of these variables may always be factored as:

$$\Pr(x_1, x_2, \ldots, x_m) = \Pr(x_1)\Pr(x_2 \mid x_1)\Pr(x_3 \mid x_2, x_1)\ldots \Pr(x_m \mid x_1, x_2, \ldots, x_{m-1}).$$

In each term, the conditioned variable (to the left of the "|") may be conditionally independent of some of the previous variables on which it is conditioned in this generic factoring. If so, these variables may be dropped from the term's list of conditioning variables (to the right of the "|") without affecting the value of the expression, leaving the factored form as:

$$Pr(x_1, x_2, \ldots, x_m) = \Pi_i Pr(x_i \mid \mathbf{Pa}_i) \text{ (DAG-based factoring of joint pdf)}$$

where $\mathbf{Pa}_i$ denotes the parents of x_i in the DAG. That is why, for example, the DAG $x \to y \to p$ allows the joint probability of the variables to be expressed as $Pr(x, y, p) = Pr(x)Pr(y \mid x)Pr(p \mid y)$. The factors in the product are typically much easier to evaluate than the original joint probability distribution. As discussed later, DAG representations of the factoring (which show the conditional independence relations among variables explicitly, by absence of arrows) also lend themselves to Gibbs sampling and other computationally efficient numerical approximation techniques. Graph algorithms also have been developed to propagate evidence through such graphs by conditioning on observations. These algorithms are distributed, i.e., calculations at each node depend only on local information, but the result is the same as if the whole joint distribution had been conditioned on the observations (using Bayes' rule) and then the resulting posterior joint distribution had been factored and represented by an updated DAG model.

Other causal interpretations of DAG models that add to the conditional independence meaning are often useful and may assist intuition, although they are not required for the theory based on conditional independence. For example, causal graphs may be used to represent causal dependency relations among quantities. As discussed by Iwasaki and Simon (1994), these relations "are based on the notion of *manipulability* of a variable by *intervention* into the system. When we say that 'variable B depends on variable A', which we will write as

$$A \to B,$$

we mean that there exists some mechanism that determines the value of B from that of A, whether it be a physical law, an actual device, or some other kind of communication path between the two variables. Thus, everything being equal, a change in A will necessarily result in a change in B."

In general DAG formulations of causal graphs, a change in A need not necessarily change the value of B. The relation between them may be stochastic or may be such that B has the same value for several different values of A. For example, if A has caused an irreversible change in B, then a further change in A may not change the value of B further. Nonetheless, interpretations of DAGs in which changes in input variables propagate among the other variables in a model along the arrows linking them may assist intuition. They are useful for many purposes in risk analysis, as well as in other areas of applied science and engineering. As noted by Iwasaki and Simon (*ibid,* p. 146) "The method for determining causality based on propagation of disturbances is generally consistent with, but somewhat less general than, the more formal method of causal ordering". The latter method, drawing on earlier econometric work on structural equation modeling, is based on dependency relations among variables in systems of structural equations (Bentler and Stein, 1992). It orders variables in such a way that earlier ones determine the values of later ones, i.e., the values of later ones can be derived from the values of earlier ones. Thus, causal orderings of variables in models can provide a further interpretation of, and motivation for, DAG graphs.

Diagrams such as Figure 1 suggest an additional meaning for causal graphs. In this figure, values generally propagate from nodes at the left (Emissions Reduction) and upper rows to nodes at the right (Excess Deaths) and lower rows (Total Cost). Changing the probability distribution of the inputs to a node is expected to change the probability distribution of its value based on the formula or conditional probability stored at that node. Moreover, this change is interpreted as causal, rather than merely reflecting statistical inference. For example, consider the link:

Concentration → *Health Damage*

This sub-model is intended to represent a causal exposure-response relation, meaning that changing exposure *Concentration* causes subsequent changes in population *Health Damage* according to a probabilistic relation that can be symbolized as Pr(*Health Damage* | *Concentration*). When the model is executed, each simulation run first determines a value for *Concentration* via Monte Carlo simulation. This value is supplied as an input to the *Health Damage* node, meaning that a value for *Health Damage* is sampled from the conditional probability distribution Pr(*Health Damage* | *Concentration*). Such a flow of sampled values from node to node along the arrows in the diagram represents a stronger concept of causation than that embodied in the conditional independence and composition relations alone. It includes a definite order in which values of variables are determined, reflecting temporal relations for the real-world phenomena being modeled as well as computational order within the model. From this standpoint, if *Health Damage* and *Concentration* were deterministically related by the formula:

Health Damage = q * *Concentration*,

then it would be correct to draw the diagram as:

Concentration → *Health Damage*

and to store the formula (*Health Damage* = q * *Concentration*) at the *Health Damage* node. It would be incorrect to draw it as:

Concentration ← *Health Damage*

and to store the formula (*Concentration* = *(1/q)*Health Damage*) at the *Concentration* node. The latter is incorrect not because it expresses wrong constraints on the joint values of the variables (the constraints are the same in both cases), nor because it misrepresents conditional independence relations among variables (it may represent them correctly). It is incorrect because it misrepresents the order in which uncertainties are resolved and the values of variables are determined.

More generally, the real-world risk processes described by influence diagrams and other DAG models evolve in time. Yet most DAG models, like that in Figure 1, do not explicitly show the role of time. (Some do, by including time-stamped variables as nodes, but graphs like Figure 1 are common.) More refined models can describe the time evolution of the processes being modeled, e.g., via systems of differential equation or by discrete-event simulation models that simulate the histories of individual entities and events. When such systems are presented and analyzed through simplified abstractions such as DAG models, details of timing in the dynamic relations among variables are lost or suppressed. Formulas and conditional probability relations at nodes typically only represent approximate long-run, steady-state, or equilibrium relations between inputs and outputs. In such cases, the directions of the arrows in a DAG model can be used to retain and indicate information about the temporal order in which changes in variables propagate and their values are determined.

Most theoretical results and algorithms for DAG models do not require that arrows indicate temporal orderings of variables. They only depend on the conditional independence and composition relations encoded in the DAG structure. But the stronger interpretation of diagrams such as Figure 1 in terms of temporal and causal orderings of their variables and *propagation* of values of variables along paths can assist intuition. It is often what human experts intend such diagrams to represent. Of course, when diagrams are learned directly from data, there is no guarantee that they will respect temporal and causal orderings unless these are enforced as constraints on DAG-learning (or DAG-construction) algorithms.

4.3 Testing Hypothesized Causal Graph Structures

Developing causal risk models from data requires qualitative and quantitative knowledge. The qualitative part can be organized and documented using causal graphs. The graphs imply statistically testable properties, as already discussed, namely: (a) Each variable (represented by a node in the graph) is *conditionally independent* of its more remote ancestors, given the values of its immediate parents; and (b) The input-output mappings represented by arrows in the graph can be *composed,* e.g., the relation between x and p in the DAG model

$$x \rightarrow y \rightarrow p$$

is the composition of the two mappings $x \rightarrow y$ and $y \rightarrow p$. [In the special case of linear relations, this means that the correlation coefficients for the pairs of variables satisfy

$$\rho_{px} = \rho_{py}\rho_{yx}.$$

The TETRAD II program (Glymour and Cooper, 1999) uses systems of constraints on correlation coefficients and partial correlation coefficients, as well as information on which correlations and partial correlations are estimated to be zero, to infer possible DAG structures from data, assuming that the relations among variables can be adequately characterized by correlations and partial (i.e., conditional) correlation coefficients.]

Both the conditional independence and the composition implications can be tested for validity if the variables involved can be measured and if adequately large and diverse samples of the values of the variables for different individuals are available. If the samples lack adequate variability (e.g., if some variables have no variance), then of course the effects of some variables on others may not be uniquely identified by the data.

Example: Composition of Conditional Probability Relations

In the DAG model:

$$X \rightarrow Y \rightarrow P$$

suppose that X, Y, and P are all binary (0-1) random variables and that **Pr(y | x)** and **Pr(p | y)** are defined as the following two matrices (probabilistic mappings):

Pr(y | x):	X = 0	X = 1
$Y = 0$	1	0.5
$Y = 1$	0	0.5

Pr(p | y):	Y = 0	Y = 1
$P = 0$	1	0.2
$P = 1$	0	0.8

(These are the two conditional probability tables that would be stored at nodes *Y* and *P*, respectively, in software packages.) Then their composition is the matrix product **Pr(p | x) = Pr(p | y)Pr(y | x)**:

Pr(p | x):	X = 0	X = 1
P = 0	1	0.6
P = 1	0	0.4

Given a probability distribution column vector for the input *X*, say π (with non-negative components summing to 1), the column vector **Pr(p | x)**π gives the corresponding probability distribution for *P*. More generally, if *X*, *Y*, and *P* have any number of levels, then the constraint implied by composition is:

$\Pr(P = \mathrm{p} \mid X = \mathrm{x}) = \Sigma_{\mathrm{y}}\Pr(P = \mathrm{p} \mid Y = \mathrm{y})\Pr(Y = \mathrm{y} \mid X = \mathrm{x})$, abbreviated
$\Pr(\mathrm{p} \mid \mathrm{x}) = \Sigma_{\mathrm{y}}\Pr(\mathrm{p} \mid \mathrm{y})\Pr(\mathrm{y} \mid \mathrm{x})$.

If *X*, *Y*, and *P* can all be measured for each individual, then the composition relation **Pr(p | x) = Pr(p | y)Pr(y | x)** can be used to test whether the model structure

$$X \rightarrow Y \rightarrow P$$

is consistent with the data. Each of the three 2 x 2 tables **Pr(p | y)**, **Pr(y | x)**, and **Pr(p | x)** can be estimated directly from the count data on *X*, *Y*, and *P* for the *N* individuals in the data set. If the product **Pr(p | y)Pr(y | x)** differs significantly from the matrix **Pr(p | x)**, then the model $X \rightarrow Y \rightarrow P$ is incorrect.

Notice that **Pr(p | x)** could be estimated from data on *X* and *P* alone, ignoring the measured values of *Y*. However, if the DAG structure of the model is correct, then including measurements of *Y* and making **Pr(p | x) = Pr(p | y)Pr(y | x)** a constraint can improve estimation of **Pr(p | x)**, as measured by criteria such as generalized least squares (GLS), maximized log-likelihood, or expected utility of decisions. Moreover, for any individual for whom *X* and *Y* have been measured and *P* is to be predicted, the prediction will in general be at least as good [as measured by the entropy of the probability distribution for *P* (Cover and Thomas, 1991) or by the expected utility of an optimized decision based on the prediction (Laffont, 1990)] if one conditions on both *X* and *Y* as if one conditions on only one of them. In general, a decision-maker never does worse (i.e., never has lower *ex ante* expected utility) by conditioning on more information (Laffont, 1990, Chapter 4). Thus, even though *P* is conditionally independent of *X* given *Y*, observations of *X* can help to improve estimation of the component relations **Pr(p | y)** and **Pr(y | x)**, and thus improve estimation of their composition, **Pr(p | x) = Pr(p | y)Pr(y | x)**. By thus reducing model uncertainty, composition relations can improve the prediction of the conditional probability distribution of *P* given an observed value of *X*.

Example: Composition and Reduction of Deterministic Models

Composition relations can be used to eliminate unobserved variables from DAG models, thus leading to more parsimonious "reduced-form" models relating observable quantities only. In general, this reduction should be done before Monte Carlo uncertainty analysis is used to quantify uncertainties about observable outputs. For example, suppose that the values of the three variables in the DAG model

$$x \to y \to p$$

satisfy the system of equations:

$$y = a + bx$$
$$p = c + dy.$$

Then the four parameters a through d may be combined to form the *reduced model* (Judge *et al.*, 1985):

$$p = a' + b'x$$

having the two reduced parameters

$$b' = bd \quad \text{and} \quad a' = (c + ad).$$

If y is a latent variable, i.e., only x and p can be observed, but not y, then only the reduced parameters can be estimated from the observations. The four original parameters (called *structural model parameters* in econometrics and path analysis) are not uniquely identifiable from the observed quantities, no matter how many observations are available.

Reducing a model by combining its equations so that only relations among observed quantities are described typically yields fewer reduced parameters that still determine the same constraints among observable quantities. In general, *uncertainty about reduced parameters is less than uncertainty about their constituent (structural) parameters.* To make this claim precise, recall that the entropy, measured in bits, of a discrete random variable X is defined as

$$\text{ent}(X) = -\Sigma_X \text{pr}(x)\log_2 \text{pr}(x),$$

where x ranges over all possible values of X and pr(x) = Pr(X = x) is the probability density assigned to value x. ent(X) may be interpreted as the expected amount of information (in bits) that will be gained by learning the value of X. For many purposes, ent(X) is a useful measure of the uncertainty about a discrete random variable (Buckley, 1985). (For a continuous random variable, the *differential entropy* is defined as

$$\text{ent}(X) = -\int p(x)\log_2 \text{pr}(x)dx$$

where pr(x) is the probability density of x and the integral is taken over the entire set of possible values of X.)

Let q be a vector of quantities (parameters or variables) in a DAG model. Let r(q) be the vector of reduced quantities (algebraic combinations or other functions of q) obtained by reducing the model. Then the uncertainty about r(q) is less than or equal to the uncertainty about q, i.e.,

$$\text{ent}[r(q)] \leq \text{ent}(q).$$

(Cover and Thomas, 1991, p. 43). Intuitively, r(q) may not distinguish among two or more values of q (i.e., r(.) may be many-to-one), so that less information is needed to determine the value of r(q) than to determine the value of q.

Monte Carlo uncertainty analyses of risk models sometimes mistakenly introduce unnecessary uncertainty in output probability distributions by not using reduced-form models. For example, in the structural model

$$y = a + bx$$
$$p = c + dy.$$

uncertainty about p might be estimated by first estimating probability distributions for x, a, b, c, and d; then sampling from the distributions of x, a, and b to obtain a distribution of y; and finally sampling from this distribution and from the distributions of c and d to obtain the estimated probability distribution of p. But this straight-forward approach introduces unnecessary uncertainty into the distribution of p by failing to exploit the fact that it depends on a, b, c, and d only through the reduced parameters $b' = bd$ and $a' = (c + ad)$. The uncertainty about (a, b, c, d), i.e., the entropy of its joint distribution, will be greater than or equal to the uncertainty about (a', b'). In general, as mentioned in Chapter 2, Monte-Carlo uncertainty analyses should be based on reduced-form models rather than on structural models.

For any two random variables X and Y, ent(X | Y) ≤ ent(X), with equality if and only if X and Y are independent. In words, *"conditioning reduces entropy"* or leaves it unchanged (Cover and Thomas, 1991, Theorem 16.1.5, p. 483). Thus, to minimize the entropy of the probability distribution for a predicted quantity, p, observed quantities that are relevant for estimating p should be included in the estimation process, even if they contain very large measurement errors. Doing so can never make the estimate of p worse (i.e., make its posterior PDF, after conditioning on observed quantities, have higher entropy) than it would be if some observations were ignored. This contrasts with recommendations from econometric procedures that seek to minimize mean square prediction error. The squared error loss criterion often prescribes that a subset of regressors should be used in preference to the complete set of regressors if some of them contain substantial measurement errors (Judge *et al.*, 1985, p. 866).

Example: Using Classification Trees to Test Conditional Independence

As previously illustrated for Campylobacter risks, classification trees can be used to find the conditional independence relations in a data set by exploiting the fact that (generically, assuming adequate variability in the sample) the dependent variable in a tree is conditionally independent of the variables *not* in the tree, given the values of the variables that *are* in the tree. For example, consider the following hypothesized causal model for lung cancer risk among women under age 40:

Beer Consumption ← *(Lifestyle)* → *Smoking Status* → *Lung Cancer.*

The model postulates an unobserved latent variable, *Lifestyle,* that explains both *Smoking Status* and *Beer Consumption.* In turn, *Smoking Status* affects the probability that the health response variable, *Lung Cancer*, will be present (1) instead of absent (0). Beer consumption has no direct causal impact on Lung Cancer in this model, despite their statistical association.

To determine whether this hypothesized model might be correct, suppose that data on women under age 40 are assembled and that they turn out to be described by the following classification tree:

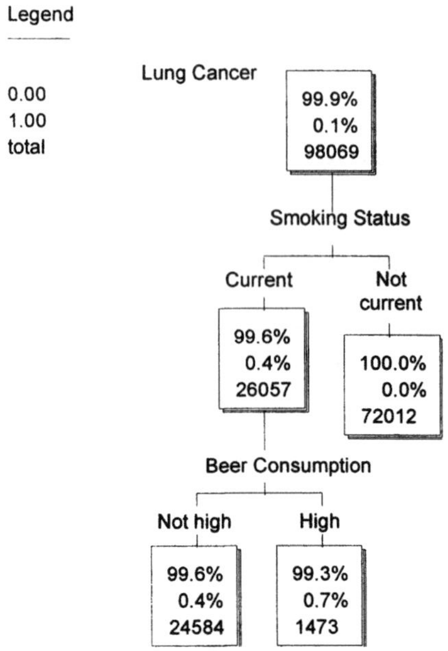

Then the hypothesized model is inconsistent with these data. If it were correct, then *Lung Cancer* would be conditionally independent of *Beer Consumption*, given *Smoking Status*. Hence, Beer Consumption would not appear below *Smoking Status* in the above classification tree.

The following alternative causal model is consistent with the data:

Beer Consumption → *Lung Cancer* ← *Smoking Status*

If *Smoking Status* enters the tree below *Beer Consumption* when *Beer Consumption* is made the first split, this would further confirm the model, suggesting that both beer and smoking are risk factors for lung cancer in this population. This illustrates how classification tree analysis can be used to help test hypothesized models and/or to help discover the correct DAG model structure from data.

Once a DAG model structure has been defined and verified by testing its implied conditional independence relations, the next step is to quantify its component input-output relations. Classification trees may also be used for this purpose. For example, for the DAG model in the preceding example:

Beer Consumption → Lung Cancer ← Smoking Status,

the above classification tree shows that the conditional probability table to be stored at the *Lung Cancer* node is as follows:

Beer Consumption	Smoking Status	Pr(Lung Cancer = 1)
Not high	Not current smoker	0.0%
High	Not current smoker	0.0%
Not high	Current smoker	0.4%
High	Current smoker	0.7%

For each combination of values of the two inputs shown in the first two columns, the third column gives the empirical estimate of the conditional probability of the output values for the node (here, just the probability of a 1 instead of a 0). Storing such a conditional probability table at each node of a DAG model provides the information needed to infer probabilities of different consequences (i.e., values at output nodes) from input values or joint probability distributions of input values.

4.4 Creating Causal Graph Structures from Data

Rather than manually generating alternative causal hypotheses and testing them for consistency with data, one may use any of several computer programs that automatically search for causal graph structures that are consistent with the composition and conditional independence relations found in multivariate data sets. Alternatively, one can also use a mix of knowledge-based constraints and automated search heuristics to discover potential causal models from data.

Example: A Partial Causal Graph Model Derived from Data

To investigate whether occupational exposure to diesel exhaust might increase the risk of lung cancer, the following partial causal graph for lung cancer risk among train workers was derived from occupational exposure estimates and follow-up mortality and cause-of-death data (Cox, 1997).

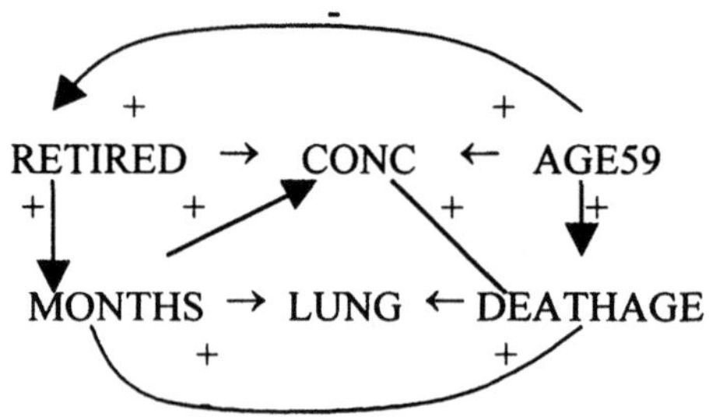

The variables are as follows: AGE59 = age of the worker in 1959; RETIRED = year that worker retired; MONTHS = estimated months of exposure to diesel exhaust; CONC = estimated average concentration of occupational diesel exhaust; DEATHAGE = worker's age at death; LUNG = 1 if worker died with lung cancer, else 0. Data are from Garshick et al. (1988). The graph is only partial, because the direction(s) of the arrows relating CONC and MONTHS to DEATHAGE are unknown and because other variables (e.g., year of death, which is determined by AGE59 and DEATHAGE) that could be added to the model are not shown.

The graph structure was derived as follows. Classification tree analysis was performed on each variable in turn, starting with LUNG, to identify the parents of each node. As usual, a node's parents were identified as those variables that (a) Appeared in every classification tree having the node as a dependent variable (typically near the top of the split list generated by the classification tree program, since a node's parents, like its children, tend to be relatively strongly associated with its value); (b) Could not be removed by splitting on another variable first; and (c) Cause all more remote ancestors to drop out of the node's tree after splitting on the parents. First the parents of LUNG were identified (i.e., MONTHS and DEATHAGE) then their parents (RETIRED and AGE59, respectively) and "siblings" (i.e., other children of parents, e.g., CONC is a sibling of LUNG) based on splits that did not disappear after conditioning on their parents), then parents of parents and siblings, and so forth. (See Cox, 1997 for details.) Starting with LUNG, which has no children, grounded the recursive identification of graph structure by successive identification of parents and siblings. This procedure revealed the dependence and conditional independence relations shown.

Orientations (i.e., directions) were assigned to some of the arrows based on directed paths. For example, the arrow from MONTHS to LUNG is implied by the fact that LUNG is conditionally independent of RETIRED given MONTHS (but is not conditionally independent of MONTHS given RETIRED). Other orientations were assigned based on knowledge of the meanings of the variables. For example, retirement limits the number of months of exposure (MONTHS). Similarly, RETIRED could affect CONC, since the year of retirement constrained the calendar years of exposure, and hence the exposure concentrations that would have been experienced, but it is unlikely that CONC causally affects RETIRED, especially since their association is positive. DEATHAGE was significantly positively associated with LUNG (i.e., the conditional distribution for DEATHAGE is significantly rightward-shifted for LUNG = 1 compared to LUNG1 = 0), and no variable could be interposed between them that would "explain away" this association, as indicated by the direct link between them. The most reasonable causal interpretation is that lung cancer is more likely to be a cause of death in old age than at younger ages. Accepting this causal interpretation establishes the arrow between them as pointing from DEATHAGE to LUNG. (The alternative interpretation that presence of lung cancer prolongs life, so that the arrow would run from LUNG to DEATHAGE, was rejected on knowledge-based grounds).

AGE59 plays a special role in causal interpretation. It not only indicates birth year, but also serves as a proxy for historical events affecting an age cohort. It is also likely that DEATHAGE involves informative censoring (threat 4). For example, AGE59 and DEATHAGE are positively associated, perhaps because among the workers who had died by 1980 ‾ a censoring condition imposed by the study ‾ those who were relatively young in 1959 died primarily because of accidents or other causes of premature death that precluded their reaching high values of DEATHAGE. Thus, the statistical relation between AGE59 and DEATHAGE might be quite different if the study were conducted later, after death times were known for most of the workers. The graph shows how to analyze the *conditional* relations among variables to determine possible causal relations, despite these potential threats.

For two links – those joining MONTHS or CONC to DEATHAGE – there was a significant statistical association between the linked variables, but no clear basis to determine the causal direction(s) of the arrows between them. Additional model-building is needed to either explain these associations in terms of other variables or to establish a causal ordering. For a fully quantitative analysis, such information is essential. To answer questions about whether a causal relation has been demonstrated between diesel exhaust exposure and lung cancer mortality risk, however, the partial model shown suffices.

Signs (+ or -) are shown on most of its arrows. These indicate signs of association between the corresponding variables that hold in *all* classification trees and sub-trees that have been conditioned on the other relevant variables, i.e., the other parents of the nodes involved. Such uniform directions of association between variables, regardless of location within a tree, need not exist in general, although it simplifies causal interpretations when they do. For example, in the tree analyses, AGE59 is negatively associated with LUNG except after conditioning on RETIRED, where the association appears to be positive. The graph suggests the following explanation. AGE59 affects LUNG only indirectly, i.e., there is no direct arrow from AGE59 to LUNG. If the level of RETIRED is held approximately fixed, then the indirect positive effect of AGE59 on LUNG via DEATHAGE dominates; otherwise, the indirect negative effect of AGE59 on LUNG via RETIRED and MONTHS, dominates.

Other examples of significant non-monotonic relations between pairs of variables include hill-shaped (single-peaked) relations when RETIRED is plotted against CONC, when DEATHAGE is plotted against MONTHS, or when MONTHS is plotted as a function of CONC. In these cases, too, the multiple oppositely signed paths in the graph can explain the observed nonlinear patterns of association. Moreover, conditioning out the effects of other variables via classification tree analysis reveals consistent directions for direct effects (e.g., RETIRED and MONTHS are positive signs for CONC, while MONTHS and DEATHAGE are negatively associated), even though the raw regression relationship between them is non-monotonic.

The partial graph is able to answer some important questions about potential exposure-risk relations. Two of the key findings are that (a) There is no evidence of a causal path from CONC to LUNG, suggesting that exposure to diesel exhaust is not a risk factor for lung cancer in this population; and (b) There is evidence for a direct causal link between MONTHS and LUNG, suggesting that something in the occupational environment other than diesel exhaust (CONC) may be a risk factor for lung cancer.

Fully automatically learning causal graph models from data is a computationally intractable (NP-hard) problem in any of several formal versions (Glymour and Cooper, 1999, Chapter 1, pp 43-44, citing earlier results by Boukaert, 1995 and Chickering, 1996.) For example, determining whether there exists a causal network with at most k arcs that represents a joint probability distribution over a set of binary variables – or finding a causal network with a minimum number of arcs -- is NP-hard, even if conditional independence tests can be performed in O(1) time. Similarly, determining whether there exists a network with at most k parents per node that has probability at least p (or finding a maximum probability network) is NP-hard. Moreover, the search spaces involved are vast. While a problem with 2 measured variables has only 3 possible DAG structures relating them ($X \rightarrow Y$, $X \leftarrow Y$, and X Y), a problem with 6 variables has over a million possible DAG structures and a problem with 10 variables has over 4×10^{18} (Glymour and Cooper, 1999, p. 43.) If latent variables are allowed, then the number of possible models that could be fit to a data set becomes infinite. It is therefore essential to focus the search for "good" models using intelligent heuristics and/or knowledge to restrict the search, e.g., by exploiting knowledge of what causal orderings are possible among variables. (For example, it is useful to know that LUNG can not be a cause of some of the other variables, such as AGE59, in the diesel exhaust example.)

The following main strategies for automatically building causal graph models from data have been extensively investigated, with encouraging results (Jordan, 1999; Cooper and Glymour, 1999; Pearl, 2000):

- *Constructive ("model-building") approaches* use the data in the data matrix D to justify creating and orienting links, starting from a set of nodes representing the variables. The preceding example of diesel exhaust and lung cancer risk illustrates one such approach.
- *Constraint-based algorithms* prune away links not justified by the data, starting from a fully connected graph, and then orient the links that remain.
- *Search methods* seek graph models that best explain the available data, according to any of several criteria.

(using computational Bayesian methods such as MCMC) to arrive at posterior probabilities for different models.

- *Bayesian Model-Averaging* (BMA) for DAG models uses multiple posterior models to make predictions, taking into account uncertainty about which model is correct.

The following paragraphs elaborate on these approaches, which form a very active area of current research.

Constraint-based algorithms for learning DAG models

Constraint-based algorithms use constraints implied by the data to identify possible DAG models. For example, the *PC algorithm* (Glymour and Cooper, 1999) starts with a complete undirected graph (i.e., all pairs of variables are linked by undirected edges) and then uses conditional independence tests to successively remove edges wherever nodes are marginally or conditionally independent. (By contrast, the classification tree approach outlined in the above example starts with no links between nodes and recursively develops them, starting from a presumed output node, e.g., LUNG.) Orientations are assigned to edges in the remaining undirected graph as follow. Any triple of nodes of the form X – Y – Z is assigned the orientation X → Y ← Z if and only if X and Z are dependent when conditioned on Y (or on any superset of variables, not including X and Z themselves, that includes Y.) Any triple of the form X → Y – Z with X and Z unconnected (by an arrow or an undirected edge, so that Z is conditionally independent of X given Y), is assigned the orientation X → Y → Z. Finally, each unoriented dependent pair X – Y with a directed path through X to Y is assigned the orientation X → Y. (Recall in the preceding example how the edge from MONTHS to LUNG was oriented.) This process of assigning orientations is repeated until no more assignments can be made. Then the algorithm stops. It may terminate with some edges having unassigned orientations (like the MONTHS – DEATHAGE link in the example.) As in the preceding example, knowledge-based constraints may also be imposed to remove edges and to determine orientations.

The two-phase approach of first eliminating edges based on conditional independence and then assigning orientations to the remaining edges wherever possible, thus turning them into arcs that are interpreted as representing direct causal relations, has been extended to allow for latent variables (i.e., hidden or unobserved variables) and possible sample selection biases. Since DAG models emphasize *conditional* relations, they can be correctly learned and used in many circumstances even if sampling is biased, so that the *unconditional* frequency distributions of the variables do not represent those in the sampled population. The resulting *Fast Causal*

Inference (FCI) algorithm (*ibid*, Chapter 6) is the basis for graph-learning algorithms found in some commercial software products.

Example: A Causal Graph Model for Campylobacteriosis

The following DAG model structure for the causal relations among variables related to risk of campylobacteriosis in the USA was derived by applying a TETRAD II type algorithm (implemented in the Bayes Net Toolbox for Matlab™, available at http://www.cs.berkeley.edu/~murphyk/Bayes/bayes.html) to the CDC case-control survey data. The input to the program is a set of correlations between pairs of variables. The output, based on a PC-like algorithm, is the following DAG model of the relations among the variables.

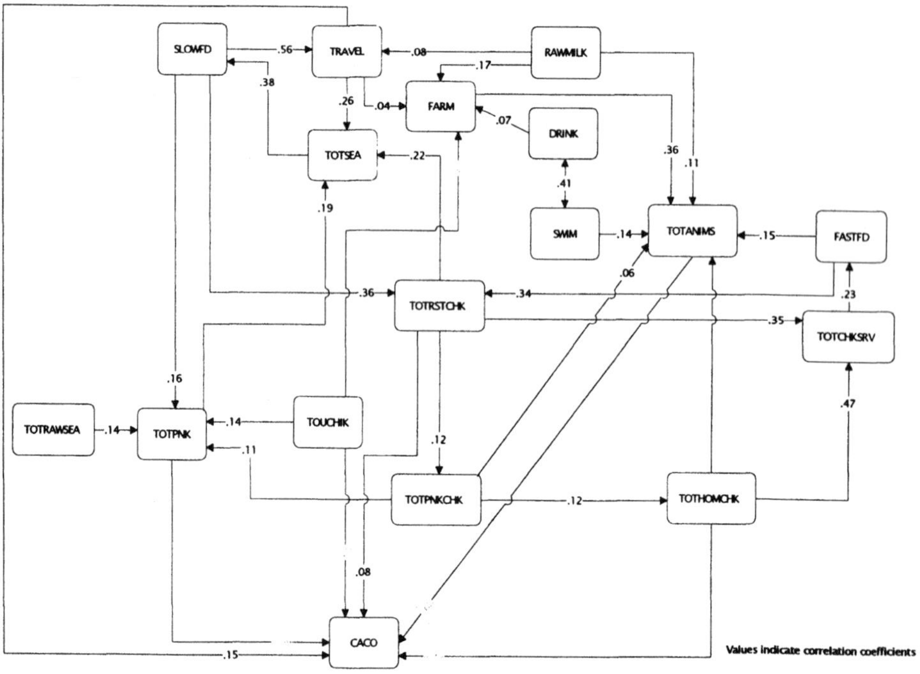

Variables: **CACO** = case-control indicator (1 = case, 0 = control), **TOTCHKSRV** - # servings of chicken (any form – any place) eaten; **TOTRSTCHK** - # of types of restaurant chicken eaten; **TOTPNKCHK** - # of types of pink chicken eaten; **TOTHOMCHK** - # of types of home prepared chicken eaten; **TOTPNK** - # types of pink meat eaten other than chicken; **FARM** - live on or visit a farm in past week; **TOTANIMS** - # of types of animals in contact with; **SLOWFD** - # of non fast food restaurant meals; **TOUCHICK** - raw chicken was touched; **SWIM** = swim in lake, river or stream; **DRINK** = drank untreated water; **RAWMILK** = drank any raw milk; **TOTSEA** = # of seafood types consumed; **TOTRAWSEA** = # of raw seafood types consumed. Time frame is 1 week for all activities.

This graph suggests that CACO is *positively* associated with foreign travel (TRAVEL) and eating restaurant chicken (TOTRSTCHK) and *negatively* associated with consumption of pink meat (non-chicken) (TOTPNK), touching raw chicken (TOUCHIK), contact with animals (TOTANIMS), and consumption of chicken at home (TOTHOMCHK). It is conditionally independent of other variables in the figure, given these six variables and based on the correlations in the data set.

4.5 Search, optimization, and model-averaging heuristics

In principle, one might search for a causal graph model that explains a given data set as well as possible according to some criterion, in hopes that it will either be the correct model with high probability (if the correct model is in the set being searched) or will be a useful approximation to it, meaning that its predictions provide an expected utility-increasing guide to action. Several algorithms have been developed to search sets of DAG models for the one(s) that "best" explain the data. Rather than judging the quality of an explanation by knowledge-based criteria (e.g., the clarity, economy, and universality of the laws invoked; the accuracy of the simplifications and approximations made; and the sparsity and testability of the assumptions), many programs rely instead on statistical criteria such as the following:

Statistical Criteria for Comparing Alternative DAG Models

- *Maximum likelihood (ML) approaches* seek DAG model(s) that make the observed data more likely (according to the model) than do any other models considered. To justify this strategy, one may seek conditions under which it will select the correct model with high probability or make predictions that, with high probability, differ from those of the correct model by at most a small amount.
- *Maximum a posteriori (MAP) methods* use Bayesian computations, discussed below, to seek a most-probable model conditioned on the data, starting from a set of candidate models with prior probabilities. Their posterior probabilities are obtained by conditioning the prior probabilities of models on the data using Bayes' rule.
- *Minimum description length (MDL) algorithms* seek DAGs and conditional probability relations at nodes that allow the available data to be reproduced using the smallest possible amount of information.
- *Minimum information-distance* methods seek to minimize the Kullback-Leibler distance between model predictions and the true joint probability distribution of the data.

It turns out that these criteria are closely related.

Let G be a hypothesized DAG structure for data set D and let β_G be its parameters, e.g., the numerical values in the probability tables stored at the nodes of G. The complete causal graph model is thus $M = (G, \beta_G)$. Let G^h be the hypothesis that exactly the conditional independence relations implied by G are true. [It can be shown that a Kullback-Leibler measure of the minimum approximation error induced by an assumed network structure will be zero if and only if G^h is true, i.e., all the independence relations implied by G hold for the true joint probability distribution (Friedman and Goldszmidt, 1999).] Then Bayes' Rule implies that the probability of graph structure G^h given data D satisfies:

$$Pr(G^h \mid D) \propto Pr(G^h)Pr(D \mid G^h),$$

i.e., "Posterior is proportional to prior x marginal likelihood" (with the proportionality constant being 1/Pr(D). This may be interpreted as a Bayesian measure of how well the network structure G^h fits the data D. If the prior probabilities of all possible DAG structures are considered equal (as an easy, though *ad hoc*, approach to creating a non-informative prior) then this criterion for choosing among DAG structures reduces to the marginal likelihood function, $Pr(D \mid G^h)$. MAP and ML estimates then coincide. They will be asymptotically equal in other cases where there is enough data so that the prior is unimportant and the marginal likelihood $Pr(D \mid G^h)$ dominates.

The Bayesian Information Criterion (BIC) approximation to the marginal likelihood is:

$$\log Pr(D \mid G^h) \approx \log Pr(D \mid G^h, \beta^*_G) - (d/2)\log N$$

where

- β^*_G = maximum-likelihood estimate of the model parameters (e.g., the probability tables or parameter values stored at the nodes of G)
- d = dimension (number of free parameters) of G at the maximum-likelihood value β^*_G. Equivalently, d = rank of the Jacobian matrix for the transformation from the "structural" parameters of the full network to the "reduced" parameters of the network relating only the observable variables; see Geiger et al., p. 463 of Jordan, 1999.
- N = sample size of the data

This criterion is easy to calculate. In particular, $Pr(D \mid G^h, \beta^*_G)$ is the probability of data D according to the complete model (G^h, β^*_G). This maybe calculated by any of several *Bayesian network inference algorithms,* such as the arc-reversal or Gibbs sampling algorithms discussed in Section 5 in the context using DAGs to draw inferences. Thus, the BIC provides a computationally practical criterion for evaluating and comparing alternative models. Moreover, it can be shown that the BIC criterion coincides with the MDL criterion (Jordan, 1999, pp. 331 and 462), even though they have entirely different motivations. Also, the BIC gives asymptotically correct estimates (for large sample sizes) of the marginal likelihood.

Approximations more accurate than the BIC have been developed for estimating the marginal likelihood. In addition to exact formulas (for the case of Dirichlet prior distributions) and computationally expensive but accurate MCMC (Gibbs sampling) algorithms, promising approaches have been developed that are intemediate in accuracy and computational requirements between MCMC and the BIC. Perhaps most notable are Gaussian approximation and Laplace approximation techniques, based on a second-order Taylor's series expansion and multivariate Gaussian approximations to $\Pr(D \mid G^h, \beta_G)\Pr(\beta_G \mid G^h)$ in the vicinity of the MLE, β^*_G (Jordan, 1999). However, the BIC criterion is relatively easy to calculate and suffices for many practical purposes.

All optimization-based search methods require methods for generating alternative models and for evaluating or comparing them. The above criteria meet this requirement. Even with the BIC as a criterion, however, searching the space of possible DAG models is impractical, making it essential to use intelligent heuristics to focus search effort. A variety of heuristics, typically involving adding or removing arrows between nodes to move from one candidate model to another, have been developed to generate alternative DAG model structures, often starting from an initial (perhaps randomly generated) set of candidate models. Simulation experiments suggest that even greedy heuristics are often quite successful in finding highly predictive models, despite the theoretical NP-hardness of finding optimal networks (Glymour and Cooper, 1999). Once the graph structure G is known, an EM algorithm can be used to identify MAP and ML estimates of the DAG model parameters, β_G, although at the risk of becoming trapped in local maxima.

Rather than relying on any single estimate, such as MAP or ML, of the causal graph model describing a data set, Bayesian model-averaging (BMA) approaches provide estimates that may be less biased and more accurate. They are especially appealing when the conditional independence relations that can be discovered in a data set do not uniquely determine the underlying DAG. If more than one DAG model is consistent with available knowledge and data, then risk calculations can be made and reported for different plausible models. BMA for DAG models provides a way to combine such conditional calculations by weighting risk estimates from different plausible models by factors reflecting their relative likelihoods. (Alternatively, if all plausible models imply the same conclusions, then the conclusions can be stated without resolving uncertainties about the correct model. "Plausible" models might be defined as those with conditional independence and composition relations consistent with the data and implying temporal orderings of variables consistent with available data and causal knowledge.)

As in other areas of statistics, BMA methods for learning DAG models and for making predictions based on them require considering multiple possible models explicitly. Let D be the n x m data matrix containing all available evidence (values of *m* variables for each of *n* individuals, with missing data and measurement errors allowed) and let S be a statement or proposition – perhaps about a conditional independence relation among variables – whose probability is to be assessed. Then the Bayesian approach quantifies the probability of S as follows:

$$\Pr(\text{S is true} \mid \text{data D}) = \Sigma_M \Pr(S \mid M)\Pr(M \mid D),$$

In principle, M indexes the set of *all* DAG models. In reality, there are almost always too many possible models to enumerate, so heuristics must be used to approximate the sum.

One of the best-known heuristics averages the predictions over *selected* models only, rather than over all possible models. Probability and parsimony are used to select the models to be included, as follows.

Occam's Window for Selective Model-Averaging

1. Create an initial set of models (e.g., initialize by including just the trivial model with no arrows between nodes).
2. Consider all one-step changes (e.g., arrow additions, deletions, and reversals that do not create cycles).
3. For each new model, m, evaluate the probability of m as $\Pr(m \mid D) = \Pr(D \mid m)\Pr(m)/\Pr(D)$. The normalizing constant Pr(D) can be ignored and Pr(m) can be taken directly from the prior (or ignored for an uninformative prior), while $\Pr(D \mid m)$ can be assessed as discussed above, e.g., using Bayesian network inference algorithms for the complete model $\Pr(D \mid G, \beta^*_G)$ where $m = (G, \beta^*_G)$.
4. Keep any new graphs that (a) are at least 5% as probable as the most probable model known and (b) Contain no more-probable subgraphs. (This is the "Occam's Window" step that eliminates needlessly complex models from consideration.)
5. Repeat steps 2 to 4 until no more graphs added.
6. Perform model-averaging using all the structures found, i.e., calculate $\Sigma_m \Pr(S \mid m)\Pr(m \mid D)$, where m ranges over all generated models.

Reference: Madigan and Raftery, 1994

Such heuristics approaches are computationally practical. Based on evaluations in simulation studies for which the right answers are known, they appear to produce usefully accurate results most of the time. Heuristic criteria such as the BIC and heuristic search procedures such as the Occam's window version of BMA are starting to make automated prediction based on causal graphs automatically constructed from data a practical reality.

5. USING CAUSAL GRAPHS IN RISK ANALYSIS

Once a complete causal graph model has been developed by hand or learned from data, it may be applied to perform several tasks essential in risk assessment and risk management. These include drawing statistical inferences, predicting probable impacts of changing input variables, optimizing decisions, and assigning shares in the blame for undesired consequences to multiple contributing probabilistic causes. This section summarizes algorithms for the first three applications, i.e., inference, prediction, and optimization. Section 6 develops a theory of risk attribution for causal graphs.

5.1 Drawing Probabilistic Inferences in DAG Models

Suppose that several variables are measured or observed for an individual and that risk assessment requires inferring the probable values of other variables, as yet unobserved. For example, the application may be to infer the probability of eventual death with lung cancer (a future variable, not yet observed) for an individual from a history of occupational exposures and individual covariates. If the joint probability distribution of all variables, Pr(X), were known, then in principle the inference problem would be solved completely by calculating

$$\Pr(X_U \mid X_K) = \Pr(X_U, X_K)/\Pr(X_K).$$

where X_U denotes the vector of unmeasured variables of interest and X_K denotes the vector of known (measured) variables. But directly conditioning a multivariate joint probability density function on observed values of some of its variables is usually impractical, both because the joint PDF Pr(X)is not known (and is typically unwieldy to represent for all possible values of the variables) and because the calculation requires computationally demanding multivariate integrations or summations to calculate $\Pr(X_K)$

DAG models and algorithms provide a solution. There is no need to invoke causal interpretations. Simply factoring a joint PDF and representing the result by a DAG model is enough to allow Bayesian inference to be performed using distributed algorithms, i.e., algorithms that operate on the information available at one node at a time. These graph-theoretic algorithms avoid the brute-force multivariate integration and concomitant combinatorial explosion entailed by direct conditioning of the full joint PDF.

Example: Direct Conditional Probability Calculation

Problem: Given a DAG model $X \rightarrow Y$ with a marginal probability distribution Pr(X) stored at X and a conditional probability table Pr(Y | X) (tabulating Pr(Y = y | X = x) for all values of x and y) stored at Y, suppose it is observed that $Y = \underline{y}$ for some individual (*Here and later we adopt the convention that observed variables are underlined.*) Then what is the posterior probability distribution for X for this individual, assuming that the DAG model applies to this individual?
Solution: The definition of conditional probability yields:

$\Pr(X = x \mid Y = \underline{y}) = \Pr(Y = \underline{y} \mid X = x)\Pr(X = x)/\Pr(Y = \underline{y})$, abbreviated
$\Pr(x \mid \underline{y}) = \Pr(\underline{y} \mid x)\Pr(x)/\Pr(\underline{y})$.

By Bayes' Rule, this is equal to:

$$\Pr(x \mid \underline{y}) = \Pr(\underline{y} \mid x)\Pr(x)/[\Sigma_j \Pr(\underline{y} \mid X = j)\Pr(X = j)].$$

Now, each of the elements on the right side can be read off from the node tables. Specifically, $\Pr(\underline{y} \mid x)$ and $\Pr(\underline{y} \mid X = j)$ come from the table Pr(Y | X) stored at node Y, while Pr(x) and Pr(X = j) come from the marginal distribution Pr(X) stored at X. Thus, all the elements required to calculate $\Pr(X = x \mid \underline{y})$ for each value of x are available in the DAG model.

Example: Arc-Reversal

Problem: Given a DAG model M = {$Y \leftarrow X \rightarrow Z$, Pr(Y | X), Pr(Z | X), Pr(X)} and observation $Y = \underline{y}$, *find* the posterior PDF for Z. (Under a causal interpretation of the arrows, this is a classical inference problem, in that an effect (Y) of X is observed and it is necessary to draw inferences about the underlying cause X and other as-yet unobserved effects Z.)

Solution: The problem can be solved in two steps, as follows:
Step 1 (Arc-Reversal): $\Pr(X \mid \underline{y}) = \Pr(\underline{y} \mid X)\Pr(X)/\Pr(\underline{y})$.
Step 2: (Forward chaining): $\Pr(Z \mid \underline{y}) = \Sigma_x \Pr(Z \mid x)\Pr(x \mid \underline{y})$

Step 2 is the usual composition formula, just as for the diagram $Y \rightarrow X \rightarrow Z$ (which explains the terminology "arc reversal" for Step 1). On its right-hand side, Pr(Z | x) comes from the node table Pr(Z | X) while $\Pr(x \mid \underline{y})$ comes from the output of Step 1. On the right-hand side of Step 1, $\Pr(\underline{y} \mid X)$ comes from the node table Pr(Y | X), Pr(X) comes from the node table Pr(X). $\Pr(\underline{y}) = [\Sigma_j \Pr(\underline{y} \mid X = j)\Pr(X = j)]$, where j ranges over all values of X, just as in the preceding example. Again, all required elements are found in the node tables specified as part of the model.

This example illustrates the essence of the *arc-reversal algorithm* for propagating evidence through DAG models (Schachter, 1990). Rather than simply propagating distributions forward from inputs to outputs, as in the *Analytica*™ engine, arc-reversal allows evidence (i.e., observations) also to be propagated backward to update beliefs about causes based on diagnostic evidence about their effects. Commercial software such as HUGIN™ (Jensen, 1996) incorporates arc-reversal and other DAG-processing algorithms to permit efficient inference in general DAG models. These and other advanced graph algorithms for inference in DAG models (e.g., Pearl, 1988; Shenoy, 1992) propagate observations through DAG models to update each node. The net result is to calculate the same posterior PDF for each variable as would be obtained by using brute-force conditioning – but without having to perform the multivariate integrations.

The idea of arc-reversal for incorporating diagnostic evidence, followed by forward propagation of prognostic (causal) information has been extended to arbitrary DAG and influence diagram models (Pearl, 1988; Schachter, 1990; Shenoy, 1992; Jensen, 1996). Once a probability model, meaning a joint PDF, Pr(X), for a vector of variables, X, has been specified and represented by a DAG model M, it can be treated as a query-answering expert system. [As usual, M = (G, β), where G is the DAG structure (nodes and arrows) and β represents the contents of the node tables.]

A user query can be defined very generally as any function, g, (possibly vector-valued) of any subset of variables in the model, say X_J.. (J is a set of index values indicating a subset of the nodes, i.e., all X_j with j in J.) To answer a user query based on some evidence, one conditions the probability model Pr(X) on the evidence. Evidence is assumed to consist of observed values of a subset of variables, X_K. An *answer* to the query is a conditional probability distribution $\Pr(g(X_J) \mid X_K)$. This may be post-processed to report various aspects of the conditional distribution, such as its mean or the conditional probability that $g(X_J)$ falls in some target set of values. Or, g itself may be defined as an indicator function having value 1 if X_J falls in a target set and 0 otherwise.

In this framework, a general algorithm for answering user queries consists of the following steps.

Expert System Query-Answering Algorithm

Inputs: A user query, body of evidence and model, $g(X_J)$, X_K, M = (G, β)
Outputs: An answer, $\Pr(g(X_J) \mid X_K)$.
Method:
1. Create a new node in the DAG to represent the query. This node may be numbered as node 0 and defined as $g(X_J)$.
2. REMOVE all nodes from the DAG except node 0 = $g(X_J)$ and the nodes in K. The resulting conditional PDF for $g(X_J)$ is the answer to the query.

The REMOVE procedure is as follows. To REMOVE node j from a DAG:

a) Order j's children (if any) so that the first one is not followed by any of its ancestors.

b) REVERSE arcs (i.e., arrows) from node j to each of its children in order, leaving j with no children. Arc reversal is accomplished by applying Bayes' Rule as in the above example. (Reversing arc between two nodes requires them to have the same parents, i.e., be conditioned on the same events, so that Bayes' Rule applies. It may be necessary to add arcs to the diagram to meet this criterion; see Owens et al., 1997.)
c) When j has no remaining children, remove it from the DAG. (The justification is that its effects on node 0 have already been propagated inward toward node 0 and it has nothing more to contribute.)

Executing this procedure calculates $Pr(g(X_J) \mid X_K)$ exactly. Many other graph operations (especially, formation of junction trees with cliques of variables as nodes), data structures, and algorithms have been devised to improve the conceptual and storage economy and the computational efficiency of inference in DAG models (e.g., Shenoy, 1992; Jensen, 1996). These advanced techniques are embedded in commercial programs such as HUGIN™ and in various research programs.

An alternative approach to DAG inference that is especially useful for large networks and networks with latent variables is based on Gibbs sampling. The core algorithm is as follows. Let $X = (X_1, X_2, \ldots, X_m)$ be the variables in the model, numbered so that no node is followed by any of its parents. (This is always possible in a DAG after a topological sort.) To quantify the PDF of X implied by a DAG model M, do the following:

Gibbs Sampling Calculation of Pr(X | M) for a Known DAG Model, M

1. Sample from $Pr(X_1)$. Call this value x_1.
2. Sample from $Pr(X_2 \mid X_1 = x_1)$. Call the result x_2.
3. Continue in this way, eventually sampling from $Pr(X_n \mid x_1, x_2, \ldots, x_{m-1})$. (Thus, X_j is sampled from $Pr(X_j \mid \mathbf{Pa}(X_j))$, where $\mathbf{Pa}(X_j)$ = parents of X_j.)
4. The values $(x_1, x_2, \ldots, x_m)$ constitute one sample from the joint PDF of X. Repeat many times to simulate the joint PDF of all variables (including latent ones).

In effect, this procedure samples from Pr(X), although each sample point is constructed by *m* draws from appropriate univariate distributions rather than by sampling once from the *m*-dimensional joint distribution (Jenzarli, 1996).

The procedure can also be used to quantify the marginal likelihood of a set of observations, D, according to a DAG model M. Suppose that $D = \{X^1, X^2, \ldots, X^n\}$, where X^i denotes the m-vector of variable values for individual i for i = 1, 2, …, n. Then, $Pr(D \mid M)$ is just the product of $Pr(X^i \mid M)$ over all individuals i in the data set, assuming that each individual can be thought of as having variable values sampled at random (and independently across individuals) from Pr(X).

To incorporate evidence into the Gibbs sampling procedure, suppose that values of some variables, X_K, are observed and that the posterior joint PDF Pr(X | X_K) is to be determined. A simple modification suffices: generate samples in exactly the same way, but reject any sample that does not match the observations in X_K. The remaining sample points will be drawn from Pr(X | X_K), as desired. This is Henrion's *probabilistic logic sampling* procedure (Cowell, 1999, p. 28.) Its computational efficiency can be low when only a small fraction of sample points are not rejected. To improve efficiency, it is desirable to sample *groups* of variables in cliques (via junction trees), which can be arranged by pre-processing the DAG (*ibid*). Then, instead of rejecting an entire sample, it becomes possible to reject only the parts that do not agree with X_K while still being able to use the rest.

A more general Gibbs sampling algorithm can be used to estimate the expected value of any function g(X) from a DAG model representing the joint probability distribution of the variables, Pr(X). In many applications, g(X) will be either an indicator function (with a value of 1 if some target condition is achieved and a value of 0 otherwise) or a utility function at the value node of an influence diagram.

Gibbs Sampling to Estimate E[g(X)]

1. Guess at the values of all unknown quantities, e.g., the X_j.
2. Update each quantity in turn by sampling a value from its conditional distribution, given the data and the current values of all other parameters. After each update, calculate g(X) using the current values of all variables.
3. Iterate Steps 1 to 3 and track the average value of g(X).
4. Check for convergence. In steady state, each full iteration (updating all variables once) is an independent random sample from the joint posterior distribution of all quantities. The sample average value of g(X) converges to the correct value $E_{Pr(X)}$ g(X)] provided that Pr(X) is irreducible (e.g., contains no 0 probabilities).

The condition of irreducibility implies that all possible configurations of X values can be sampled.

Gibbs sampling is especially valuable when some variable values are missing for some individuals and/or when there are latent variables in the DAG model. Since it is possible to sample from conditional probability distributions for variables in the DAG when their values are missing in the data set, the procedure can still generate samples from Pr(X) even when components of X are missing for some or all individuals.

In summary, DAG inference algorithms either calculate exactly or approximate by Monte Carlo simulation the same results that would be obtained by conditioning the full prior joint distribution for all model quantities on all observations using Bayes' rule, and then extracting the posterior distribution for quantities of interest by "marginalizing them out", i.e., integrating over all other quantities (Shenoy, 1992). However, DAG inference algorithms are practical for applied work and are available in commercial decision analysis software packages such as HUGIN™ (Jensen, 1996) as well as in a variety of academic and research software. By contrast, straightforward Bayesian computation of the posterior PDF by direct application of Bayes' rule is often computationally unmanageable because it requires multivariate integration or summation over a combinatorially explosive number of possible joint values of model quantities. DAG inference algorithms thus make Bayesian inference computationally practical in many applications where the conceptual framework is attractive.

5.2 Applications of DAG Inference in Risk Assessment

Inference algorithms for DAG models provide a unified framework for addressing some of the most important challenges in statistical risk assessment. These include:

a) Modeling and correcting for errors and uncertainties in exposure estimates and responses;
b) Dealing with unobserved or imperfect sources of inter-individual variability and heterogeneity; and
c) Controlling for the effects of unmeasured confounders.

For example, consider the following DAG model structure:

$$\text{Model DAG:} \quad \begin{array}{ccccccccc} \underline{X} & \leftarrow & X & \rightarrow & Y & \leftarrow & Z & \rightarrow & \underline{Z} \\ & & & & \downarrow & & & & \\ & & & & \underline{Y} & & & & \end{array}$$

Here, $\underline{X}$ denotes measured or recorded exposures and X denotes true exposures; $\underline{Y}$ is measured or recorded response while Y is actual response; and Z is measured or recorded covariates while Z is the true (but perhaps unknown) vector of covariates affecting responses. Z may include uncertain parameters (e.g., of pharmacokinetic and dose-response models) as well as variables such as age and sex and latent variables such as frailty or susceptibility. (If Z contains unmeasured confounders, then simply add an arrow from Z to X and the following discussion still applies.) Stored at node $\underline{X}$ is the conditional probability table $\Pr(\underline{X} \mid X)$, and similarly for $\underline{Y}$ and $\underline{Z}$. There is no need to adopt the path diagram convention of writing

$X \rightarrow \underline{X} \leftarrow \text{error}_X$,

for example, since $\Pr(\underline{X} \mid X)$ already contains the probabilistic relation between true and measured exposures.

In this framework, a key challenge for statistical inference if the model parameters are unknown is to *observe* measured values $\underline{X}$, $\underline{Y}$, and $\underline{Z}$ (which may contain measurement errors and be poor proxies for X, Y, and Z, respectively) and then to *infer* the true exposure-response relation $\Pr(Y \mid X, Z)$, taking into account any effects of heterogeneity (Z) and any unobserved confounders (if there is an arrow from Z to X). This problem may be solved (given sufficiently numerous and variable data) using the techniques mentioned in Section 4, e.g., by using an EM algorithm to construct ML or MAP estimates of the parameters β for the model $M = (G, \beta)$. A second inference problem when M is already known is to predict Y from $(\underline{X}, \underline{Z})$ (if no measurements on Y are available) or to estimate the true response Y from observations of $(\underline{X}, \underline{Y}, \underline{Z})$ (if noisy response data $\underline{Y}$ are available). In either case, this inference problem can be addressed by propagating the observed values $(\underline{X}, \underline{Y}, \underline{Z})$ through the DAG model M using arc reversal, Gibbs sampling, or other algorithms.

To apply DAG inference to practical risk assessment problems, it is necessary to supply (or to extract from data) a conditional independence model with quantified conditional relations for each uncertain quantity (node), determining its probable value from the values of its parents. In addition, the prior distributions for input nodes (e.g., population exposure frequency distributions) must be given. If these inputs are uncertain, then misspecifying input assumptions may bias the results (Richardson and Leblond, 1997).

The steps required to apply DAG inference methods to estimate exposure-response relations are as follows (Richardson and Gilks, 1993):

1. Represent conditional independence relations among quantities by a DAG model such as $\underline{X} \leftarrow X \rightarrow \underline{Y} \leftarrow Z \rightarrow \underline{Z}$. (In this example model, responses are observed without error, i.e., $\underline{Y} = Y$.)
2. Quantify the following conditional relations:
   - A *measurement error model:* $X \rightarrow \underline{X}$. The notation: "$X \rightarrow \underline{X}$" represents that relation $\Pr(\underline{X} \mid X)$. This model is usually derived from a validation study or from repeated measurements. In the *Berkson formulation* of measurement error models, the specified relation is $\underline{X} \rightarrow X$, i.e., the probability distribution of true exposure given measured ones. In effect, the Berkson formulation has already performed arc reversal. If information is available on covariates, then the link $Z \rightarrow \underline{Z}$ describing measurement errors in Z must also be specified.

- A *response model:* X → <u>Y</u> ← Z, i.e., Pr(Y | X, Z). This is also called the disease model. It describes how probability of response varies as a function of exposure and covariates. Recall that unknown dose-response parameters may be included in Z as latent variables. Thus, specifying the disease model does *not* require resolving uncertainties about all unknown quantities. Those are addressed by the inference algorithms after the model is specified.
- *Exposure model:* X. This specifies the marginal probability distribution for X. (In general, there may be an arrow from Z to X if individual characteristics in Z affect exposure patterns. Z → X

3. *Propagate evidence* (<u>X</u>, <u>Y</u>, <u>Z</u>) through the graph to obtain the joint posterior distribution of all quantities.

Meeting the input requirements is the main challenge for this DAG inference approach. Its main strengths are that (a) The entire posterior distribution of any model variable(s) can be estimated as precisely as desired by MCMC sampling; and (b) It uses all available data, including data on health outcomes, to obtain the best possible probabilistic estimates of individual exposures and their effects. Richardson and Gilks (1993) describe an application to risk estimation when exposure estimates are based on job-exposure matrices and survey data.

This section has referred throughout to DAG models of the joint probability distribution Pr(X) of values of variables describing *individuals*. However, the same technology can be reinterpreted and applied to model risks in *populations* by reinterpreting Pr(X) as the joint frequency distribution of variables in the population. Pr(X_U | X_K) then describes the joint frequency distribution of the variables X_U in within the sub-population satisfying description X_K. DAG representations of Pr(X) can then be used to characterize population frequency distributions of risk in various sub-populations.

5.3 Using DAG Models to Make Predictions

Causal graphs can be used to predict the probable consequences of different combinations of input variables. A simple propagation algorithm using a forward-chaining process with Monte-Carlo sampling at nodes suffices. This approach is embodied in several popular commercial software products for Monte Carlo risk and uncertainty analysis such as @RISK™, Crystal Ball™, and *Analytica*™ (Buede, 1998). It is very useful for applications such as predicting the effects of different emission levels (or other inputs) on the probability distributions of lives lost (and other consequences), as in the example of Figure 1, *provided* that the underlying model gives a usefully accurate description of how the world works, i.e., of how changes in inputs propagate to determine probable changes in other variables. This is an important proviso. When satisfied, it allows the modeler to gloss over philosophical perplexities about the meaning of causation and the logic of counterfactuals and manipulation in probabilistic causal predictions. Instead, input values are simply sampled and propagated and the results interpreted as probability distributions for derived variables brought about (or "caused") by given configuration of inputs.

Forward propagation of probability distributions is conceptually simple, although the Monte Carlo techniques used often incorporate sophisticated variance-reduction and sampling techniques to boost efficiency. Suppose that in the model $\{X \rightarrow Y, Pr(Y \mid X), Pr(X)\}$ the input X is a random variable. Then the probability distribution for Y induced by Pr(X) is:

$$Pr(y) = \Sigma_x Pr(Y = y \mid X = x) Pr(X = x) \text{ for each possible value, y, of Y.}$$

Rather than calculate these probabilities explicitly, a Monte Carlo approach samples values of X from Pr(X) and then samples corresponding values of Y from Pr(Y | X). Repeating many times and tracking the relative frequency of each value of Y that appears gives an approximation to Pr(Y). The approach can be extended to large networks by repeatedly propagating sample values from inputs to derived variables, as described previously in the box on "Gibbs Sampling Calculation of Pr(X | M)". Now, if risk management actions change the values (or marginal distributions) of one or more inputs, then the Monte Carlo propagation procedure will produce new conditional distributions for the derived variables.

The Monte Carlo simulation procedure for predicting probable consequences of input configurations (or of probability distributions of input values) by propagating them through a causal graph may be summarized as follows. (This box also recapitulates key aspects of the causal graph modeling framework for ease of reference.)

Monte Carlo Prediction Procedure for Causal Graphs

1. **Pre-Processing:** At each node of the causal graph, store the conditional probability distribution of the node variable, given each distinct combination of values for its parents. (Recall that classification trees can help to automate this step when DAG models are constructed from data.)
2. **Inputs:** Specify the combination of input values, one value for each input node, for which the joint probability distribution of output variables is to be predicted. (If some of the inputs are random variables, then repeatedly sample input values from their distributions and follow the rest of the procedure.) Input nodes are those with only outward-directed

arrows. To assess the causal impact of an intervention, the value(s) of the corresponding node(s) should be set to their new values under the proposed intervention. If an intervention sets the values of some non-input nodes, then the DAG structure should be modified by deleting the links determining the probable values of these nodes from the values of their parents, thus turning them into input nodes whose values may be set (Pearl, 2000).

3. **Propagation:** Propagate the input values through the causal graph to obtain the joint probability distribution of the other (non-input, i.e., derived) variables. Propagation is accomplished as follows. Begin with all derived nodes marked as "unevaluated" and each input node marked as "evaluated". Until all nodes are marked evaluated, do the following: Choose any unevaluated node, say j, all of whose parents have been evaluated. Use Monte-Carlo simulation to sample repeatedly from the conditional value distributions of j's parent nodes, and, for each such sampled set of input values, sample from the conditional probability distribution for j, given that input combination, as determined by the information stored at j in step 1. Repeat until the conditional value distribution of j has been determined with adequate precision (using Monte-Carlo sampling and stopping rules). Store the conditional value distribution for j and mark j as "evaluated".
4. *Stop when all of the variables or output node(s) of interest have been evaluated.* The conditional value distributions of these variables, given the input values (including any that have been set by intervention), constitute the causal graph's prediction for the probable consequences of the selected input values.

The task in Step 1 of representing the graph structure and storing probabilistic input-output relations as conditional probability distributions at each node may be accomplished using available software packages for creating and manipulating influence diagrams such as *Analytica*™ (see Figure 1) or *Hugin*™ (e.g., Jensen, 1996). Bayesian model averaging can combine predictions from multiple models if the correct one is unknown.

The concept of *setting* the values of some variables to model effects of interventions has provoked confusion and controversy, although many find the idea very intuitive (Pearl, 2000). For example, in the DAG model $(G, \beta) = \{X \rightarrow Y \rightarrow Z, \Pr(X), \Pr(Y \mid X), \Pr(Z \mid Y)\}$, one can imagine directly manipulating Y without changing X, thus apparently violating the causal theory embodied in the model. To avoid such conceptual difficulties, it is useful to adopt the influence diagram convention in which some nodes are explicitly identified as decision nodes or choice nodes, and all manipulations consist of choosing the values of these variables from specified choice sets.

5.4 Decision-Making and Optimization

Influence diagrams were invented to support decision-making. To do so, a causal graph DAG is augmented with *decision nodes* (conventionally shown as squares, such as "Emissions Reduction" in Figure 1) and a *value node* (e.g., the "Total Cost" node in Figure 1) that is a deterministic function of outcomes representing consequences of interest. The value node typically represents a utility function for outcomes (see Chapter 5), and we will assume this. Originally, influence diagrams were developed in the context of DAG representations of joint probability distributions, with no formal causal interpretation. However, common practice (Pearl 2000) and common sense suggest that the DAGs used to relate decisions to their consequences should be constructed and interpreted as causal graphs, with choices at decision nodes propagating along arrows to affect outcome probabilities and ultimately the probability distribution of the value node.

Decision nodes have a precedence order reflecting temporal precedence. In principle, each decision node has arrows into it from all of its predecessor decision nodes, indicating that information about all previous decisions is assumed to be available when any decision is made. (In practice, some of these arrows may be omitted to avoid clutter where previous decisions are judged to be irrelevant ones for current ones.) More generally, arrows into a decision node reflect the information available when that decision is made. Associated with each decision node is a set of alternative decisions or *acts* that can be taken there. A *decision rule* maps the information entering each decision node into a choice of act at that node. The goal of decision analysis is to identify decision rules that will maximize the expected value of the value node, i.e., the expected utility of the outcomes. Chapter 5 justifies choosing expected utility as the objective function to be maximized.

The key to solving the decision problem is to change the DAG to express the value node directly as a function of the decisions made, rather than only as a function of its parents (which typically include chance and deterministic nodes). Figure 1 illustrates this idea using loss instead of utility as the objective function – an inessential change. By expressing "Total Cost" as a function of "Emissions Reduction", the graph makes clear that an emission reduction of about 0.7 is optimal [As often happens in applications, the optimised value of the objective function is fairly insensitive to the exact value of the decision parameter, illustrating the "flat maximum" principle (von Winterfeldt and Edwards 1986).] In general, once the value node has been expressed as a deterministic function of the decision variables, optimization techniques can be used to select the optimal decisions.

A technique for expressing expected utility as a deterministic function of decision variables is readily available: simply eliminate chance nodes (and deterministic nodes as special cases) using the REMOVE procedure already discussed for the Expert System Query-Answering Algorithm. Doing so allows all evidence to be propagated toward the value node until, when the procedure is finished, what is left is just the value node with arrows pointing into it from the decision nodes that affect it (Barlow, 1998, Chapter 10). This is precisely what is needed in order to optimize the decisions and thus solve the decision problem. The algorithm is as follows:

Solving Influence Diagrams with Decisions

1. Eliminate all nodes (except the value node) that do not point to another node. (These are called "barren nodes").
2. If one or more nodes point into the value node, do the following:

a) If there is a decision node that points into the value node, and if all other nodes that point into the value node also point into that decision node, remove the decision node by policy determination, i.e., optimizing it. Remove any nodes (other than the value node) that no longer point to any other node. Go back to step 2.

b) If there is a chance node that points into only the value node, remove it (by averaging, i.e., taking the expected value of the value node with respect to it). Go back to step 2.

c) Find a chance node that points into the value node and not into any decision, such that reversing one or more arcs pointing into that chance node (without creating a cycle) makes the chance node point into only the value node. Reverse those arcs. Go back to step 2.

Source: Owens et al., 1997

"Optimizing out" the decisions, i.e., eliminating decision nodes by replacing them with the optimized values of the decision variables, is also referred to as *policy determination* (Owens et al., 1997).

Example: A Bayesian Decision Analysis

This example illustrates how to combine Bayesian inference and decision-making in a simple example decision problem. Figure A shows the causal graph for the decision problem. It has a single decision node, represented by x_2.

Figure A: A risk management example requiring inference, extrapolation, and decision-making in a model with uncertain input values

$$\underline{x_1} \rightarrow \underline{y_1} \rightarrow \underline{P_1} \rightarrow \underline{n_1} \leftarrow \underline{N_1} \quad \text{High Occupational Exposure}$$

$$\uparrow \quad \uparrow$$

$$b \quad q \; \} \text{ risk model parameters}$$

$$\downarrow \quad \downarrow$$

$$x_2 \rightarrow y_2 \rightarrow P_2 \rightarrow n_2 \leftarrow \underline{N_2} \quad \text{Low Public Health Exposure}$$

The observed quantities in this model are as follows:

- x_1 = vector of dose levels to which a cohort of workers was exposed
- N_1 = vector giving the number of workers in each dose group
- n_1 = a vector giving the number of workers in each dose group that contracted a disease
- N_2 = size of a currently exposed public population
- x_2 = standard specifying the maximum allowed dose to which the public population may be exposed (a decision variable).

The remaining quantities are not observed. They include y_1 and y_2, the internal doses created by administered doses x_1 and x_2, respectively; P_1 and P_2, the response probabilities due to these doses; *b* and *q*, the parameter vectors that determine the internal dose from the administered dose and probability of response from the internal dose, respectively; and n_2, the uncertain number of cases of the disease that will occur in the exposed public population.

If the standard x_2 had already been fixed and n_2 observed, then the inference problem might be how to combine the evidence from the two studies, 1 and 2, to determine the probable values of the unobserved quantities. This would be a DAG inference problem. In Figure A, however, such pooling or meta-analysis is unnecessary, since only the data from the x_1 study is available. Instead, the inference problem is to use the evidence contained in (x_1, n_1, N_1) to estimate *b* and *q* and to determine the dose-response relation between x_2 and P_2. The decision problem is then to set a level for x_2 that will minimize expected loss, given all relevant information about P_2.

For simplicity and concreteness, suppose that *b* is a scalar and that the node formulas are as follows:

- $y_i = bx_i$ for i = 1, 2
- $P_i = 1 - \exp(-qy_i)$ for i = 1, 2
- $\text{Loss} = (v/x_2) + wn_2$

where v and w are value weights reflecting a preference or value trade-off between the costs of control (assumed to be proportional to $1/x_2$)and the loss due to cases of the disease. Let the empirical evidence be the following data set:

data = {$n_1 = 0$ for $x_1 =$ 0 ppm-years and $N_1 = 200$ workers;
$n_1 = 10$ for $x_1 = 100$ ppm-years and $N_1 = 1000$ workers}

Then the expected loss associated with decision x_2, given the evidence contained in this data set (i.e., after conditioning on the observations) is

$$E(\text{Loss} \mid \text{data}) = (v/x_2) + wE(n_2 \mid \text{data}) = (v/x_2) + wN_2E(P_2 \mid \text{data}),$$

where

$$E(P_2 \mid \text{data}) = 1 - E[\exp(-bqx_2) \mid \text{data}]$$

Choosing x_2 to minimize the expected loss is the same as choosing it to minimize the simpler expression

$$(1/x_2) + kE(P_2 \mid \text{data})$$

where $k > 0$ is defined as $k = (wN_2/v)$. All relevant value trade-offs in this example can be reduced to selection of a value for k.

The term $E(P_2 \mid \text{data})$ may be quantified via Bayes' rule if a prior joint PDF for the reduced parameter *bq* is specified. (Note that it is impossible to estimate *b* and *q* separately from the data. Therefore, assigning a PDF to each independently and then sampling their values would introduce spurious uncertainty into the model. Using the reduced parameter *bq* avoids this difficulty.) A perhaps more natural, but logically equivalent, alternative is to define a prior PDF for P_1 and update it by conditioning on the data. The posterior PDF for P_1 thus obtained induces a unique corresponding posterior PDF for *bq*, since the two quantities are related by the transformation

$$bq = -\ln(1 - P_1)/x_1.$$

If the prior PDF for P_1 is the maximum-entropy prior, i.e., the uniform density between 0 and 1 then, after observing the data, the posterior PDF for P_1 has a beta-distribution with parameters $(n_1 + 1)$ and $(N_1 - n_1 + 1)$. Its mean value is:

$$E(P_1 \mid \text{data}) = (n_1 + 1) / (N_1 + 2) = 11/1002$$

(DeGroot, 1970, p. 160). The posterior PDF for P_1 implies a posterior PDF for P_2, for any choice of the decision variable x_2. Namely, the formula

$$P_2 = 1 - \exp(-bqx_2) = 1 - \exp[\ln(1 - P_1)(x_2/x_1)],$$

transforms the beta PDF for P_1 into a pdf for P_2 by a change of variables.

To minimize the scaled expected loss function

$$(1/x_2) + kE(P_2 \mid \text{data})$$

by appropriate choice of x_2, it is necessary to be able to compute the expected value of P_2 for each choice of x_2. This can be done conveniently using influence diagram tools such as *Analytica*™.

Using this software, and assuming that k = 1000, the example can quickly be solved for the optimal value of x_2. The optimal action is found to be $x_2 = 3$ ppm-years, with expected loss 0.67. Making the standard too relaxed or too stringent by a factor of 10 (i.e., setting $x_2 = 30$ ppm-years or $x_2 = 0.3$ ppm-years, respectively) increases the expected loss 5-fold in each case, to about 3.36, and shifts the cumulative loss distribution function rightward.

The Bayesian decision-analytic approach illustrated in this example requires a prior PDF for uncertain quantities as an input. Its defenders have developed several responses to the criticism that the need to specify a prior PDF injects undesirable subjectivity into the analysis. One defense is that use of subjective prior information is not necessarily undesirable. A second, pragmatic defense is that, when there are adequate data, the posterior PDF is not sensitive to the specific prior selected. Finally, there have been numerous attempts to cope with "vague priors", e.g., by using sets of prior PDFs, by relaxing the requirement that prior beliefs be represented by additive probability measures (as in super-additive Dempster-Shafer belief functions), by selecting maximum-entropy priors as defaults, and so forth (Dubois *et al.,* 1992). Powerful "coherence" arguments support the use of conventional Bayesian prior PDFs – for example, on the grounds that beliefs can be updated in a dynamically consistent fashion only if they are capable of being represented by a Bayesian prior PDF (Epstein and Breton, 1993).

6. ATTRIBUTABLE RISKS IN CAUSAL GRAPHS

Attributable risk calculations play a key role in retrospective risk assessment. They have been proposed for use in determining legal liability and compensation award amounts, estimating public health burdens from activities such as smoking, and deciding when to trigger regulatory risk management interventions based on monitoring data of risks attributable to bacteria in food. Yet, how to define and calculate attributable risks when multiple factors affect risk has been far from clear. This section introduces a new theory of attributable risk calculations in causal graph models that overcomes key limitations of traditional approaches and that better ties attributable risk calculations to causality instead of to statistical associations.

6.1 Why is Risk Attribution Hard?

Formulas for calculating attributable risks are a mainstay of basic epidemiology. For example, if case-control data are summarized in a 2 x 2 table that classifies individuals as exposed or not and as cases or not, then the attributable risk associated with exposure can be immediately calculated. In the following table:

	Exposed	Not Exposed
Case	a	b
Control	c	d
	a + c	b + d

the *attributable risk* (AR) is defined as the disease rate in exposed people minus the disease rate in unexposed people, i.e.,

$$AR = [a/(a + c)] - [b/(b + d)].$$

The population attributable risk (PAR) is defined as:

PAR = fraction of cases associated with exposure = $(N_1 - Np_0)/N_1$

where

- N = sample population size = a + c + b + d
- N_1 = number of cases = a + b
- p_0 = disease rate in the unexposed population = b/(b + d)

So, PAR = (AR) x (prevalence of exposure to risk factor in the population)
= (total cases – expected cases based on rate among unexposed)/(total cases)

Thus, at first it seems easy to calculate and interpret attributable risk. It is commonly referred to as the risk "due to" or "caused by" exposure and is interpreted as the proportion of cases (for PAR) or as the probability of disease (for AR) that would be prevented if effects of exposure were eliminated. However, in general, such interpretations are unjustified.

When diseases are caused by multiple factors, there is no necessary relation between attributable risk, as traditionally defined by the above formulas for AR and PAR,, and the risk that would disappear if exposure were removed. Indeed, Simpson's paradox shows that a positive population attributable risk (PAR) may be calculated from a 2 x 2 table even in cases where preventing exposure would *increase* risk. More generally, attributable risk calculations do not produce numbers that can successfully be interpreted as the proportions of blame for an undesired outcome that should be allocated to exposures (or to other risk factors).

Example: Over-Attribution of Risk to Multiple Risk Factors

Suppose that eating restaurant chicken is a risk factor for diarrhea from Campylobacter if and only if (a) The chicken is contaminated with Campylobacter prior to cooking; and (b) The chicken is undercooked and served pink on the inside. To highlight the conceptual issue as starkly as possible, suppose that the dose-response relation is such that an individual will become ill if and only if she eats undercooked (pink) contaminated chicken. Thus, the probability of illness can be tabulated as a function of *Contamination* and *Cooking,* as follows:

	Cooking	
Chicken Contamination	Well cooked	Pink
Not Contaminated	0	0
Contaminated	0	1

To keep the arithmetic simple, suppose that 40% of a certain population of size 100 (perhaps at a summer camp) is described by the (Contaminated, Pink) cell and that

20% of the population is described by each of the other three cells. Then the 2 x 2 table for the *Cooking* factor would be as follows:

	Exposed (Pink chicken)	Not Exposed (Well cooked chicken)
Cases	40	0
Controls	20	40

The attributable risk for the *Cooking* factor is the illness rate among exposed people minus the illness rate in unexposed people, i.e.,

$$AR_{cooking} = [a/(a + c)] - [b/(b + d)] = (40/60) - (0/40) = 0.67.$$

Symmetrically, the attributable risk for the *Chicken Contamination* risk factor is also 0.67. But these two ARs total to 1.34, i.e., to more than 100% of the risk being attributed. Such over-attribution (allocation of more than 100% of the total risk among the factors that contribute to causing it) can be arbitrarily great if there are many risk factors. Nothing in the definition of AR for individual factors prevents over-attribution or assures consistency of risk attribution among multiple factors in the sense of requiring the shares of risk allocated to different factors to sum to 100% of the total risk being allocated. This must be corrected in any theory of multivariate risk attribution that seeks to consistently allocate risk to multiple factors.

Example: Over-Attribution of Risk to Factors in a Causal Sequence

Consider the following causal chain:

$$V \rightarrow W \rightarrow X \rightarrow Y \rightarrow p.$$

Suppose that p is the probability of an adverse event and that V, W, X, and Y are all 0-1 variables satisfying the structural equations:

$$p = Y; \quad Y = X; \quad X = W; \quad W = V.$$

Each event in the sequence occurs (i.e., each variable becomes 1) if and only if its predecessor occurs, i.e., the chain describes a domino effect. If each individual in a sample described by this process is classified as a case ($p = 1$) or control ($p = 0$) and is cross-classified as exposed to X (i.e., $X = 1$) or as unexposed to X (i.e., $X = 0$), then the resulting 2 x 2 table can be used to calculate an attributable risk for X. However, exactly analogous calculations can be calculated for all the other elements in the chain preceding p, i.e., for V, W, and Y, as well as for X Thus, if the attributable risk for X is 0.35, for example, then the attributable risks for V, W, and Y must also be 0.35. Thus, the total attributable risk assigned to these factors would be 4 x 0.35 = 1.4, well over 100%. By making the chain longer, the total risk attributed to different factors can be made as large as desired. If compensation awards, financial liability for cleaning up waste sites, or other risk management decisions are based on such attributable risk calculations, more than 100% of the total blame (or cost, damage, etc.) may be allocated to various factors that are correlated with each other due to participation in the same causal sequence. Intuitively, it might seem more sensible to allocate all of the blame in this example to the input variable, V, rather than to any of its consequences in the causal chain leading from V to p.

Example: Attributable Risks for *Campylobacter*

The World Health Organization (WHO) and other authorities have called for data on the number of human *Campylobacter* illnesses per year in different countries that are attributed to consumption of chicken. But attribution is not purely data-driven. It requires risk-accounting judgments. For example, how much of the risk of human campylobacteriosis from eating undercooked chicken should be attributed to the chicken *vs.* to cooking and kitchen hygiene practices? How much, if any, should be attributed to processing that failed to eliminate (or even amplified) the microbial load of *Campylobacter* in chicken carcasses before they arrived at the kitchen? Should microbial load of Campylobacter be attributed only to the chicken that carries it or to the sources that originally contaminated the chicken (e.g., untreated drinking water, droppings from wild birds, insect vectors, etc.)? Should any share in risk be attributed the immune status of victims if it renders them susceptible to doses of *Campylobacter* that might not have affected someone else? Should 100% of the risk be allocated to each step at which sufficient care could eliminate *Campylobacter* risk? Or should the total amount of risk allocated to different causes be constrained to sum to no more than 100% of the total risk being allocated? Such questions cannot be answered by collecting more data. They are methodological rather than empirical. The answers to such conceptual questions should be driven by the intended *use* of attributable risk calculations – e.g., to guide or trigger actions or to report risk attributed to risk management decisions, events, conditions, or personal attributes and behaviors.

Example: Smoking-Attributable Mortality

An often-cited number is that cigarette smoking in the United States kills over 400,000 people per year. For example, the Centers for Disease Control (CDC) web site asserts that "Each year, more than 400,000 Americans die from cigarette smoking. In fact, one in every five deaths in the United States is smoking related.." (http://www.cdc.gov/tobacco/research_data/health_consequences/mortali.htm) But this statistic does *not* imply that if smoking were eliminated (or had never begun in the population), there would be 400,000 fewer deaths per year. For, some of those who die from smoking-associated diseases in any year have diets, alcohol consumption patterns, exercise habits, occupational exposures, and socioeconomic factors that are associated with both cigarette smoking and with increased mortality rates. The 400,000 number is based on differences in mortality rates between smokers and non-smokers – the traditional definition of AR. But the entire difference is almost certainly not due entirely to smoking. Some of these deaths might have occurred even if the victims had not smoked (but did have their other risk factors). Thus, attributable risk calculations are implicit in the reporting of crucial public health numbers, such as smoking-attributable mortality statistics. It is important to make explicit the assumptions embedded in such attributable risk numbers. For the particular example of the 400,000 deaths per year statistics, critics have argued that controlling for confounders would greatly reduce this number, while supporters claim that adjusting for confounders makes little difference, or even increases estimated attributable risk. The debate then turns to issues such as controlling for residual confounding (i.e., to confounding that remains despite attempts to control for it), effects of omitted or unmeasured confounders, age-specific hazard rates, and other important technical details. However this debate is eventually resolved (if at all) the fact remains that the reported statistic of 400,000 smoking-attributable deaths per year does not have a clear, unique interpretation in terms of the deaths per year that would be prevented by eliminated smoking. Yet, this is the type of information that would be most useful for many public health initiatives. A more useful theory of risk attribution should be able to specify the change in probability distribution for deaths per year attributable to changes in input values. Causal graphs provide a natural framework for such calculations.

Example: Attributable Risks for Two Synergistic Factors

Suppose that workers occupationally exposed to a particular factor, Factor 1 (e.g., radiation, coal dust, diesel exhaust, an industrial chemical, etc.) have a greater incidence rate of a given health effect (e.g., lung cancer or asthma) than do otherwise similar workers who are not exposed. Factor 1 interacts synergistically with a second factor, namely, tobacco smoke, as shown in Table A.

Table A: Lung cancer incidence rates for four combinations of two factors

		Factor 1	
		Absent	Present
Factor 2 (Smoking)	Nonsmokers	10% →	20%
		↓	↓
	Smokers	15% →	40%

The natural background rate in the population over a suitably defined time period, is 10%: this is the incidence rate in the absence of either factor. When Factor 1 alone is present, the incidence rate climbs to 20%. The attributable risk due to exposure to Factor 1 alone is 10%, the difference in incidence rates between exposed and unexposed populations. This increment is the *marginal risk* contributed by Factor 1 when Factor 2 is absent. Similarly, exposure to Factor 2 (smoking) alone produces an attributable risk of (15% - 10%) = *5%*. Finally, exposure to both factors gives an incidence rate of 40%, for a total increase over the no-exposure baseline of (40% - 10%) = 30%. How much of the 30% increase in incidence rate should be attributed to Factor 1 and how much to Factor 2?

Since Factor 1 alone produces only a 10% increment in incidence rate, whereas the two factors together produce a 30% increment, Factor 2 might be held responsible for the difference, which is 20%. This corresponds to calculating the attributable risk for each factor along the "upper path" in Table 1 (i.e., the path from the upper left-hand cell through the upper right-hand cell to the lower right-hand cell). The total attributable risk of 30% is then allocated with 1/3 (= 10%) being attributed to Factor 1 and 2/3 (= 20%) being attributed to Factor 2. But this approach fails to determine a *unique* allocation risk, since one could equally well calculate attributable risks along the "lower path" (from the upper left-hand cell through the lower left-hand cell to the lower right-hand cell), arguing that smoking alone causes only a 5% increase in incidence rate, so that the remaining difference of 25% between the incidence rate when both factors are present (= 40%) and the incidence rate when neither is present (= 10%) must be attributable to Factor 1. Equivalently, 25% is the *marginal risk* contributed by Factor 1 when Factor 2 is present. Now only 1/6 (instead of 2/3) of this difference is attributed to smoking, and 5/6 (instead of 1/3) is attributed to Factor 1.

If a worker sues his employer because of a health effect associated with Factor 1, and if that worker also smokes, then it may become important to find a rational, defensible procedure for determining how much of the blame for the health effect should be attached to Factor 1, and how much to the worker's decision to smoke (Landes and Posner 1983). Such a procedure might also prove important in establishing an equitable compensation plan for employees contracting the health effect, perhaps as part of a partially employer-subsidized insurance plan, whether or not lawsuits are involved.

The situation in Table A arises in many contexts where a risk results from the synergistic interaction of multiple factors. If an industrial plant's emissions produce health, ecological, and materials-damage effects whose magnitudes depend on the emissions of nearby plants and traffic, for example, then it may be useful for setting discharge permit fees and plant-specific cleanup costs to know how much of the total effect is attributable to each point source. A common regulatory principle is that each source should pay a cost sufficient to offset the reduction in benefit that is properly attributable to it. But this principle and related principles of social cost-benefit analysis cannot be implemented in practice until a satisfactory concept of attributable risk has been developed. Thus, the question of risk attribution in the presence of multiple causes or factors is of more than merely philosophical interest; it may be very useful as one input in the allocation of social costs and of responsibility for jointly produced effects. It is reasonable for public decision makers to look to risk analysts for methodological advice on the attribution of risks.

6.2 Principles for Risk Attribution

What principles must a satisfactory attributable-risk formula for multiple risk factors formula satisfy? Suppose that "risk" is represented by a 0-1 chance node, R, in a causal graph or causal influence diagram. In this section, the causal graph is interpreted so that causes precede effects in the partial ordering of nodes induced by the DAG and effects of changes in inputs are assumed to propagate through the graph as indicated by the arrows. The goal of risk attribution is to allocate responsibility for the risk event if it occurs (i.e., if R = 1) among other nodes in the graph. Potential desiderata for a risk attribution method of formula include the following:

Principles for Risk Attribution

1. ***Full allocation:*** A satisfactory attributable risk formula should allocate exactly 100% of the risk being analyzed among the factors that cause it. (Thus, the risks attributed to the different causes should not sum to more or less than the total risk being allocated.)
2. ***Marginality.*** The risk attributed to a variable is calculated from its marginal contributions to risk, i.e., from the incremental contributions that it makes to risk when the values or levels of other variables are held fixed. (Its marginal contributions to risk may depend on the values or levels of other risk factors. For example, in the above example of synergy between smoking and Factor 1, the marginal contributions for Factor 1 were 10% along the upper path and 25% along the lower path, and thus depended on whether smoking was present or absent.) This marginality principle is implied by the following stronger condition:

2b *Monotonicity.* Suppose that DAG models G and G' are identical except that a variable X in G is replaced with a variable X' in G' that makes a marginal contribution to risk, i.e., to Pr(R = 1), that is at least as great as X's for *every* configuration of values of the other variables. (In other words, X' always has a marginal risk at least as great as that from X.) Then X' is assigned at least as much attributable risk in G'' as X is assigned in G. Marginality also suggests that a variable that makes zero marginal contribution to risk for *all* configurations of the other variables should have zero risk attributed to it. This may be stated as another principle:

2c *Zero risk attributed to irrelevant variables.* If the probability distribution for the risk variable, R, is independent of the value of a variable Y for all configurations of values of the remaining variables, then Y is assigned zero risk. For example, in the DAG model $X \rightarrow R \leftarrow Y$, if $\Pr(R \mid X, Y) = \Pr(R \mid X)$ for all possible values of X and Y, then zero blame for R would be attributed to Y. Note that this principle implies:

2d *Allocation to causal predecessors:* A satisfactory attributable risk formula should allocate 0 risk to any variable that is not a causal predecessor (parent or ancestor) of the risk node. For example, in the model $Y \leftarrow X \rightarrow R$, where R is the risk node, zero blame for R should be assigned to (i.e., attributed to) Y.

3. ***Impartiality.*** Variables that affect risk equally are assigned equal attributable risks. (This may also be stated as: How variables are labeled or numbered does not affect the risk attributed to them. All that matters is how they contribute to risk.)

Mathematical analysis shows that these principles imply a unique method for calculating attributable risks in certain situations. For example, suppose that all variables are 0-1 variables, representing binary risk factors that may be present (exposed) or absent (unexposed) for each individual. $X_j = 1$ for an individual exposed to risk factor j and $X_j = 0$ for an individual not exposed to j. Assume that risk is a non-decreasing function of these variables, any of which may be present or absent:

$$\Pr(R = 1 \mid X_1, X_2, \ldots, X_m) = f(X_1, X_2, \ldots, X_m)$$

Then there exists only one method for allocating non-negative attributable risks among the variables that is consistent with the previous principles.

Attribution Formula for Binary Risk Factors

Suppose that the risk node in a causal graph model has *m* parents, that all *m* of these variables are binary (1 = present, 0 = absent), that all configurations of values are possible, and that risk is a non-decreasing function of these *m* variables. Let $X = (X_1, X_2, \ldots, X_m)$ be a specific configuration of values for the *m* parents. Define the *fair share* of risk for parent j in configuration X as its *average marginal contribution* (AMC) to risk in configuration X. That is, it is the average incremental increase in risk when X_j is increased from $X_j = 0$ to $X_j = 1$, averaged over all orders (i.e., permutations) of the non-zero variables in X. Denote the fair share assigned to parent j by $AR_j(X)$.

THEOREM: $AR_j(X)$ is the unique measure of attributable risk that satisfies the principles of Full Allocation, Marginality (or Monotonicity), and Impartiality. In other words, these principles imply that attributable risk coincides with fair share of risk, for each causal factor.

The proof is based on reinterpreting the axioms and representation and uniqueness theorem for the Shapley value in games of fair cost allocation (Cox, 1985), updated to use the axioms of Young (1994.)

Example: Two Synergistic Risk Factors (Cont.)

The theory just outlined for binary risk factors suffices to solve for attributable risks in the preceding synergy example for *Factor 1* and *Smoking*. Again, consider the 2 x 2 table for individual risks as a function of factor combinations (Table A below). For a smoker exposed to Factor 1, the configuration of interest is $(X_1, X_2) = (1, 1)$. The DAG representation has as its graph

Factor 1 → Risk ← Smoking,

i.e., the Risk node has Factor 1 and Smoking as its two parents. The fair shares of risk for these two factors and for configuration X = (1, 1) are:

Table A: Lung cancer incidence rates for 4 combinations of 2 factors

		Factor 1	
		Absent	Present
Factor 2 (Smoking)	Nonsmokers	10%	20%
	Smokers	15%	40%

$AR_1(X) = (0.10 + 0.25)/2 = 0.175$
$AR_2(X) = (0.05 + 0.20)/2 = 0.125.$

Thus, $AR_1(X) + AR_2(X) = 0.30$, the total excess risk in configuration (1, 1) compared to (0, 0), as required by Full Allocation. The fair share solution generalizes this approach, i.e., the average marginal contribution solution, to any number of binary variables.

Attribution for Binary Variables in a Causal Graph

The average marginal contribution (AMC) principle can be extended to solve more general attribution problems. For example, to attribute risks to variables other than the parents of the risk node, it suffices to (a) Calculate the AMC for each variable by averaging only over all orderings of variables that are consistent with the DAG constraints (and with the configuration for which risk is to be attributed); and (b) Applying a chain rule that allows *fair share fractions* to be multiplied along chains. The fair share fraction or fair attributable fraction, abbreviated AF, is defined as the fair share of risk attributed to a factor divided by the total risk being attributed. In general, fair share fractions are multiplied along chains and summed over the chains entering a node, thus generalizing ideas from path analysis (Bentler and Stein, 1992) beyond the context of linear models.

Example: Attribution of Risk in a Causal Sequence

Consider the sequence:

$X \rightarrow Y \rightarrow R$

with model parameter values
$E(R \mid Y = 0) = Pr(R = 1 \mid Y = 0) = 0.2$
$E(R \mid Y = 1) = 0.6$
$E(Y \mid X = 0) = 0.2$

$E(Y \mid X = 1) = 0.4$
$E(X) = Pr(X = 1) = 0.5$

Then the risk of R attributed to Y (or, more precisely, to the event Y = 1) in the configuration (X, Y, R) = (1, 1, 1) is: $AR_Y = (0.6 - 0.2) = 0.4$. Thus, of the total risk, $E(R \mid X = 1, Y = 1) = 0.6$, 2/3 is attributed to Y. The fraction 2/3 may be called the *fair share fraction* or the *fair attributable fraction* of R = 1 due to Y = 1, denoted as $AF_{RY} = 0.67$. Likewise, the risk of Y attributed to X = 1 is (0.4 – 0.2) = 0.2, giving a fair attributable fraction of $AF_{YX} = 0.5$ for the risk of Y = 1 due to X = 1. The chain rule then states that $AF_{RX} = (AF_{RY})(AF_{YX}) = 0.33$. Of the total risk of $E(R \mid X = 1, Y = 1) = 0.6$, the amount $(AF_{RR})(0.6) = (1 - AF_{RY})(0.6) = 0.2$ is due to background (or to R's own propensity to happen in the absence of other factors) and $(AF_{RY})(0.6) = 0.4$ is due to Y. Of the 0.4 total that is due to Y, $(AF_{YY})(0.4) = (1 - AF_{YX})(0.4) = 0.2$ is due to Y itself (i.e., to Y's propensity to occur even without X present) and the remaining 0.2 is due to X. Thus, the final allocation of risk of R = 1 is: 0.2 to background, 0.2 to Y, and 0.2 to X. These final shares correspond to *final attributable fractions* (abbreviated FA) of $FA_{RR} = FA_{RX} = FA_{RY} = 0.33$. The final attributable fraction FA_{Rj} is defined and interpreted as the proportion of the risk of R = 1 that is attributed solely to factor j and not to any of its ancestors. Thus, $FA_{RY} = (AF_{RY})(1 - AF_{YX}) = (0.67)(0.5) = 0.33$, for example.

In a more complex model such as

$$X \rightarrow Y \rightarrow R \leftarrow W \leftarrow V,$$

factors AF_{RY} and AF_{RW} would be calculated using the AMC formula and then allocated along the chains leading into R using the chain rule to obtain the FA_{Rj}.

Example: Over-Attribution of Risk in a Causal Sequence (Cont.)

The chain rule resolves the problem of risk attribution for the model

$$V \rightarrow W \rightarrow X \rightarrow Y \rightarrow p \text{ with node formulas } p = Y = X = W = V.$$

In the configuration p = Y = X = W = V = 1, the calculation is: $AF_{PY} = AF_{YX} = AF_{XW} = AF_{WV} = 1$, corresponding to $FA_{pY} = FA_{pX} = FA_{pW} = 0$, $FA_{pV} = 1$.

Attribution for Continuous Variables

Most causal graph models useful in risk analysis involve variables with more than two possible values. This section shows how to extend the AMC solution concept to continuous variables. In many applications, it is convenient to allow the factors X_j to be continuous rather than binary. For example, a continuous X_j may be used to denote the *level of exposure* to Factor j. (Extensions of AMC to ordered and unordered categorical variables can also be envisioned but will not be developed here.)

To extend the previous discussion of attributable risk to this continuous case, it is necessary to make some additional assumptions and to modify the Impartiality axiom. Let *R(X)* now denote a risk assignment function, mapping nonnegative *m*-vectors of variable values, $X = (X_1, X_2, \ldots, X_m)$, onto the nonnegative reals, so that $R(X) \geq 0$ is a numerical index of risk (e.g. the probability of an undesirable event, the expected number of fatalities in an accident, or the average number of incidents per year, depending on the context.) It is now convenient to assume that R(0) = 0 (perhaps after a shift of origin to remove a constant "background" risk rate) and that *R*(.) has continuous first partial derivatives. Assume that *R*(.) is nondecreasing in each of its arguments: increasing the level of any factor never decreases (but may increase) the resulting risk level. Thus, the risk index, $R = R(X)$, is a smoothly increasing (or at least nondecreasing) function of X, assigning zero risk to zero exposure.

It is now necessary to consider the *units* in which variables are measured – a problem not arising in the discrete formulation. Specifically, *the choice of units should not be allowed to influence the attribution process:* measuring a factor in *mg* instead of μg should leave the total risk share attributed to it unchanged (and hence should scale up the *per-unit* risk attributed to it by 1000). Per-unit risks will be denoted by coefficients, $a_i = a_i(R(.),X)$, where the attribution function *a*(., .) must satisfy the equation

$$a(R(.),x) \bullet x = R(x). \qquad \text{(Full Allocation Constraint)}$$

The total risk allocated to Factor *i* by *a* is just

$$r_i = a_i X_i$$

and the full allocation constraint requires that $r_1 + \ldots + r_n = R$, the total risk to be attributed.

A final complexity introduced by continuous factor levels is that it may in some cases be possible to conceptually *subdivide* total exposure to a factor into parts. The per-unit risks attributed to these parts should then be the same, since they all refer to the same factor. For example, if the risk of a storage tank rupture depends on the number of gallons stored, then the risk per gallon should be the same for all gallons in the tank, given the total number of gallons stored. Similarly, if the risk of Campylobacteriosis depends only on the total colony-forming units (CFUs) of Campylobacter ingested in a meal, then the risk per CFU attributed to ingested Campylobacter should be the same for each CFU, given the total number of CFUS ingested.

These considerations extend the concept of Impartiality and are summarized in the following axiom:

Principle 3a: ***Impartiality for Continuous Factors***. (a) The risk share allocated by a satisfactory risk attribution function to variable i, namely $r_i = a_i X_i$, does not depend on the units used to measure the factor levels: rescaling units of X_i rescales a, reciprocally, so that $r_i = a_i X_i$, is unaffected. (b) If the risk level R depends on X_i and X_j only through their *sum*, $X_i + X_j$ (so that X_i and X_j act as if they were really just two parts of the "combined factor" $X_i + X_j$.), then $a_i = a_j$ (i.e., the "parts" are assigned the same per-unit risk rate, which can also be thought of as the risk for their sum).

A compact mathematical expression of this axiom is given in Billera and Heath (1982, p. 34), under the name "Aggregation Invariance". The foregoing follows Mirman et al. (1983) in dividing it into two parts for clarity of interpretation

Attribution Formula for Continuous Variables

THEOREM: There is exactly one risk attribution function that satisfies Impartiality for Continuous Factors and an additivity condition (that the allocation of a sum of risks must equal the sum of their allocations; see Cox, 1985). It is given by $a^*(R(.), X)$, where a_i^* the per-unit risk attributed to X_i, is defined by

$$a_i^* = \int_{0 \le t \le 1} R^i(tX)\, dt \qquad \text{(Continuous AMC)},$$

the average marginal contribution to risk from variable X_i as exposure increases uniformly from 0 to X. Here, $R^i(.)$ denotes the marginal risk function for X_i i.e., the partial derivative of R(.) with respect to its ith argument, $\partial R/\partial X_i$.

PROOF. This is a reinterpretation of the Shapley value for continuous ("non-atomic") games. The extension of the Shapley value to the continuous case was adapted to the context of cost allocation in Billera and Heath (1982) and Mirman et al. (1983). The mathematical proof of Theorem 2 as a result in functional analysis is given in detail in Billera and Heath (1982).

Example: Risk Attribution for Logistic Regression Models

To illustrate the application of the continuous AMC formula, consider the popular logistic dose-response relation, $R(X) = \Pr(\text{Response} \mid X_1, X_2, \ldots, X_M) = [1 + \exp-(\beta_1 X_1 + \ldots + \beta_M X_M)]^{-1} - 0.5$. Here, the probability of a health effect, given exposure profile ($X_1, \ldots, X_m$), is taken as the measure of risk. Differentiating and performing the integration in the continuous AMC formula gives the fraction of the total risk, R(X), attributed to variable i, namely $a_i{*}X_i/R$ is just $\beta_i X_i/((\beta_1 X_1 + \ldots + \beta_m X_m)$, or $\beta_i X_i/(\beta \bullet X)$, for short. This provides one conceptual approach to answering such traditionally perplexing problems in tort law liability, as: "An individual who is exposed to a carcinogenic substance at his work place develops lung cancer, but he also smokes five packs of Camels a day – what is the incremental contribution of exposure to the carcinogen?" Exposures to each factor can be entered into a prior dose-response relation (assumed to have continuous first partial derivatives, etc.) for the individual involved (if it is known), and relative contributions assessed as in the above example.

Example: Attribution of Risk Externalities

N industrial plants release stack gases to air. The resulting health risk to humans, on some appropriately defined scale, is given by $R(X) = (\alpha_1 X_1 + \ldots + \alpha_N X_N)^2$, where X_i is the release rate from Plant i. (We choose a squared loss function simply to illustrate synergistic interaction among risk factors.) What is the relative risk share attributable to Plant i? Using the continuous AMC formula), the proportion of the total risk attributable to Plant i, namely $\alpha_i{*}X_i/R(X)$, is given by $\alpha_i X_i / \alpha \bullet X$, as in the previous example.

Example: Attribution of Externalities with a Threshold

A challenging methodological problem in risk analysis to date has been how to account for "risk externalities," such as those that arise when routine emissions from electric power plants combine with emissions from other sources, e.g., other power plants or local traffic, to produce cumulative pollutant concentrations that pose a threat to human health or safety. For example, suppose that three neighboring industrial plants labeled *A, B,* and *C* produce 1, 2, and 3 "units" of pollution, respectively, and that the average health damage (a risk to the exposed population) resulting from *x* units of pollution can be expressed as

$$H(x) = \begin{cases} 0 & \text{if } x \leq 5 \\ (x-5)^2 & \text{if } x \geq 5. \end{cases}$$

For the conceptual purposes of this discussion, it does not much matter how the units of pollution and health effect are defined, although clearly this would be important in practice. The important aspect is that the increment in health damage resulting from one more unit of pollution depends on how many units are already present. $x = 5$ is the threshold level at which health effects begin. The total risk in this situation is $H(1 + 2 + 3) = 1$; it is attributable to sources *A, B,* and *C,* which play the role of the "factors" in the general analysis above. Each plant owner can truthfully argue that its emissions alone would produce no health damage, but together the damage is produced, and the problem of attributing the risk of magnitude $H(6) = 1$ among the three plants arises.

If emission levels are treated as continuous variables, then the continuous AMC formula implies that each plant is assigned a risk share proportional to its emission level, so that $(r_A, r_B, r_C) = (1/6, 2/6, 3/6)$. On the other hand, if emissions are discrete (with each plant being either present or absent), then all three sources are assigned the same attributable risk, namely $FA_j = 1/3$, $j = A, B, C$. In particular, the attributable risk from Source *C* is equal to the attributable risk from Source *A*, even though Source *C* produces three times as much pollution. This is because the sources play symmetrical roles in creating the total risk to be attributed: any one of them could eliminate all of the risk by shutting down.

If the output of pollution from Plant *C* had been 5 units instead of 3 units, how would the attributable risks assigned to the three sources change? In the continuous case, attributable risk is still allocated in proportion to emissions levels so that the new attributable risks are $(r_A, r_B, r_C) = (9/8, 18/8, 45/8)$. Note that the risk assigned to *each* plant increases, even though *C* is the only one that has increased its emissions. For the discrete case, the new total risk of H(8) = 9 is assigned as follows: 11/6 to Source A, 20/6 to Source *B*, and 23/6 to Source *C*, for a total of 54/6 = 9. Of the 8 point *increase* in H(x) brought about by C's additional emissions, 1.5 points (18.75%) are attributed to Source A, 3 points (37.5%) are attributed to Source B, and 3.5 points (43.75%) are attributed to Source C.

Why should the risk attributed to Sources *A* and *B* be increased just because Source *C* has a greater pollution output? Shouldn't *C* be held responsible for all of the extra risk? The answer given by the AMC principle is an unequivocal "no": Sources *A* and *B* are assigned a substantial share (56.5%) of the risk increment in question. The reason and justification is that *A*, *B*, and *C* are linked via the "externality" of pollution: the presence of *A* and *B* magnifies the effects of C's emissions, or, equivalently, reduces the amount of pollution that *C* can produce without causing H(x) to exceed a specified level. When *C* has a greater level of pollution associated with its operation, therefore, *A* and *B* are largely responsible for the resulting increase in health risks, H(x). The reciprocal nature of responsibility for private outputs which are linked by synergistic externalities has been a familiar theme in environmental economics since it was first clearly brought out by Coase (1960) in the context of the reciprocal obligations implicit in property rights. A similar analysis applies to the case of a new plant that adds its pollutants to an already existing background level produced by established sources. Unless the increment in health risks from this new plant is *independent* of the background level (implying a linear relation between source emissions and resulting risks), not all of the new risks will be attributed to the new source.

It may be tempting to conclude that the AMC principle offers a method for scientifically allocating social responsibility for joint effects. This is not so. Suggestive as the formulas are, they only allocate risk to factors, not to socially responsible parties. For example, Source *A* in the above example could as well represent a natural background source as a man-made one, in which case the share of risk attributed to it would carry no implication of social responsibility; or it could be a source such as automobile traffic, in which responsibility is so diluted as to become practically meaningless. Thus, the fact that part of a risk is *attributable* to a given source does not mean that there is a corresponding social *responsibility* for that risk share. Moreover, figuring out how much of a total risk is attributable to each source does not tell which sources are best able to reduce it. Differential control costs across sources, which are crucial to cost-effective social programs for risk control, do not in any way influence the calculation of attributable risk.

Either formula can, however, be applied in such a way that all risk is allocated among the subset of factors for which ownership and responsibility are meaningful. Indeed, the risk analyst can attribute 100% of the risk to *any* desired subset of factors that may be present simply by deleting from consideration factors that are not to have risk attributed to them, and incorporating their effects into the risk assignment function, R (.). In the above example, if Source A represented natural background emissions (e.g. volcanic emissions), and if it were decided that risk should not be attributed to such natural hazards, then it could be deleted from the list of sources to which risk is allowed to be attributed by replacing the risk assignment function H(x) with the revised function

$$H'(x) = \begin{cases} 0 & \text{for } x \leq 4, \\ (x - 4)^2 & \text{for } x \geq 4, \end{cases}$$

where x is now the sum of the pollution outputs from Sources B and C only.

The ability to choose among formulations, by choosing which factors are to be considered candidates for attribution and which are not, means that ethical and social considerations must come into play in determining "attributability." If Factor 2 in the smoking synergy example (Table A above) had been gender (coded as 0 = male, 1 = female) instead of smoking (0 for no smoking, 1 for smoking) for example, then attributable risk could be calculated separately for males and females (0.2 and 0.3 respectively) with Factor 1 being considered the only risk factor that is relevant, e.g. in determining compensation liability. Thus, social and ethical considerations could prohibit a formulation in which gender is a candidate for attribution, with a risk of only 0.25 being attributed to Factor 2, and a risk of 0.15 being attributed to "being female," although the analogous formulation would be acceptable for "smoking."

Example: Attributing Campylobacter Risk to Multiple Causes

The approach to attribution of risk proposed so far allows the risk of infection in the following graph to be attributed between the two continuous factors *CFUs in bought chicken* and *Cooking preparation care.*

CFUs in bought chicken → CFUs in ingested chicken ← Cooking/preparation care
↓
risk of infection

The risk of infection is 100% attributed to *CFUs in ingested chicken.* Responsibility for *CFUs in ingested chicken* can be allocated between *CFUs in bought chicken* and *Cooking preparation care* using the continuous AMC formula with *CFUs in ingested chicken* as the dependent variable (instead of risk). (To apply the formal framework, the variable representing cooking care must be oriented so that greater values correspond to greater risk, i.e., it is really a "carelessness" variable. For example, this variable might represent the fraction of CFU in bought chicken that is *not* eliminated by cooking and preparation prior to serving.) Risk of infection is then allocated between these two input sin the same proportion as responsibility for *CFUs in ingested chicken.*

Note that this attribution result may not have any clear implications for risk management, however. The key question for rational risk management is: Which acts (from among a set of alternative decision options) lead to the most desirable joint probability distribution of outcomes, e.g., illnesses, chicken prices, and control costs? No theory of attributable risks addresses this range of concerns. On the other hand, an influence diagram analysis could address this risk management decision problem directly. More generally, attributable risk calculations are seldom useful for guiding prospective risk management decisions. Decision-analytic techniques such as Monte-Carlo prediction and influence diagram analysis are designed for this purpose and should be used instead of attributable risks. On the other hand attributable risk calculations may have some value in retrospective risk management decisions about compensation or liability if the principles leading to the fair share of risk and the AMC solution are considered compelling desiderata.

Example: More Realistic Attribution Problems for Campylobacter

This example discusses other aspects of risk attribution that arise in more complex DAG models.

Cross-contamination of other foods by contaminated chicken is a possible route for Campylobacter to reach ingested food. This possibility is reflected in the following DAG:

CFUs from chicken → CFUs from other foods ↓ ↓ cooking care→ total CFUs ingested → *risk of infection*

Suppose that attention is focused on the relation between the two italicized variables *CFUs from chicken* and *risk of infection.* The proportion of the risk of infection that should be attributed to CFUs from chicken may be found by applying the chain rule and by summing over the paths entering each node. Presumably, *risk of infection* is 100% attributable to *total CFUs ingested.* Blame for *total CFUs ingested* is allocated between *CFUs from chicken* and *CFUs from other foods* in proportion to their magnitudes (by Impartiality for Continuous Factors), but then part of the blame for *CFUs from other foods* is attributed to *CFUs from chicken,* again based on their actual physical contributions. CFUs from chicken that grow on other food are still considered attributable to chicken based on the AMC logic and the fact that contamination of chicken precedes cross-contamination in this model (so that te latter cannot occur without the former. This is an example of how the DAG model constrains the set of orderings of variables that must be considered in the AMC calculation.)

In the next model, *immune status* and *age* are shown as factors that can affect *susceptibility,* which is a latent variable. If a policy decision is made not to attribute risk to these factors, then the *(susceptibility)* node and its parents can be deleted and the conditional distribution of *risk of infection* given *total CFUs ingested* can be updated by averaging over the different levels of *susceptibility* weighted by their relative frequencies.

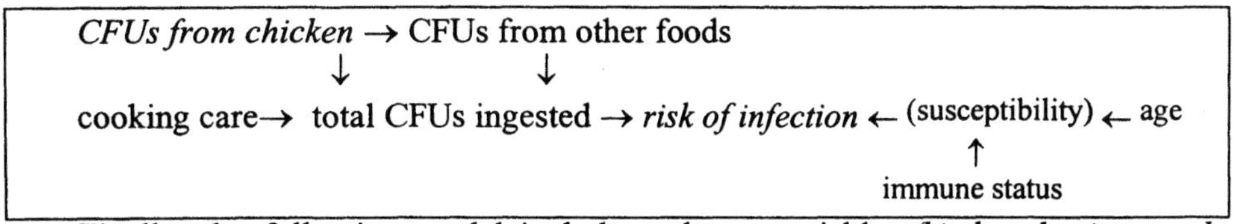

Finally, the following model includes a latent variable, *(kitchen hygiene and food preparation care)* as well as cross-contamination.

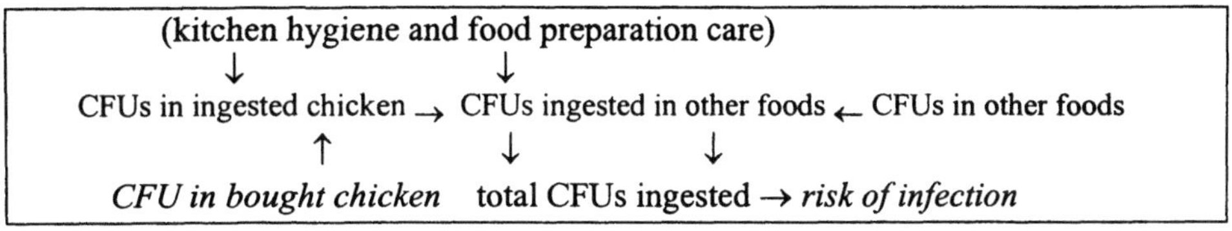

The proportion of *risk of infection* (or, equivalently, of *total CFUs ingested*) that should be attributed to *CFU in bought chicken* can be calculated using the AMC and chain rule principles. However, if it is only desired to calculate how the probability of *risk of infection* will change if a given reduction in *CFU in bought chicken* is accomplished, then a discrete-event model (Chapter 2) or Monte-Carlo based prediction will be more appropriate than an attributable risk calculation.

7. CONCLUSIONS

The causal graph modeling approach presented in this chapter provides a uniform conceptual and algorithmic framework for accomplishing some of the most important goals of quantitative risk analysis. These include:

- *Representing and consolidating causal knowledge* about how the world works in connecting risk management actions to their probable consequences. Causal graphs and causal influence diagrams provide an explicit, documented description of how changes in some variables (including decision variables) are hypothesized to propagate along causal paths (albeit probabilistically in most cases) to change the probability distribution of outcomes.
- *Testing and refining causal hypotheses and models.* Conditional independence tests and composition tests can be used to determine whether hypothesized causal graph models are consistent with available data. If not, they will usually reveal where the relations implied by the DAG model are violated and where new arcs need to be added or old ones need to be re-oriented.

- *Learning possible causal patterns from data.* Causal graph learning algorithms do not stop with identifying statistical associations, but rather start with them and then attempt to build full DAG models that are consistent with the data. Although some version of the DAG-learning problem are NP-hard, simple search heuristics seem to be effective in discovering useful models. The Occam's Window heuristic and the BIC criterion enable BMA ideas to be applied in many cases with practical amounts of computational effort.
- *Inferring* probable true exposure-effect relation from observations with errors and missing data (e.g., for true exposures and effects.)
- *Estimating* effects of unobserved heterogeneity and confounders using DAG inference algorithms. These apply modern computational statistics algorithms (such as the EM algorithms, Data Augmentation algorithms, and Gibbs Sampling or other MCMC algorithms) to solve important statistical inference problems.
- *Predicting* probable consequences of decisions via Monte Carlo simulation and *optimizing* decisions using influence diagram algorithms.
- *Attributing* risks/allocating blames for undesired outcomes (or exposures) to their possible contributing causes, as represented in a causal graph model.

Advances in theory and algorithms continue to be made in each of these important areas. However, the present state-of-the-art of causal graph modeling already appears to make it promising as a reasonably rigorous, comprehensive, and computationally practical framework within which to address many of the analytic challenges of quantitative risk analysis.

Chapter 5

Individual Risk Management Decisions

1. INTRODUCTION

The goal of applied risk assessment is usually more than just to identify hazards or to quantify health risks for individuals and populations from different exposures. A successful quantitative risk assessment (QRA) usually quantifies the probabilities of different consequences for various alternative *decisions*. It provides information to improve action. While QRA may also produce intermediate information, such as decision-to-exposure or exposure-to-response-probability relations, its main goal is to assess the risks (determined by consequence probabilities) and uncertainties for risk management decision alternatives that decision-makers must choose among. In this context, information from a QRA has value if it helps to identify or make decisions that increase the probabilities of more-preferred consequences and reduce the probabilities of less-preferred outcomes.

This is the first of three chapters dealing with how QRA information can and should be used to improve choices among risk management options. Closely related topics include how to use QRA information to identify useful and effective ("undominated") risk management options; and pitfalls in describing and communicating risk information. These topics are introduced through examples and via summaries of experimental evidence.

Normative and descriptive theories of decision-making under risk are very different. Normative theories often lose their prescriptive power when the realities of decision-making behavior are taken into account. For example, real decision-makers may not be able to supply the coherent preferences and beliefs assumed or required by prescriptive theories. This chapter and chapters 6 and 7 present findings from experimental economics and the psychology of risk perceptions, risk preferences, and risky choices. These findings make it clear that *any* normative theory of risk management, if it is to be of practical value, must deal with gaps and paradoxes in human thought and behavior. Chapter 7 introduces some of the practical options made available by multicriteria decision-making (MCDM) and optimization.

The theoretical developments and experimental data covered in these chapters have strong implications for how information about health risk can and should be communicated to achieve various ends effectively. As discussed in Chapter 8, goals of risk communication range from warning individuals about potential hazards (perhaps to limit liability) to empowering them with balanced information to support better-informed decision-making to trying to manipulate their perceptions and behaviors (e.g., in attempts to reduce teen smoking or to promote the use of sun blocks). The findings on risk perception and risk communication presented in this chapter and Chapters 6 and 7 can help to achieve all of these goals more effectively.

Risk communication is often better modeled as a participatory political process than as a one-way information transfer from a source to one or more sinks. In *analysis-deliberation* models of risk management decision-making, those who care about the results of a risk management decision raise and deliberate about issues that concern them, with risk assessment being drawn on to provide relevant facts, data, and analyses to inform the deliberative part of the process. Other models of risk communication and of participatory risk management processes take into account the incentives of participants to truthfully reveal what they know and to attempt to manipulate collective choice processes for their own ends. These complexities can be illuminated by game-theoretic models, as developed in Chapter 8.

2. VALUE FUNCTIONS AND RISK PROFILES

In most applications, "risk" refers not only to the probability or frequency of a hazardous event, but also to the severity of its consequences. Hazard functions are useful for expressing the frequency – or, more precisely, the expected number of occurrences per unit time – for an event. To represent information about the likely severity of its consequences, however, it is necessary to introduce another component into the description of risk. That is one purpose of this section. A second purpose is to introduce the idea of a *risk profile,* i.e., the complementary CDF of consequence severity. Most of this chapter and the next then focus on techniques for comparing and choosing among risk profiles. This allows one to compare and choose among prospects or risk management actions that generate the risk profiles.

Suppose, then, that there is a set of possible *consequences* of a hazardous event. Call this consequence set C, with generic member c. Depending on the application, consequences might describe the number of people hurt or killed in an accident, the number exposed to an injurious chemical in the event of a leak, the number of quality-adjusted life-years (QALYs) lost in a population from a disease, and so forth. It is not necessary that consequences be measured on a single scale. For example, a vector of costs, economic benefits, environmental changes, and health effects may represent a single consequence. Or, a consequence may be represented by a set of such vectors, one for each member of the potentially affected population.

Not all potential consequences are equally preferred. Some are considered "less severe" than others, meaning that the less severe consequences are strictly preferred to more severe ones. Although relative severity is often easy to judge for a single-attribute consequence (e.g., number of lives, life-years, or QALYs saved or lost), it may be harder to compare multi-attribute consequences. Chapter 7 introduces techniques to simplify and structure comparisons among prospects with multi-attribute consequences. The techniques in this chapter are more general. They evade the difficulties of making value trade-offs among consequences with multiple incommensurable dimensions by assuming that well-defined preferences exist, without examining how they are (or should be) formed.

Preferences among consequences will be represented by a *value function, v*. This is a function that assigns numbers to consequences. It does so in such a way that preferred consequences are assigned higher numbers while equally preferred consequences are assigned the same number. For a given set of potential consequences, the most preferred one is typically assigned a value of 1, while the least preferred (or "most severe") one is assigned a value of 0. Values of other consequences are measured on the scale calibrated by these two end-points, as explained further below. The notation $v(c)$ denotes the *value* of consequence c. When consequences are evaluated using a single attribute as a criterion, that attribute itself, e.g., number of deaths per year from a hazardous activity, may be an appropriate value function (apart from any rescaling performed to satisfy the convention of lying between 0 and 1.)

If $v(c)$ is a value function, then so is $f(v(c))$, for any increasing function f with $f(0) = 0$ and $f(1) = 1$, such as $v^2(c)$. In some settings, a natural severity scale exists (e.g., lives lost in an accident) for which twice as much is considered twice as bad. Specifying values for the endpoints of the value scale (namely, 0 for 0 loss and 1 for some loss at least as large as the greatest one to be considered) then determines the values of all other consequences. Treating "value" itself as an attribute allows many of the ideas that follow to be extended directly to more complicated settings.

From this perspective, the output of a consequence analysis may be expressed as a cumulative distribution function (CDF) for the single attribute called "value". Let F denote the CDF for the value of the consequence of a hazardous event. It is defined as follows:

$F(s) = Pr[v(c) \leq s]$ for all s, $0 \leq s \leq 1$, and for all $c \in C$ (risk profile, CDF)

Reorienting the axis to show the probability that the consequence severity from an event will be at least s, for all s between 0 (least severe, most valued consequence) and s = 1 (most severe, least preferred consequence) produces a *risk profile* for that event. (In some of the older literature, risk profiles are also called "frequency-severity" curves or "F-N" curves, the latter name deriving from applications in which risk profiles represented the expected annual frequency of accidents that would cause at least "N" deaths or other units of damage. Such curves typically conflate the risk information from multiple events, hazard functions, and consequence severity probability distributions, however.) Consequences in particular application domains may have a clear unit of severity, such as number of lives lost, acres burned, dollars spent to accomplish a mission, and so forth. Then, it is common practice to present risk profiles with severity expressed in its natural units, rather than normalizing to the 0-1 scale used for more abstract discussions.

Since CDFs for consequence values contain exactly the same information as risk profiles for consequence severities, with the possible exception of a normalizing constant that maps natural units to a 0-1 scale, we will henceforth use the generic term *risk profile* and the symbol F to refer to either.

For simple situations, an alternative notation is also widely used. If there is a small list of n possible consequences with a single natural attribute (e.g., lives or dollars) for evaluating them, then the following *prospect notation* may be used:

$$(p, v) = (v_1, p_1; v_2, p_2, \ldots, v_n, p_n) \qquad \text{(prospect notation for values)}$$

This notation summarizes the probability density function (PDF) for the uncertain prospect by listing the probability p_j of receiving a consequence with value v_j, for each possible consequence $j = 1,2, \ldots, n$. The consequences are usually numbered in order of increasing (or non-decreasing, if there are ties) values. If the consequences are specified in natural units on a single attribute scale (e.g., life-years or dollars), rather than being expressed on a value scale, then we will use the following notation:

$$(p, c) = (c_1, p_1; c_2, p_2, \ldots, c_n, p_n) \qquad \text{(prospect with natural units)}$$

Here, the pair (c_j, p_j) denotes the consequence and probability of the j-th possible outcome. This is called a *single-attribute prospect*. If each consequence is described instead by a vector of attributes (e.g., life-years lost *and* dollars spent), then the resulting *multi-attribute prospect* may be denoted by this same formula, but with each c_j now interpreted as a vector (i.e., an ordered list) of consequences.

In summary, a single event that produces an uncertain consequence may be represented by a risk profile, *F*. In some simple cases, it may also be represented as a prospect *(p, v)* or *(p, c)*. Table 1 summarizes the steps needed to calculate the risk profile.

Table 1: Steps to Calculate a Risk Profile

1. *Bound the consequence set.* Identify the set of potential consequences to be considered.
2. *Assess a value function for the consequences.:* Assess the value (or severity) of each potential consequence identified in Step 1. It is conventional (but not necessary) to scale the values of the consequences to run from 0 to 1. With this normalization, define *severity = 1 – value.*
3. *QRA:* Assess the probability of each potential consequence identified in Step 1 using a QRA model. (If the set of consequences is continuous, then assess a probability density function for consequences.)
4. *Calculate the risk profile.* From the information in Steps 1 and 2, prepare a risk profile showing the probability that the severity of the consequences will be at least *s*, for each *s* between 0 and 1. This information may also be presented in a cumulative distribution function for value showing the probability that the value of the consequences is no greater than *s*, for each *s* between 0 and 1.

Each step presents problems and challenges that can be addressed by appropriate techniques. For example:

- *Bounding the consequence set.* Possible consequences may be listed explicitly if there are only a few of them and they are known. More often, the full set of possible consequences is uncertain. The possible consequences or outcome scenarios to be considered are specified implicitly, e.g., by specifying constraints on the possible values of the variables describing them. Or, possible consequences may be randomly sampled using a computer simulation model that generates particular outcome scenarios – defined as combinations of values of variables in the model – that can occur according to the assumptions and constraints embedded in the model. Although there is no perfect general solution to the problem of identifying all and only the consequences that might occur, simple techniques mentioned in Chapter 2, such as event tree analysis, fault-tree analysis and knowledge-based forward chaining, as well as more complex simulation modeling, can help to systematically identify the most plausible potential outcomes and avoid oversights.
- *Assessing value functions.* When potential consequences are complex and described by multiple incommensurate attributes or evaluated by

multiple criteria, it may not be clear how they should be ordered in terms of relative severity. Different judges may rank them differently. In this case, formal techniques can help to clarify preferences and value trade-offs. Multicriteria decision-making (MCDM) and multattribute utility theory (MAUT), discussed later in this chapter and in Chapter 7, respectively, can help to elicit coherent subjective value functions from individual decision-makers or stakeholders. They can also identify partial orderings of consequences that all judges would agree with (or at least all judges with roughly compatible preferences, e.g., for lower costs and fewer deaths).

- *QRA and consequence probability modeling.* As discussed in Chapter 2, assessing the probabilities of different consequences can require detailed scientific, engineering, and applied mathematical models, as well as applied probability and statistics. The details vary greatly across application areas, with the mathematics required to model fires and explosions and their consequences, for example, having little in common with the mathematics used to model the AIDS epidemic. Despite this diversity, many consequence models and their strengths, limitations, and uncertainties can be discussed within the unified graphical "meta-modeling" framework of DAGs discussed in Chapter 4. These can show the structure of model calculations by which values (or probability distributions) for output quantities are derived from those for input quantities, while abstracting from the details of the calculations.
- *Calculating risk profiles.* If the total set of outcome scenarios to be considered is too large to list explicitly, then random sampling via simulation may be the best that can be done. In this case, the true risk profile can only be approximated by repeatedly sampling consequence scenarios, evaluating their severities, and estimating the fraction of scenarios that exceed different severity levels. Importance-weighted sampling and other special techniques can be used to help obtain accurate simulation-based estimates of probabilities of rare events in Markov processes and other stochastic systems. This is currently an area of rapid progress in operations research methodology.

Despite these challenges, the steps required to calculate a risk profile can often be carried out with little difficulty or ambiguity using the types of engineering, statistical, and causal models surveyed in Chapters 2-4.

Risk profiles play a crucial role in risk assessment. They allow expertise in comparing and evaluating risk profiles to be separated from the expertise required to create an accurate risk profile. The former requires principles of decision analysis, as elucidated in this chapter and the next. The latter typically requires subject matter expertise and modeling skills of the types mentioned in Chapters 2-4. Thus, the risk profile is a summary and communication device that allows the numerical results of a QRA to be handed over to risk management decision-making in such a way that the decision-maker need not have the deep subject-matter and modeling expertise often required to create the risk profile via QRA.

3. RATIONAL INDIVIDUAL RISK-MANAGEMENT VIA EXPECTED UTILITY (EU) THEORY

The dominant prescriptive paradigm for how rational individuals should make decisions is expected utility (EU) theory. It provides the following normative process for decision-making.

Steps in Rational Decision-Making via EU Theory

1. Identify the set of alternative *decision options* to be considered
2. Identify the set of potential *consequences* to be considered.
3. Assess the N-M *utility* of each consequence. (This is discussed below.)
4. Quantify the *probability* of each consequence if each decision option is selected. This information yields a *risk profile* for each option.
5. Eliminate options with dominated risk profiles. (This optional step is described further in Chapter 6, e.g., using various dominance concepts.)
6. Choose the option whose risk profile gives *maximum expected utility*.
7. Present recommendations (undominated and EU-maximizing options) and *sensitivity analyses* showing impacts on decisions of changing values, assumptions, probabilities, risk attitudes (Chapter 6), and information. Graphical displays such as spider- and tornado-diagrams (Clemen, 1996; see http://www.palisade.com/html/ptree.html for software) present sensitivity analyses in formats that many decision-makers find useful. They allow one to visually compare multiple factors to see which ones most affect the optimal decision.

Steps 1-6 are for an individual decision-maker. Step 7 (presentation and sensitivity analysis) is important when EU-based decision analysis is used as a basis for recommending actions to others, such as the analyst's client or a risk manager who is relying on decision analysis to help clarify choices. For excellent introductions to the basics of EU decision analysis and applications, see Raiffa (1968), Keeney and Raiffa (1974), and Clemen (1996). Pannell (1997) surveys sensitivity analysis techniques.

This version of the EU procedure applies when there is a small number of decision options and a small number of consequences. It assumes that the scope of the decision problem is clear, i.e., it is known (a) Who is the relevant decision-maker (d.m.); (b) What are the alternative decision options available to choose among; and (c) What are the potential consequences to be considered in making the decisions. There is no ambiguity about what actions are possible and allowed (e.g., decision-making authority is not in question). Moreover, the formulation presumes that criteria and preferences for evaluating consequences are clear and stable enough so that a well-defined utility function can be assessed (see below). Finally, Step 4 requires that it be possible to complete a quantitative risk assessment (QRA) to give consequence probabilities for different decision options. If these pre-conditions are not satisfied, then the EU theory framework may not be applicable. Approaches to decision-making that may be useful even when EU theory is not directly applicable are considered at the end of the chapter.

3.1 EU Decision-Modeling Basics

The traditional EU framework conceptualizes a decision problem as consisting of the following components:

- A *consequence function*, $c: A \times S \to C$ where A = set of alternative *acts* or decision options to be considered; S = the set of *states of the world* or *scenarios* (representing events that the decision-maker cannot directly choose or control); and C = a set of possible *consequences*. The notation $c(a, s)$ denotes the consequence if act a is selected and state s occurs.
- A *utility function* u: $C \to [0, 1]$, where 0 represents the utility of the least-preferred consequence in C and 1 represents the utility of the most-preferred consequence in C. For any other consequence in C, define $u(c)$ as the probability of receiving the consequence with utility 1 (and otherwise receiving the consequence with utility 0) that makes one indifferent between this lottery and receiving c with certainty. Of course, for this definition to make sense, one must assume that such an indifference probability can be uniquely defined. This and other assumptions of EU theory are listed below.
- A *conditional probability measure*, $Pr(c \mid a)$, specifying the probability of each consequence in C if each act a in A is selected. (This is the definition of $Pr(c \mid a)$ if C is finite or countable. More generally, it specifies the probability density of any Borel-measurable subset of C if a is selected. Moreover, strictly speaking, acts may not be events, so that the conditional notation "|" for events is being abused. However, it is convenient and intuitive enough to warrant continued use.)

Usually, $Pr(c \mid a)$ is thought of as being derived by summing or integrating the state-dependent consequence of an act, weighted by the state probability, over all states, i.e.,

$$Pr(c \mid a) = \int_{s \in S} c(a, s)Pr(s)ds \quad \text{(calculating consequence probabilities)}$$

This calculation can be extended to probabilistic consequence functions, say

$Pr(c \mid a, s)$. (stochastic consequence model)

The calculation is: $Pr(c \mid a) = \int_{s \in S} Pr(c \mid a, s)Pr(s)ds$

$Pr(c \mid a)$ for each consequence and each act is obtained from a QRA model. Given $Pr(c \mid a)$ and $u(c)$, the expected utility of an act a in A is given by:

$$EU(a) = E_{Pr(c \mid a)}[u(c)] = \int_{c \in C} u(c)Pr(c \mid a)dc.$$

For finite or countable consequence sets, the integral is replaced with a sum; for mixed discrete and continuous consequences, it is replaced by a Lebesgue-Stieltjes integral. Even if $Pr(c \mid a)$ cannot be solved for exactly, it can often be approximated from an underlying model of act-consequence relations via Monte-Carlo simulation. In any case, the notation

$E_{Pr(c \mid a)}[u(c)]$ or $EU(a)$

indicates expected value of the utility function $u(c)$ when consequence probabilities are described by the measure $Pr(c \mid a)$.

It is often useful to formulate risk management models that clearly distinguish between *acts,* summarizing everything that the d.m. can do, choose, or control to affect the consequence, and *states,* which summarize everything else that affects the consequence. The DAG model is:

$$\begin{array}{ccccc} a & \rightarrow & c & \leftarrow & s \\ & & \downarrow & & \\ & & u & & \end{array}$$ (act-state decision model)

To calculate expected utilities, it suffices to know the conditional probability of c given a and s and to know the state probabilities (for any a, if choice of a affects state probabilities).

The canonical case in which states are defined so that their probabilities do not depend on the act and in which acts and states uniquely determine consequences is sometimes called *normal form* analysis. Allowing an arc from a to s and interpreting $a \rightarrow c \leftarrow s$ to mean that only the conditional *probability* of each consequence is determined by the act-state pair a and s increases the flexibility of the decision model without significantly complicating calculation of expected utility. Influence diagrams allow these extensions.

3.2 Decision-Making Algorithms and Technologies

Other well-known decision formalisms are also useful for specifying the relations among acts, states, and consequences. They provide templates for structuring and representing knowledge, quantifying key relations, and organizing the concepts and data needed to support effective decision-making. Table 2 lists the most common ones and shows how they may be interpreted as special cases of the act-state-consequence framework.

Table 2: Decision Modeling Frameworks

Framework/ Representation	**Act**	**State**	**Consequence**	**Optimization algorithms**
Payoff matrix (Raiffa, 1968)	row	column	Payoff in cell	Eliminate dominated rows, then choose row to maximize expected utility
Normal Form • *Consequence model c(a, s)* • *State probability model p(s | a)* • *utility model u(c)*	a = act (an input to the con-sequence model)	s = state (input to consequence model)	c(a, s) = consequence model (a matrix or bivariate function)	Mathematical programming: $\max_{a \in A} \Sigma_c u(c)p(c \mid a)$ s.t. $p(c \mid a) = \Sigma_s c(a, s)p(s \mid a)$
Decision tree (Raiffa, 1968)	Decision at each choice node	Outcome at each chance node	Payoffs at tips of tree	Averaging out and folding back (simple dynamic programming)
Influence diagram (Chapter 4 and Clemen, 1996)	Decision at each choice node	Outcome at each chance node	Value at value node	• Arc reversal • MCMC (Bielza *et al.*, 1999)
Markov decision process (MDP); Semi-Markov decision process (Hazen and Pellisssier, 1996)	Choice of act in each state	Transition time and next state at each visit to a state	Reward per unit time in states and/or at transitions; transition rates among states	• Value iteration • Policy iteration • MCMC
Stochastic optimal control	Decision rule (maps observed history to control trajectory)	Evolution of state trajectory, given control trajectory	Function of state and/or control trajectory	Stochastic dynamic programming (Bertsekas and Tsitsiklis, 1996; Philbrick and Kitanidas, 2001)

If the number of possible acts and the number of possible states are both small, then the decision-relevant information can be summarized in a table, called a decision table or *payoff matrix,* showing the utility of the consequence from each act-state combination. Such an array for three acts and four states would look as follows:

Payoff matrix for a decision problem with 3 acts and 4 states:

		States			
		s_1	s_2	s_3	s_4
Act	a_1	u_{11}	u_{12}	u_{13}	u_{14}
	a_2	u_{21}	u_{22}	u_{23}	u_{24}
	a_3	u_{31}	u_{32}	u_{33}	u_{34}
		p_1	p_2	p_3	p_4
		State probabilities			

In general, if |A| = number of acts and |S| = number of states, let **U** denote an |A| x |S| normal-form payoff matrix in which row i and column j contains the utility of consequence c_{ij}, say, u_{ij}. Let **p** be a column |S|-vector of state probabilities. Then the expected utilities for the acts are just the elements of the column |A|-vector **Up**, i.e., the sums of $u_{ij}p_j$ over all states, j, for each act, i. The decision algorithm implied by EU theory is to calculate **Up** and to choose the act corresponding to its largest element.

Decision tables have the advantage of being simple to use and understand. Their major limitation is that they do not represent detailed causal information, and hence the method by which consequences or values are associated with act-state pairs is reduced to a black box that may not be trusted by those affected by the decision. Decision tables are most useful for representing static decision problems in which no new information arrives over time. They do not clarify the dynamics of sequential decisions.

Decision trees (Raiffa, 1968; Clemen, 1996), with which we assume familiarity, explicitly represent the time sequences of acts and events leading to different consequences. Probabilities at chance nodes are conditioned on the histories of decisions (i.e., choices at decision nodes) and events (i.e., resolutions of uncertainties at chance nodes) along the entire path leading from the root of the tree up to the chance node of interest. A decision tree may be interpreted as a simple form of causal model showing how decisions affect probable consequences. In the "act-state-consequence" framework, a single act for an entire decision tree model consists of a choice of what to do at *each* decision node in case one arrives there.

When the numbers of possible acts and states are large, efficient ways of storing decision problem information must be used. Decision trees have long been used for this purpose, but they are often inefficient for large problems. They are most useful for small dynamic models with finite sets of decision options and events available at each choice or chance node, respectively. They quickly become unwieldy when there are many combinations. For example, consider the easy problem of choosing an order in which to inspect N sites to minimize the expected cost of finding an object hidden at one of them. The probability that the object is at site j is p_j and the cost of inspecting site j is c_j. A decision tree approach requires $N!$ leaves, one for each order, even though many of the sub-trees are identical and thus, in principle, need only be solved once. The optimal solution (examine sites in decreasing order of p_j/c_j) is much simpler than such a large tree with $N!$ branches might suggest.

Influence diagrams (IDs) and similar DAG models (Chapter 4) provide alternatives to decision trees. In typical ID software, nodes of different types are used to represent not only decisions and chance variables, as in decision trees, but also deterministic functions, stochastic functions, and value or utility functions. IDs often provide compact representations of large decision problems compared to decision trees, especially when the trees are very symmetric (i.e., have many copies of identical sub-trees). Influence diagrams also reveal conditional independence relations among the variables in a decision model, thus clarifying the logic of the decision problem. They can incorporate the results of detailed causal models and QRA in the conditional probability relations stored at nodes. Special graph-theoretic techniques have been created to improve the economy of graphical representations of decision situations and the computational efficiency of influence diagrams, as well as various generalizations and hybrids of ID and decision tree ideas (e.g., Shenoy, 1992; Jordan, 1998).

For large and complex risk models, calculation of $EU(a)$ and of the optimal act can often be accomplished efficiently via Monte-Carlo simulation. States are sampled from $Pr(s \mid a)$, consequences are sampled from $Pr(c \mid a, s)$, and expected utilities are obtained as the sample mean of $u(c)$. Markov Chain Monte Carlo (MCMC) algorithms have been developed to solve for optimal (EU-maximizing) acts even in large and complex decision problems represented by influence diagrams (Bielza *et al.*, 1999). Samples drawn from a specially constructed joint probability distribution for $Pr(a, s)$ [derived from $Pr(s)$, $u(c)$, and $Pr(c \mid a, s)$] make it computationally tractable to identify the expected utility maximizing act.

DAG models and algorithms have revolutionized decision analysis technology, both for knowledge representation and for computation of optimal (EU-maximizing) decisions. However, they apply best to situations in which conditional probability relations are known explicitly and can be stored at the network nodes (preferably as small tables for discrete variables, or as conditional means and variance-covariance matrices for continuous variables modeled by Gaussian influence diagrams.)

By contrast, if only the rates of stochastic transitions among states are known as functions of the decision variables, then the probable consequences over time of different decisions are only implicitly determined. In this case, Markov and Semi-Markov decision processes (e.g., Hazen and Pellisssier, 1996) and stochastic optimal control models, including stochastic dynamic programming (SDP) models, may be the most natural decision-modeling frameworks. In *adaptive* SDP (ADP) models, the effects of decisions on learning the correct model must be balanced against their effects on payoffs (Bertsekas and Tsitsiklis, 1996). These models are discussed further at the end of the chapter.

In ADP models, the effects of decisions on probability distributions of payoffs over time (which arise from durations spent in states, transitions among states, or both) need not be quantified explicitly in order to calculate optimal decisions. Instead, the best policy, i.e., the decision to make in each state to maximize average reward per unit time or other objectives, based on observed information to date, can be solved for from knowledge of how decisions affect costs and transition rates among states. Even if these are unknown, adaptive procedures can still learn approximately optimal policies by intelligent trial and error, provided that there is adequate opportunity for experimentation (Bertsekas and Tsitsiklis,1996; Das et al., 1999). When experimentation is too costly, however – e.g., because the costs of trial or the potential costs of errors are too high – then QRA models such as those in Chapter 2 become useful or even essential for simulating the relation between decisions and their probable consequences without incurring the costs of real-world trial and error.

A limitation of dynamic programming and stochastic optimal control approaches is that their objective functions typically lack solid utility-theoretic foundations. Criteria such as average reward per unit time or expected discounted reward from some number of periods (possibly infinite) are typically used without explicit justification in terms of the properties that preferences can or should be assumed to satisfy. Indeed, fully dynamic generalizations of EU theory have only started to be developed (see e.g., Hazen and Pellisssier, 1996, for dynamic utility theory for semi-Markov decision processes). In general, the objective function to be used for evaluating control policies for stochastic processes is not known. The theory of preferences for such processes (Chapter 7) requires further development. Meanwhile, dynamic programming and other optimization methods typically assume that preferences are separable across time periods or make other simplifying assumptions that may lack strong theoretical foundations.

Example: Risky Choices Over Time (Cubitt and Sugden, 2001)

Two frequently proposed normative principles for choices made over time are *separability* and *timing independence* (or *dynamic consistency*) of preferences. As explained by Cubitt and Sugden (2001):

> "The principle of *timing independence* requires that the action which an agent takes, when faced with a choice in a dynamic problem, and the action that he would have committed himself to, if he had committed at the start of the problem to some action to be taken at this point, are the same. ... The principle of *separability* asserts that, at any point in a dynamic problem, the history of how that point was reached is irrelevant to the choice made there: on this view,

> choice is forward-looking and unaffected by the contingencies that can no longer occur. ... Separability is the justification for the folding-back and dynamic programming approaches to single-agent dynamic decision problems: timing independence (combined with reduction [see Section 3.2]) is the justification for the practice of analyzing such problems in normal form."

Many practical computational techniques, such as Bellman's Principle of Optimality for dynamic programming, take such principles for granted. Yet, some real choices violate them (see Chapter 7). In experiments where individuals choose at different times the number of optional rounds to play in a repeated monetary gambling setting with some initial compulsory rounds, the pattern of observed decisions implies preferences that violate at least one of these two principles (Cubitt and Sugden, 2001). Preferences among choices depend on the recent history of gains and losses, perhaps because subjective attachments to recently-acquired balances form in ways that violate the normative principles. Whatever the reason, real decisions commonly violate timing independence, separability, and closely related principles (e.g., the economic principle that "sunk costs" should be disregarded in optimizing choices going forward; and the "reduction principle" that only final probability distributions for consequences should matter in choosing among risky prospects).

Now, imagine trying to explain to someone *why* he or she *should* comply with principles such as separability and time-independence. Is there any objective basis for claiming that these or other suggested normative principles should be compelling when untutored preferences violate them? As explored in Section 3.2 and subsequently, a powerful line of argument is that *if* one wants one's decisions to be consistent with certain axioms (e.g., that higher probabilities of preferred outcomes are preferred, other things being equal), *then* principles such as timing independence and separability are logically implied. One might object that actual preferences are primitive and that prescribing principles that they should satisfy in order to be logically consistent with desirable axioms does not provide a constructive approach for modifying them. To this, decision analysts riposte that preferences for *consequences* are indeed taken as primitive, but preferences for *acts* or choices should be constructed, based on normative principles, from preferences for consequences and beliefs about the QRA relation between acts and their probable consequences. EU theory develops this approach in detail.

3.3 Optimization Modeling for Risk Management Decisions

Operations research approaches to most decision problems use formal optimization models and methods. Health and safety risk management decisions are no exception. Table 3 summarizes some key areas of health and safety risk analysis to which optimization models have been applied.

Formal optimization models sometimes lead to striking insights and clarification of risk management decision problems. However, they are limited in many health risk management applications by the fact that the objective function to be optimized is not clear or necessarily agreed to by all those participating in or affected by the decisions.

Table 3: Optimization Formulations of Risk Management Problems

Problem	Decision variables	Typical Constraints	Typical Objective Functions
Noxious facility location	Select one among N candidate sites for a hazardous facility	Facility must be put somewhere	Maximum exposures; QALYs lost
Hazardous material (HAZMAT) transportation	Routes/schedules for shipments (i.e., what shipments to send where, how, & when) (Erkut & Verter, 1998)	Shipments must arrive by due dates; environmental justice constraints; transportation network constraints	Population exposure from the shipments
Production process selection	Select a process (e.g., a fuel cycle for electricity production)	Meet production requirements. Sub-processes must be technically compatible.	Average QALYs lost per unit produced
Pollution abatement technology selection	What technologies to deploy when and where.	Must choose some process. Sub-processes must be compatible.	Reduce measured concentrations to target levels
Project selection and prioritization	Which project(s) to work on next; resource allocation	Limited resources available to work on projects	Discounted net benefits from finished projects
Risky project management	Abandon now, wait some more, or invest to improve	Value of project follows a diffusion process	Discounted net value from finished projects
Optimal inspection and screening	When to check (e.g., for breast cancer) and what tests to perform		Inspection cost + Cost of non-detection
Product testing and inspection (Loch et al., 2001)	Test battery design; optimal sequencing, intensity, & stopping of product testing	Budget, precedence constraints among possible tests; test characteristics	Cost of testing + cost of errors
Optimal maintenance policies	When to inspect and when to replace components		Inspection cost + failure costs + replacement costs
Hazardous waste site investigation	Sampling, screening, stopping, and action decision rules. (See Chapter 3 case study.)	Allowed form of decision rules.	Costs of investigation + costs of action + residual risks
Hazardous waste site remediation	Where to place wells and pumps; pumping rates (Tucciarelli et al., 1998)	Keep pollutant plume in ground water away from drinking wells	Costs of equipment

Example: Optimal Investment in a Risky Product R&D Project

Problem Setting: Suppose that a new product (e.g., a new drug) is being developed via an expensive R&D project with periodic reviews. At each review, management must decide whether to abandon, continue or correct/improve (by an additional investment) the current approach. The value of the project ultimately depends on such uncertain factors as market acceptance and sales of the product, cost and time consumed in completing the project, and the actual level of safety and effectiveness (generically called "performance" or "quality") delivered by the final product. It also depends on the performance level required by regulators and/or consumers. This requirement may be uncertain as the R&D is being done.

As R&D continues, the expected performance of the product drifts up or down due as uncertainties are resolved, tests are completed, and so forth. Assume that the expected payoff from launching a product of performance level x is:

$$V(x) = m + F(x)(M - m)$$

where m is the profit margin (possibly 0) if the product fails to meet the required performance level and M is the profit margin if the product does meet or exceed the required performance level. F(x) is the probability that the required performance level is no greater than x. V(x) is assumed to be an S-shaped (i.e., convex-concave, increasing) function of product performance level, x.

Decision Problem: Assuming that the performance drift is unbiased (i.e., it follows a martingale process with zero mean), what is the optimal policy determining when to continue, when to improve, and when to abandon?

Solution: This problem can be formulated and solved via SDP. (For details of such an SDP model, see Huchzermeier and Loch, 2001. These authors also interpret the solution in terms of the *option-value of managerial flexibility,* referring to the economic value of being able to wait and decide what to do after some relevant uncertainties have been resolved.)

The optimal policy is characterized by two *control limits,* as follows. The optimal decision in period t is to *abandon* the project if expected quality (i.e., performance level) is below some minimum acceptable level, A(t); to *continue* if the it is above some upper bound $B(t) \geq A(t)$; and to *improve* if the expected quality is between these two control limits. The numerical values of the control limits A(t) and B(t) may be solved for via the dynamic programming recursion.

The qualitative properties of the solution are informative. For example, increasing the variability in the required performance level decreases the option value of the project and may reduce the set of states (i.e., expected performance levels) in which it is worthwhile to invest in improvement. Such useful qualitative insights (as well tractable closed forms for the optimal policy, such as the control-limit form illustrated here) emerge fairly often from SDP optimization of dynamic decisions under uncertainty.

Example: Objective Functions for HAZMAT Transportation Risks (Boffey and Karkazis, 1995; Erkut and Verter, 1995, 1998)

Problem Setting: Suppose that hazardous materials (HAZMATs) must be sent through a transportation network (e.g., trucked via a network of roads) from origins where they are produced or stored to destinations where they are consumed or disposed of. Different links in the network have different hazard rates for accidents per vehicle-mile (perhaps for different types of vehicles and loads) and have different population sizes potentially affected by an accident if one occurs there. Moreover, the potentially affected population may vary by time of day and by day of week. The transportation network must be used to move HAZMATS from origins to destinations with enough speed so that unacceptable backlogs at the origins and/or shortfalls at the destinations do not accumulate at any place in the network.

Decision Problem: How should each shipment be scheduled and routed so as to accomplish the goal of moving all HAZMATS from their origins to their destinations to satisfy scheduling requirements and constraints on risk metrics while minimizing costs, remaining risks, or a convex combination of these criteria?

Discussion: Erkut and Verter (1998) examined the routes recommended by a commercial software package for solving HAZMAT optimal-routing problems for a single origin-destination pair. The software has a database of about 500,000 miles of road in the continental U.S. It gives the length and surrounding population density for each road segment and allows accident and release probabilities to be determined by road type. The optimization program considers an impact radius that may depend on the material being considered. (However, this radius ignores prevailing wind direction and other local meteorological and terrain information). The program solves a network optimization problem to find routes that minimize any of several user-selected criteria. The pre-packaged criteria include: travel distance, number of people exposed per trip, accident probability per trip, incident probability per trip (an incident is an accident that leads to a release of the HAZMAT) and a "societal risk" metric. Societal risk is defined as the expected number of people impacted per trip. It is calculated as the sum over road segments in a route of miles per segment x accident probability per mile x conditional probability of a release given an accident x population density in the area x impact area (= πr^2, where r is the impact radius for the HAZMAT being considered.)

Empirical evaluation of the optimized paths for different criteria on various problems revealed that (a) The paths selected by different criteria are usually not similar (as measured by quantitative indices of similarity based on extent of overlapping edges); and (b) Paths optimized for one criterion (e.g., minimizing number of people exposed, which tends to select back roads) typically perform poorly as measured by other criteria (e.g., accident probability, which tends to be much higher per mile on back roads as compared to highways.) Routes that minimize the "societal risk" metric are on average 1.7 times longer than the shortest path and are 3.5 times more likely to lead to an accident and release (i.e., to an "incident") than routes that minimize incident probability.

When shipments from multiple origins to multiple destinations are considered, concern for risk equity may prompt additional constraints (Erkut and Verter, 1995). A typical example of such a constraint is that no individual should be exposed to more than a certain total probability per year of exposure to (or adverse impact from) the HAZMAT flows. Constraints motivated by concern for "environmental justice" may specify that people of specified types (typically defined based on demographic characteristics), or communities with certain percentage compositions of people, should have their exposures or expected adverse impacts kept below specified target levels. Such additional constraints define interesting classes of transportation network routing and scheduling optimization problems (Boffey and Karkazis, 1995; Erkut and Verter, 1995).

Example: Objective Functions for Siting a Hazardous Facility

The problem of where to place a single noxious facility has been addressed by a large literature using techniques ranging from multiattribute utility theory (Chapter 7) and multi-criteria decision-making to public choice and negotiation theory (Chapter 8). Possible objectives include choosing the location to:

a. Maximize the minimum distance between the facility and some protected set or region (e.g., of surrounding towns) (Boffey and Karkazis, 1995)
b. Minimize the sum of the exposures or effects expected from the facility (*ibid*).
c. Minimize the sum of the compensation payments required so that everyone feels at least as well-off after the facility is built as before.
d. Minimize the sum of the amounts that individuals would be willing to pay to have the facility moved somewhere else.

Algorithms for optimizing (a) and (b) have been studied for facilities located on transportation networks as well as facilities located in the plane. The joint problem of locating multiple facilities and routing hazardous shipments having these facilities as origins or destinations remains a promising area for research (Boffey and Karkazis, 1995; Erkut and Verter, 1995).

Criteria (c) and (d) raise some practical difficulties that emphasize the difference between what variables are *intended* to represent in the theoretical framework of an economic or decision analysis and what they sometimes *do* represent to those affected by the decisions being analyzed. Jenkins-Smith and Kunreuther (2001) report that offers to compensate those adversely affected by a facility (namely a prison, an incinerator, a landfill, or a high-level nuclear waste repository) may actually reduce the acceptability of the facility to respondents. More accurately offering large financial compensations may reduce the acceptability of a proffered facility-location/risk mitigation/community compensation package. Compensation may be perceived as a bribe or as being inappropriate. It may also perhaps be perceived as signaling that the mitigation measures are inadequate. Similarly, as discussed in detail later, stated willingness-to-pay (SWTP) amounts are often better indicators of political attitudes and perceptions about rights and justice than of preferences for consequences. Such considerations make it difficult to operationalize constructs (c) and (d), as elicited values are confounded by factors other than preferences for different locations.

In practice, sites for major hazardous facilities are seldom or never selected using optimization models alone. The decision processes almost always involve multiple stakeholders, potentially liable parties, and decision-makers. Such situations are analyzed via game-theory models in Chapter 8.

Example: Meeting Air Pollution Management Goals

Environmental goals such as targeted reductions in urban and regional levels of nitrogen oxides (NOx), ozone, and hydrocarbons, can often be achieved by many different mixes of control measures. For example, improvements in vehicle technologies (e.g., catalytic converters), fuel control measures (e.g., mandatory introduction of cleaner-burning fuels), inspection and maintenance measures, and driver incentives (e.g., use of HOV lanes) can all affect emissions and therefore concentrations of pollutants.

The optimization problem of finding a minimum-cost bundle of controls (one per region) and a set of global control measures (e.g., regulations that apply to all regions) to achieve the required pollution abatement targets can be formulated as a large-scale integer program (Degraeve and Koopman, 1998). It can be solved using a column-generation heuristic. Policy insights from applying this optimization model to air pollution management in Europe include: (a) Rebalancing the air quality policy mix to make more use of economic instruments will provide a more cost-effective approach to meeting air pollution goals; and (b) Using economic incentives to drive technical changes in vehicle technology instead of relying on centralized command-control enforcement of regulatory standards better addresses the strong geographic differences in air quality within the European Union.

This example illustrates how optimization models can help to clarify the efficiency of different bundles of policy instruments – and different institutional frameworks -- in achieving public health and/or environmental goals.

As these examples and numerous others (e.g., Table 3) show, optimization modeling can be a valuable part of risk management. Indeed, the EU framework explicitly justifies selection of an act that maximizes expected utility, thus inviting the use of optimization methods to identify the EU-maximizing act. However, the objective function to be maximized – the expected utility function – may not be easy to define in all situations. Moreover, if there are multiple decision-makers or multiple affected parties, then neither the EU framework nor optimization modeling necessarily provides an appropriate framework for analysis. This chapter and Chapters 6 and 7 examine specific functional forms for utility functions, while Chapter 8 examines alternative decision-making frameworks for use when multiple stakeholders participate in or are affected by decisions.

3.4 Axioms for EU Theories

To follow the prescriptions of EU theory, it must be possible to assign utilities to consequences. This is guaranteed if (a) Consequences can be unambiguously rank-ordered from most-preferred to least-preferred (with ties allowed); and (b) The "canonical lottery" giving probability p of utility 1 and probability $1 - p$ of utility 0 can be uniquely solved for the value of p

that renders it indifferent to receiving any specified consequence, *c*, with certainty. [Recall that this value of *p* is defined as the NM utility *u(c)*.]

Many more-or-less normatively appealing sets of axioms imply EU theory (Kreps, 1988). Perhaps the best-known is the following.

Axioms for Expected Utility (EU) Theory

Preference Notation: Let R be a preference relation (interpreted as "is at least as preferred as") among lotteries, i.e., objective probability distributions or probability measures over the set of consequences. Let P be the corresponding strict preference relation, "is strictly preferred to". Thus, for any two lotteries a and b, aPb if and only if not(bRa).
Mixture Notation: (a p c) is the compound lottery giving lottery a with probability p and lottery c with probability 1 - p.

EU Representation Theorem: Suppose that preferences among lotteries satisfy the following four axioms. Then they can be represented by expected utility, i.e.,

aRb if and only if $EU(a) \geq EU(b)$

where EU(x) = expected utility of lottery x. (Thus, EU(x) is the expected value in lottery x of u(c), where u(c) is the NM utility of consequence c.)

Axioms:

- *Reduction: Only probabilities of consequences matter:* Any two choices that create the same consequence probabilities (i.e., the same probability measures or CDFs) are indifferent to each other. This is sometimes called the *reduction principle* or reduction axiom.
- *Weak order:* R is transitive, reflexive, and complete. (Thus, if aRb and bRc, then aRc; aRa; and any two lotteries can be compared by R.)
- *Independence:* aRb if and only if (a p c) R (b p c) for any three lotteries a, b, and c and any probability $0 < p < 1$.
- *Continuity:* If aPb & bPc, then, ∃p in (0, 1), (a p c) is indifferent to b.

References: Kreps, 1988; Quiggen, 1993

EU theory gives prescriptions for "rational" decision-making in the exact and limited sense that it is the only procedure that satisfies these axioms. To the extent that they are accepted as normatively compelling criteria for rationality, EU theory is justified.

Example: The Reduction Principle as a Normative Axiom

Consider a decision problem in normal form, represented by a table with rows corresponding to possible acts that a d.m. must choose among and columns representing alternative possible states that may occur. Each state is assumed to have a known occurrence probability. A "state" may be thought of as a description, or set of facts, about the world that is detailed enough so that it fully determines the consequence of the d.m.'s decision. The number in row i, column j of the table indicates the payoff or reward to the d.m. if she chooses act i and if state j occurs. It may be denoted symbolically by $c(a_i, j)$.

In the following example, there are two possible acts (labeled a_1 and a_2) and six equally probable possible states, labeled S_1 to S_6 with probabilities of 1/6 each, shown below them. The table shows payoffs in dollars; thus, the payoff if act a_1 is selected and state S_3 occurs is \$3.

state:	S_1	S_2	S_3	S_4	S_5	S_6
Pr(s):	1/6	1/6	1/6	1/6	1/6	1/6
$c(a_1, s)$:	1	2	3	4	5	6
$c(a_2, s)$:	2	3	4	5	6	1

The Reduction Axiom implies that acts a_1 and a_2 must be equally preferred, since both give the same PDF for consequences (namely, 1/6 probability for each possible outcome.) Yet, is it truly "irrational", or normatively unacceptable, to prefer act a_2 in these circumstances, even though it has a 5/6 chance of giving a better outcome than act a_1? According to the Reduction Axiom, the answer is yes. Most normative decision theories require that each act be evaluated only on the basis of its *own* probability distribution over consequences, without regard for what would have happened had a different act been chosen. Exceptions include *state-dependent utility theories* and the *disappointment* and *regret* theories mentioned at the bottom of Table 4. However, these are most often viewed as descriptive rather than prescriptive theories.

4. EU THEORY CHALLENGES AND ALTERNATIVES

Many generalizations and extensions of the EU Representation Theorem have been established by weakening the axioms that imply it. Each axiom has been challenged as not necessarily being normatively compelling. Each has been relaxed in various ways, producing many variants on EU theory (e.g., Fishburn, 1989).

Some of the most important generalizations of EU theory hold for the *single-attribute* case in which the consequence set, *C*, is an interval of the real line (as for monetary gains or losses, life durations,or QALYs). Then, the axiom of Independence can be replaced by a smoothness condition in which the utility function for consequences is allowed to vary smoothly depending on the lottery being evaluated (Machina, 1982; Quiggin, 1993, Chapter 12). Table 4 outlines other modifications and extensions of EU theory. Chapter 6 examines single-attribute utility theory (SAUT) and specific functional forms for SAUT utility functions in greater detail.

The following sections summarize other conceptual challenges to the EU axioms and framework.

4.1 Cognitive Heuristics and Biases Violate Reduction

The *Reduction Axiom* often does not describe real behaviors. In addition to depending on consequence probabilities, choices and preferences among prospects may also depend on a psychological reference point, as in Cumulative Prospect Theory. This reference point, in turn, may depend on how a prospect is described – e.g., in terms of lives saved or lives lost, using words or using numbers, using tables or using graphs, with uncertainties presented in sequential stages or all at once, and so forth. It may also depend on how information about probabilities is revealed over time, with the reference point shifting as partial information is acquired. It may depend on how values for consequences are elicited (e.g., by asking for willingness to pay, willingness to sell, indifference probabilities in canonical lotteries, preference orderings, or other information.) Although it is easy to accept the principle that preferences among prospects *should* not depend on which logically equivalent description is used ("framing") or on which methods for determining values is selected ("elicitation procedure"), the fact that they sometimes *do* depend on these choices challenges normative theory (McFadden, 1999). Insisting that preferences *should* be robust and consistent, if in fact they are not, does not necessarily help risk managers. Deciding what to do when preferences for consequences (or actions) are highly sensitive to framing and elicitation biases requires new principles, since most existing normative theories (Table 4) assume that stable preferences for consequences exist.

Example: Framing Affects Preferences for Health Measures

The following "Asian Flu" example illustrates how strongly framing can affect preferences and choices among risky prospects.

Table 4: Alternative Theories for Discrete Choices and Consequences

Name of Theory	Representation	Key Axioms, Ideas, and Comments
NM EU theory (von Neumann and Morgenstern) *Note:* Acts a and b are interpreted as probability measures or distributions on a consequence set.	Act a should be preferred to act b (i.e., aPb) iff V(a) > V(b), where $V(a) = EU(a) = E_a[u(c)] = \Sigma_c Pr(c \mid a)u(c) = \Sigma_s Pr(s \mid a)u[c(a, s)]$ u(c) = N-M utility of c (a number between 0 and 1)	Only consequence probabilities are relevant in choosing acts. Axioms of Weak Order, Independence, and Continuity imply this EU representation. Preferences for acts are derived from preferences for consequences and from consequence probabilities.
Weighted linear utility (Chew and MacCrimmon, 1979)	aPb if and only if (i.e., iff) u(a)w(b) > u(b)w(a), where u and w are linear functionals and w is a positive weighting function	Satisfies Weak Order and Continuity, but weakens Independence axiom. Reduces to EU theory if w is constant.
Skew Symmetric Bilinear (SSB) (Fishburn, 1989)	aPb iff $\phi(a, b) > 0$, where $\phi(a, b) = E_{ab}[\phi(x, y)] = \Sigma_{x \in C}\Sigma_{y \in C}\phi(x, y)Pr(x)Pr(y)$ where $\phi(x, y)$ is skew-symmetric, [i.e., $\phi(x, y) = -\phi(y, x)$], and bilinear, i.e., linear in each argument.	Implied by axioms of Continuity, Convexity, Symmetry (Fishburn, 1989). Implies stochastic dominance but allows non-transitive preferences, preference reversal, and violations of Independence. Generalizes weighted utility (SSB with transitivity reduces to weighted utility theory). Reduces to NM EU theory if $\phi(a, b) = u(a) - u(b)$ where u is linear.
Implicit weighted utility (Chew, 1989)	aPb iff V(a) > V(b), where $V = V(X) = \Sigma_i p_i u(x_i, V)$ for any prospect X.	Weakens SSB. Replaces Independence axiom of EU with the following Betweeness axiom: If aPb, then aP(a p b) for any p strictly between 0 and 1.
Disappointment and regret theory (Bell, 1982; Loomes and Sugden, 1982)	aPb if and only if $E_{Pr(s)}\phi[(c(a, s), c(b, s)] > 0$, where Pr(s) is a probability density on the set of states, S, and ϕ is an SSB utility function on C x C.	Roughly, aPb if the expected regret from choosing a is less than that from choosing b. A typical regret function is: $\phi(x, y) = v(x) - v(y) + k[v(x) - v(y)]$, $k > 0$. Allows preference cycles (violating Weak Order) and violations of Reduction Axiom and stochastic dominance.
Gul's **disappointment** theory (Gul, 1991)	aPb if V(a) > V(b), where $V(X) = E[U(X)] - \beta E\{U[CE(X)] - U(X) \mid X < CE(X)\}$	β = coefficient of disappointment aversion. Theory preserves stochastic dominance, reduces to EU if $\beta = 1$

Cumulative Prospect Theory (CPT) (Tversky and Kahneman; Chateauneuf and Wakker, 1999)	aPb if V(a) > V(b), where $V(X) = \Sigma_x\pi[p(x)]v(x)$. π = nonlinear *weighting function* for probabilities. v = *value function* for gains or losses around a reference point. It is typically asymmetric.	CPT preserves stochastic dominance and describes many of the observed violations of EU theory. It reduces to EU theory if $\pi(p) = p$ and if $v(x) = U(x)$ where U is the NM utility function. CPT is applied by a trade-off consistency axiom together with Reduction, Weak Order, Stochastic Dominance, and Continuity.
Rank Dependent Expected Utility (RDEU) theory (Qiggin, 1993)	aPb if V(a) > V(b), where $V(X) = \Sigma_i u(x_i)[g(p_1 + p_2 + \ldots + p_i) - g(p_1 + p_2 + \ldots + p_{i-1})]$	Reduces to EU theory if $g(p) = p$. Can explain Allais Paradox and Preference Reversal (q.v.). It implies Stochastic Dominance.
Lottery-Dependent Utility Theory (Becker and Sarin, 1987)	aPb if V(a) > V(b), where $V(X) = \Sigma_i p_i u(x_i, c_F)$ and $c_F = \Sigma_i p_i h(x_i)$	Assumes only stochastic dominance, weak ordering, and continuity.

Asian Flu Framing Example (Kahneman and Tversky, 1984)

Subjects were asked to suppose that a disease (e.g., a new strain of Asian flu virus) is expected to kill 600 people if there is no intervention. 152 subjects were asked to choose between the following two public health interventions (prospects), which represent alternative vaccination strategies, one certain and the other experimental and uncertain.

A: Save 200 lives.

vs.

B: Save all 600 lives with probability 1/3, else save 0 lives with probability 2/3.

Another 155 subjects were asked to choose between the following two prospects, which are logically identical to A and B:

C: 400 people die.

vs.

D: 0 people die with probability 1/3, else all 600 die with probability 2/3

Descriptions A and B refer to lives saved, while the logically equivalent descriptions C and D are phrased in terms of lives lost. Both sets of descriptions present exactly the same factual information. Nonetheless, 72% of the subjects receiving the first pair of descriptions preferred A to B, while 78% of the subjects receiving the second pair of descriptions preferred D over C. Thus, *communicating risk information in terms of gains (lives saved) or in terms of losses (lives lost) significantly affects the perceived relative desirability of different risk management choices.*

The pattern of choices in this example clearly violates the Reduction Axiom. It has been confirmed in many other experiments. Prospect Theory explains this violation as follows. Presenting lives saved constitutes a *positive frame,* i.e., subjects are invited to consider how much may be gained by the two different decisions. Option A describes a sure gain while option B describes an uncertain one with the same expected value. Therefore, subjects who are risk-averse for gains choose A over B. Presenting the same information in terms of lives lost constitutes a *negative frame,* with subjects being asked to consider how much may be lost by their decisions. Option C presents a sure loss (with respect to a different implicit reference point, namely, 0 lives lost, compared to the implicit reference point of 0 lives saved in Option A. The facts are the same, but the reference point is not.) Option D represents an uncertain loss with the same expected value as Option C. Therefore, a subject who is risk-seeking for losses will choose D over C. This is what is observed.

In general, Prospect Theory stipulates that the value function v(x) is asymmetric for gains and losses around the reference point, being steeper for losses than for gains for most decision-makers. This property is called loss-aversion. Moreover, v(x) is typically concave for gains and convex for losses, suggesting diminishing marginal utility for gains and diminishing marginal disutility for losses. Therefore, logically equivalent descriptions that shift the reference point from which gains and losses are measured can stimulate different responses as subjects form preferences for the alternative decision options.

The practical significance of framing effects in private health and medical decisions has been demonstrated many times. For example, Mareteau (1989) found that positive or negative framing of risk information significantly affects the advice that medical students would give to patients about which medical procedures to undergo (or which procedures the students themselves would choose) in response to different medical problems. How risks are framed can also affect which surgical procedures physicians and patients elect. Physicians' choices of treatments can be significantly affected by whether treatment risks are presented as survival rates or as (logically equivalent) mortality rates. For example, the fraction of doctors favoring radiation over surgery for treatment of lung cancer can be more than doubled by presenting relevant data using mortality instead of survival (McNeil, Pauker, and Tversky, 1988, reviewed in Camerer, 1995.)

Personal choices about health protection are also affected by framing. For example, Detweiler *et al.* (1999) showed that brochures promoting sunscreen use more effectively induced beach-goers to use sunscreen if they were written to emphasize a gain frame instead of a loss frame. Thus, risk communications designed to change perceptions or to manipulate behaviors can take advantage of framing effects.

Such findings raise the question of what frame(s) should be used to present information about risks – whether in medical, public health, personal health, financial, or engineering risk management. *Presenting risk information without influencing perceptions and choices may be impossible,* since some way to present the information must necessarily be chosen. However, presenting the same factual information about risks in different ways, e.g., using multiple frames and perhaps multiple presentation modes (such as text, tables, graphs, various visual metaphors, comparisons to other risks, and so forth) may help to reduce decision biases. Bias, in this context, refers to a systematic effect on decision-making that would change if different, logically equivalent descriptions of risk information were used. The following examples illustrate other types of biases.

Example: Presentation Biases Can Affect Choices

When choices are presented to a decision-maker, the format of the presentation may systematically affect (i.e., bias) choices. For example:

- Important decisions can turn on details of how risk information is presented. For example, using the word "survival" (instead of "mortality" or "death") in summarizing statistical risks of different cancer treatment therapies can greatly affect individual choices among alternative therapies (O'Connor, 1985).
- Numerous studies have examined the impacts of tables, graphs, icons, and other displays and shown how differences in display formats affect how people respond to risk information. For example, Elting et al. (1999) showed 34 physicians tables, pie charts, bar graphs, and icon displays containing hypothetical data from a clinical trial and asked them to decide whether to continue the trial or stop for an unplanned statistical analysis. Correct decisions were made more than twice as often with icon displays than with bar graphs and almost twice as often when data were negatively framed rather than positively framed in tables (Elting et al., 1999).
- Formal risk assessments can have different impacts on decisions depending on how their conclusions are summarized and displayed. For example, summarizing mortality risks in terms of deaths per million people in a population, deaths per million dollars of benefit produced, deaths per facility, deaths per facility-year of operation, and so forth invite different risk-benefit comparisons (Crouch and Wilson, 1992, cited in Slovic, 1999).
- Tabulating normal-form decision problem information horizontally instead of vertically (i.e., putting acts in columns and states in rows instead of the reverse) can reduce or eliminate regret effects. As discussed later, *regret effects* are based on correlations among the payoffs from different acts. Since such correlations play no role in calculating the consequence probabilities and the expected utility of each act, the Reduction axiom and EU theory, respectively, prescribe that they should be irrelevant for decision-making. In practice, however, regret effects can significantly affect choices. Reformatting risk information to hide or obscure correlations among payoffs from different acts – for example, by presenting the consequence PDF or risk profile for each act

separately, without showing states – can diminish or eliminate regret effects (Camerer, 1995, pp 655-6).

- In a decision table in normal form, merely splitting one state into several sub-states (with same outcome and the same total probability) can affect choices among acts, in defiance of the Reduction Axiom (*ibid*). This bias can be avoided by dropping the normal form (act-state-consequence) presentation and instead showing only the risk profile for each act, i.e., its CDF (or 1 minus its CDF) for consequences.
- Including "irrelevant" alternatives (acts or rows) in a normal-form decision table can change choices among the other acts, although the probabilities of consequences for each act – and hence its expected utility – do not depend on what other acts are displayed. For example, adding an inferior alternative (a new row) to a payoff table can make some attributes more salient in comparing the other alternatives, leading to a change in choice, even though the new alternative is selected (Simonson and Tversky, 1992, reviewed in McFadden, 1999).
- Presenting probabilities of wins and losses for *each play* of an iterated gamble may lead to different preferences for accepting a sequence of the gambles than would presenting the implied probability distribution for *total* gains and losses (Benartzi and Thaler, 1999). For example, only 63% of subjects would accept a hypothetical sequence of five 50-50 chances at winning \$2000 or losing \$500, and yet 83% of subjects would accept the corresponding "aggregated" probability distribution of total gains or losses (*ibid,* p. 374, citing Redelmeier and Tversky, 1992). Similarly, a decision-maker who weighs losses 2.5 times more strongly than gains, i.e., whose value function is $v(x) = x$ for $x \geq 0$, $v(x) = 2.5x$ for $x < 0$, would accept the prospect (\$400, 0.25; \$100, 0.5; -\$200, 0.25), since its expected value of 25 is positive. This same decision-maker might reject each of two identical prospects described by (\$200, 0.5; \$-100, 0.5), since each has a negative expected value to him, i.e., -25. Yet, the two gambles together have the same risk profile as the first prospect. In practice, decision-makers often consider gambles independently (sometimes described as *narrow framing*), neglecting the portfolio in which they are embedded. Presenting risk information in segregated rather than aggregated terms plays to this weakness and can affect choices. Langer and Weber (2001) show that for low-probability, high-loss gambles, the aggregated presentation (i.e., showing the PDF for total gains or losses) may cause a sequence of gambles to be *less* attractive than the segregated presentation giving win and loss probabilities for each play.
- Presenting risky options in terms of event trees that show the sequential resolution of uncertainties also commonly leads to violations of the Reduction Axiom. For example, consider a choice between the following two prospects (Kahneman and Tversky, 1979):
  (A) Win 0 with probability 0.75; else move to a second stage and win \$3000 with certainty.
  (B) Win 0 with probability 0.75, else move to a second stage and win \$4000 with probability 0.8 and 0 with probability 0.2.

Most people choose A over B, even though they would choose (0.2, 4000) over (0.25, 3000). Yet, logically, A reduces to (0.25, 3000) and B reduces to (0.2, 4000)! The reason that A is preferred to B when the prospects are presented as two-stage gambles is that, in comparing A to B, subjects tend to cancel the common first stage (0.75 probability of winning 0) and to compare (1, 3000) to (0.8, 4000), with the former being preferred. This artificial *isolation* of non-common portions of prospects is a common cognitive process that violates Reduction. Other similar processes, such as *segregation* of a prospect into a sure outcome and a gamble relative to it, are also well documented; see the review by McFadden (1999).

- Presenting non-monetary benefits before monetary ones in discussing with a community whether to accept a proposed noxious facility there (e.g., a prison, incinerator, landfill, or high-level nuclear waste repository) may make the proposal more acceptable. Discussing monetary benefits first may be perceived as an attempt to bribe or inappropriately buy agreement or to compensate for unstated deficiencies in risk management measures (Jenkins-Smith and Kunreuther, 2001). While the previous examples emphasize how presentation affects cognitive processing of risk information, this one suggests that moral judgments and evaluations (and assumptions about the intent or motivation of those presenting risk information) can also be triggered differently by different presentation orders.

Examples like these suggest two different reactions. One is that regret, disappointment or elation, anxiety and hope during the sequential resolution of uncertainties, and so forth should be considered legitimate components of total psychological consequences received, and therefore should be included in the modeling of consequences. Multiattribute utility theory (MAUT) models, discussed in Chapter 7, might then be used to capture these extra consequence dimensions. By contrast, many of the above examples assume that the consequences of an act refer (or should refer) only to actual outcomes, and not to feelings of regret or other psychological aspects based on a display showing what would or might have happened had a different act been chosen.

An alternative response is to assert that presentation mode biases reflect psychological illusions that truly should be rejected as guides to action. Rearranging the presentation of risk information should not affect choices among actions, and choices that rest on such grounds should be reconsidered. From this perspective, it is significant that most of the effects just described apply when subjects are asked to state their preferences for acts or prospects. But, EU theory and many other normative models prescribe separating assessment of preferences for *consequences* from preferences for *acts*. Preferences for acts are not expressed directly, but instead are derived by formal calculations of expected utilities from preferences for consequences and beliefs about probable consequences of acts. This decomposition helps to protect against presentation mode biases.

In summary, experimental evidence reveals some disadvantages of failing to follow the normative recipe for deriving preferences for acts from consequence probabilities and consequence values (e.g., utilities). Aspects of information presentation that most decision-makers agree should play no role in principle can powerfully affect holistic preferences for or choices among acts in practice. Decision procedures that enforce normative principles such as the Reduction Axiom helps to prevent decisions from being affected by such incidental features of information presentation.

Example: Anchoring, Embedding, and Other Response Mode Biases Violate the Reduction Axiom for Decisions About Public Projects

Much as framing and presentation biases let the presentation of *risk information* affect choices among risk management actions, so the presentation of *response options* used in surveys to elicit preferences (or beliefs) can also affect the answers expressed. For example, asking whether the value of a proposed public project exceeds a certain stated amount may lead to significantly more people saying "yes" than would be implied by the distribution of values obtained from open-ended responses (Green et al., 1996, cited in McFadden, 1999, p. 87.) People seem to anchor their estimates of values on the stated amount and to adjust their valuations insufficiently around this anchor. (A similar phenomenon, also called anchoring, has been found in studies of subjective beliefs about uncertain quantities.) As an example of anchoring bias in elicitation of values for environmental risks, Kahneman et al. (1999) reported that the mean stated willingness-to-pay (SWTP) to save 50,000 birds annually ranged from $20 (with a $5 anchor in the design of the survey) to $143 (with a $400 anchor.)

A different anchoring bias has been found in surveys of the economic value assigned by respondents to bundles of environmental programs (e.g., oil spill prevention, preservation of salmon stock in the Pacific Northwest, protection of migratory birds, and so forth.) The monetary value assigned to the first in a sequence of goods serves as an anchor for the rest. The value assigned to the first good is typically much larger than for subsequent goods. In this case, *absence* of a specified anchor or reference point (called a "modulus") for the monetary response scale leads to the first good and value being assigned that role. The importance of such reference points is seen from the facts that (a) monetary values assigned to individual environmental goods typically sum to far more than the values that would be assigned to the set of goods evaluated jointly; and (b) placing goods with higher monetary values first in the sequence inflates the value assigned to all the goods in the sequence (Payne et al., 2000).

Other *response mode biases* include the effects of response categories, scales, or sequences of questions used to elicit judgments about economic (or other) values on the elicited values (McFadden, 1999). As a striking example, evaluations of environmental, health, and safety benefits decrease dramatically when stated willingness to pay (SWTP) values are elicited not by a direct question, but by embedding a specific good (e.g., protecting a particular species) in a more general category (e.g., environmental protection programs). Allocating the SWTP for a category to specific items can lead to stated values that are orders of magnitude less than those obtained when the SWTP for each item is assessed in isolation. For goods such as protection against oil spills off the coast of Alaska, an average SWTP of $85 elicited by direct questioning falls to $0.29 when an embedding design is used (Kahneman et al., 1999, p. 217). In summary, *the magnitudes and ordering of response options can significantly affect the responses elicited,* even when logically equivalent options are presented. Specifically, whether the perceived benefits of a public project, such as a health, safety, or environmental risk management measure, outweigh its perceived costs and uncertainties can depend dramatically on how responses are elicited.

Responses about the perceived values of proposed projects are often taken to imply preferences among acts. For example, a decision may be made to undertake a project if and only if its perceived value outweighs its cost. Or, a choice from among several competing public projects may be made by selecting the one with the greatest perceived value, or by selecting all those with sufficiently great perceived benefit-cost ratios, as determined from survey responses. To the extent that response mode biases affect the perceived values for prospects (but without affecting their consequence probabilities or the perceived values of different consequences), and hence affect inferred choices or preferences among acts, they violate the Reduction Axiom. Moreover, as discussed later under Preference Reversal, responses about perceived values may contradict preferences for choices.

Finally, the use of hypothetical questions to elicit responses may induce biases. For example, in studies of contingent valuation methods (CVM) evaluating willingness to pay (WTP) for health, safety, and environmental benefits, the hypothetical WTP responses often exceed corresponding real WTP values, even when there is no incentive to strategically misrepresent true WTP values. However, correcting for subjective uncertainty about stated WTP values using a 0-10 certainty scale can reduce or eliminate this bias (Johannesson et al., 1999).

Example: Stated Willingness-to-Pay (SWTP) Values May Be Insensitive to Consequence Probabilities and Values

Kahneman *et al.* (1999) and other investigators have assembled experimental evidence strongly suggesting that stated willingness-to-pay (SWTP) values for proposed public projects more typically reflect *attitudes* rather than *preferences*. Thus, SWTP values are based on many considerations other than probability distributions for consequences. Among the well-documented anomalies for SWTP values are the following.

- **Insensitivity of SWTP to consequences.** The literature reviewed by Kahneman et al. (1999) shows that varying the number of lives saved, or of non-fatal injuries prevented, or of environmental ills prevented, by proposed risk-reduction programs has little or no effect on their SWTP values. For example, (a) The SWTP of UK respondents for a program to reduce non-fatal road injuries increased by only 29% when the number of prevented injuries was increased by 300%. (b) The SWTP of residents to protect 57 wilderness areas in four western states was only 1.3 times the SWTP to protect a single wilderness area.

(c) Increasing the number of lives saved by a factor of 10 has no impact on SWTP in some experiments. (d) A 600-fold increase in the mortality risk reduction benefits of a hypothetical water treatment program produced only a 4-fold increase in SWTP.

- **Insensitivity of SWTP to probabilities.** Hammitt and Graham (1999) present several examples showing that SWTP for health risk reductions is inadequately sensitive to the magnitudes of the reduction. For example, they found no change in SWTP to reduce the risk of food contamination by Salmonella bacteria when the size of the postulated risk reduction was varied ten-fold. Moreover, careful attempts to communicate risks by using verbal analogies as well as numerical probabilities had little effect on this insensitivity to the magnitude of risk reduction. These results confirm and extend previous findings showing that stated WTP amounts are not sensitive indicators of economic preferences.

Findings such as these suggest that SWTP is often not an effective way to elicit economic values for prospects involving risks to health, safety, or the environment. Survey-based methods such as contingent valuation that ask respondents to directly express their willingness to pay for risk reductions may give results that are inappropriate guides to risk management policy. Assessing utilities or other measures of preference for consequences may give more sensitive, robust, and useful results (Kahneman et al., 1999).

Example: Elicitation Procedure Biases Violate Procedure Invariance

Even someone who accepts the principles of EU theory may be unable to apply them to obtain clear results. Hershey and Schoemaker (1995, cited in Camerer, 1995) showed that NM utilities elicited by different methods vary systematically. For example, one way to assess NM utilities is to vary probabilities in a canonical lottery to obtain the probabilities at which the canonical lottery and the outcome being evaluated are equally preferred. A different method, when the payoffs are in dollars, is to ask for the amount that, if received with certainty, would be indifferent to a given lottery. (This is a "certainty equivalent" method.) The utility of the certain amount then should be the same as the expected utility of the lottery. However, Hershey and Schoemaker show that such different procedures lead to different numerical utilities for the same outcomes. Logically, however, they should produce identical results, since each is designed to uncover the NM utilities.

More generally, the idea that the value of a quantity (an NM utility, for example) should not depend on the procedure used to measure it is called the principle of *procedure invariance.* Another example of how this principle may be violated involves different utilities elicited by systematically increasing the probability (starting from 0) in a canonical lottery until indifference is obtained compared to systematically decreasing it (starting from 1) until indifference is obtained. When the procedure used to assess utilities affects the utility values obtained so much the optimal (EU-maximizing) act changes, the prescriptions of EU theory are no longer clear – even to someone who wishes to act according to its prescriptions.

Example: Preference Reversal

A dramatic violation of procedure invariance that also illustrates an important response mode bias is the notorious *preference reversal* phenomenon (originally identified by Slovic and Lichtenstein, 1968; see Camerer, 1995 for a review.) When asked to *state a dollar value* (specifically, a minimum selling price) for a prospect (p, x), subjects tend to state values that are strongly correlated with the potential payoff x. On the other hand, when asked to *choose* among prospects, the same subjects tend to make choices that are strongly associated with the probability of winning, p. As a result, many subjects will assign a higher dollar value to a low-probability, high dollar value bet, but actually choose a higher-probability, lower dollar value bet. This pattern has been observed with simple laboratory experiments (e.g., using a 0.99 probability to win $4 and a 0.01 chance to lose $1, compared to a 0.33 probability to win $16 and a 0.67 chance to lose $2. Many subjects would choose the former but bid more for the latter.) It has been replicated using real bets of various sizes and using different types of prizes.

In summary, the preferences revealed by choices among simple prospects may contradict the expressed preferences obtained by asking people how much they would be willing to pay for the prospects (or how much they would be willing to sell the prospects for if entitled to them). Large reversals discrepancies between choosing and pricing have been documented for environmental values such as cleaner air, as well as for consumer goods. The direction of the reversal can change when the stakes are sufficiently large. Reversals increase dramatically when probabilities are re-stated to make them more difficult to process (e.g., by rewriting 0.9 as 513/570.) Much of the explanation for preference reversal appears to stem from a psychological tendency to weight an attribute of a prospect more heavily when it is compatible with the response scale (e.g., pricing is more compatible with the payoff attribute x rather than the probability attribute p) (Camerer, 1995).

4.2 Other Violations of EU Axioms

Weak ordering has been challenged on the grounds that transitivity of indifference, and even of strict preference, need not hold. For example, any two consecutive stimuli in a series may not be noticeably different, and yet the first and last elements may be very noticeably different, with one being clearly preferred to the other. The very startling but thoroughly confirmed phenomenon of preference reversal, in which people assign a higher monetary value to financial lottery A than to financial lottery B, and yet prefer B to A, may cast doubt on the exact meaning of preference orderings. A generalization of EU theory called *skew-symmetric bilinear (SSB) utility theory* (Table 4) allows for both preference reversal and violations of transitivity, but still implies the existence of most-preferred prospects, satisfaction of stochastic dominance principles and several other desirable properties of EU theory.

Example: Multiattribute Choices May Violate Transitivity

In experimental studies, subjects often seem to treat probabilities and consequences as two different attributes or dimensions of a risk, rather than combining them according to one of the formulas in Table 4. Psychological processes used to compare multiattribute alternatives, such as eliminating those that are not sufficiently attractive on one attribute before considering other attributes, can lead to choices and behaviors that are highly anomalous from the standpoint of normative theory. For example, Loomes, Starmer and Sugden (1991, cited in Camerer, 1995) showed that about 17% of subjects in a U.K. experiment chose (0.6, 8) over (0.3, 18) and (1, 4) over (0.6, 8), but also chose (0.3, 18) over (1, 4), where all consequences are in pounds. This non-transitive preference pattern is consistent with the larger probability being the basis for choice when payoffs are close, but not when they are as different as 18 and 4.

Continuity is dropped in theories that, in effect, allow some outcomes to be infinitely preferred to others. More precisely, preferences in these theories are represented by utility vectors rather than by utility numbers, with the *lexicographic ordering* of vectors (i.e., based on the first component in which they differ) representing the non-continuous preference ordering. Lexicographic preference orderings have been extensively studied and axiomatized by Fishburn and colleagues (e.g., LaValle and Fishburn, 1996).

Independence is the most-often-challenged axiom. Not only do many observed choices and behaviors violate it, but its normative status has been questioned. However, decision-makers that violate the independence axiom must also violate other axioms with considerable normative appeal. One such axiom is *dynamic consistency,* which asserts that a rational decision-maker will not plan to do one thing at a future node in a decision tree (if it is reached) and then do something else if that node actually is reached (LaValle and Wapman, 1986). Decision-makers whose preferences among gambles violate independence can be induced to exhibit dynamic inconsistency, and to accept gambles in which they lose with probability 1 (Quiggin, 1993, Chapter 9). Despite this rationale, actual behaviors and choices in certain situations do violate independence.

Example: The Certainty Effect (Allais Paradox) Violates Independence

Empirically, many people prefer \$3000 with certainty to the lottery (\$4000, 0.8, \$0) (i.e., 80% probability of \$4000 but nothing otherwise) and also prefer (\$4000, 0.2, \$0) to (\$3000, 0.25, \$0). Such preferences violate Independence. The first preference implies that $u(\$3000) > 0.8u(\$4000)$ while the second implies that $0.2u(\$4000) > 0.25u(\$3000)$, which is equivalent to $0.8u(\$4000) > u(\$3000)$. This type of violation is consistent and robust: many similar examples exhibit it. It is often called the *certainty effect,* since it may be interpreted as indicating that decision-makers over-weight the utility of prizes received with certainty. It is also called the *Allais paradox*, after its discoverer. Any normative or descriptive theory that permits the certainty effect must drop the Independence axiom.

Example: Russian Roulette and Nonlinear Weighting Functions

The following example (due to R. Zeckhauser, cited in Wu and Gonzalez, 1996) illustrates another type of certainty effect. Suppose that you must play Russian roulette but can purchase the removal of one bullet from the gun. Would you pay as much to reduce the number of bullets from 4 to 3 as to reduce the number of bullets from 1 to 0? Most people would pay more to reduce the risk from 1/6 to 0 than to reduce it from 4/6 to 3/6, suggesting that they do not treat probabilities linearly (Wu and Gonzalez, 1996).

5. SUBJECTIVE PROBABILITY JUDGMENTS AND SUBJECTIVE EXPECTED UTILITY (SEU)

In the theory of decision analysis for a single decision-maker, a more general result than EU theory, called *subjective expected utility* (SEU) theory, has attracted even greater attention than EU theory. While EU theory applies to decision-making in which acts are represented by objective lotteries, i.e., by known probabilities for consequences, SEU theory applies when the probabilities of consequences are unknown and must be derived from subjective beliefs. The former context is often described as decision-making under *risk,* while the latter is referred to as decision-making under *uncertainty.* Decision-making under risk is the primary focus in this book, as we will rely on risk assessment models to provide estimates of probabilities. This section briefly introduces SEU, which is of dominant importance in decision theory.

There are many versions of SEU. One of the best-known is the "horse-lottery" (or Anscombe-Aumann) version. Its key constructs and results are summarized in the box below. This approach to SEU assumes that objective lotteries exist. An older approach, due to Savage, derives the existence of both a well-behaved (finitely additive) subjective probability measure and an NM utility function, such that the EU representation holds, from seven axioms without assuming objective lotteries. The axioms (Weak Order, the Sure Thing Principle, State Independence, Comparative Probability, Nontrivial acts, and two technical conditions guaranteeing that there enough random events to allow needed comparisons) derive probabilities for events from preferences for bets conditioned on the events. However, the axioms are not entirely empirically verifiable, even in principle. The Savage approach continues to be a useful point of departure for SEU and its generalizations (e.g., Nehring, 1999), but many risk analysts and decision-theorists find the Anscombe-Aumann framework simpler and more useful.

Anscombe-Aumann Subjective Expected Utility (SEU) Theory

Define a *horse-lottery* to be a function mapping states of nature to objective lotteries over consequences. In a horse lottery, the choice of an act does not affect state probabilities. Each act-state pair determines probabilities of consequences. But, the state probabilities may be uncertain. (Many horse lotteries, including real racetrack bets, map states of nature, i.e., the winning horse, to deterministic consequences.)

Suppose that preferences satisfy the three EU axioms and the following one:

State-Independence: Let g and h be two horse lotteries giving the same prizes (lotteries) in all states except for state s. Then gPh if and only if the lottery given by g in state s is preferred to the lottery given by h in state s. (Assume that the d.m. cares which lottery is given by state s, i.e., the subjective probability of s is not zero.)

SEU Theorem (Anscombe and Aumann): These four axioms imply a unique set of subjective state probabilities and a unique NM utility function (normalized to have values of 0 and 1 for the least-preferred and most-preferred consequences, respectively) for consequences such that preferences for horse lotteries are represented by their SEUs.

(*Note:* In general, the subjective probabilities are unique only if the utilities are state-independent, as EU theory implies.)

References: Kreps, 1988; Quiggen, 1993

All versions of SEU share a concern with quantifying probabilities (or other measures of uncertainty, such as interval-valued probabilities) based on personal degrees of belief and/or preferences for lotteries with prizes conditioned on the occurrences of different states. These have been axiomatized in many ways, leading to a rich variety of generalizations of SEU theory (Fishburn, 1989). In many health risk analysis applications, however, probabilities of states and consequences are not obtained from unanalyzed beliefs, but rather from explicit statistical and quantitative risk assessment (QRA) models.

When probabilities of consequences are unknown and must be assessed judgmentally, numerous errors, biases, and violations of SEU may occur. For example:

- *Heuristics and Biases.* Subjects may select, weigh, and combine information for decision-making in ways that are inconsistent with the axioms and principles of probability theory. Table 5 lists well-documented heuristics and biases that affect judgments about probabilities, likely causes, and uncertain quantities.

Table 5: Heuristics and Biases in Judgments Under Uncertainty

Heuristic/Bias	**Explanation**
Anchoring	Judged probabilities, frequencies, or values remain too close to suggested values (e.g., cues provided in the problem statement)
Availability • Primacy and recency • Saliency • Status quo bias	Background information is under-weighted compared to readily retrieved information. • Initial events and most recent events are most easily recalled and hence receive excessive weight • Information that seems relevant at the moment is over-weighted compared to other information • Historical experience is more vivid and is weighted more heavily than hypothetical alternatives
Base rate neglect	Base rates are neglected or underweighted in many situations (unless they are salient or interpreted causally). Instead, judgments of representativeness (q.v.) can inappropriately overwhelm background information about base rates.
Calibration and Over-confidence	Assessed probabilities are not accurate. For example, events judged to be impossible in some experiments happen about 20% of the time, while events judged to be certain happen only about 80% of the time. Calibration for groups of events tends to be better than for individual events. (Camerer, 1995)
Credulity/ superstition	Evidence that supports patterns and causal explanations for coincidences is too readily accepted (McFadden, 1999, Table 1)
Confirmation Bias	Evidence that supports previously formed hypotheses (especially about likely causes or diagnoses) is over-weighted compared to other evidence. (See also calibration/over-confidence) In medical settings, this can lead to *premature closure* of alternative diagnostic hypotheses.
Conjunction fallacy	Physicians, statisticians, and other expert and lay subjects often consider a conjunction of events (e.g., patient was exposed *and* exposure caused observed symptoms) to be *more* probable than any of the individual components of the conjunction. This violates probability theory.
Conservatism	Sample information that reveals likelihoods is often underweighted. Many subjects, including various experts, are poor intuitive Bayesians. They do not reason correctly about diagnostic evidence/probabilities.
Focal effects	Recalled or stated values are binned into approximate categories.
Disjunction ("irrational prudence")	Although a decision-maker will take the same action whether or not an event occurs, he may still prefer to wait for the uncertainty to be resolved.
Framing	Presentation of data (e.g., in ascending or descending order) can affect judgments of likelihoods or estimated values for uncertain events or quantities (e.g., the geometric mean of the data), respectively.

Gambler's Fallacy	Belief that random fluctuations will tend to occur to cancel out previous unusual patterns and restore the representativeness (q.v) of the sample.
Hindsight bias	Belief that whatever happened was inevitable or predictable in advance
Illusion of control	Subjects perceive that skill can affect the outcomes of chance events.
Law of Small Numbers	Subjects assume that even small samples are representative of the populations from which they are drawn. Since representativeness (q.v.) is not sensitive to sample size, they tend to gather too little data and to over-generalize from small samples.
Regression to the mean	Many subjects expect that observed patterns will continue (i.e., that the observations are representative of the whole data-generating process), rather than expecting unusual fluctuations to be followed by a return to more usual levels ("regression to the mean").
Representativeness	Pr(X | E) is estimated by how "representative" X is of E, ignoring or being inadequately sensitive to base rates for Pr(X) and Pr(E) and violating the probability rule, Pr(X | E) = Pr(E | X)Pr(X)/Pr(E).

Tversky and Kahneman applied the term *heuristics and biases* in the 1970s and 1980s to describe how people *do* make probability judgments and decisions under uncertainty. This often differs from all normative models of how such judgments and decisions *should* be made. The surveys by Camerer (1995), McFadden (1999), and Piattelli-Palmarini provide details, examples, and discussions of additional heuristics and biases. Examples include: Assigning too much weight to small samples, ignoring or under-weighting background information, confusing the representative-ness of a sample with its probability (or letting high conditional probabilities create unwarrantedly high unconditional ones), assigning causal interpretations to coincidences, being too confident in subjective estimates, over-weighting recent or readily retrieved information, and expressing numerical probabilities that violate probability theory (McFadden, 1999).

- *Ambiguity Aversion.* Most decision-makers are averse to uncertain (or "ambiguous") probabilities – so much so that they may let the certainty or objectivity of the probabilities affect their preferences among risky prospects (the Ellsberg Paradox.) They prefer prospects with known probabilities. Many subjects will pay a substantial premium (e.g., 10%-20% of the expected value of a prospect) to eliminate ambiguity about probabilities, replacing vague subjective estimates with crisp objective ones (Camerer, 1995). Such *ambiguity aversion* can create reluctance to participate in uncertain prospects and may help to explain why betting is not more frequent in a world where complete agreement on subjective probabilities is rare. However, SEU theory does not allow for a preference for "true" (or "crisp" or "objective") probabilities over subjective ones. Generalizations of SEU theory that do allow for ambiguity aversion have led to joint representations of preferences and beliefs in which decision-makers select acts that *maximize the minimum expected utility with respect to an entire set of prior probabilities for states.* This reduces to EU theory when there is only one prior, as for decision-making under risk. This "maximin EU" representation has been axiomatized for both Savage acts (Casadesus-Masanell et al., 2000) and the Anscombe-Aumann framework (Gilboa and Schmeidler, 1989.)

- *Non-additive probabilities.* Judged probabilities are often described by sub-additive probability measures, i.e., mutually exclusive, collectively exhaustive alternatives are assigned subjective probabilities that sum to less than 100%. Alternatives to probability theory, such as Dempster-Shafer belief functions and fuzzy measures of belief, may express judgments better than probabilities in some situations. Table 6 lists some alternatives to probability theory for representing and reasoning with uncertainty.

Table 6: Non-Probabilistic Methods for Reasoning Under Uncertainty

Technique	**Basic Ideas**
Analogy and similarity	Assume that "similar" causes have similar effects. (Definition and measurement of "similarity" is often problematic.)
Conditional logics and "rough" inference	Partition data points into classes of mutually "indiscernible" points based on having the same attributes values. Draw rough inferences by conditioning inferences on these partitions.
Default logics	Assume that "usual" conditions hold unless contradictions indicate that some assumptions otherwise.
Dempster-Shafer belief functions; upper and lower probabilities	Use interval-valued and/or non-additive probability measures to generalize conventional probability theory. Allows less than 100% of belief to be committed when sufficient information is unavailable.
Deontic logics	Reason about what one *should* do
Fuzzy logic and inference	Chains of implications are as strong as their weakest links, where the strength of a link (implication) is indicated by a continuous variable.
Inductive logic programming	Generalize from examples; obtain a small set of rules that describes observed cases
Modal logics of knowledge and belief	Modal operators such as K ("It is known that"), N ("It is necessary that…"), "It is not known that it is not true that…" and so forth may prefix other statements
Nonmonotonic logics.	Adding new facts or evidence can cause previous inferences or suppositions to be retracted.
Possibilistic logic	Formulas have numerical or linguistic degrees of truth (i.e., truth is indicated by a continuous variable rather than a binary variable). See fuzzy logic.
Truth Maintenance Systems	Identify and propagate justifications for all assertions. When a contradiction is encountered, retract minimal set of assumptions needed to restore consistency.

Reference: Sombe, 1990.

These are only some of the best-developed approaches to representing and reasoning with incomplete, possibly incorrect and inconsistent information, knowledge, and evidence. Many other theories specifying how to form, express, and revise beliefs based on data and knowledge have been developed in detail and tested in computer programs in the sub-discipline of artificial intelligence (or artificial intelligence and statistics) devoted to uncertainty. As yet, however, these approaches have had limited impact on the theory or practice of risk analysis. Risk analysis continues to be dominated by probability methods, and many theorists have argued that probability theory is uniquely sound and useful for practical applications. However, the practical value of analogy, similarity, approximate reasoning, and other non-probabilistic modes of inference in risk assessment (e.g., in QSAR applications and hazard analysis in complex systems) is undeniable.

SEU theory has been generalized to allow for non-additive subjective probability measures, with nonadditive SEU calculated via Choquet summation or integration. Axiomatic treatments of such generalized SEU theories have been advanced, largely in the mathematical economics literature (e.g., Nehring, 1999). Although the thrust of much of the theoretical literature is that beliefs that cannot be described by probability theory violate one or more normatively desirable axioms, the fact remains that real beliefs (including judged probabilities) often do not satisfy the laws of probability.

Example: Interaction of values and beliefs.

A key aspect of SEU and its normative generalizations is that preferences for consequences should not affect judgments about their probabilities. Wishful thinking is not allowed in models of rational decision-making. Yet, in practice, beliefs about the magnitudes of risks from various hazards tend to be negatively associated with their perceived benefits, and information about benefits or values systematically affects beliefs about likely consequences (Slovic, 1999).

Some generalized SEU theories can describe ambiguity-aversion and many other features of real decision-making. However, SEU remains attractive as a normative theory, mainly because behavior that violates it also violates axioms that many people want to use to guide their decisions.

Example: Incoherent Expert Probabilities (Moskowitz and Sarin, 1983)

An expert on energy utilities believes the following:

- Pr(utility A's oil-fired capacity will be restricted within the next 10 years) = 0.9
- Pr(conservation will be undertaken at A within 10 years) = 0.5
- Pr(conservation | restriction) = 0.7

These judgments will be abbreviated as: Pr(restriction) = 0.9, Pr(conservation) = 0.5, Pr(conservation | restriction) = 0.7. Given these judgments, *what is the value of Pr(restriction | conservation)?* The answer is readily calculated from the definition of conditional probability, as follows:

Pr(restriction | conservation)
= [Pr(conservation | restriction)Pr(restriction)]/ Pr(conservation)
= (0.7)(0.9)/(0.5) = 1.26.

The expert's subjective beliefs are inconsistent with all possible causal models, since probabilities should never be able to exceed 1. The phenomenon of *incoherent subjective probabilities,* i.e., probabilities that do not satisfy the axioms of probability theory, occurs fairly frequently in practice.

Some suggested tips for practical assessment of subjective probabilities follow. For more thorough discussions, see Kahneman et al., 1981.

Problem in Eliciting Subjective Probabilities	**Suggested Solutions**
Incoherent/inconsistent probabilities	• Elicit *joint* probabilities rather than marginal probabilities (Moskowitz and Sarin, 1993) • Draw influence diagrams to clarify dependencies (see Chapter 4)
Over-confidence	• Avoid anchoring (e.g., do not provide anchoring cues) • Check calibration of estimates and feed back the results to the expert.
Availability and representativeness, heuristics and biases	• Decompose the assessment. Check conditional as well as marginal probabilities • Formulate explicit models
Fallacies, e.g., gambler's fallacy, regression to the mean, assumption of statistical independence	• Be careful, especially about judgmental "causal" reasoning. Use formal methods.

Example: Overconfidence about Perceived Cause-of-Death Frequencies

In the following table, each cause of mortality in the right column is more frequent than the corresponding cause in the left column (e.g., death from measles is more common than death from fireworks accidents.)

Confident Answer	**Correct Answer**
Pregnancy, abortion, and childbirth, combined	Appendicitis
All accidents	Heart attacks
Homicide	Suicide
Fireworks accidents	Measles
Suicide	Diabetes
Breast tumor	Diabetes

Source: Piattelli-Palmarini, 1994

Yet, most people believe the reverse of the patterns shown and are very confident in these mistaken beliefs. The combination of incorrect beliefs and high confidence illustrates one form of over-confidence. Benjamin et al. (2001) point out that costs and relevance of information help to explain why people of different ages differ in how well they estimate cause-specific mortality rates. For example, a survey of college students showed that they tend to overestimate deaths per year from rare causes and to underestimate deaths per year from more common causes when asked about cause-specific mortality rates in the general population. This confirmed earlier research. But estimates of mortality rates in their own age group were nearly unbiased, suggesting that the respondents were relatively well informed about risks that were relevant for them.

Example: Intuitive Bayesian Inference is Unreliable

Suppose that a clinical test designed to diagnose a specific illness comes out positive for a particular patient. The test has a sensitivity of 79%, i.e., Pr(test is positive | illness) = 0.79. It has a false positive rate of 10%, i.e., Pr(test is positive | no illness) = 0.10. It has a false negative rate of 21%, i.e., Pr(test is negative | illness) = 0.21, and Pr(test is negative | no illness) = 0.9. The illness affects 1% of the population in the same age group, having the same gender, etc. as the patient, i.e., the prior probability of illness for this patient before the test, based on all available information, is 0.01. Now, given the evidence from the test, what is the posterior probability of illness?

The correct answer to this question can be calculated from Bayes' Rule:

$$\Pr(\text{illness} \mid \text{test positive}) = [\Pr(\text{test positive} \mid \text{illness})*\Pr(\text{illness})]/\Pr(\text{test positive})$$

where the denominator is given by:

$$\Pr(\text{test positive}) = \Pr(\text{test positive} \mid \text{illness})\Pr(\text{illness}) + \Pr(\text{test positive} \mid \text{no illness})\Pr(\text{no illness}) = (0.79)(0.01) + (0.10)(0.99) = 0.1069$$

So,

$$\Pr(\text{illness} \mid \text{test positive}) = \Pr(\text{test positive} \mid \text{illness})*\Pr(\text{illness})/\Pr(\text{test positive}) = (0.79)(0.01)/(0.1069) = 0.0079/0.1069 = 7.4\%.$$

However, *intuitive answers, even among U.S. hospital doctors, typically exceed 50% and are often close to 79%!* The compatibility of the positive test result with the hypothesis that the patient has the illness has a much stronger impact on judgments than the fact that the illness is relatively rare (Piattelli-Palmarini, 1994, p. 81). This is an example of base rate neglect, mixed with a tendency to confuse causes with diagnostic correlations.

Example: Presentation Can Affect Probability Judgments about Causes

Subjective probability judgments about probable causes and diagnoses are often unreliable. One reason is that intuitive Bayesian inference is notoriously inaccurate, as in the previous example. A different reason is that probability judgments can be very sensitive to details of how questions are posed or how subjective probabilities are elicited. For example, asking physicians or other experts to assess the probabilities of several different causes of observed symptoms can lead to very different answers for the probability of a specific cause, depending on the amount of detail used in listing other possible causes. Tversky and co-authors have developed and found experimental evidence for a model of subjective probability formation called *support theory*, in which the judged probability of a hypothesis is based on the perceived support for it compared to an alternative hypothesis (e.g., Rottenstreich and Tversky, 1997). One prediction from support theory that has been experimentally confirmed with physicians and others is that the judged support for a hypothesized cause is increased when the description of the hypothesis is "unpacked" into a more detailed, explicit description (Redelmeier et al., 1995). This is so even though the logical and statistical support for it are unaffected by the change of description. The effect is strong enough so that a disjunction may be judged *less* likely than its disjoint components, in violation of probability theory (*ibid*). This is one of many examples of how descriptions can strongly influence judged probabilities.

Example: Factors Affecting Perceptions and Beliefs About Risks

Perceived mortality risks (e.g., beliefs about frequencies of death from different causes and hazards), perceived risks from technology hazards, and beliefs about the frequencies of illnesses, deaths, and injuries from many other hazards are shaped by factors that would not play a role in purely "rational" models of belief formation. For example, the sex and gender of a subject are powerful predictors of perceived risks from many sorts of hazards (Slovic, 1990).

- *Risk denial/unrealistic optimism:* Most people believe that they personally are subject to smaller-than-average risks – a belief that must be wrong a high percentage of the time. To a lesser extent, they believe the same about their families. For example, in a large Swedish survey, respondents rated perceived risks to self, family members, and members of the general population on a 0 to 6 scale. They rated risks from 15 hazards ranging from alcohol, AIDS and traffic accidents to industrial pollution and traffic exhaust to radon, lightning, and the green house effect and ozone depletion. The mean perceived risk from each hazard was smallest for personal risk, next smallest for risks to other family members, and largest for risks to the general population (Sjoberg, 2000).
- *Psychometric factors.* Since a seminal 1978 paper by Fischhoff and co-workers, many studies of risk perceptions have shown that a few underlying factors (i.e., linear combinations of the original survey variables) account for a significant portion of the variance in responses. Three factors, often interpreted as New/Unfamiliar Risk, Dread, and Number Exposed have been used to help interpret risk rankings of various hazards. Frequency and severity of adverse

consequences, which are used in most technical definitions of risk, do not necessarily play a decisive role in determining risk perceptions. Indeed, they are only two among many attributes, including voluntariness, intentionality, equity, familiarity, identifiability of victims, threats specifically to children, media attention, and so forth that jointly determine judgments about risk acceptability and extent of public concern. Interestingly, perceptions about these "contextual" attributes may also affect perceptions about the "technical" attributes of frequency and severity (Slovic, 1999).

- *Individual types, attributes, and concerns.* The Cultural Theory of risk perception pioneered by Douglas and Wildavsky, 1982, postulates that different "types" of people (e.g., egalitarians, individualists, hierarchists, fatalists) focus on different types of hazards. Which hazards are perceived as being most threatening may be better predicted by the respondent's type than by the technical attributes (e.g., frequency and severity) of the risks. Although the theory in its original form may be overly simple and have limited predictive power (Sjoberg, 2000), empirical research has shown conclusively that attributes of respondents are strongly correlated with stated beliefs about risks. Judged risks from a very wide range of different hazards are well predicted by factors such as the respondent's gender, race, political worldview, emotional affect (e.g., liking or disliking for different technologies based on their benefits), and affiliation (e.g., industry vs. academia) (Slovic, 1999). This is true for experts, e.g., toxicologists, as well as for lay persons. For example, about 30% of white males with relatively high educations and household incomes and mainly conservative political views tend to perceive the risks from 25 sources (ranging from storms and floods to pesticides or bacteria in food to nuclear waste to radon in homes) as being "slight" far more often than other groups. General social, cultural, and political attitudes strongly affect judgments and beliefs about risks, as well as beliefs about other complex issues.

These findings show that attitudes that affect evaluations of consequences also strongly influence beliefs about likelihoods of consequences. This type of interaction between beliefs and values is incompatible with SEU's prescription for a separate, neutral evaluation of probabilities apart from utilities, followed by arithmetic combination of these distinct elements. Instead, *a responder's own attributes (e.g., income, political party) are often strong predictors of how he or she will perceive the risks from different hazards.* This contrasts with models of objective (observer-independent) formation of beliefs based on analysis of data in which all people with access to the same information draw the same conclusions about probabilities, regardless of their personal preferences and attributes.

In summary, plentiful experimental evidence has established the following two important conclusions that limit the direct application of EU and SEU theory.

- Subjective judgments of probabilities are often unreliable and/or incoherent (e.g., implying probabilities for mutually exclusive, collectively exhaustive events or hypotheses that do not sum to 1).
- The methods used to present information and to elicit utilities and probabilities can significantly affect the numerical values elicited.

These findings suggest that simply asking people to express judged probabilities and utilities and then using the expressed values to calculate expected utilities of actions may not lead to effective recommendations of decisions.

We believe that the preceding findings, and the analogous findings that EU does not hold in practice, provide strong motivation for normative theories of risk assessment and risk management. Good prescriptive theories are needed precisely *because* people do not intuitively separate assessments of facts from judgments of values, or combine these elements in ways that always lead to good (e.g., undominated) decisions. The framework of risk analysis and prescriptive decision theory, including assessment of risk profiles for actions and comparisons of profiles based on EU principles, can do much to overcome cognitive limitations and illusions. Because judged probabilities are often untrustworthy, even where lives and medical treatments are at stake, it is desirable to use formal statistical or mathematical models to document what is known and what is uncertain and to estimate probabilities of events or hypotheses based on these models and on available data. (See Chapter 3.) Then, when models have been used to create risk profiles and to characterize uncertainties, it is desirable to use formal decision and risk management methods to help choose among the actions that generate different risk profiles. How to do so without requiring utilities to be assessed with unrealistic precision and coherence is the subject of the following sections.

6. BEYOND SEU: ADAPTIVE DECISION-MAKING WITH UNKNOWN OR INCOMPLETE MODELS

The consequence function $c: A \times S \rightarrow C$ for a normal-form model of a decision requires that the set of possible states, S, be known. This may be unrealistic in many cases. For example, a perennial challenge in engineering risk assessments, especially for new, complex technologies and systems, is that all potentially important failure modes and consequences may be difficult or impossible to anticipate. While formal fault tree analysis, event trees, and other systematic methods can be used to identify and help quantify many failures, some unenvisioned possibilities may remain. Then, the state set S and the consequence function c are only partially known. In medical decision-making, too, consequences may be very important but hard to visualize and the probabilities of different outcomes from choices may be so uncertain that useful subjective point estimates may be unattainable.

Even when states or their probabilities are unknown, it may still be possible to identify decision-rules that will lead to optimal (EU-maximizing) behaviors. Several ideas and algorithms from the intersection of artificial intelligence (AI) and operations research have produced useful methods for identifying optimal, near-optimal, or asymptotically-optimal decision-making with unknown states and/or unknown state probabilities. Among the most promising approaches are the following.

- *Simulation-optimization.* A discrete-event simulation (DES) model *implicitly* represents knowledge about the relations among decisions and consequences. This makes it unnecessary to explicitly identify all of the relevant states and their probabilities, thus avoiding a potential combinatorial explosion. The DES model encodes local knowledge about the probability distributions and transition rates for states of individual objects and entities, expressed as functions of their own histories. The simulation integrates this local knowledge to give a global model of the probabilistic relation among inputs and output histories. Running the simulation may reveal unanticipated failure paths or patterns emerging from the interactions of the parts of the modeled system that were not recognized by its designers. Careful sampling of the set of possible input combinations, e.g., using importance sampling techniques, can help to identify and quantify the probabilities of rare events (Shahabuddin, 1994; http://www.informs-cs.org/heibib.html).
- *Simulation-optimization* algorithms, which are often based on stochastic gradient search or stochastic approximation methods (Wei, 1997), adaptively optimize the inputs to a DES model to keep the probabilities of undesired outputs acceptably small and/or to optimize a performance evaluation function. (For recent algorithms and applications, see e.g., http://faculty.uscolo.edu/gosavi/websites.html; and Andradóttir, 1996.)
- *Reinforcement learning and adaptive dynamic programming* techniques (Bertsekas and Tsitsiklis,1996; Das et al., 1999; Russell and Norvig, 1995, Chapter 20; Sutton and Barto, 1998) allow the rewards associated with different act-state pairs to be learned directly by trial and error. The state set S and act set A must be known, but the state transition probabilities, given acts, may initially be unknown. In Markov Decision Processes (MDPs) and Semi-Markov Decision Processes (SMDPs), modifying the probability of choosing each act in each state to reinforce (i.e., make more probable) acts that have yielded greater average awards will eventually lead to an optimal decision rule. Algorithms have been developed to find optimal or approximately optimal decision rules for discounted reward and average reward per unit time criteria. For

example, the one-step *temporal-difference Q-learning* algorithm uses Monte Carlo simulation to estimate the expected value of taking action a in state s. The resulting number, denoted by $Q(a, s)$, is updated after each transition from state s_t to state s_{t+1} via the formula:

$$Q(a_t, s_t) \leftarrow Q(a_t, s_t) + \alpha[r_{t+1} + \beta\, max_a Q(a, s_{t+1}) - Q(a_t, s_t)]$$

Here, r_{t+1} is the observed reward from the transition and α and β are parameters selected to assure convergence. The motivation is that in steady state, when all estimated Q values are correct and Q-maximizing acts are always selected, the optimal Q values are fixed points of this updating rule. Sutton and Barto (1998) provide details and Russell and Norvig (1995) compare the performance of Q-learning (CPU time, speed of convergence to optimal decision rule) to exact and approximate Adaptive Dynamic Programming (ADP) algorithms. They show that ADP algorithms, although more computationally intensive, learn more quickly and accurately on some test problems. Das et al. (1999) give an efficient reinforcement-learning algorithm for solving SMDP problems.

- *Population of learning agents:* Easley and Rustichini (1999) introduce a related learning scheme in which neither the states nor their probabilities need ever be known. The goal is to learn which action generates the highest expected utility by trial and error. Axioms of symmetry, monotonicity, independence, and exchangeability (essentially stating that preference orderings for actions depend only on their observed payoff frequencies and are increasing functions of these observed payoffs) are used to show that the EU-maximizing action will almost surely be selected in the long run. It is assumed that in each period, the decision-maker observes (after deciding what to do) the payoffs that she *would* have received had each action been selected – perhaps because she is part of a large population of agents experimenting with different actions. Reinforcement learning, too, requires refining estimates of act-state payoffs by considering the estimated rewards that would have been received if an estimated optimal act had been selected. Thus, although disappointment and decision regret arguably play no role in "rational" EU decision-making with known decision problem specifications, considering the gap between achieved rewards and estimated potential rewards does help to learn EU-maximizing actions or decision rules in settings where states and/or probabilities are not explicitly known.
- *N-arm bandit problems.* In an N-arm bandit problem, N alternative acts are available, but the probable consequences of each act, and hence its probable value, may be uncertain. The available information about the probability distribution of rewards (i.e., "payoffs") from each act is the history of payoffs that actually result when it is tried. Clinical trials of

new drugs and treatments provide an example, with each act corresponding to a choice of treatment for a case. For such problems, it is necessary to balance *exploration* (i.e., search for higher-payoff acts, such as more effective treatments) against *exploitation* of previously studied acts (i.e., selecting and receiving rewards from known high-payoff acts.)

A large literature addresses optimal and approximately optimal strategies for N-arm bandit problems and their generalizations. Dynamic programming and, more recently, polyhedral combinatorics, have led to a profound theory of simple (index-based) decision rules related to conservation laws (Crosbie and Glazebrook, 2000; Bertsimas and Nino-Mora, 2000). These decision rules allow one to calculate a single numerical score (called a *Gittins index* or *priority-index*) for each act, based on the history of payoffs obtained from it so far. The optimal policy (for various criteria such as average or discounted reward) consists of always choosing the act with the highest index. Based on the outcome of the selected act, its index may change. Algorithms for calculating Gittins indices have been developed in the literature on stochastic dynamic programming and the sub-literature on multiarmed bandit problems. Moreover, Sutton and Barto (1998) show that even relatively simple rules that balance exploration against exploitation, without seeking more sophisticated optimization, often perform fairly well.

The multiarmed bandit problem has been generalized in various ways, e.g., to the class of *restless bandits* in which the probability distribution of payoffs from each act may be evolving over time (Bertsimas and Nino-Mora, 2000). In a clinical trial setting, such drift can occur if the efficacies of different treatments change as the microorganisms being targeted develop resistance.

Example: Allocating Clean-Up Efforts to Randomly Deteriorating Sites

Suppose that N hazardous waste sites or underground storage tanks may deteriorate stochastically over time. Each of the N locations corresponds to a potential remediation project, and each potential project is in exactly one of a finite number of distinct *physical states* at any moment (e.g., containment is intact; containment is breached but no leak has yet occurred; leak has occurred but has not yet reached ground water; leak has reached ground water but contaminant plume is still within site boundary, contaminant plume is no longer spreading but is being reduced by ongoing remediation efforts, etc.). More generally, an *information state* for a project is a probability distribution for its physical state. Work on a project and/or the passage of time may cause its information state to change, even if its physical state does not. We will use the generic term "state" to refer to information states and/or physical states.

At the start of each planning period (e.g., month, quarter, or year), exactly M < N projects must be worked on, where M is determined by the number of available contractors, crews, etc. Each project undergoes stochastic transitions among states between periods and earns penalties based on which physical state it is in. For example, states in which contamination enters drinking water may carry large penalties, while states in which the contamination is contained and kept away from public drinking supplies may carry small or zero penalties. Both the penalties and the transition rates among states may depend on whether the project is being worked on. Even if a project is not worked on, stochastic transitions may carry it from a low-penalty state to a high-penalty state, i.e., the site may deteriorate over time.

The decision problem is to select a decision rule (sometimes called a *Markovian policy*) that determines the current action to take (i.e., which projects to work on in each planning period) as a function of the current information states to minimize the total expected discounted penalty over an infinite time horizon. [The decision rule may be randomized and may depend on time as well as on project states, but it can be shown that the optimal solution can be achieved by deterministic policies that do not depend on time (Bertsimas and Nino-Mora, 2000, Theorem 1).]

This is an example of a *restless bandit problem,* i.e., one in which projects may change even while they are not worked on. Such problems can be formulated as linear programs that are PSPACE-hard in general (although polynomially solvable for the special case of the multiarmed bandits, for which Gittins index policies are optimal.) However, index-based policies appear to provide excellent heuristics for approximating optimal solutions even when they are not optimal (*ibid*).

Approaches to "model-free" learning of optimal or near-optimal policies in MDPs and SMDPs are among the most active areas of research in machine learning, dynamic programming, and AI methods for decision analysis (http://www.cs.brown.edu/people/lpk/rl-survey/rl-survey.html). Much current research is focused on *approximate* ADPs with value functions approximated by simulation and then smoothed by splines, neural nets, classification trees, MARS, and so forth (Bertsekas and Tsitsiklis, 1996; Das et al., 1999; Chen et al., 1999). The granularity of the approximations is often adaptively selected to reduce computational effort except where better approximations make an important difference. Interpolation of continuous, smooth cost-to-go values can also greatly improve computational performance (Philbrick and Kitanidas, 2001; Chen et al., 1999). Breakthroughs have resulted in the practical solution of large-scale stochastic dynamic optimization problems using local approximations.

Reinforcement learning and related techniques depend heavily on feedback from the environment to teach a decision-maker which acts have the greatest payoffs in different situations. In many risk analysis applications, however, such reliable feedback is not available. The adverse events of interest may be very rare and the impacts of decisions on their occurrence rates may be difficult or impossible to ascertain from data alone. Therefore, risk analysis and risk management decision-making pose special challenges to machine-learning methods. To meet this need, QRA and simulation models (e.g., DES models, fault tree and consequence analysis models, or compartmental flow models) are commonly used to predict the probabilistic relation between acts and consequences. These models enable simulation-optimization or other algorithms to calculate optimal decisions without requiring hard-to-obtain, expensive feedback from real-world trial and error. However, the resulting decisions are only optimal with respect to the risk models used. Model validation and treatment of remaining model uncertainties, as discussed in Chapters 3 and 4, can therefore be especially important for effective risk-management decision-making.

7. CONCLUSION

In summary, this chapter has introduced EU theory, which justifies the use of optimization techniques for decision-making when the objective function (i.e., the expected utility as a function of decision variables) to be optimized is known. It has also surveyed some limitations and extensions of EU theory. Chapter 6 turns to techniques for identifying specific functional forms for the utility function in the case of single-attribute utility theory (SAUT), i.e., for decision situations where consequences can adequately be summarized by a single number. Chapter 7 extends the analysis to multi-attribute consequences and utility theory (MAUT) and introduces models of decision-making over time. Chapter 8 considers risk management decision-making for multiple decision-makers and stakeholders.

Chapter 6

Choosing Among Risk Profiles

1. INTRODUCTION

To solve risk management problems using the expected utility (EU) framework, one must first represent different decision options by their risk profiles using QRA modeling methods such as those in chapters 2-4. Then, it is necessary to choose the option corresponding to the most-preferred risk profile. This chapter covers methods for choosing among risk profiles – or, equivalently, cumulative probability distribution functions (CDFs) – when the consequences of interest can be summarized and compared using a single numerical attribute. This is not necessarily a severe restriction, as the attribute may be a synthetic one such as "net benefit" or "value" that summarizes preferences for consequences. The results in this chapter can be used to construct explicit utility functions for single-attribute consequences so that the expected utilities of prospects can be calculated. They also allow some pairs of prospects to be compared even without defining an explicit utility function, using various dominance criteria.

In many applications, the set of possible consequences has the structure of an interval of the real line, with preferences increasing (or decreasing) as values increase. This defines the setting for *single-attribute* utility theory (SAUT). For example, financial decisions often have outcomes that can be summarized as gains or losses of money. Health risks may affect the probability distribution of quality-adjusted life years (QALYs), and some decision-makers may consider this single numerical attribute an adequate summary of health impacts. Personal health and safety choices may affect the probability distribution of one's own QALYs or remaining life span, and again this single numerical attribute may be deemed an adequate summary of consequences to guide decision-making.

When the consequence set can be represented as a compact interval of the real line, with higher values preferred to lower ones, normative theories become more specific and powerful than in the discrete case. An important benefit of SAUT is that specific functional forms for the NM utility function can be derived from apparently reasonable, or at least interesting, assumptions. A second benefit is that the EU representation and some important generalizations can be derived from different, perhaps more appealing axioms than in the case of discrete consequences. Finally, it is possible to define new concepts of risk for SAUT, such as certainty equivalents and measures of risk aversion, that exploit the continuity of the consequence set.

2. BASIC EXPECTED UTILITY (EU) THEORY FOR SINGLE-ATTRIBUTE CONSEQUENCES

For simplicity, suppose that the consequence set is scaled to run from 0 to 1, i.e., C = [0, 1], where 0 is the least-preferred consequence and 1 is the most-preferred consequence. A generic consequence in this set will now be denoted by x rather than c, with $0 \leq x \leq 1$. The scale is conventionally oriented so that larger values of x are preferred to smaller ones. For example, each value of x between 0 and 1 might represent the proportion of a population that survives exposure to a hazard without developing an adverse health effect.

In this setting, *acts* are represented by probability density functions (PDFs) or their cumulative probability distribution functions (CDFs) on the consequence set [0, 1]. Let F and G denote two such CDFs, or acts, with corresponding PDFs of f and g, respectively. The expected utility of act F (for a utility function u) is given by:

$$EU(F) = E_F[u(x)]$$
$$= \int_{0 \leq x \leq 1} u(x)dF(x) = \int_{0 \leq x \leq 1} u(x)f(x)dx \text{ if F is absolutely continuous.}$$

Here u(x) is the *von Neumann-Morgenstern (NM) utility function* giving the NM utility of each consequence, *x*, usually normalized to lie between 0 and 1. Since *x* acts as a variable of integration in the evaluation of expected utility, we sometimes abbreviate $E_F[u(x)]$ as simply $E_F(u)$. The notation EU(F) is also used when the utility function is clear from context. EU(.) is the *expected utility functional* mapping distributions to numbers between 0 and 1. (A "functional" is a mathematical function that takes other functions as arguments. A functional assigns a unique number to each function in its domain. Expected values and variances of probability density functions are examples, as are definite integrals, certainty equivalents, and medians.)

The EU representation theorem states that:

Act F should be preferred to act G if and only if $E_F(u) > E_G(u)$

This EU representation theorem can be derived from axioms of Reduction, Weak Order, Continuity, and Independence introduced in Chapter 5. However, other axioms also suffice for it in the SAUT setting that are not available in the discrete-consequence setting, as will now be illustrated.

2.1 Certainty Equivalents

The consequence set in SAUT has mathematical properties (such as weak order and order density) of the real line. These make it possible to be more parsimonious in stating normative assumptions that allow an NM utility function representation of preferences for acts. In fact, the single assumption that a well-defined certainty equivalent functional exists suffices.

Let CE(.) be a functional that maps CDFs over [0, 1] into numbers in the interval [0, 1]. CE(.) is interpreted as the *certainty-equivalent functional* that assigns to each CDF the deterministic consequence to which it is indifferent. (Of course, there may be no such function when consequences are discrete.) Suppose that CE(.) satisfies the following two axioms:

- *Deterministic consequences:* Let δ(x) denote the degenerate distribution putting probability 1 on x. For any deterministic consequence x in [0, 1], CE[δ(x)] = x.
- *Betweenness:* If F and F' are any two CDFs on [0,1] such that CE(F) = CE(F'), then for any p in [0,1] and any other CDF on [0, 1], say, H, CE[pF + (1 - p)H] = CE[pF' + (1 - p)H].

These two axioms are necessary and sufficient for the existence of an NM utility function over the consequence set [0, 1]. They imply (and are implied by) the existence of a unique NM utility function, u(x), normalized, as usual so that u(0) = 0 and u(1) = 1, that is a continuous, monotonically increasing, function of x (Chew, 1989). The possibility of deriving the EU representation for SAUT from the preceding two axioms illustrates the power of imposing the mathematical structure of the real line on the consequence.

The utility function and the certainty-equivalent functional are simply related. The utility of the certainty equivalent of any probability distribution is just the expected utility of that distribution. Either can be derived from the other. In symbols,

$$u[CE(F)] = EU(F) = E_F[u(x)], \text{ for any CDF, F, on } [0, 1].$$

The certainty equivalent functional has the property that CE[pF + (1 - p)H] lies between CE(F) and CE(H), for any two CDFs F and H over [0, 1]. [Note that pF + (1 - p)H is a CDF that is a weighted average (or convex combination) of the CDFs F and H; it is not the probabilistic mixture (F,p,H).] Moreover, if F lies everywhere to the right of G, then CE(F) is at least as great as CE(G). For other axiomatic derivations of EU theory and generalizations in the SAUT context, see Chew, 1989.

2.2 Risk Attitudes, Risk Aversion, and Prospect Theory

A *prospect* in SAUT refers to a random variable taking values in the consequence set, often represented by the interval [0, 1]. Each prospect may be represented by a corresponding CDF. The difference between the expected value and the certainty equivalent of a prospect is its *risk premium.* It reflects the decision-maker's attitude toward risk. A d.m. is said to be *risk-averse, risk-neutral,* or *risk-seeking* if her certainty equivalent for prospects is less than, equal to, or greater than its expected value, respectively. Thus, a risk-averse d.m. will pay a positive risk premium (in units of x) to replace an uncertain prospect with its expected value.

Example: Risk-Aversion May Depend on the Prospect

Tversky and Kahneman and other prominent investigators have studied certainty equivalents for simple prospects of the form (x, p), offering $x with probability p and nothing ($0) otherwise. A common pattern of findings, illustrated in the following table, is that many decision-makers are risk-averse for most gains and low-probability losses, but risk-seeking for most losses and low-probability gains.

	Gain	**Loss**
Low probability	Risk seeking, e.g., CE(100, 0.05) = $14	Risk averse, e.g., CE(-100, 0.05) = -$8
High probability	Risk averse, e.g., CE(100, 0.95) = $78	Risk seeking, e.g., CE(-100, 0.95) = $-84

Source: Tversky and Wakker, 1995, Table 1 (based on Tversky and Kahneman, 1992). Numbers are median certainty equivalents for several subjects.

This pattern has been observed across different cultures (e.g., Chinese and American) and different experimental conditions. It appears to be a robust feature of decision-making psychology for risky choices.

Example: Certainty Equivalents in Prospect Theory

According to Prospect Theory (Kahneman and Tversky, 1979) and Cumulative Prospect Theory (CPT) (Tversky and Kahneman, 1992), the subjective value of a simple prospect (x, p) is given by a product $\pi(p)v(x)$, where $\pi(p)$ is a nonlinear *weighting function* for probabilities and v(x) is a *value function* for gains and losses. Gains and losses are measured with respect to a reference point, sometimes interpreted as an *aspiration level,* that may be affected by how the decision problem is described or "framed". In this framework, the experimentally observed pattern of certainty equivalents in the above table can be interpreted as showing that typical weighting functions *overweight small probabilities* (e.g., 0.05) and *underweight large ones* (e.g., 0.95). This pattern has been widely observed. Probabilities of 0.25 and smaller are typically overweighted and probabilities of 0.5 and greater are typically underweighted (Bleichrodt and Pinto, 2000). For example, an increase in probability from 20% to 25% is often perceived as less important than an increase from 0% to 5% (or from 95% to 100%), even though all three changes are equally important from the standpoint of single-attribute expected utility calculations (Tversky and Wakker, 1995). Nonlinear weighting functions and value functions appear in many descriptive and some normative models of decision-making.

Empirically, $\pi(p)$ is typically "inverse S-shaped", i.e., concave for probabilities less than about 0.40 and convex (upward-curving) for larger probabilities (Wu and Gonzalez, 1996). The shape of the probability-weighting function may be different for different outcome domains, consistent with empirical evidence that decision-makers appear to be more risk-averse for life duration than for money (Wakker and Deneffe, 1996). Numerically, subjective evaluations of prospects of the form (x, p) are often well approximated by parametric models such as:

$$v(x)\pi(p) = (x^{\alpha})p^{\beta}/[p^{\beta} + (1-p)^{\beta}]^{1/\beta} \qquad \text{(Parametric CPT Model)}$$

in which $v(x) = x^{\alpha}$ is the value function and $\pi(p) = p^{\beta}/[p^{\beta} + (1-p)^{\beta}]^{1/\beta}$ is the weighting function. The power function form $v(x) = x^{\alpha}$ has empirical support for life duration (Bleichrodt and Pinto, 2000). The conceptual basis for such Cumulative Prospect Theory (CPT) models is discussed below. Values of β between 0.56 and 0.71 (and values of α between about 0.5 and 0.8) are realistic for several experimental data sets (Wu and Gonzalez, 1996). The two-parameter model $\pi(p) = \delta p^{\beta}/[\delta p^{\beta} + (1-p)^{\beta}]^{1/\beta}$ has also been used successfully by several investigators (reviewed in Bleichrodt and Pinto, 2000). Models such as this suggest that interpreting certainty equivalents and apparent risk attitudes may require some subtlety, since they are confounded with risk *perceptions* (mediated by π and v) that may not be obvious based on the objective data alone.

A two-stage procedure called the *Trade-off Method* can be used to elicit utility functions even when probabilities are distorted by a weighting function (Wakker and Deneffe, 1996). The first stage constructs a "standard sequence" of consequences equally spaced in terms of utilities, using a sequence of hypothetical lotteries with which the probability p of winning [and hence any $\pi(p)$] are constant. The second stage uses the standard sequence of utilities to construct a standard sequence of probabilities equally spaced in terms of probability weighting (Abdellaoui, 2000;

Bleichrodt and Pinto, 2000). This technique allows both v(x) and π(p) to be assessed in practice. Prospects with multiple possible consequences can then be evaluated and ranked via the formula $\Sigma_i v(x_i)\pi(p_i)$ where i indexes the possible consequences of the prospect. CPT predicts that individual preferences for prospects follow the ranking implied by the values of $\Sigma_i v(x_i)\pi(p_i)$.

Example: Apparent Risk Attitude may Change with Concreteness of Problem Description (Eckel, 1999)

It is sometimes difficult to define unique certainty-equivalents and risk attitudes from individual responses to questions, as illustrated by the following experimental finding (Eckel, 1999). One set of subjects was asked to write down a certainty equivalent (CE) for the abstract prospect (x, p) = ($10, 0.9). The average value was CE($10, 0.9) = $9.18. A second set of subjects was asked to provide CEs for the following concrete problem. Each subject is to choose on egg from among twenty in a basket containing two "bad eggs" and 18 "good eggs". Good eggs may be sold for $10 and bad eggs for $0. The chances of selecting a good egg are 18/20 = 0.9. Thus, this situation amounts to a monetary prospect ($10, 0.9). However, for the egg gamble, CE($10, 0.9) = $8.73, significantly less ($p < 0.03$) than the $9.18 for the abstract gamble. Moreover, subjects appear to be risk-seeking for the abstract gamble (CE > expected value of $9) but risk-averse for the eggs gamble.

Example: Certainty Equivalents May Vary with Interpretation

The following observation provides additional evidence that well-defined certainty equivalents for prospects may not exist. Many subjects who would refuse to exchange a large but unlikely loss for a small but certain one when both prospects are called "gambles" accept the exchange when the first prospect is called a "risk" and the second "insurance". For example, significantly more subjects prefer a loss of $10 with certainty to the prospect (-$1000, 0.01) when it is called an insurance premium than when it is not (Hershey and Schoemaker, 1980, cited in Camerer, 1995.) Thus, *how prospects are labeled may affect preferences among them.*

Example: Certainty Equivalents and Risk Attitudes for Life-Years

Prospect theory has been applied to interpret certainty equivalents for gambles involving remaining years of life. Verhoef *et al.* (1994) asked groups of 30 women to give certainty equivalents for two-outcome gambles with equal probabilities of longer or shorter remaining life spans. Short expected remaining lifetimes led to risk-seeking attitudes (i.e., certainty equivalents greater than expected values), while longer expected remaining lifetimes led to risk-averse attitudes. These data can be interpreted in terms of Prospect Theory. The women's choices are consistent with aspiration levels for remaining lifetimes that declined as they aged. They were willing to gamble to obtain longer lives when expected remaining lifetime was short compared to the aspiration level, but were averse to gambling when expected remaining lifetime was long compared to the aspiration level. Such changing risk attitudes could help to explain changes in willingness to undergo risky surgical procedures at different ages (Verhoef *et al.*, 1994).

Simple analysis (based on Jensen's inequality) proves that *risk aversion corresponds to concavity of the NM utility function.* If the d.m. is globally risk-averse, i.e., risk averse at all levels of x from 0 to 1, then u(x) is concave in x over this interval. Indeed, the following qualitative global properties are all logically equivalent.

Five Equivalent Qualitative Characterizations of Global Risk-Aversion

- The d.m. is risk-averse for attribute x.
- The d.m. prefers to receive the expected value of any 50-50 lottery to playing the lottery.
- The d.m. prefers to receive the expected value of *any* prospect to the prospect.
- u(x) is concave in x (e.g., "hill-shaped"; slope decreases as x increases).
- The risk premium of any non-degenerate lottery is positive.

Source: Keeney and Raiffa, 1976, Chapter 4.

How much of a desirable attribute a risk-averse d.m. would be willing to give up to replace an uncertain prospect with its expected value may depend on how much of the attribute the d.m. has to start with. Thus, risk attitude, as reflected in the risk premium, is in general a local property that varies with x. The following *local risk aversion function* gives a quantitative measure of local risk aversion for twice-differentiable utility functions:

$$r(x) = -u''(x)/u'(x) \quad \text{(Arrow-Pratt local risk aversion function)}$$

where primes denote derivatives. The numerator u"(x) reflects the local curvature of the utility function, which is negative if the utility function is locally concave (i.e., risk-averse). The denominator u'(x) makes the ratio r(x) invariant if the utility function is re-normalized and translated so that utilities take values over a different interval from [0, 1]. (The conventional choice of 0 and 1 to represent the NM utilities of the least- and most-preferred consequence is arbitrary. Any other interval could be used equally well, much as either Fahrenheit or Centigrade scales may be used with equal validity to measure temperature. Therefore, invariance of concepts under re-normalizations of the utility scale are required for the concepts to have meaning independent of the particular units chosen in describing them.) Finally, the minus sign preceding the ratio u"(x)/u'(x) makes r(x) positive for risk-averse utility functions and negative for risk-prone ones, meaning those with negative risk premiums and convex (upward-curving) utility functions. For the risk-neutral utility function, u(x) = x, it is clear that r(x) = u"(x) = 0.

r(x) has desirable properties as a measure of risk aversion. For example, larger values of r(x) imply larger risk premiums at x for *any* risky prospect (Keeney and Raiffa, 1976, Theorem 4.12). Therefore, a d.m. with a uniformly larger local risk aversion function may be said to be *more risk-averse* than a d.m. with a smaller local risk aversion function. Similarly, r(x) is an increasing function of x if and only if the risk premium for any risky prospect is an increasing function of x.

Risk aversion has been defined here for a single attribute. There is no requirement that a d.m. who is risk-averse for one attribute should be risk-averse for all others. Moreover, in reality, decision-makers often care about gains and losses relative to a starting position, although the EU paradigm prescribes that only probabilities of final consequences should matter.

While EU theory allows risk-averse utility functions, it also imposes strong constraints on the degrees of risk aversion that a d.m. can exhibit for gambles involving stakes of different sizes. In particular, an EU maximizer who rejects moderate-stakes gambles must also reject certain larger-stakes gambles. Many of these implications seem unrealistic. For example:

- A d.m. who rejects a 50-50 chance of gaining \$110 or losing \$100 (starting from any wealth level) must also reject a 50-50 chance of losing \$1000 or gaining *any* sum of money, no matter how large.
- A risk-averse d.m. who rejects a 50-50 chance at gaining \$105 or losing \$100 at wealth levels less than \$350,000 must also reject a 50-50 chance to win \$635,670 or lose \$4,000, starting from a wealth of \$340,000.

The proof of these startling results (Rabin, 2000) rests on the fact that risk aversion to moderate-stakes gambles implies that the marginal utility of money must decline unrealistically quickly.

3. INTRINSIC VALUE AND EXPONENTIAL UTILITY

This section introduces the important but somewhat subtle concept of *intrinsic value* for consequences. Intrinsic value is intended to reflect strength or intensity preferences. Its main benefit is that, when preferences for consequences are measured on an intrinsic value scale, a single parametric family of NM utility functions – exponential utility functions – emerges as natural for representing preferences among risky prospects. This family allows a clear separation of the objective and subjective components of risk in the SAUT setting and provides a model for how risk and subjective attitudes toward it can be defined and quantified.

The mathematical and philosophical underpinnings for this analysis run deeper than its simply stated results might suggest. They include the ideas that (a) Meaningful measurement scales must leave key relations (e.g., the ordering of differences in values) invariant under allowed transformations – indeed, this is the defining property of a "meaningful" scale in measurement theory (Narens and Luce, 1990); and (b) This type of measurement scale may be applicable to preferences (Luce and Narens, 1990; French, 1988 based on work by Dyer and Sarin, 1979). These references provide the groundwork for defining and using units of measurement for consequence values that can express differences as well as orderings. Such cardinal measurement of preferences for consequences can dramatically simplify and clarify SAUT methods and results.

Is it meaningful to say that an NM utility function is concave? A potential difficulty is that the curvature of the function depends on how the attribute axis is defined. If u(x) is linear, for example, then $u(x^2)$ and $u(x^{½})$ are not. Yet, x and x^a for any $a > 0$ contain exactly the same ordinal information about the preference ordering of consequences on the interval C = [0, 1]. Hence, there appears to be a danger that the same d.m. could have a utility function that is linear, convex, or concave on [0, 1], depending on an arbitrary choice of x, x^2, or $x^{½}$ as the attribute for which preferences are studied. This would make risk-attitude depend on which representation is chosen for the attribute, rather than purely a property of the d.m.

A simple solution to this potential problem is to recognize that there is more to the mathematical structure of most consequence sets than just the ordering of consequences. Single-attribute consequences are usually measured in "natural" units, such as life-years, time-to-tumor (in years), or QALYs. What makes these units (but not their squares or their square roots) natural is that their sums and differences are well defined and meaningful (Narens and Luce, 1990; Luce and Narens, 1990). Arbitrary order-preserving transformations of the x axis are *not* allowed because they do not preserve the ordering of sums and differences. Thus, a sequence of consequences that are equally spaced before such a transformation would not be after it. (Changes of scale units, e.g., from life-years to life-months, are still allowed by this criterion, but they leave concavity or convexity of the utility function unchanged. Moreover, even this degree of freedom is removed if the endpoints of the consequence axis are set equal to 0 and 1.)

When the scale of the x axis is constrained by the requirement that differences have meaning, it may be possible to draw more specific conclusions about the mathematical form of the utility function. For example, suppose that consequences are measured in units such that *equal increments are valued equally,* independent of the base level to which the increment is added. (Thus, for any four levels of the consequence attribute, a, b, c, and d, the inequality $(a - b) > (c - d)$ holds if and only if the d.m. prefers a to b at least as strongly as she prefers c to d. For a full discussion of what, if anything, this quaternary relation means, see French, 1988, Section 9.3). Then, under additional technical assumptions, the NM utility function must be either:

$u(x) = x$, for $0 \leq x \leq 1$ (Linear utility function)
or
$u(x) = (1 - e^{-kx})/(1 - e^{-k})$ for $0 \leq x \leq 1$ (Exponential utility function)

(The denominator here normalizes the utilities to lie on a scale from 0 to 1. If this conventional normalization is dropped, then the NM utility function can be simplified to $u(x) = b - ae^{-kx}$ where $a > 0$ and b are arbitrary constants.) k is the *coefficient of relative risk aversion.* It is the value of the local risk aversion function, $r(x) = k$, which is a constant independent of x for exponential utility.

As a slight generalization, suppose that v(x) is a value function that assigns numbers to consequences so that differences in the numbers reflect strengths or intensities of preferences (i.e., so that $v(A) - v(B) > v(C) - v(D)$ if and only if A is preferred to B more strongly than C is preferred to D. See French, 1988, for a discussion of what this may mean.) On such a scale, equal increments in the consequence attribute are equally valued. Values measured on such a scale are sometimes called *intrinsic values, cardinal values,* or *measurable values.* They are uniquely defined (with the usual normalization $v(0) = 0$ and $v(1) = 1$) under relatively mild assumptions.

The intrinsic value function for monetary consequences is usually assumed to be a smoothly increasing concave function of x, thus exhibiting "diminishing marginal utility", in classical economics terminology. (The "utility function" of classical microeconomics is not an NM utility function, but might well be a measurable value function.) Let v(.) be a measurable value function mapping consequences expressed in natural units (e.g., in dollars) to an intrinsic value scale on which equal increments are equally valued, regardless of the value level to which they are added. Then the corresponding exponential NM utility function mapping consequences expressed in natural units to NM utilities is just:

$u(x) = (1 - e^{-kv(x)})/(1 - e^{-k})$ (Exponential utility of intrinsic value)

In short, intrinsic value may itself be treated as the single attribute to which SAUT is applied. The resulting NM utility function will then, under some technical assumptions, be either a linear or an exponential function of intrinsic value. Empirically, analysis of farmer preferences for a single monetary attribute (price of potatoes) show that exponential forms for both u(x) and v(x) are reasonable and fit the data better than power functions, while the exponential form for u(x) as a function of v(x) fits the data better than a power function form or a linear form, i.e., $u(x) = v(x)^k$ (Smidts, 1997).

Several different sets of assumptions imply the exponential form. For a risk-averse decision-maker with a smooth (twice-differentiable) increasing NM utility function on the consequence interval, the conditions in Table 1 are all mutually equivalent. One or more of these conditions may be considered normatively compelling once consequences have been expressed on an intrinsic value scale. If so, exponential utility functions should be used to choose among prospects.

Table 1: Equivalent Characterizations of Exponential Utility Functions

- *Strong Risk Independence:* Let X and Y be two risky prospects (random variables) on C = [0, 1]. If w + X is preferred to w + Y for *some* level of w, then w + X is preferred to w + Y for *all* levels of w. (When the attribute is money, w is sometimes interpreted as the initial wealth level. Then, this condition says that the preference ranking of prospects does not depend on the initial level of wealth.)
- *Risk Premium Independence:* The d.m.'s risk premium for any risky prospect (i.e., any random variable on C = [0, 1]) is independent of the initial level of x. (It depends only on the prospect.)
- *Certainty Equivalent Independence:* The certainty equivalent of any risky prospect is independent of the initial level of x. Thus, if X is the prospect and w is any initial level of x, then CE(w + X) = CE(X) + w.
- *Delta Property for Certainty Equivalents:* If a constant, w, is added to every possible outcome of a prospect X, then the certainty equivalent of the new prospect thus formed is CE(X) + w. (This is sometimes called the "delta property", due to Pfanzagl, 1959.)
- *Equal Buying And Selling Prices:* For any prospect X and attribute level w, the d.m. is indifferent between w + CE(X) – X and w + X – CE(X).
- *No Buying-Price/Selling Price Reversals:* The ranking of prospects or of information sources based on their certainty equivalents (i.e., "selling prices") can never disagree with their ranking based on buying prices. (This assumes the d.m. is risk-averse. Otherwise, the risk-neutral utility function u(x) = x is possible.)
- *Exponential Utility:* $u(x) = (1 - e^{-kx})/(1 - e^{-k})$ for $0 \leq x \leq 1$. (If the [0, 1] normalization is dropped, this may be simplified to $u(x) = e^{-kx}$.)

References: Dyer and Jia, 1997; Hazen and Sounderpandian, 1999

These equivalent conditions and the measurable value scale approach provide two alternative routes to exponential utility. A third approach is based on careful consideration of what it means to add consequences. Harvey (1990) analyzes preferences for consequences defined on an open interval. Any two consequences can be composed to form a third one via a continuous group operation (i.e., an associative binary operation with an inverse and identity), interpreted as addition of attribute levels. This implies the existence of a strictly increasing continuous function v mapping attribute levels to numerical values in such a way that the value assigned to the sum of two attribute levels is the sum of their values. (In symbols, there is a continuous function $v : C \rightarrow R$ with $v' > 0$ and such that $v(x + y) = v(x) + v(y)$, where + is the continuous group operation for combining attribute levels, and + is ordinary addition of real numbers.) If preferences satisfy the EU representation, then the following conditions are equivalent:

- If consequence x is preferred to the prospect (x + y, p, x − y) (i.e., the prospect giving x + y with probability p and x − y otherwise) for *some* level of x, then this preference holds for *all* levels of x.
- $u(x) = e^{[kv(x)]}$ if $k > 0$, $u(x) = -e^{[-kv(x)]}$ if $k < 0$, $u(x) = v(x)$ if $k = 0$. Thus, if the d.m. is risk-averse, the utility function is exponential.

In summary, many different conditions imply that the utility function must be exponential. Yet, it is clear that none of them need hold for an attribute like money. For, the initial level of monetary wealth may well affect the relative desirability of different prospects. In this sense, the prospects do not have *intrinsic* values. Instead, their values typically depend on the level of initial wealth. However, when consequence values are measured on an intrinsic value scale, it seems much more reasonable to suppose that the value of each prospect should be independent of the context in which it occurs. The various conditions leading to an exponential utility function express the informal notion of context-independence of value or preferences in different precise ways. Thus, while exponential utility might not be reasonable for dollar outcomes, it seems much more compelling for attributes such as lives saved or QALYs that might be expected to be measured on an intrinsic value scale.

4. NON-EXPONENTIAL SAUT UTILITY FUNCTIONS

For consequences that are not measured on intrinsic value scales, other parametric families of utility functions may be justified. One family that has emerged as especially important in axiomatic treatments is the linear-plus-exponential family, consisting of utility functions of the following form:

$$u(x) = x - be^{-cx}, \qquad \text{(linear-plus-exponential utility)}$$

where b and c are positive constants. Several axiomatic routes lead to this destination. For example, the following two statements are equivalent (Dyer and Jia, 1997, based on earlier work by D.E. Bell):

1. *Risk Independence Condition:* For any two single-attribute random variables X and Y with the same expected value, if w + X is preferred to w + Y for *some* value of w, then w + X is preferred to w + Y for *all* values of w.
2. The NM utility function is one of the following:
   - Quadratic utility: $u(x) = x^2$
   - Linear utility: $u(x) = x$
   - Exponential utility: $u(x) = 1 - e^{-kx}$
   - Linear-plus-exponential utility: $u(x) = x - be^{-kx}$

where b and k are constants. (These representations do not make the conventional [0, 1] normalization for u, and so they may be rescaled and translated arbitrarily.) Assuming that the d.m. is risk-averse eliminates linear utility, while assuming that more of x is always preferred (at least weakly) to less eliminates quadratic utility. Hence, with these assumptions, risk independence implies exponential or linear-plus-exponential utility.

An alternative route to linear-plus-exponential utility for monetary prospects is via the following axioms (Pedersen and Satchell, 1997, citing earlier work by D.E. Bell.)

An Alternative Characterization of Linear-plus-Exponential Utility

Suppose that a d.m.

- Prefers more of attribute x to less
- Is risk-averse at all levels of x
- Is decreasingly risk-averse at all levels of x (i.e., the risk premium for a risky prospect decreases as initial wealth increases)
- Has risk aversion approaching zero as initial wealth increases without bound
- Switches preferences between any two prospects at most once as wealth increases. (Note that requiring *zero* switches leads to exponential utility.)
- Has preferences represented by a smooth (infinitely differentiable) NM utility function, u(x).

Then the d.m.'s utility function can be represented as follows:

$u(x) = x - be^{-kx}$ for positive constants b and k.

Proof: Pedersen and Satchell, 1997, http://ideas.uqam.ca/ideas/data/Papers/fthcambap9735.html

The key intuition here is that as a d.m.'s initial wealth increases, he or she may be willing to pay smaller risk premiums to replace prospects with their certainty equivalents. Hence, a relatively risky prospect with a high expected value may become preferred to a less risky one with a lower expected value if starting wealth is increased sufficiently. However, increasing starting wealth still further would not cause this switch in preferences to be reversed. This provides a plausible motivation for a one-switch (i.e., linear-plus-exponential) family of utility functions for monetary prospects.

Additional justifications for linear-plus-exponential utility include the following. It is the *only* family of increasing risk-averse utility functions (i.e., with $u' > 0$ and $u'' < 0$) for which any (and all) of the following three properties holds (Bell, 1995):

- It is always preferable to resolve a larger uncertainty about initial wealth rather than a smaller uncertainty before choosing among risky prospects.
- The NM expected utility of a prospect depends only on initial wealth, the expected value of the prospect, and some measure of the prospect's risk.
- If two prospects switch in preference as initial wealth increases, then they do so in favor of the prospect with the higher expected value.

Still another line of justification for the linear-plus-exponential form is a condition stating that an increase in initial wealth always decreases the risk premium for one prospect relative to another (Gelles and Mitchell, 1999).

In financial settings, it is common practice to consider risks of portfolios of prospects, in which an investor may take shares of different sizes in different prospects. The monetary return received from a prospect is proportional to the investor's share in it. If preferences between shares in any two prospects with non-negative outcomes and having the same expected value are the same regardless of the sizes of the shares (i.e., independent of the constant of proportionality determining return), then one is led to NM utility functions of the following forms:

- $u(x) = \log_a(x)$
- $u(x) = x^a$
- $u(x) = x + b\log(x)$
- $u(x) = x + bx\log(x)$
- $u(x) = x + bx^a$

where a and b are constants (Dyer and Jia, 1997).

These parametric families of utility functions, as well as the linear-plus-exponential family, have been much studied by financial risk theorists. Although their axiomatic basis is mathematically interesting, their relevance for evaluating health, safety, and environmental risks is unclear. For example, the idea of an initial level of "wealth", i.e., of the desirable attribute x, seems sensible for decisions involving monetary outcomes. But it is not obvious how to interpret it for health outcomes. As previously discussed, such outcomes may be *intrinsically* valued, with a prospect representing a random gain or loss having the same value independent of the context (i.e., the base level and/or other gains or losses) in which it occurs. Then, exponential utility is more appropriate than these alternative utility functions.

Parametric NM utility functions for health consequences such as QALYs, often emphasize exponential utility functions. However, a major theme of SAUT is that many comparisons of prospects in terms of desirability and/or risk do *not* require specific parametric utility functions to be known. This is explored next.

5. OBJECTIVE COMPARISONS OF RISK PROFILES

The theory of exponential utility summarized above provides a basis for comparing risk attitudes of decision-makers. One exponential-utility d.m. is *more risk averse* than another if and only if he has a greater coefficient of relative risk aversion (i.e., a greater value of k). This suggests the following way to compare the risks of different prospects. For any two random variables X and Y defined on the same intrinsic value scale, define X to be *riskier than* Y (for exponential utility d.m.s) if and only if *all* risk-averse exponential-utility d.m.s have a larger risk premium for X than for Y. The notation $X\ R_{exp}\ Y$ will indicate this relation between prospects. After illustrating some possible applications of this comparative risk ordering, we will consider how to generalize it to cases where exponential utility functions cannot necessarily be assumed.

Example: Comparing Risks of Normal Distributions by their Variances

Suppose that X is a normally distributed random variable with mean E(X) and variance Var(X). The risk premium for X for a d.m. with an exponential utility function is calculated as follows:

$$\text{Risk premium} = E(X) - CE(X) = E(X) - x^*,$$

where

$$u(x^*) = EU(X) = \int_{[-\infty,\infty]} u(x)dF(x) = = \int_{[-\infty,\infty]} u(x)f(x)dx.$$

In this example, $u(x) = (1 - e^{-kx})/(1 - e^{-k})$ and f(x) is the PDF of the normal random variable X. (The normalization of u(x) to u(0) = 0, u(1) = 1 is not required here, since a normal random variable is defined on the open interval [-∞, ∞]. Using $u(x) = e^{-kx}$ or any equivalent NM function, i.e., one of the form $a + be^{-kx}$ for arbitrary constants b > 0 and a would work just as well.) Hence,

$$u(x^*) = [1/(1 - e^{-k})]\int_{[-\infty,\infty]}(1 - e^{-kx})f(x)dx = [1/(1 - e^{-k})][1 - \int_{[-\infty,\infty]}e^{-kx}f(x)dx].$$

Now, this last integral is known. It is essentially the moment generating function of the normal distribution (Ross, 1970). Its value is $\exp\{-kE(X) + Var(X)k^2/2\}$. Thus,

$$u(x^*) = \{1 - \exp[-kE(X) + Var(X)k^2/2]\}/(1 - e^{-k}),$$

Comparing this to the utility function $u(x) = (1 - e^{-kx})/(1 - e^{-k})$ reveals that

$$x^* = CE(X) = E(X) - (k/2)Var(X). \quad \text{(CE of normal distribution)}$$

Hence, the risk premium for a normally distributed prospect is:

$$E(X) - CE(X) = (k/2)Var(X). \quad \text{(Risk premium for normal distribution)}$$

It follows that one normal distribution is riskier than another (according to the definition of R_{exp}) if and only if it has a greater variance.

Example: Comparing the Risks of Compound Poisson Processes

Suppose that X(t) is the total amount of damage accumulated in a compound Poisson process over a time interval of length t. Accidents arrive at rate λ and each accident is assumed to cause a statistically independent random amount of damage, described by a variable Y measured on an intrinsic value scale for damage. The scale is oriented so that larger amounts are preferred, e.g., by measuring damage in negative numbers. Thus, in this example, the conventional 0-1 normalization of the consequence scale is not used. The utility function for damage is assumed to be exponential. The certainty equivalent for X(t) satisfies the equation:

$$u(x^*) = EU[X(t)] = E\{\exp(-kX(t)]\} = \int_{[-\infty,\infty]} e^{-kx}f(x)dx$$

where f is the PDF of X(t). From the moment generating function for a compound Poisson process (Ross, 1970), this integral evaluates to $\exp\{\lambda t[E(e^{-kY}) - 1]\}$. Comparing this expression to $u(x^*) = e^{-kx^*}$ reveals that the certainty equivalent of X(t) is:

$$CE[X(t)] = x^* = (\lambda t/k)[1 - E(e^{-kY})]. \quad \text{(Certainty equivalent for process)}$$

Hence, the risk premium is:

$$\text{Risk premium} = E[X(t)] - CE[X(t)] = \lambda tE(Y) - (\lambda t/k)[1 - E(e^{-kY})]$$

For the special case where Y is normal with mean E(Y) and variance Var(Y), the term $E(e^{-kY})$ is just $e^{-k[E(Y) - Var(Y)(k/2)]}$ and the risk premium becomes:

$$\text{Risk premium} = \lambda tE(Y) - (\lambda t/k)[1 - e^{-k[E(Y) - (k/2)Var(Y)]}].$$

If $k[E(Y) - (k/2)Var(Y)]$ is numerically small, then the approximation $1 - e^{-x} \approx x$ for small x may be used to simplify this expression further:
Small-approximation risk premium $= \lambda tE(Y) - (\lambda t/k)(k)[E(Y) - (k/2)Var(Y)] = (\lambda tk/2)Var(Y)$.

In this special case, *one compound Poisson process is riskier than another if and only if it has a higher value of $\lambda Var(Y)$.* The product of the frequency of accidents and the variance in severity (i.e., damage per accident) provides a numerical measure of risk for such stochastic processes.

Compound Poisson processes are especially convenient to analyze and interpret because, intuitively, they generate risks and values at a constant rate per unit time. The certainty equivalent per unit time for such a process is therefore well defined and may be used to compare alternative compound Poisson processes. The preceding analysis shows that it is given by the following explicit formula:

$$\text{CE per unit time for compound Poisson process} = (\lambda/k)[1 - E(e^{-kY})].$$

When Y is deterministic and small, this reduces to approximately λY, which is independent of k. Thus, in this special case, all decision-makers with exponential utilities can agree on the ranking of alternative processes in terms of their certainty equivalents per unit time, without regard for personal risk attitudes. In this limited but important sense, one such process is *objectively* preferable to another if it produces a smaller average damage per unit time.

Unfortunately, CE per unit time is not well-defined for more general renewal processes. For example, consider a renewal process in which the time between renewals (e.g., accidents) is uniformly distributed between 0 and 1 year. The hazard rate for occurrence of the next renewal epoch approaches infinity whenever the elapsed time since the previous renewal approaches 1. (Recall that the hazard function for such a process is

$$h(t) = f(t)/[1 - F(t)] = 1/(1 - t) \text{ for } 0 \leq t \leq 1$$

where t is the time since the previous renewal.) Thus, the certainty equivalent per unit time may change over time, increasing with time since the previous renewal and then dropping back to its original level whenever a renewal occurs. Preferences among such processes may well depend on subjective attitudes toward anxiety and suspense, preferences for the timing of uncertainty resolution, and aversion toward (or preference for) bunching of adverse consequences in time. A criterion other than CE per unit time is needed to compare and choose among such processes.

These examples illustrate a particular approach to "objective" definition and measurement of risk that can be generalized to settings where exponential utility cannot be assumed. The steps are as follows:

1. Identify conditions under which *all* risk-averse decision-makers with a certain class of preferences agree on how to rank a set of prospects from least risky to most risky. The most commonly considered class of preferences consists of those that can be represented by increasing risk-averse utility functions.
2. Assign numbers (risk premiums in the preceding examples) to prospects in such a way that prospects ranked as being more risky receive higher numbers. (Often, the prospects in question will belong to a parametric family, as for normal distributions or Poisson processes, and the numbers representing the ordering of risks may have a natural interpretation in terms of parameters of the prospects.)

Then all decision makers in the specified class will agree that the numbers represent the risk ordering of prospects, and hence provide at least an ordinal measure of risk. This measure is objective in the limited but useful sense that it is not necessary to know details of a d.m.'s preferences to compare the risks of different prospects. (For example, variances can be used to compare risks of normal distributions for all d.m.s with exponential utility functions without needing to know individual values of the risk aversion coefficient k.) If the numbers assigned to prospects enjoy additional properties, such as that the risk of the sum of two random variables or prospects is the sum of their risks, then cardinal measurement of risk may be possible.

5.1 First-Order Stochastic Dominance (FSD)

Suppose that every d.m. to be considered prefers more of attribute x to less. For example, x might indicate the mean time between accidents or fatalities in a Poisson arrival process, or the sum of remaining QALYs in a population. What, if anything, does this agreement on preferences for *consequences* imply about agreement on preferences for *prospects*? ("Prospects" are defined here, as usual, as random variables on the set of consequences.) Expected utility theory provides a precise answer.

One prospect is called *stochastically larger* than another if the first is at least as likely as the second to exceed *any* specified level of the consequence attribute. Thus, the CDF for the first lies to the right of the CDF of the second, possibly touching at one or more points. When more is preferred to less, this relation between prospects (or the CDFs representing them) is also called *first-order stochastic dominance,* abbreviated FSD. In symbols, this definition is:

X FSD Y if and only if $\Pr(X > a) \geq \Pr(Y > a)$ for all a.

The notation "X FSD Y" is read "X dominates Y by first-order stochastic dominance." (Second-order and higher-order dominance relations are defined later.) In terms of CDFs, the FSD relation becomes:

$F_X(a) \leq F_Y(a)$ for all a,

where F_X is the CDF for X and F_Y is the CDF for Y. Thus, a d.m. who chooses F_X instead of F_Y (or, equivalently, X instead of Y) is at least as likely to obtain a consequence of at least x as a d.m. who makes the reverse choice, for *all* levels of x.

FSD is the one and only specific relation for comparing prospects that all d.m.s with increasing NM utility functions must agree on. The following result is fundamental (e.g., Quiggin, 1993; Ross, 1996):

Characterizations of First-Order Stochastic Dominance (FSD)

The following are equivalent:

1. X FSD Y (i.e., $\Pr(X \leq a) \leq \Pr(Y \leq a)$ for all a. Equivalently, $F_X(a) \leq F_Y(a)\ \forall a$)
2. The expected utility of X is at least as great as the expected utility of Y for *every* increasing utility function, i.e., $EU(X) \geq EU(Y)$ for all monotonically increasing u(x).
3. $Y = X - Z$, for some random variable Z such that $\Pr(Z \geq 0) = 1$.
4. There are two random variables X* and Y* defined on the same probability space as X and Y and with the same CDFs as X and Y such that $\Pr(X^* \leq Y^*) = 1$.

Sources: Quiggen, 1993, Chapter 3; Ross, 1998; Shaked and Shanthikumar, 1994, Chapter 1. The equivalence of 1-3 was proved by Hadar and Russell, 1969.

The first two conditions show that all EU decision-makers who prefer more of x to less should agree on the ranking of any prospects that are ordered by FSD, i.e., by the relation "lies to the right of" among their CDFs. The third condition gives an additional interpretation based on the fact that the difference of two CDFs ordered by FSD can be interpreted as the CDF of another prospect.

Example: Comparing Probability Distributions by FSD

The first member of each of the following pairs dominates the second by FSD:

a) *U[a, b] FSD U[c, d] if a > c and b > d:* Here, the notation is that U[a, b] denotes a random variable uniformly distributed between a and b, where a > b. (To see this, note that the CDF of U[a, b] lies to the right of the CDF for U[c, d] except when both are 0 or 1.)
b) *$N(\mu, \sigma^2)$ FSD $N(m, \sigma^2)$ if $m < \mu$ i.e., normal distributions with the same variances are ordered by their means.* The normal distribution with mean μ and variance σ^2, denoted by $N(\mu, \sigma^2)$, dominates $N(m, \sigma^2)$ by FSD whenever $m < \mu$. (To see this, note that the first distribution is a rightward-shifted version of the second.)
c) *B(n, p) FSD B(n', p') if (n, p) ≥ (n', p'):* A binomial distribution with parameters n and p, denoted by B(n, p) (i.e., the sum of n independent, identically distributed Bernoulli random variables, each having a value of 1 with probability p and a value of 0 with probability 1 - p) dominates by FSD a binomial distribution with parameters n' and p', where n ≥ n', p ≥ p'. In words, the binomial distribution B(n, p) is stochastically increasing in each of its two parameters.
d) *Poisson random variables are ordered by their means:* A Poisson random variable with mean λ dominates by FSD any Poisson random variable with mean λ' < λ. In other words, a Poisson random variable is *"stochastically increasing"* in its mean.
e) *Exponential random variables are ordered by their means:* An exponential distribution with mean 1/λ' dominates by FSD an exponential distribution with mean 1/λ, where λ' < λ.

These comparisons have natural intuitive interpretations. For example, part (d) implies that in a Poisson arrival process, the number of arrivals by any fixed time is stochastically increasing in the average arrival rate, λ. Part (e) implies that the exponentially distributed time between successive arrivals in a Poisson process is stochastically decreasing in the mean arrival rate.

That a decision-maker should prefer FSD-dominating prospects to FSD-dominated ones is often considered fundamental for rational decision-making – even more basic and defensible than EU theory. Indeed, FSD may be used as an axiom to help derive either EU or more general preference theories for single-attribute consequences. For example, the following five axioms, including FSD, imply (and are implied by) EU theory in the SAUT setting. They hold if and only if preferences among prospects can be represented by corresponding expected values of an NM utility function. Generalizations of EU theory are obtained if Axiom 5 (Independence) is replaced with weaker conditions while retaining FSD and the other axioms.

Axioms (Including FSD) for EU Theory in the Single-Attribute Setting

1. *Reduction* (i.e., sufficiency of CDFs, or "Only CDFs matter for decision-making"): If prospects X and Y have the same CDF, then they are equally preferred.
2. *Weak order:* Prospects can be weakly ordered (i.e., ranked with ties allowed) by preference. Equivalently, the binary relation "is at least as preferred as" between prospects is complete, transitive, and reflexive.
3. *FSD:* If X FSD Y then X is at least as preferred as Y.
4. *Continuity:* The preference relation "is at least as preferred as" is continuous (in the topology of weak convergence. Roughly, this means that if X is strictly preferred to Y, then all prospects with CDFs sufficiently close to the CDF for X will be preferred to all prospects with CDFs sufficiently close to Y's. More accurately, for all X, the set of all prospects that are at least as preferred as X is closed in the topology of weak convergence of CDFs.)
5. *Independence:* If F, G, and H are three CDFs over the consequence interval and F R G, then for any w in [0, 1], [wF + (1 – w)H] R [wG + (1 – w)G].

Reference: Quiggin, 1993, Chapters 3 and 9. See also Kreps, 1988.

Example: Rank-Dependent Utility Implies FSD

A psychologically realistic description of choice under risk must account for the fact that people tend to systematically over-weight some probabilities (especially, small ones) compared to the prescriptions of EU theory. A generalization of EU theory that accommodates this and other departures from EU is *rank-dependent expected utility* (RDEU) theory (Quiggin, 1993). In RDEU, preferences among prospects are represented as follows. The prospect

$$X = (x_1, p_1; x_2, p_2; \ldots, x_n, p_n) \qquad \text{(discrete prospect)}$$

that gives consequences $x_1, x_2, \ldots, x_n$ (numbered so that preferences for x_j increase with j) with probabilities $p_1, p_2, \ldots, p_n$, respectively, is assigned a rank-dependent utility of:

$$\text{RDEU}(X) = \sum_{j=1}^{n} \pi_j v(x_j) \qquad \text{(RDEU representation of preferences).}$$

In this representation, v(.) is called the *value function* and the π_j are called *decision weights* (following terminology of Tversky and Kahneman, 1992). The RDEU decision weights are derived from the outcome probabilities via the following formula:

$$\pi_j = w(CDF(x_j)) - w(CDF(x_{j-1})) \qquad \text{(decision weights formula)}$$

Here, w(.) is a strictly increasing *weighting function* from [0, 1] to [0, 1] with w(0) = 0 and w(1) = 1. The weighting function maps probabilities to transformed values that allow phenomena such as over-weighting of small probabilities to be explicitly described. CDF(j) is defined as the probability of obtaining an outcome no better than the jth one, i.e.,

$$CDF(x_j) = \sum_{i=1}^{j} p_i.$$

(If j = 1, so that j – 1 = 0, then define $CDF(x_{j-1}) = \sum_{i=1}^{0} p_i = 0$.)

RDEU theory is considerably more flexible than EU theory, to which it reduces in the special case of a linear weighting function, w(p) = p. Nonlinear weighting functions can yield preferences that violate EU theory by exhibiting the Allais paradox, preference reversals, and other anomalies (Quiggin, 1993, Chapter 10.) Despite its flexibility, however, RDEU theory also imposes some sharp constraints on allowed patterns of preferences. Most importantly, *RDEU implies FSD,* i.e., prospect X is preferred to prospect Y by all decision-makers who obey RDEU if and only if X FSD Y (Quiggin, 1993). Thus, RDEU is an example of a coherent theory of decision-making under risk that obeys FSD but that is less restrictive than EU theory. Indeed, RDEU implies Axioms 1-4 above and is implied by them in conjunction with various weaker conditions than Independence. This equivalence is developed in the next example.

Example: Cumulative Prospect Theory (CPT) Implies FSD

Additional psychological realism can be gained by modifying EU further to allow not only for nonlinear weighting of probabilities, but also for the fact that people often focus more on gains and losses compared to a reference point than on probabilities of absolute outcomes. To this end, it is useful to decompose each risky prospect into its gain and loss parts. For any prospect X, define X^+ as the prospect formed from X by replacing all loss outcomes (compared to the *status quo*) with the *status quo* outcome. Similarly, define X^- as the prospect formed from X by replacing every gain by the *status quo*. *Cumulative prospect theory* (CPT) holds if preferences among prospects can be represented by two RDEU functions, one for gains and one for losses, as follows: Prospect X is at least as preferred as prospect Y if and only if CPT(X) ≥ CPT(Y), where

$$CPT(X) = RDEU^+(X^+) + RDEU^-(X^-).$$

Here,

a) $RDEU^+(.)$ is an RDEU function for gains, with some weighting function w^+;
b) $RDEU^-(.)$ is an RDEU function for losses, with some weighting function w^-; and

c) There is a value function v: $X \rightarrow R$ with $v(0) = 0$, where outcome 0 represents the *status quo*.

To describe empirical observations, the value function is often assumed to be s-shaped, i.e., concave (risk-averse) for gains and convex (risk-seeking) for losses.

CPT theory describes real decision-making more accurately than EU theory. It allows for, and even predicts, several types of violations of the normative axioms of EU theory. However, it, too, implies FSD. Indeed, CPT holds with a continuous value function if and only if

a) Axioms 1-4 above hold; and
b) A weaker condition than Independence called *Tradeoff Consistency,* also holds.

Under these conditions, the value function is uniquely determined up to multiplication by a positive constant, and the weighting functions w^+ and w^- are uniquely determined (Chateauneuf and Wakker, 1999).

Trade-off consistency (TC) applies to any four consequences z, w, x, and y that are all gains or all losses. For concreteness, suppose that $z > w$ and $x > y$. Then TC requires that, if the preference difference between consequences x and y is greater than the preference difference between consequences z and w, then the preference difference between z and w is not greater than the preference difference between x and y. Preference difference is defined as follows. x is preferred to y more strongly than z is preferred to w (i.e., the preference difference between x and y is greater than that between z and w) if and only if there are two lotteries, L_1 and L_2, with the same probabilities but different possible outcomes, such that:

- L_1 with x as one of its possible outcomes is at least as preferred as L_2 with y as one of its possible outcomes.
- If z is substituted for x in L_1 and w is substituted for y in L_2, then L_1 is not preferred to L_2.

In notation,

$$(c_1, p_1; c_2, p_2; \ldots; x, p; \ldots; c_n, p_n) \text{ R } (d_1, p_1; d_2, p_2; \ldots; y, p; \ldots; d_n, p_n)$$

but

$$(d_1, p_1; d_2, p_2; \ldots; w, p; \ldots; d_n, p_n) \text{ P } (c_1, p_1; c_2, p_2; \ldots; z, p; \ldots; c_n, p_n)$$

For example, suppose that (100, 0.3; 1, 0.7) R (110, 0.3; 0, 0.7) but (110, 0.3; 10, 0.7) P (100, 0.3; 9, 0.7). Then the preference difference between 1 and 0 is greater than the preference difference between 10 and 9. TC then requires that the preference difference between 10 and 9 cannot be greater than that between 1 and 0. In conjunction with axioms of Sufficiency, Weak Ordering, Continuity, and FSD, TC is necessary and sufficient for CPT. Moreover, if TC is strengthened by requiring it to apply whether or not z, w, x, and y have the same sign (i.e., all gains or all losses), then Axioms 1-4 together with this "strong TC" condition imply RDEU (Chateauneuf and Wakker, 1999).

In summary, FSD can be used as one of a small set of axioms that justify the EU, CPT, or RDEU theories, depending on whether Independence, TC, or Weakened TC, respectively, is included as an axiom. The main logical relations among different SAUT theories discussed so far may be diagrammed as follows. (Arrows here indicate logical implications, i.e., each theory or condition is implied by the conjunction of the items pointing into it.)

Relations among preference independence conditions and representations

TC ← strong TC ← Indep. ← Risk Indep. ← Strong Risk Indep.
↓ ↓ ↓ ↓ ↓
Axioms 1-4 → CPT → RDEU → EU → linear + exponential utility → exponential utility

This diagram is slightly over-simplified, in that, to obtain the exponential or linear-plus-exponential families, some additional assumptions (e.g., decreasing risk aversion, differentiability) are also required, as explained previously. Moreover, the diagram does not reveal that the alternative sets of axioms obtained by different weakenings or strengthenings of Independence are *necessary* (implied by) as well as sufficient for the alternative representations. Nonetheless, it clearly shows that Axioms 1-4 above, including FSD, are of fundamental importance for EU theory and several of its generalizations and specializations.

Example: Generalized Smooth Utility Implies FSD (Machina, 1982)

A different direction for generalizing EU theory is to allow the utility function to depend not only on the consequences being evaluated, but also on the probability distribution for consequences. The key construct is the *local utility function at F,* denoted by u(x ; F). This assigns utilities to outcomes x in an interval of the real line, but allows the utility of each x to depend on the CDF, F, of the risky prospect being evaluated. If preferences depend smoothly on F in an appropriate technical sense (formalized by Frechet differentiability of the functional representing preferences for prospects), then preferences among prospects satisfy FSD if and only if u(x ; F) is non-decreasing in x for all CDFs F. This property can be proved without requiring Independence, which is replaced by the smoothness condition (Machina, 1982). The resulting "generalized smooth utility theory" reduces to EU theory if u(x ; F) is the same for all F. However, it is in general not compatible with RDEU theory, which violates the smoothness assumption (Quiggin, 1993).

Example: FSD Implies that Risk is Not Variance

The FSD relation among prospects (or their CDFs) has an interesting implication for possible quantitative definitions of risk. Suppose that risk is defined as the information about a prospect that a rational (EU-maximizing) d.m. must have, in addition to knowledge of its expected value, in order to calculate its certainty equivalent. In symbols, this becomes:

$$CE(X) = f[E(X), R(X)],$$

where f is some function and R(X) is the risk of prospect X. Under some additional assumptions, this formulation leads to linear-plus-exponential utility functions and to explicit formulas for R(X) (Dyer and Jia, 1997). Before this was known, a common proposal in the early days of financial risk analysis was that R(X) might be the variance of X. This resulted in *mean-variance decision theories,* some of which are still sometimes used in financial portfolio risk management. We shall now show that such theories are, in general, inconsistent with the FSD axiom.

Recall that, for the special case of normally distributed prospects and exponential utilities, the certainty equivalent functional is:

$$CE(X) = f[E(X), Var(X)] = E(X) - (k/2)Var(X)$$

where $k > 0$ is a coefficient indicating the extent of risk aversion. Attractive as this formula may be on grounds of simplicity and interpretability, it is not satisfactory in general for non-normal distributions. For example, the following two prospects have the same means and the same variances, and yet there is no reason to suppose that a d.m. should assign them the same certainty equivalent, especially since A involves no potential losses compared to the *status quo*, while B involves a possible large loss.

A: ($1M, 0.1; $0, 0.9)

B: ($0.2M, 0.9; -$0.8M, 0.1)

More generally, the assumption that preferences among prospects depend only on their means and variances can be shown to conflict with the FSD axiom. For, suppose that, to the contrary, well-defined mean-variance indifference curves do exist, such that

a) All prospects with the same mean and variance are equally desirable;
b) For any given level of variance, prospects with higher means are preferred to ones with lower means;
c) For any given mean value, prospects with lower variances are preferred to ones with higher variances; and
d) For any two prospects X and Y with X FSD Y, X is at least as preferred as Y, and X is strictly preferred to Y if the CDF for X lies strictly to the right of the CDF for Y.

To show that these properties are inconsistent, consider any two distinct mean-variance pairs, say (m_1, v_1) and (m_2, v_2), that lie on the same indifference curve, i.e., that are equally preferred. It is easy to construct a one-parameter family of prospects, totally ordered by FSD, that includes members corresponding to (m_1, v_1) and (m_2, v_2). This implies that the prospect corresponding to one of these mean-variance points dominates by FSD the prospect corresponding to the other. But then one should be strictly preferred to the other, contradicting the postulate that they are on the same indifference curve.

As a specific example of the required one-parameter family of prospects, take the shifted binomial family:

$$a + B(n, p)$$

with mean and variance $m = a + np$ and $v = np(1 - p)$, respectively. Given two mean-variance points (m_1, v_1) and (m_2, v_2), the following system of four equations in four unknowns $(a, n, p_1, \text{and } p_2)$ can be solved for the values of (a, n, p_1, p_2):

$$a + np_1 = m_1$$
$$a + np_2 = m_2$$
$$np_1(1 - p_1) = v_1$$
$$np_2(1 - p_2) = v_2$$

(For simplicity, we will only consider points such that the resulting values of p_1 and p_2 are between 0 and 1, although similar constructions can be completed when this does not hold.) In general, one of p_1 and p_2 will be greater than the other. But, the family $a + B(n, p)$ is stochastically increasing in each of its three parameters, and specifically in p when a and n are held fixed. This provides the desired construction.

Simpler counter-examples can be constructed for special cases. For example, consider the indifference curve passing through the origin, with $(m, v) = (0, 0)$. This curve, which must slope upward from the origin, would divide the set of mean-variance pairs into those that are better (below the curve) and worse (above it) than the status quo of (0, 0), based on mean-variance values only. Suppose that it passes through some point (m_2, v_2) with $m_2 > v_2$. To show that FSD is violated, i.e., that the existence of the hypothesized indifference curve is inconsistent with FSD, it is only required to solve the system:

$$a + np_1 = 0$$
$$a + np_2 = m_2$$
$$np_1(1 - p_1) = 0$$
$$np_2(1 - p_2) = v_2$$

Taking $p_1 = 0$, $a = 0$ implies the solution: $p_2 = 1 - (v_2/m_2)$, $n = m_2/p_2$. But, the second prospect, $B(n, p_2)$, clearly dominates by FSD the first prospect, $B(n, 0)$, since $p_2 > 0$. Similar counter-examples can be constructed for other cases to show that *assuming that preferences among prospects depend only on their means and variances is inconsistent with FSD* (Borch, 1969). Only if the prospects considered are restricted, e.g., to normal distributions or to other location-scale families, does mean-variance preference theory give coherent results. In these cases it agrees with EU theory and respects the FSD partial ordering of prospects (Quiggin, 1993, Chapter 3.)

Many attempts have been made to define measures of risk based on moments of distributions. Any such attempt is bound to violate FSD or other axioms leading to EU theory, however. It can be shown that for *any* preference relation based only on the moments (mean, variance, skewness, kurtosis, etc.) of random variables, there are many pairs of random variables in which EU prescribes the opposite of the moment-based preference relation. This is true for a broad class of SAUT utility functions (Brockett and Kahane, 1992). Therefore, general-purpose quantitative measures of risk that are consistent with EU theory cannot be based only on means, variances, or other moments of the distributions corresponding to prospects.

Example: Risk-Value Model Is Consistent with FSD

Quantitative definitions of risk in SAUT settings can be obtained from EU theory, although they cannot be expressed in terms of the variance or other moments alone. To this end, Jia and Dyer (1996) introduced a *risk-value model* of the form:

$$V[CE(X)] = V[E(X)] - \phi[E(X)][R(X) - R(0)] \text{ (risk-value model).}$$

It is constructed from (a) An increasing *value function*, V(.); (b) A risk-value *tradeoff factor* $\phi[E(X)] > 0$ that may depend on E(X); and (c) A *standard measure of risk* of the form:

$$R(X) = -E\{u[X - E(X)]\}, \qquad \text{(standard measure of risk)}$$

where u is an NM utility function. Given these three elements, what a d.m. needs to know in order to choose among prospects is just their means and their risks, E(X) and R(X).

This risk-value model is implied by the Risk Independence axiom and the assumptions of EU theory. Moreover, V(.) and R(.) are unique up to choice of origin and scale and ϕ is unique up to choice of scale. If EU theory holds, then $V(x) = u(x)$ and $\phi(x) = u''(x)/u''(0)$. In this case, the model becomes:

$$EU(X) = u[E(X)] - \phi[E(X)][R(X) - R(0)] \text{ (EU risk-value model).}$$

Additional assumptions (e.g., risk aversion, increasing smooth utility function) imply that u(.) must be linear-plus-exponential or exponential. Explicit formulas for R(X) can then be developed for various parametric families of CDFs (Jia and Dyer, 1996). Thus, risk-value theory provides a satisfactory approach, consistent with FSD and EU theory, to many of the issues that mean-variance theory was formerly used for, such as providing explicit, interpretable formulas for risk.

Risk-value theory can be applied even if EU theory is not assumed. The components V(.), R(.), and ϕ(.) can then be modeled independently. ϕ(.) represents some important aspects of perceived risk, while R(.) continues to represent a standardized measure of risk. The resulting framework is flexible enough to encompass some fairly rich descriptive theories, including some that violate FSD (Dyer and Jia, 1997; Inman *et al.*, 1997). It can also be used to explain many violations of EU observed in practice and in experimental settings.

Example: Some Experimentally Observed Choices Violate FSD

Normative decision theories, including several that imply FSD, prescribe how decision-makers *should* behave in order to be consistent with principles which, upon reflection, they wish to adopt. Such theories are potentially valuable guides to action precisely because real decision-makers do not always behave in accord with their prescriptions in the absence of formal analysis. For example, most d.m.s agree (perhaps after reflection) that they want their decisions to satisfy the FSD axiom, meaning that they would not want to choose prospect Y over prospect X if X FSD Y. Yet, in practice, d.m.s do sometimes inadvertently choose dominated prospects. Formal decision analysis can help to reveal and thus prevent such mistakes.

Psychologists studying individual choice behavior under uncertainty have demonstrated consistent violations of the FSD axiom in laboratory experiments. One such experiment, reported in the pioneering work of Tversky and Kahneman (1981), is as follows. Suppose that you must choose one A-lottery (A1 or A2) *and* one B-lottery (B1 or B2) from the following two pairs:

A1: Gain $240 for sure. *vs.* A2: 25% chance to gain $1,000 (else, $0)

and

B1: Lose $750 for sure *vs.* B2: 75% chance to lose $1,000 (else, $0)

In other words, you must choose exactly one of the four pairs (A1, B1), (A1, B2), (A2, B1), (A2, B2). *What do you choose?* Do you choose A1 or A2, and do you choose B1 or B2? When this question is asked to many subjects, it turns out that many prefer (A1, B2) to (A2, B1). However, this preference violates FSD, and is thus "irrational" if FSD is accepted as a principle of rational decision-making. For, the combination (A1, B2) amounts to a prospect that offers a loss of -$760 (= $240 - $1000) with probability 0.75 and a gain of $240 otherwise; while the prospect (A2, B1) offers a loss of -$750 (= -$750 + $0) with probability 0.75 and a gain of $250 (= $-750 + $1000) otherwise:

(A1, B2) = 75% chance of losing -$760, else win $240

(A2, B1) = 75% chance of losing -$750, else win $250.

Clearly, (A2, B1) dominates (A1, B2) by FSD. Thus, all subjects who choose (A1, B2) over (A2, B1) violate FSD.

Tversky and Kahneman (1981) suggest why framing this concurrent choice problem as two separate choices leads many decision makers to choose a dominated alternative. Separating the components of the decision this way leads to different reference points and choices than integrating them into a choice among alternative outcome probability distributions. Such examples occur commonly in real decision-making by doctors, lawyers, engineers, and others involved in risk management decision-making (Piatelli-Palmarini, 1994). These consistent patterns of irrational (e.g., FSD-violating) choices in experimental and real settings help to motivate formal decision analysis as an aid to help avoid choosing dominated prospects.

Birnbaum and Navarette (1998) present additional examples of systematic violations of FSD. For example, they report that most people choose A = ($6, .05; $91, .03, $99; .92) over B = ($6, .02; $8, .03; $99, .95). In other words, they choose the prospect corresponding to column A in the following table to the prospect corresponding to column B, thus violating FSD.

Data leading to FDS violations

	A	B
Pr $6	.05	0.02
Pr $8		0.03
Pr $91	.03	
Pr $99	.92	.95

Arraying the payoff information in a table of CDFs (or complementary CDFs, i.e., risk profiles) such as the following one would presumably make the FSD relation more obvious and perhaps help to reduce such violations. However, no descriptive theory that implies FSD, such as RDEU or CPT, can explain such systematic violations of FSD in actual choices.

Same data arrayed in cumulative format:

	A	B
Pr at least $6	1	1
Pr at least $8	.95	.98
Pr at least $91	.95	.95
Pr at least $99	.92	.95

Example: Prospect Theory Violates FSD

The original version of Prospect Theory (Kahneman and Tversky, 1979), which preceded Cumulative Prospect Theory (CPT), was intended to offer a psychologically realistic description of decision-making under risk. Its prominent features included:

- An *Editing phase,* consisting of stages of *coding* (which establishes reference point for gains and losses), *combination* (pools identical outcomes), *segregation* of riskless component (e.g., (300, p, 200) is segregated as the sum of the riskless 200 and the lottery (100, p, 0)), and *cancellation* of common outcomes.
- An *Evaluation phase* in which each outcome probability p is assigned a *decision weight* $\pi(p)$. (Typically, $\pi(p) + \pi(1 - p) < 1$ and low probabilities are over-weighted compared to their true values.) The evaluation phase also assigns a *value number* v(x) to each x (measured as a deviation around the reference point). The value function v(x) is typically concave (risk-averse) for gains, convex (risk-seeking) for losses, and steeper for gains than for losses.

Kahneman and Tversky showed that Prospect Theory could account for a variety of ways in which real decision-makers violate the prescriptions of EU theory in experiments. However, it also implied that some decisions would necessarily violate FSD. One of the main motivations for Cumulative Prospect Theory was to retain the descriptive realism of Prospect Theory (including its decision weights and value functions) while satisfying the normative principle of FSD.

Example: Some Theories of Disappointment and Regret Violate FSD

Several descriptive theories of decision-making and risk perception have been developed that help to explain and predict violations of FSD, as well as other observed violations of EU theory. Most involve some form of aversion to *disappointment* in the consequence received and/or aversion to *regret* for the action chosen if it turns out that another actions would have yielded a preferred consequence. Some *state-dependent utility theories,* which allow the utility function for consequences to depend on the state of the world when the consequence is received, also lead to violations of FSD.

The following modification of Example 2.1 captures the qualitative intuition behind several theories that allow for violations of FSD. Again, there are six equally probable states, S_1 to S_6, and two possible acts, a_1 and a_2. Now, however, a_1 FSD a_2.

state:	S_1	S_2	S_3	S_4	S_5	S_6
Pr(s):	1/6	1/6	1/6	1/6	1/6	1/6
$c(a_1, s)$:	1.001	2	3	4	5	6
$c(a_2, s)$:	2	3	4	5	6	1

Yet, a d.m. who preferred a_2 to a_1 when they had identical CDFs (thus violating Axiom 1 for SAUT EU theory, i.e., that only CDFs matter) might still have this preference even though a_1 now dominates a_2 by FSD. In both cases, the rationale is that a_2 has a 5/6 chance of yielding a higher payoff (if states 1-5 occur) than the one that a_1 would have yielded for the same state. Aversion to disappointment or to regret if the lower payoff is received when a higher payoff could have been obtained by choosing a_1 might prompt one to choose a_2 over a_1, even though a_1 has greater expected utility, expected value, and FSD ordering. While choices made to avoid disappointment or regret rather than to maximize expected utility often have some psychological plausibility, people who make them will consistently tend to receive less-preferred outcomes than people who ignore state-by-state comparisons and simply choose the prospects with the dominant CDFs. In this sense, such behavior is mal-adaptive or irrational.

Not all models of disappointment and regret aversion violate FSD. For example, suppose that the psychological value of a risky prospect depends only on its expected value and on the expected "disappointment" or "elation" from receiving its outcome. (Here, the difference between the prospect's expected and actual values is defined as the "disappointment" in the outcome if this difference is positive and as the "elation" if it is negative.) A risk-value model that is algebraically equivalent to such a disappointment model is:

$$CE(X) = E(X) - (k/2)E[|X - E(X)|]. \quad \text{(Disappointment risk-value model)}$$

Here, k is the difference in the value weights for disappointment and elation and $E[|X - E(X)|]$ is the absolute deviation of X (Inman, 1997; Jia, Dyer and

Butler, 2001, building in part on work of Bell, 1985). A d.m. with preferences described by this model will be risk-averse if $k > 0$ and will satisfy FSD if $|k| < 1$. Thus, FSD is consistent with, but not implied by, the disappointment risk-value model. Moreover, the model can explain some well-known violations of EU theory, including the common ratio effect and the Allais paradox (Jia *et al.*, 2001).

Other specific models have been proposed to describe disappointment-averse decision-making. A well-known one is the Loomes and Sugden (1986) model:

$$CE(X) = E(X) - E[D(X - E(X))] \qquad \text{(Loomes-Sugden, 1986 model)}$$

Here, D is a smooth (continuously differentiable) function, convex for $x > E(X)$ and concave for $x < E(X)$, with the skew-symmetry property $D(x) = -D(-x)$. Also, the preceding risk-value models make the *expected value* of a prospect the reference point or aspiration level with respect to which elation or disappointment are calculated once the actual outcome is received. Other models of disappointment make the *expected utility* or the *certainty equivalent* the reference point (Loomes and Sugden, 1986; Gul, 1991), and other choices for the aspiration level, such as the *status quo* reference level used in CPT, can be envisioned.

Disappointment models tend to be far more descriptively realistic than normative theories such as EU. They often involve estimating only one or a few additional parameters. For example, the axiomatic model of Gul, 1991, represents preferences as a weighted sum of elation and disappointment expected utilities. A single weight, $\beta \geq 0$, interpreted as a "coefficient of disappointment aversion" is its sole parameter. The usual representation of the (disappointment-adjusted) expected utility of a risky prospect for a disappointment-averse individual has the form:

$$E[U(X)] - \beta E\{U[CE(X)] - U(X) \mid X < CE(X)\} \qquad \text{(Gul, 1991 model)}$$

where U is a conventional NM utility function (increasing and concave). In effect, the individual attaches some extra disutility to outcomes that are less than a prospect's certainty equivalent (Palacios-Huerta, 1999). This model can explain the Allais paradox and some other observed violations of EU theory, contains EU theory as a special case (when $\beta = 0$), and is the simplest model with these features, as explained and proved by Gul, 1991.

Despite the greater descriptive realism of disappointment and regret models, many of them allow the possibility of violating FSD (as, indeed, real choices sometimes do) for some values of their parameters. Since FSD is perhaps the most compelling of all normative axioms for choosing among risky prospects, such disappointment-aversion and regret-aversion theories are truly descriptive, rather than prescriptive.

5.2 Extensions of FSD

The theory and practice of comparing distributions by FSD have been extended in several directions. Among the most important are the following.

- *Algorithms to calculate FSD.* Determining whether F FSD G can be computationally challenging for arbitrary CDFs F and G, although it is easy for special cases (e.g., if F and G are CDFs of normally distributed prospects). Numerical algorithms for calculating FSD relations have been developed in quantitative financial analysis and operations research. For example, Fishburn and LaValle (195) present efficient algorithms for the case in which all outcomes lie in a grid of evenly spaced points (e.g., for monetary outcomes rounded to the nearest cent), which is realistic for many practical applications.
- *Multivariate stochastic dominance relations* have been defined and characterized (Shaked and Shathikumar, 1994, Chapter 4). These are discussed later in the section on multiattribute utility theory.
- *FSD for prospects based on perceived risks.* Recall that prospect theory and cumulative prospect theory (CPT) both postulate that subjects choose among prospects not by considering the actual probabilities of different outcomes, but by considering nonlinearly transformed probabilities. A transformation, $\pi(p)$, is applied to probabilities, p, before expected values are calculated. $\pi(p)$ is monotonic and non-decreasing, with $\pi(0) = 0$ and $\pi(1) = 1$. Because of this transformation, subjects are sometimes described as "over-weighting" or "under-weighting" objective probabilities as they form preferences among prospects. Levy and Wiener (1998) prove that FSD is preserved under arbitrary monotonic nondecreasing transformations, so that if two CDFs can be compared by FSD based on their true probabilities, then they can also be compared by FSD (and the comparison will be the same) using the transformed probabilities.
- *FSD and related orderings for temporal random variables.* Suppose that prospects X and Y are two random variables describing the waiting times until occurrence of an undesirable outcome (e.g., death for a person, failure of a system) if two different actions are taken. Thus, the single attribute in question is the time until the risk event occurs. The definition for X FSD Y, i.e., that X is stochastically larger than Y if $\Pr(X > a) \geq \Pr(Y > a)$ for all a, still holds with time as the attribute of interest. Let X_t denote the additional life time from t on, given survival until time t. Then X has an *increasing failure rate (IFR)*, i.e., its hazard rate function

$h_X(t)$ is increasing in t, if and only if X_t is stochastically decreasing in t. X has a *decreasing failure rate* (DFR), i.e., $h_X(t)$ decreases with t, if and only if X_t is stochastically increasing in t. Moreover, $h_X(t) \geq h_Y(t)$ for all $t \geq 0$ if and only if Y_t FSD X_t for all $t \geq 0$. This relation defines the *hazard rate ordering* between random variables X and Y. It can be used to prove that if m(t) is the renewal function (i.e., mean number of renewals by time t) of a renewal process with a DFR interarrival time denoted by random variable T, then:

$$t/E(T) \leq m(t) \leq [t/E(T)] + \{E(T^2)/2[E(T)]^2\} - 1$$

For these and other results, including a *likelihood ratio ordering* that implies the hazard rate ordering, see Ross (1996, Chapter 9) or Shaked and Shanthikumar (1994).

- *Indirect stochastic dominance (ISD) and preference for flexibility* (Nehring, 1999). Suppose that an act taken today induces a probability measure on subset of acts (i.e., "opportunity sets") to choose from tomorrow. Then, even if preferences among tomorrow's acts are uncertain (e.g., because they depend on information not yet received), a new relation of *indirect stochastic dominance* (ISD) may still be defined among today's acts. If *f* and g are two acts available today, then *f ISD g* if and only if, for *any* weak preference ordering of tomorrow's acts, the opportunity set from *f* is at least as likely as the opportunity set from g to allow at least any given level of preference to be achieved. (If the mapping from today's acts to tomorrow's opportunity sets is deterministic, then *f* induces a larger opportunty set than *g*, i.e., ISD corresponds to set inclusion in this special case.) Under Savage-like axioms (of Weak Order, Sure-Thing Principle, State-Independent Preferences, Comparative Probability, and an Archimedean axiom), non-trivial preferences for today's acts satisfy ISD if and only if there is a subjective probability measure and a utility function such that *f* is at least as preferred as *g* if and only if $EIU(f) \geq EIU(g)$. Here, *EIU(f)* denotes the *expected indirect utility* of *f*, i.e., the expected utility if *f* is chosen today and then the most-preferred act in the opportunity set resulting from *f* is chosen tomorrow.
- *Comparing stochastic processes.* The idea of stochastic ordering by FSD has been extended to allow certain types of stochastic processes to be compared. Important examples include:
  - SIR epidemic models (Shaked and Shanthikumar, 1994)
  - Failure counting processes induced by alternative maintenance policies in reliability systems (*ibid*, Chapter 15)
  - Failure counting processes induced by alternative resource allocations in complex reliability systems (*ibid*, Chapter 16)

  - Renewal processes and renewal-reward processes (Ross, 1996)
  - Some Markov decision processes (*ibid*, Chapter 10)
  - Multi-arm bandit decision resource allocation processes (Magnac and Robin, 1999). A dynamic version of FSD that allows for learning (by Bayesian conditioning) about the characteristics of multiple competing projects is shown to imply that undominated projects are also optimal (in the sense of Gittins indices for stochastic dynamic programming) under certain conditions.
- *Hypothesis testing about FSD.* Statistical tests for comparing uncertain distributions in terms of FSD based on samples from them are discussed in Chapter 8 of Shaked and Shanthikumar, 1994, Statistical inference for stochastic ordering, by B. Bhattacharya and R.L. Dykstra. These authors show how to greatly improve on traditional tests such as the Kolmogorov-Smirnov test.

In summary, the concept that "more is better" for a single-attribute consequence set leads to useful partial orderings for (a) Probability distributions over consequences; and (b) Alternative stochastic processes that generate consequences over time. FSD and its extensions can be used to partially order alternative risk management actions that are represented by probability distributions or stochastic processes over consequences. Moreover, the resulting partial orderings do not require detailed knowledge about preferences (beyond more being preferred to less) and are even somewhat robust to psychological distortions of objective probabilities.

6. HIGHER-ORDER STOCHASTIC DOMINANCE AND RISK DEFINITIONS

A possible definition of risk for single-attribute prospects is as follows: The "risk" of a prospect is the information about it that a rational d.m. must have, in addition to its expected value (and his or her own risk attitude), in order to calculate its certainty equivalent. As discussed previously, this definition leads to linear-plus-exponential NM utility functions in the EU framework and to the standard measure of risk in risk-value models. An alternative, implicit definition of risk is: The "risk" of a prospect is reflected in the risk premium that a risk-averse decision-maker would be willing to pay to replace it by its expected value. Any risk-averse d.m. would sacrifice some amount of a desirable attribute (the risk premium) to replace an uncertain prospect with its expected value. Doing so may be interpreted as paying some of the desirable attribute to reduce "risk", and hence may be a basis for figuring out what "risk" means. This attempted approach to defining risk is not circular – although it certainly comes close to defining risk as "that which risk-averse people avoid" – because risk-aversion has already been defined in terms of concavity of the NM utility function. The consequences of this approach are explored next.

First-order stochastic dominance (FSD) can be used by any d.m. to partially order singe-attribute prospects, provided that preferences for consequences increase (or decrease) with the amount of the attribute. For d.m.s whose preferences satisfy additional conditions, such as risk-aversion, additional stochastic dominance relations can be defined. These enable additional comparisons among prospects. *Second-order stochastic dominance* (SSD), *third-order stochastic dominance* (TSD) and higher-order dominance relations for comparing CDFs may be defined as follows. (Notice that the definition refers only to the CDFs, and not to their underlying random variables.)

Higher-Order Stochastic Dominance Relations

Let F and G be two right-continuous CDFs. Define the *n-th order difference relation* between F and G for n = 2, 3, ... recursively, as follows:

$D^1 = G - F$

$D^{n+1}(t) = \int_{[-\infty,\, t]} D^n(s)ds$ for n = 1, 2, ...

Then the *n-th order strict stochastic dominance relation* $>_n$ between G and F is defined as follows:

$G >_n F$ if and only if $G \neq F$ and $D^n(t) \leq 0$ for all t.

First-order, second-order and third-order strict stochastic dominance (abbreviated FSD, SSD and TSD) are defined as the special cases of $>_n$ for n = 1, n = 2, and n = 3, respectively.

THEOREM (e.g., Shaked and Shanthikumar, 1994): For all right-continuous CDFs F and G and for any positive integer n = 1, 2, ..., the following two conditions are equivalent:

a) $G >_n F$

b) $E_G(U) > E_F(U)$ *for all functions U with continuous derivatives of orders 1 through n that alternate in sign, with* $U'(x) > 0$ *for all x in the support of F or G, i.e., all x with* $F(x) > 0$ *or* $G(x) > 0$.

Here, $E_F(U)$ denotes the expected value of the function U(x) with respect to F, i.e., it is defined as follows: $E_F(U) = \int_{(-\infty,\, \infty]} U(x)dF(x)$. Recall that, for absolutely continuous CDFs, $E_F(U) = \int_{(-\infty,\, \infty]} U(x)f(x)dx$, where $f(x) = F'(x)$ is the PDF corresponding to F. Condition (b) can also be stated as follows:

EU(X) < EU(Y)

For *all* increasing, concave (if n > 1), sufficiently smooth (n-times continuously differentiable) functions U with derivatives of alternating signs on the support of F or G, where X and Y are random variables having F and G for their respective CDFs.

Each order of stochastic dominance implies those that follow it; thus:

(G FSD F) → (G SSD F) → (G TSD F) → ...

where "(G TSD F)", for example should be read as "G dominates F by strict third-order stochastic dominance" and arrows denote implication. Some pairs of CDFs that cannot be compared by lower-order stochastic dominance relations can be compared by higher-order stochastic dominance relations.

Of these higher-order stochastic dominance relations, SSD best illuminates the implications of risk-aversion. The key results are as follows.

Characterizations of increasing risk (Rothschild and Stiglitz, 1970)

For any two single-attribute prospects X (with CDF F) and Y (with CDF G), the following conditions are equivalent:

1. All non-satiating, risk-averse expected-utility d.m.s prefer X to Y. (In this sense, Y is *riskier than* X.) That is EU(X) $\geq$ EU(Y) for all increasing, concave functions U, i.e., for all utility functions with $U'(x) \geq 0$ and $U''(x) \leq 0\ \forall x$.
2. F SSD G.

If E(X) = E(Y), then (1) and (2) are equivalent to each of the following:

3. Y = X + Z, where Z is a random variable such that E(Z | X) = 0 and Y is equal in distribution to X + Z. ("X is equal to Y plus noise".)
4. Y is obtained from X by a finite sequence of simple mean-preserving spreads (or as the limit of an infinite sequence of simple mean-preserving spreads). A simple *mean-preserving spread* (MPS) of a random variable moves probability mass out of an interval to both sides, so that the CDF of the original random variable crosses the CDF of the new ("spread") random variable once from below.

Reference: Quiggin, 1993, Section 6.6; Gotoh and Konno, 1999.

Intuitively, one prospect is riskier than another with the same expected value if its probability distribution is more "spread out" around the mean. The idea of a mean-preserving spread makes this intuition more precise. If both prospects are normally distributed with equal means, then the one with the smaller variance will be preferred by all risk-averse decision-makers, based on the SSD criterion. In this sense, the one with the smaller variance may be defined as less risky than the one with the greater variance. However, for non-normal random variables, variance is not necessarily an adequate measure of risk.

Example: SSD and Risk-Aversion in a Normal Form Decision Problem

A d.m. must choose between two acts, a_1 and a_2. The consequence of choosing act j if state s occurs, denoted by $c(a_j, s)$ for j = 1, 2, is shown in the following table.

State s:	s1	s2	s3	s4
Pr(s)	0.1	0.2	0.3	0.4
$c(a_1, s)$	\$1	\$2	\$3	\$4
$c(a_2, s)$	\$3	\$3	\$3	\$3

Specifically, if act 2 is chosen, then \$3 is received with certainty, while if act 1 is chosen, then the possible payoffs range from \$1 (with probability 0.1) to \$4 (with probability 0.4), depending on the state. *Which act should the d.m. choose?*

Solution: In general, the answer depends on the d.m.'s attitude toward risk. She must decide whether she prefers \$3 with certainty to the uncertain prospect (potentially more profitable but also potentially less profitable than \$3) corresponding to act 1. Clearly, neither act dominates the other by FSD, so the d.m. cannot choose between the two acts based on FSD alone. However, higher-order stochastic dominance relations (specifically, SSD) can help.

A d.m. is said to be *risk-averse for gains* if she prefers to receive the expected value of any non-negative random variable to receiving a realization of the random variable (assuming that she prefers more of the outcome to less.) If G and F are any two CDFs over non-negative returns (i.e., CDFs with non-negative supports) having the same mean, then *a d.m. who is risk-averse for gains will always prefer G to F if G SSD F.* This provides enough information to solve the problem, if we assume that the d.m. is risk-averse for gains (as most d.m.s are). The normal-form information can be re-cast as the following dominance-calculation table.

c	\$1	\$2	\$3	\$4
F(c)	0.1	0.3	0.6	1.0
G(c)	0	0	1	1
D^1	-0.1	-0.3	0.4	0
D^2	-0.1	-0.4	0	0

Here, F and G are the consequence CDFs representing acts 1 and 2, respectively. The first- and second-order difference relations between them are computed in the two last rows. Although FSD is not satisfied (since the D^1 values include a positive number, 0.4), SSD is satisfied: G SSD F. Hence, any d.m. who is risk-averse for gains will prefer G to F and should choose act 2 ($3 with certainty) instead of act 1 (the risky prospect).

Discussion: In this example, SSD held although FSD did not. Hence, a conclusion about which act is preferred could be drawn for any d.m. known to be risk-averse for gains, but not for a d.m. whose preferences are only known to be increasing in the outcome (i.e., dollar payoff). More generally, if SSD as well as FSD fail to hold, then TSD and higher-order stochastic dominance criteria can be examined, but comparisons based on such criteria are only relevant for d.m.s with increasingly constrained preferences and risk attitudes. Thus, FSD is relevant for all d.m.s with monotonic preferences for the outcome; SSD is relevant for the subset of d.m.'s with risk-averse monotonic preferences; TSD is relevant for the more restricted subset of d.m.'s satisfying "decreasing absolute risk aversion" (DARA), and so forth.

Example: Mean-Variance Analysis and SSD for Normal PDFs

If all prospects are represented by normal distributions over a desirable attribute, then all risk-averse decision-makers will agree with the partial ordering of prospects implied by mean-variance dominance. In other words, prospects with higher (or equal) means and smaller variances are preferred by all risk-averse decision-makers. Thus, SSD can be reduced to comparing means and variances of distributions, for normally distributed prospects. However this result does *not* generalize to arbitrary distributions, since no moment-based definitions of risk can be consistent with EU theory in general (Brockett and Kahane, 1992).

Example: Comparing Renewal and Poisson Processes by SSD

Suppose that $\{N_F(t), t \geq 0\}$ is a renewal process with interarrival time T having distribution F. Suppose that F is NBUE (i.e., new better than used in expectation, meaning that $E(T - a \mid T > a) \leq E(T)$ for all $a \geq 0$), with mean μ. Then $N_F(t)$ SSD $N(t)$ for all t, where $\{N(t), t \geq 0\}$ is the Poisson process with the same mean interarrival time, μ. Conversely, if F is NWUE (new worse than used in expectation), then $N(t)$ SSD $N_F(t)$ for all t (Ross, 1996, Chapter 9.) This suggests that the intensities of Poisson processes can be used to define numerical "equivalents" for renewal processes in much the same way that canonical lotteries are used to define the utilities of CDFs.

Example: Bernoulli Variables Maximize Risk for a Given Mean

Suppose that X is a random variable taking values of the unit interval [0, 1]. For example, X may be the uncertain probability of an adverse event. Then X SSD Y, where Y is a Bernoulli random variable with the same mean as X, i.e., Y has value 1 with probability E(X) and 0 otherwise (*ibid*, p. 455). Thus, all risk-averse decision makers will prefer a non-Bernoulli distribution on [0, 1] to a Bernoulli distribution with the same expected value. Intuitively, the reason is clear: a Bernoulli random variable is a mean-preserving spread (obtained by moving probability mass to the two extremes of the interval [0, 1]) with respect to any non-Bernoulli random variable having the same mean.

6.1 Extensions of SSD

The use of SSD to compare risks of prospects has been extended in several important ways, as follows.

- *Generalized smooth utility theory.* Machina (1982) showed that the relation between risk aversion and SSD extends to generalized smooth utility theory. Every d.m. for whom all local utility functions U(x ; F) are concave (i.e., risk-averse) functions of x will prefer F to G if F SSD G, and conversely (Quiggin, 1993.)
- *Extension to RDEU.* Quiggin (1993, Section 6.6) extends the relation between risk aversion and SSD to Rank-Dependent Expected Utility (RDEU) theory. He showed that all RDEU risk-averse decision-makers prefer prospect X to prospect Y if and only if Y is derived from X by a special type of mean-preserving spread (a "monotone spread") that can be interpreted as adding a risky variable to X or as shifting probability mass from the center of a probability distribution toward its tails.
- *Extension to utility functions that are not globally risk-averse.* Muller (1998), building on important papers by Landsberger and Meilijson published between 1990 and 1994, generalizes the idea of a mean-preserving spread to a *mean-preserving dispersal (MPD)*. This formalizes the single-crossing property for CDFs. The set of all utility functions that can be compared by MPD includes utility functions that are consistent with the observation that some decision-makers buy both insurance and lottery tickets. They also include utility functions that are risk-averse only to losses and utility functions that are averse only to sufficiently large risks.
- *Extension to S-shaped utility functions.* For decision-makers who are risk seeking for losses and risk averse for gains (i.e., who have "S-shaped" utility functions), the concept of SSD can be generalized as follows. F dominates G for all d.m.s with S-shaped utility functions if and only if

  $$\int_{[y,\, x]}[G(s) - F(s)]ds \geq 0 \text{ for all pairs } x > 0 \text{ and } y < 0$$

  This dominance relation is preserved under transformations $\pi(p)$ that are concave for gains and convex for losses (Levy and Wiener, 1998).
- *Mean-lower semi-standard deviation as measures of risk.* Ogryczak and Ruszczynski (1999) show that prospects that are undominated when compared based on their *means and mean-lower semi-standard deviations* are also undominated by SSD. This provides a computationally practical way of identifying many SSD-undominated prospects and motivates the use of mean-lower semi-standard deviations as measures of risk. (The mean-lower semi-standard deviation measure of risk is defined as: $\{\int_{[-\infty,\, E(X)]}[x - E(X)]^2 dF_X(x)\}^{1/2}$.)

- *Risk comparisons using TSD.* Gotoh and Konno (2000) extend the work of Ogryczak and Ruszczynski to provide computationally efficient methods for identifying prospects that are undominated by TSD. They prove that prospects that are undominated by an easily computed mean-lower semi-skewness criterion are also undominated by TSD.
- *SSD for prospects based on perceived risks.* Levy and Wiener (1998) prove that SSD is preserved under monotonic concave transformations of the outcome probabilities. Thus, for example, prospects that can be compared by SSD using true probabilities can also be compared (and the comparison will be the same) using the subjectively transformed probabilities or "decision weights" $\pi(p)$ of Prospect Theory and Cumulative Prospect Theory.
- *Comparing stochastic processes by SSD.* Some stochastic processes can be compared by SSD. For example, in a branching process, the variability of the size of the *n*th generation increases if the number of offspring per individual is more variable. For the same mean number of offspring per individual, increasing variability in the number of offspring increases the likelihood of extinction at each generation. Here, the relation "more variable" between random variables with the same mean is equivalent to SSD (Ross, 1996).
- *Calculating dominance relations:* Fishburn and LaValle (1995) give numerical algorithms for calculating SSD and versions of higher-order stochastic dominance relations when consequences are discretized to lie on a one-dimensional grid.

7. CONCLUSIONS

The concepts and techniques reviewed in this chapter provide ways to compare risk profiles and to eliminate those that should not be preferred. Many of these comparisons can be made without extensive information about subjective preferences and risk attitudes. For example, exponential utility functions reduce all subjective aspects of decision-making to a single parameter, interpretable as a coefficient reflecting degree of risk aversion. Given the value of this parameter for a d.m., the certainty equivalent and risk premium can be calculated for any prospect, i.e., for any risk profile or CDF defined over the single-attribute consequence set. Various measures of stochastic dominance are even less restrictive, enabling single-attribute prospects (and some stochastic processes) to be compared based on limited qualitative properties, such as that less of an adverse consequence is preferred to more, that higher probabilities of preferred consequences should be preferred, or that the decision-maker has risk-averse preferences. Such comparisons also suggest constructive definitions of "risk", insofar as risk can be defined as something that must be known, in addition to expected values, to allow prospects to be compared and/or to allow their certainty equivalents to be calculated.

This chapter has also reviewed findings from psychology and laboratory investigations of how real people do make decisions under risk. A striking finding is that peoples' choices often violate the normative principles and axioms that justify EU theory and its generalizations. This discrepancy provides one of the motivations for normative theory. The theory is needed because, without it, people tend to make "mistakes" (e.g., by choosing options leading to dominated probability distributions of consequences). The theory clarifies exactly why such choices can and should be considered to be mistakes, e.g., because they make preferred outcomes less probable. However, some of the most fundamental principles of decision-making, (including FSD) are respected by descriptively realistic theories such as RDEU and CPT, even when these violate the more stringent prescriptions of EU theory.

Single-attribute utility theory may be taken as a paradigmatic example of a successful normative theory for choice among risky prospects. It is well-enough developed so that its conceptual foundations can be crisply stated via a few clearly interpretable axioms. Different sets of axioms that lead to the same representations of preferences for consequences and to the same prescriptions for actions have been articulated. The implications of different axiom sets and their relaxations for are well understood. Moreover, the ways in which real behaviors deviate from the prescriptions of the theory have been clarified by several decades of ingenious experiments. Useful descriptive theories such as CPT and RDEU have been developed that specialize to normative models under special conditions (e.g., when the probability-weighting function is specialized to $\pi(p) = p$.)

Despite these successes, the theoretical edifice of SAUT is limited by the fact that it applies only to prospects that can be represented by univariate risk profiles or CDFs. Prospects with more complex representations, such as decision trees with consequences received and/or uncertainties resolved at different times, do not necessarily fit within this framework. Comparison of other stochastic processes also falls outside the scope of SAUT, with the exception of some important special cases (e.g., comparison of Poisson processes.) Consequences described by multiple attributes do not directly fit into the SAUT framework, either.

These limitations are traditionally avoided in applied decision analysis by making simplifying assumptions that bring complex prospects within the scope of SAUT. Time-structured prospects such as decision trees or other stochastic processes are commonly replaced with probability distributions (represented by CDFS or risk profiles) for *terminal asset position,* i.e., the total wealth as of a certain date by which the consequences of all decisions are assumed to have worked themselves out. Multi-attribute consequences are often replaced by single-attribute quantities – e.g., equivalent monetary values or more abstract value numbers constructed with the help of a value function – by "pricing out" the non-monetary attributes.

While such simplifications extend the reach of SAUT and facilitate economic risk-cost-benefit analysis and decision analysis for prospects with financial consequences, they are less applicable to hazards involving changes in health status and life duration. The next chapter therefore considers how the paradigm of normative decision theory can be extended from prospects represented by risk profiles to prospects in which time and value trade-offs among different consequence dimensions play essential roles.

Chapter 7

Multiattribute, Multi-Person, and Multi-Period Risks

1. INTRODUCTION

Human health hazards, ranging from pollutants in air or water to microbes in food, usually threaten the health of many people simultaneously. Typically, different people are unequally exposed to the health hazards arising from a decision affecting exposures to such hazards, raising questions about fairness and equity in the distribution of risks from the decision – or, more generally, in the distribution of its costs, risks, and benefits. Most hazards can affect several aspects of health, such as the frequency, duration, type, and intensity or severity of exposure-related illnesses. Different people, including people of different ages, may value these aspects of health effects differently.

Even for one person, the value placed on probable health effects occurring at one age, as assessed from the perspective of a different (typically, younger) age, may reflect individual time preferences and attitudes toward delayed risks. If some of those potentially affected by a decision live in the future, then uncertainty about their future preferences and risk attitudes may complicate the ethics of current decision-making. Each of these complexities – multiple health attributes, multiple people affected, and delays in resolution of uncertainties and/or in receipt of consequences from current decisions – suggests possible extensions of the basic EU framework for rational decision-making.

This chapter examines approaches for extending EU theory to multiple attributes, stakeholders, time periods, and generations. It also includes an approach to choosing among Markov or semi-Markov transition processes with multiple health states. None of these extended theories is yet as well-developed as single-attribute utility theory, but a great deal of insight has been developed in each area based on theoretical analyses and experimental evidence on how people think about and make decisions for prospects that involve these complexities. The partial progress made so far highlights challenges for advancing operations research methodology to support health risk management decisions.

2. MULTIATTRIBUTE UTILITY THEORY (MAUT)

Many prospects have outcomes that are most naturally described by several attributes. For example, the outcome of a risky medical procedure may be extra years of life gained *and* the quality of those years, leading to a two-attribute summary of potential consequences. A risky financial prospect that costs \$c and that offers a probability p of winning \$x (and a probability 1 – p of yielding \$0) may be represented as a three-attribute vector, (c, p, x). A patient's health status at any time may be summarized by several attributes, such as indicators of sensation, mobility, emotion, cognition, self-care, pain, and fertility (Feeny et al., 1995; Torrance et al., 1996).

Similarly, decisions about public projects such as construction of a power plant or incinerator may result in costs, risks and benefits to members of a population. These cost-risk-benefit combinations may be viewed as multiple attributes for each person affected. If each such combination consists of *m* attributes, then the outcome for a population of *N* people may be considered an *mN*-attribute vector. If other attributes, such as measures of environmental degradation, other non-financial costs, community impacts (both positive and negative), and so forth are needed to describe all the impacts of concern, then they can be included in the attribute vectors used to describe outcomes. In short, describing consequences by multiple attributes provides a great deal of flexibility and opportunity for realism. Keeney and Raiffa (1976) offer practical advice on how to choose a "good" (adequately complete, non-redundant, well-defined, operationally measurable) set of attributes for describing outcomes of projects.

2.1 Basics of Multiattribute Value and Utility Theory

In general, suppose that the consequence set C consists of vectors of attribute levels, with each consequence represented by an n-tuple $(c_1, c_2, \ldots, c_n)$. In other words, C can be decomposed or factored as a Cartesian product of n single-attribute consequence sets:

$C = C_1 \times C_2 \times \ldots \times C_n$ (Multiattribute consequence set).

A primary goal of multiattribute utility theory (MAUT) is to use this mathematical structure of the consequence set to develop simple expressions for consequence utilities.

A technique other than direct elicitation of utilities for n-vectors of consequence attributes is needed. Direct elicitation is undesirable because psychological heuristics, such as considering only one or a few attributes at a time, can lead to unreliable results for directly elicited preferences. Indeed, expressed preferences for multiattribute consequences can be so sensitive to incidental details of information presentation and cognitive processing that psychologists and econometricians sometimes represent them by *random utility* models (Fischer et al., 2000). Different choices among the same deterministic prospects may be made on different trials, and random utility models attempt to describe and predict choice probabilities. Principles for such *probabilistic choice models,* such as stochastic versions of transitivity, have been explored (French, 1988), but lie outside our focus. Rather, we seek to construct deterministic utility functions that represent the idealized preferences that may be hypothesized to underlie actual preference data, with its frequent inconsistencies and apparently random elements. We require an approach that will also overcome the practical difficulty that the number of attribute value combinations for which utilities must be assessed to support useful interpolation of a multivariate utility function may be overwhelming.

Two main routes lead to simple expressions for multiattribute utility functions. One focuses on *multiattribute value functions.* A value function assigns a numerical value to each outcome vector $c = (c_1, c_2, \ldots, c_n)$. The value assigned to c is denoted by v(c). "Value" is then treated as a new, synthetic attribute. A utility function can be developed for value using methods for single-attribute utility function assessment (Keeney and Raiffa, 1976, Chapter 4). The utility of a consequence vector c is just the utility of its value, i.e., u[v(c]. An example of a utility function derived via this route (not normalized to run from 0 to 1) might be:

$$u(c_1, c_2, \ldots, c_n) = 1 - \exp[-k(w_1c_1 + w_2c_2, \ldots + w_nc_n)].$$

Here, the multiattribute value function is additive:

$$v(c_1, c_2, \ldots, c_n) = w_1c_1 + w_2c_2, \ldots + w_nc_n \qquad \text{(Additive value function).}$$

The non-negative attribute *value weights* w_j may be scaled to sum to 1 if all attributes are desirable. The utility function is $u = 1 - e^{-kV}$, where k reflects risk attitude.

The second approach to constructing simple multivariate utility functions requires verifying (or assuming) that various *independence conditions* hold for different attributes or subsets of attributes. These conditions imply specific functional forms for the multivariate utility function, without using value as an intermediate construct. The strategy is one of divide and conquer. The first step is to separate attributes into mutually independent subsets such that preferences for attribute levels within one subset can be assessed without knowing levels of attributes in other subsets. (Thus, all substitute and complement effects hold within, rather than between, subsets.) Next, a component utility function is assessed for each subset. Finally, the components are aggregated to form the total utility function. Table 1 lists some of the most important independence relations among attributes underpinning current multiattribute value and utility theory.

In describing a decision problem, it is often possible to find many different sets of attributes, each of which suffices to describe the outcomes of proposed alternatives.

Table 1: Independence conditions for MAUT utility assessment

- ***Mutual preferential independence*** **(MPI)** Attributes $C_1 \ldots C_n$ are said to be *mutually preferentially independent* for a respondent if his preferences between any two distinct attribute *n*-tuples depend only on the attribute levels that differ between the two bundles and not on the attribute levels that are the same in both bundles (if any). In other words, his marginal rate of substitution between any two attributes does not depend on the levels of the other attributes (Keeney and Raiffa, 1976.)
- ***Mutual utility independence*** (MUI). Attribute C_i is said to be *utility independent* of the remaining attributes for a respondent if her preference ordering for probability distributions over C_i does not depend on the levels of the remaining attributes. Multiple attributes are said to be *mutually utility independent* (MUI) of each other if every subset is utility independent of the complementary subset, meaning that preferences among joint distributions for a subset of attributes do not depend on the levels of attributes not in that subset (Keeney and Raiffa, 1976).
- ***Difference independence*** (Dyer and Sarin 1979). Attribute C_i is said to be difference independent (DI) of the remaining attributes if the preference difference between any two attribute bundles that differ only in the level of C_i does not depend on the common levels of the remaining attributes. Similarly, attribute C_i is *weakly difference independent* (WDI) of the remaining attributes if the *ordering* of preference differences among pairs of attribute bundles that differ only in the level of component C_i depends only on the levels of C_i and not on the common levels of the remaining attributes. (Here C_i denotes both the *i*th attribute and its level. The concept of preference difference was previously defined in terms of whether the change in going from one attribute bundle to another is more or less preferred than the change in going from a third attribute bundle to a fourth.)

- ***Additive independence*** (Fishburn, 1965, cited in Keeney and Raiffa, 1976). Attributes $C_i, \ldots, C_n$ are said to be *additively independent* (AI) for a respondent if his preferences for a probability distribution over attribute levels depend only on the *n* marginal probability distributions of each attribute separately and not on their joint distribution.
- ***Zero-condition*** (Miyamoto et al., 1998). Attribute C_i satisfies the *zero-condition* if it has a level, z_i, such that all n-tuples with $C_i = z_i$ are mutually indifferent. If this level is denoted by 0, then all *n*-tuples (c_1, c_2, …, c_n) with $c_i = 0$ are mutually indifferent. For example, in a two-attribute prospect (x, p) that gives reward x with probability p and 0 otherwise, it is natural to assume that all levels of p are indifferent for x = 0. Then, attribute x satisfies the zero-condition. Similarly, all levels of x may be indifferent if p = 0. Then, p satisfies the zero-condition.
- ***Standard gamble invariance*** (SGI) (Miyamoto et al., 1998). Attribute C_i satisfies SGI if the certainty equivalent of a two-outcome gamble over attribute C_i does not depend on the levels of other attributes, provided that none of them is fixed at its 0 level.

The set that is selected should ideally be *complete,* in that descriptions of outcomes in terms of those attributes alone provide a sufficient basis for choosing among them, and *irredundant* (or minimal), in that different attributes measure different constructs.

Selecting attributes that satisfy the sorts of independence conditions in Table 1 will help meet the latter criterion, and will greatly simplify subsequent analysis and interpretation. Moreover, practical experience suggests that it is usually possible to find such a set of attributes, given sufficient thought. Keeney and Raiffa (1976) discuss techniques for efficiently verifying various independence conditions. Ideal attributes should also be *clearly defined* (e.g., their operational definitions should be known), well-understood, and *direct measures* of the outcomes that people care about, rather than remote causal antecedents (i.e., "proxy attributes") for those outcomes (Fischer, 1989).

The independence conditions yield immediate simplifications for assessing utility functions and relating them to value functions. For example, if MPI and AI hold, and if DI holds for some attribute, then under weak technical assumptions, the value and utility functions must coincide, i.e., v(c) = u(c) (Dyer and Sarin 1979). Under other conditions, u(c) may be obtained from v(c) by the transformation

$$u(c) = [1 - e^{-kv(c)}]/(1 - e^{-k})$$

where k is a coefficient of relative risk aversion that is positive for risk-averse respondents, negative for risk-seeking ones (Currim and Sarin 1984). When u(c) and v(c) are identical, or when u(c) can be simply obtained from v(c), techniques for value function assessment, such as conjoint assessment techniques, can also be used to assess utility functions. Then it is good practice to use both approaches to check on the validity of the results.

The main significance of the independence conditions is that they allow the multivariate NM utility function $u(c_1,\ldots,c_n)$ to be decomposed as a function of n component NM utility functions $u_1(c_1),\ldots,u_n(c_n)$. Moreover, the decomposition usually has a simple functional form. Typically, the multivariate utility function is a sum or product of the component utility functions. (Multi-linear aggregation formulas based on sums of products can also arise.) Table 2 summarizes some of the most important and widely applied decomposition results.

An important feature of the multiplicative aggregation rule is that it allows for possible substitute or complement effects among attributes. Specifically, if the scaling constant K satisfies $K < 0$, then the attributes are substitutes and the respondent is said to display *multivariate risk aversion.* Conversely, if $K > 0$, the attributes are complements, and the respondent is multivariate risk-seeking. In either case, multiplicative aggregation is required. Only if $K = 0$ (multivariate risk neutrality, no preference interaction among attributes) is the additive aggregation rule justified. Except where otherwise noted, the basic reference for these results and for details on how to calculate the scaling constants K and K_i, i = 1,2,......., n, is Keeney and Raiffa (1976), which also gives advice on what to do if independence conditions can not be established. (The most common solution is to rethink the choice of attributes used to describe the problem, e.g., by re-aggregating more primitive attributes into different high-level attributes.) The extension to measurable value functions is due to Dyer and Sarin (1979). French (1988) provides a useful textbook treatment.

The results in Table 2 are for NM utility functions and EU theory. However, Miyamoto and Wakker (1996) provide a powerful and concise extension of MAUT to many more general theories, including Savage's Subjective Expected Utility (SEU) theory, state-dependent SEU, Prospect Theory, Rank-Dependent Expected Utility Theory (RDEU), and Choquet-expected utility for non-additive subjective probability measures. This generalization is based on the assumption that attributes can be selected so that more is preferred to less for each attribute and MUI holds for acts rank-ordered by dominance. Then, MUI implies additive or multiplicative aggregation functions for composing multivariate preference functions from attribute-specific ones in each of these normative decision theories. That the zero-condition and SGI imply a multiplicative utility function (given the assumptions of EU theory) is a more recent result of special reference in health risk assessment, as discussed in later examples.

Table 2: Multiattribute value and utility decomposition theorems

- *MPI + regularity conditions imply additive value function.* If the attributes are mutually preferentially independent (MPI), and some technical conditions hold, then $v(c_1, c_2, \ldots, c_n) = v_1(c_1) + \ldots + v_n(c_n)$ (Krantz et al. 1971). Moreover, if the $v_i(.)$ are measurable value functions and difference independence (DI) holds for at least one attribute, then v is also measurable. Value functions that are not measurable – i.e., for which value differences have no significance – are sometimes referred to as conjoint models.
- *AI + regularity conditions imply additive utility function.* If the attributes $C_1, \ldots, C_n$ are additively independent (AI) and some other mild technical conditions hold, then $u(c_1, \ldots, c_n) = u_1(c_1) + \ldots + u_n(c_n)$.
- *MPI + WDI imply additive or multiplicative value function.* If attributes $C_1, \ldots, C_n$ are MPI, and WDI holds for at least one attribute, then:

$$v(c_1 \ldots c_n) = \frac{\Pi_i [1 + KK_i v_i(c_i)] - 1}{K} \quad \text{if } K_1 + \ldots + K_n \neq 1\text{; else}$$

$$v(c_1 \ldots c_n) = K_1 v_1(c_1) + \ldots + K_n v_n(c_n) \quad \text{if } K_1 + \ldots + K_n = 1.$$

Here, Π_i denotes the product over all components (i.e., from i = 1 to n), and all the value functions, both the components $v_i(.)$ and the aggregate v(.) are scaled to run from 0 to 1. The scaling constants K and K_i are chosen accordingly.

- *MPI + UI imply additive or multiplicative utility function.* If $C_1 \ldots C_n$ are MPI, and if utility independence (UI) holds for at least one attribute, then utilities are aggregated additively or multiplicatively. That is, the above result for value functions holds for utility functions, with u(.) and $u_i(.)$ replacing v(.) and $v_i(.)$.
- *Zero-condition + SGI imply multiplicative utility function.* If U is an NM utility function for consequences in $C = C_1 \times C_2 \times \ldots \times C_n$ and if attributes C_1 through C_{n-1} have zero levels, then SGI holds for attributes C_1 through C_{n-1} if and only if U has a multiplicative decomposition, $u(c_1, \ldots, c_n) = u_1(c_1) \times \ldots \times u_n(c_n)$. (Miyamoto, Wakker, et al., 1998)

2.2 Some Practical Aspects of MAUT

Practical techniques for assessing single-attribute component value functions and utility functions in simple situations, meaning situations where preferences are relatively clear and consistent, are widely available (Keeney and Raiffa 1976; French, 1988; Clemen, 1996). Component value or utility functions for each attribute can often be assessed fairly rapidly and reliably by first establishing a few qualitative properties (such as concavity or risk aversion) and then assessing a few points – generally no more than three – that indicate the overall shape of the function. After a component utility function has been estimated, it is verified and adjusted as necessary through a series of diagnostic questions that confirm its qualitative properties. A second opportunity for correction and verification is afforded when the whole utility function is checked for predictive validity, following aggregation of the component utility functions to form the multiattribute utility function. Such checking is an essential part of MAUT applications.

Although simple approaches work well in simple situations, successful application of MAUT modeling to difficult choices requires considerable care, especially in choosing the attributes used to describe a problem. For example:

- *Proxy attributes,* defined as indirect measures of the fundamental attributes that ultimately carry value, tend to be systematically over-weighted compared to the fundamental attributes. For example, pollution level (a proxy for harm done by pollution) tends to be over-weighted compared to the values inferred from preferences for its fundamental (non-proxy) impacts, e.g., on cost and health. So, it is desirable to express consequences in terms of fundamental attributes if possible. Otherwise, the proxy attribute bias may lead to substantially sub-optimal decisions (Fischer, 1989)..
- *Level of detail:* Parts of a decision problem that are described in more detail tend to receive disproportionate weight in decision making. This is true for attribute trees used to describe consequences, as well as for probability judgments (*ibid,* p. 492; Rottenstreich and Tversky, 1997). Hence, the level of detail used in elaborating an attribute hierarchy can create severe presentation mode biases that affect decisions.
- *Assessment methods:* Different methods have been developed and compared for assessing the weights in multiattribute value and utility functions. They do not necessarily give the same answers. Each method (e.g., value ratio, swing weighting, tradeoff and pricing out approaches) has difficulties. Presentation, elicitation, and response mode biases can strongly affect results (Clemen, 1996; Barron and Barrett, 1996; Borcherding et al., 1991). Preference reversal illustrates the strength and robustness of such effects when only two different dimensions – sizes

and probabilities of monetary gains or losses – are involved. On the other hand, using rank-ordered weights and avoiding the (perhaps spurious) precision required for numerical weights can often be surprisingly effective for determining the best multiattribute alternative (Barron and Barrett, 1996).

In summary, practical applications of MAUT must overcome presentation mode biases, elicitation biases, and response mode biases similar in ubiquity and severity to those encountered in single-attribute and discrete-alternative decision analysis. Considerable ingenuity has been devoted to making multiattribute value theory and MAUT useful. Successful applications involving health, safety, and environmental risks have been completed (Clemen, 1996). But the key challenge – that people often do not seem to *have* stable, well-defined preferences that are independent of the techniques used to elicit and express them – is at least as much a problem for MAUT as for SAUT, especially when difficult choices involving non-commensurable consequence dimensions must be made (Fischoff and Cox, 1986).

3. APPLICATIONS OF MAUT TO HEALTH RISKS

This section further develops the theory and summarizes several applications of MAUT, emphasizing consequences involving health, safety, and life. The theory also applies equally well to environmental, financial, and other consequence dimensions as well.

3.1 Using MAUT to Develop Health Status Indicators

MAUT-based studies have clarified the structure of human preferences for health attributes, both when the respondents' own health, or pain and suffering, are involved, and when some individuals (e.g., physicians or public policy makers) must make judgments involving the health states of others. These studies show the value of the MAUT framework for addressing difficult questions involving risks to life and health. For example, Torrance, Boyle, and Horwood (1982) used four aggregate attributes (relating to physical function, role function, social-emotional function, and health problems) to classify overall health states. They assessed both value and utility functions over each attribute for nearly 90 subjects to use in a cost/social-utility analysis of neonatal intensive care. Utility-assessment techniques for such health applications proved capable of eliciting usefully precise and reliable results. (This paper also considers several other aspects of the *social-utility* approach to evaluation of health effects, discussed in Chapter 8.)

The Health Utilities Index Mark 2 (Torrance et al., 1996) extends this approach for assessing health-related quality of life. It uses seven attributes (based on sensation, mobility, emotion, cognition, self-care, pain, and fertility) and a scoring formula (based on the MAUT decomposition theory discussed next) to quantify health status. A random sample of parents expressed preferences for health states lasting for a 60-year lifetime, starting at age 10. Value functions were elicited using visual analogue scaling and risk attitudes were elicited using standard gambles (with death or health as the end-points) for each attribute separately. The justification for this approach is provided by MAUT theory. Again, the results seemed consistent and robust enough to provide useful measures of health status for many applications.

Additional applications of MAUT assessments to valuation of human health include modeling the roles of risk attitudes and time preference in determining individual utility functions for health effects distributed over time (Gafni and Torrance 1984). Krischer (1980) gives an annotated bibliography of over 100 articles and reports published between 1970 and 1980 relating to single-attribute and multiattribute utility assessments, with applications to valuing health effects ranging from sore throat to cancer.

3.2 Independence Conditions and QALYs

MAUT theory has long provided a theoretical foundation for the use of quality-adjusted life-years (QALYs) to summarize health-related outcomes. Originally, Pliskin, Shepard, and Weinstein (1980) applied MAUT to health status and anginal pain. They defined a class of MAUT models for health status and life years and assessed the tradeoffs between longevity and relief from anginal pain that would produce equal utility. They assessed relevant value and utility functions for several individuals and used EU theory to derive conditions under which the use of expected QALYs as health status indices could be reconciled with the requirements of "rational" (EU) decision making. They assumed that life duration and health quality are MUI and that, for each health state, there is a number q such that Y years in that health state is judged equivalent to qY years in perfect health. Then, under certain regularity conditions, the utility of Y years in the health state is proportional to a power function of Y, with a constant of proportionality that depends on the health state.

Miyamoto et al. (1998) dramatically simplified and generalized the expected utility foundations of QALYs by introducing the zero-condition independence condition and combining it with SGI. They recognize that a health outcome in which a person lives for T years in chronic health state Q and then dies may be viewed as a two-attribute consequence. (The health state indicator Q, in turn, might be defined and measured using a MAUT model such as the Health Utilities Index Mark 2 mentioned above.) The traditional QALY model may then be expressed as the following two-attribute NM utility function:

$$u(Q, T) = v(Q)T \qquad \text{(Classical QALY model)}$$

In this classical expression for QALY, V(Q) adjust for health quality and utility is assumed to be linear in duration – not necessarily a plausible restriction. In this context, the zero-condition implies that all pairs $(Q_1, 0)$ and $(Q_2, 0)$ are equally valued, for any two health states Q_1 and Q_2, when the duration is zero. To take risk-aversion for duration of life into account, the classical QALY model must be generalized as follows:

$$u(Q, T) = v(Q)W(T) \qquad \text{(Generalized QALY model)}$$

where W is the function that values life duration and $W(0) = 0$. The theoretical justifications for these expressions and a useful corollary, as proved by Miyamoto et al., are as follows:

For an expected-utility (EU) decision-maker, the classical QALY model holds if and only if the zero-condition and risk neutrality hold. The generalized QALY model holds if and only if the zero-condition and SGI hold. In the generalized QALY model, a d.m. who is risk-averse for positive health states (i.e., states for which longer life durations are preferred) must be risk-seeking for negative health states (those for which shorter life durations are preferred), consistent with the predictions of Prospect Theory.

Elegant and parsimonious as this theory is, Miyamoto et al. note that SGI and the generalized QALY model are violated in some settings, such as when someone wants to live for a certain amount of time (e.g., to see a specific event) and then die, or when a d.m. will trade off life duration for improved health quality only if life duration is sufficiently long.

3.3 Money Values for Reductions in Risk to Life

MAUT yields useful theoretical insights into the monetary value of health and safety benefits. However, not all of the predictions of MAUT theory describe monetary evaluations, and there appear to be many psychological aspects of actual monetary evaluations that would not be expected or predicted by MAUT theory or by more general EU theory.

Example: A Two-Attribute Theoretical Model of Statistical Life Value

Weinstein, Shepard, and Pliskin (1980) used a two-attribute model to study the tradeoffs between dollars or asset position (the first attribute) and risk to life (the second). They assumed that

a) Life is preferred to death for any asset position;
b) The marginal value of an increase in assets is greater in life than in legacy;
c) A lottery giving a 50% chance at death with no legacy and a 50% chance at life with asset position *x* is preferred to death with legacy *x*, for any *x*; and
d) Individuals are risk-averse for assets.

They showed that EU theory and these assumptions imply these conclusions:

- The dollar value of a reduction in risk of dying, as measured by WTA or WTP, depends not only on the size of the risk reduction but also on the size of the baseline risk being reduced. Specifically, a risk reduction of a given magnitude is more valuable in terms of both buying and selling prices when it is made in a larger initial risk level. For example, a risk reduction from 0.4 to 0.3 would be more valuable, under the above assumptions, than a risk reduction from 0.3 to 0.2. (Note, however, that the psychological certainty effect implies that a reduction from 0.1 to 0 may be more valued than either, since it replaces uncertainty with certainty. Thus, the deductions from EU theory may not describe real preferences or behaviors.)
- The selling price per expected life saved (i.e., the selling-price value of a statistical life saved to its owner) is greater for larger risks than for smaller ones. The WTA value for reducing a risk from 0.03 to 0.02 is less than one tenth of the value of reducing it from 0.3 to 0.2. However, the WTP, or buying price, per expected life saved may be greater in the former case, in contradiction to the WTA price.
- Societal WTP, defined as the sum of individual WTP values, will tend to be greater for a health or safety regulation that saves one life with certainty than for a regulation that saves ten lives with probability 0.1 each.

Perhaps the most important overall conclusion from this research is that *a unique dollar value per expected life saved, even as valued by the owner of the life, may not exist.* Instead, this value will depend on the initial level of the risk being reduced. Thus, for small risks ranging over several orders of magnitude, one would expect to find a range of dollar values for statistical lives saved (i.e., dollars per expected life saved.) Therefore, it would be senseless to search for a universally applicable dollar value per expected life saved, even for small risk reductions.

This example illustrates the types of qualitative implications for health and safety measures that can be drawn from MAUT. It has several noteworthy features. First, the data requirements needed to draw useful qualitative conclusions are relatively mild, having to do primarily with the qualitative properties of individual risk attitudes. Second, some challenging policy questions involving health and safety trade-offs can be at least partly addressed by relatively simple analytic models.

This example also illustrates one way to bridge the gap between utility-based and money-based measures of individual preference. Namely, a MAUT function having monetary position as one attribute is assessed and used to define indifference curves or tradeoffs between money and traditionally hard-to-price attributes such as risks to life. The resulting tradeoff rates establish dollar equivalents for the hard-to-price attributes with respect to the underlying MAUT model. (Of course, unlike true market prices, such preference-based WTP or WTA figures in general apply only to the individual whose preferences are being assessed. There is no equilibrating market mechanism to equate the tradeoff rates for different individuals.) Keeney and Raiffa (1976) discuss this strategy for converting from MAUT to dollar representations of preferences under the name "pricing out." In practice, the technique is only useful when, as in the above example, there are few attributes and simplifying qualitative assumptions about preference tradeoffs can be made.

Despite the theoretical insights that MAUT produces in this example, real stated WTP and WTA numbers often do *not* reflect rational economic (expected-utility/MAUT) trade-off preferences. As previously mentioned, they often appear to reflect attitudes rather than preferences, and hence can be remarkably insensitive to the size of the risk reductions or other benefits achieved (Kahneman et al., 1999).

Example: Costs per Statistical Life Saved Differ Widely

Empirically, the average amounts actually spent per expected life-year saved by different life-saving interventions range from under $20,000/life-year for medical interventions to over 100 times that amount (for pollution and toxin control) to over $10 billion/life-year for some interventions (Tengs et al., 1995). For example, Viscusi and Hamilton (1999) estimate that the cost per cancer death averted at 130 Superfund sites is at least $11.7 billion/death averted (based on EPA's risk estimates) and more probably over $100 billion per cancer death averted (based on more realistic risk estimates.) Only a small part of this enormous range in dollars spent per statistical life saved can be explained by EU-based modeling of WTP for risk reductions (Rabin, 2000). Moreover, the upper end of the range may be difficult or impossible to justify on rational grounds. For example, Chapman and Hariharan (1994) estimate that "health and safety regulations, which cost more than $12.2 million per life saved, are likely to kill more people through increased poverty than they save directly." If this is true, then relatively expensive approaches to risk reduction may create more harm than they remove.

Example: Wage-Risk Revealed Values of Statistical Lives Saved

The implicit dollar value per statistical life saved inferred from willingness-to-accept (WTA) compensation studies of the relation between wages and hazard rates in labor markets spans a smaller range, from about $0.6M to about $16M per statistical life saved, in 1990 dollars (Viscusi, 1992, Table 4-1). Of this roughly 30-fold range of values, perhaps a factor of 2 to 3 can be explained by heterogeneity in the values that workers assign to their own lives. Workers in the highest-risk quartile of occupations put an apparent value on their (statistical) lives, i.e., on reductions in fatality risks, about 1/2 to 1/3 as great as workers in the least-risky quartile of occupations (*ibid,* Table 3-6). (However, the mortality risks they bear are more than 10 times as great on average.)

Other sources of differences in apparent WTP or WTA per statistical life saved may include different *attitudes* toward risks of different types (e.g., risks borne voluntarily and with information vs. risks not disclosed) and differences in *perceptions and beliefs* about risks. Recall that most people believe that they have lower-than-average risks, especially if they believe that they have some control over outcomes; hence, they may accept smaller compensating wage differentials than they would require if they believed that average risk rates applied to them.

Outside the context of occupational risks, WTA and WTP dollar figures for risk increases and decreases show even more dramatic variations, many of them apparently driven by attitudes and considerations apart from simple economic (EU- or MAUT-style) tradeoffs. This is especially true in surveys of dollar equivalents for changes in risk.

Example: WTP *vs.* WTA Asymmetry for Reducing Health Risks

Most consumers would not pay a dollar extra per can of a household product (e.g., insecticide) to reduce probability of inhalation or skin poisoning by 5/10,000, from 10/10,000 to 5/10,000. Yet, most find an increase in risk of even 1/10,000 (e.g., from 15/10,000 to 16/10,000) to be unacceptable at any price. The minority who would accept a riskier product for a sufficiently lesser price would require several dollars of price reduction to accept an incremental risk that they would only pay at most a small fraction of a dollar to eliminate (Viscusi, 1992, Table 8-1). Similarly, individuals who would pay $800 (median stated WTP) to inoculate themselves against a disease with a 0.001 chance of death to which they had hypothetically been exposed would demand about $100,000 in compensation to accept this same risk (Casey, 1995 citing Thaler, 1983).

EU theory allows WTP and WTA to differ by relatively small amounts, based on an income or wealth effect, but it cannot explain such large differences (*ibid,* p. 981; Rabin, 2000). Instead, the large excess of WTA over WTP may have more to do with attitudes and beliefs (e.g., "Products should be safe and I should not have to pay extra for product safety") than with EU or MAUT theory. They asymmetry between WTA and WTP amounts may also be magnified by the *endowment effect,* i.e., the well-documented inclination of decision-makers to value the *status quo* and to avoid changes (in any direction) more than would be expected under a pure EU-maximizing model. However, this effect cannot explain all observed differences between WTA and WTP. Psychological encoding rules similar to those in prospect theory, e.g., describing when and how potential gains and losses are combined and incorporated into a buyer's or seller's perceived status quo, may explain the rest of the difference (Casey, 1995).

Evidence from survey-based (contingent valuation) studies of stated WTP for road safety programs reveals the large extent to which attitudes, rather than trade-off calculations, appear to drive respondent's answers and hence the imputed value of statistical lives saved. Stated WTP for fatalities prevented in this context is remarkably insensitive to the *number* of fatalities prevented per unit time. Instead, it seems to reflect a perception that safety improvement is a good thing (independently of its magnitude) and that one should therefore be willing to spend enough on it to achieve a sense of moral satisfaction, but not enough to seriously disrupt normal saving and expenditure patterns (Beattie et al. 1998). Such reasoning has very little overlap with the assumptions of MAUT decision-making and rational economic trade-off calculations.

Stated WTP for statistical life-years saved can be very small if the life-years in question occur in advanced age, when quality of life is expected to be relatively low.

Example: Steep Discounting of Life-Saving in Old Age

Johannesson and Johansson (1996) posed the following question to 2013 Swedes aged 18 to 69:

> "The chance for a man/woman of your age to become at least 75 years old is P percent. On average, a 75-year old lives for another ten years. Assume that if you survive to the age of 75 years you are given the possibility to undergo a medical treatment. The treatment is expected to increase your remaining length of life to 11 years. Would you choose to buy this treatment if it costs $x and has to be paid for this year?"

The proportion of respondents who said yes (and the subset of these – averaging 47% of them – who said they were confident about their affirmative answer) was studied as a function of the parameters x and P and the current age of the respondent. P% ranged from 0.6 to 0.8 based on the age and sex of the respondent, while the cost was varied over six values spanning a 50-fold range (using values of 100, 500, 1000, 5000, 15000, and 50000 Swedish Crowns. $1 = about 7.5 Swedish Crowns in August, 1995 when the survey was conducted.) In this context, the average WTP for this potential increase in life expectancy was only $400-$1500, with an implied value per statistical life saved of roughly $30,000 to $110,000.

Perceived poor quality of life in advanced age, as well as modest discounting of future health benefits (at a rate estimated as between 0.3% and 3.4%), were possible explanations for these relatively low values. The proportion of respondents who expressed willingness to pay for the statistical life extension varied from 53% at the lowest cost level to 9% at the highest. Thus, it appears that many individuals may have had a zero or near-zero WTP to extend expected life duration past age 75 from 10 years to 11 years.

Other studies with Swedish populations have shown a strong preference for life-saving programs devoted toward younger people. For example, on average, saving one 30-year-old is judged equivalent to saving eleven 60-year olds (*ibid*, p. 171). Whether or not they are appropriate for policy purposes, such judgments suggest that neither the number of statistical lives saved by an intervention nor the number of statistical life-years saved adequately summarizes its impacts. The age distribution for the years saved (or lives saved) and perhaps the quality of those years are also needed to determine the perceived value of the life-saving intervention.

It has struck many observers as paradoxical that society routinely spends tens to hundreds of millions of dollars – and sometimes much more – on life-saving interventions that command a WTP of only a few million dollars or less. Chapter 8 explores some possible reasons for this discrepancy based on models of multi-party decision-making. Other reasons are found in the attitudes and psychology of individual decision-makers, especially where environmental decisions are involved.

Example: Attitudinal Factors in WTP for Environmental Regulations

A survey study of expressed willingness to vote for costly environmental risk regulations (accelerating overhaul of gasoline station storage tanks) suggests several factors that may contribute to large estimated WTPs for regulations with relatively high costs and small health benefits (Johnson, 1999). These include:

a) Belief that the regulations will not impose heavy costs on those who vote for them (i.e., that someone else – such as industrial polluters – will pay);
b) Willingness to pay to enforce perceived public property rghts to environmental goods;
c) Satisfaction from a perception that a polluter has been forced to pay. Some people are willing to support measures that will cost them a small amount and provide relatively small benefits if doing so forces large companies to spend large amounts of money in the name of environmental quality.

A key conclusion from this study was that individual WTP may depend on who else pays how much.

Subramanian and Cropper (2000) show that many people stongly prefer risk reduction measures involving industrial pollution control to other risk reduction measures (e.g., smoking education programs and other public health) even when the latter save 50 to 100 times more lives and cost less. This appears to reflect a very high relative value placed on other perceived benefits of industrial pollution control, such as cleaner air and reduced acid rain and ozone depletion, held by a substantial fraction (over 20%) of respondents.

Patt and Zeckhauser (2000) provide evidence of a *bias toward action* when action may lead to demonstrable gains (e.g., driving estimated risk below some "bright line") and non-demonstrable losses of comparable magnitude. Such action bias can lead to preferences for interventions that have zero or negative net benefits if taking action conveys a positive value.

3.4 Perceived Risk of Risk Profiles

Chapter 6's discussion of single-attribute utility theory (SAUT) examined two different approaches to defining risk: as the information that a decision-maker must have about prospects, in addition to their expected values, to choose among them based on EU; and as the aspect of a prospect (or its CDF) that risk-averse decision-makers would pay a premium to avoid. A third approach is more empirical. Subjects are asked to compare prospects using the binary comparative relation "is at least as risky as" between prospects. The answers are used to define and model how perceived risk is determined by objective aspects of the prospects, much as a psychologist might use judgments of perceived relative size or distance to define and model how these perceptions depend on various factors. Typically, axioms are used to define classes of functions that might represent the "at least as risky as" relation, and the consistency of the resulting functions with experimental data is assessed.

Table 3 summarizes several models for perceived risk that have been developed this way. Some of them can be clarified using multiattribute utility theory (MAUT), even though the prospects being compared are single-attribute prospects. For example, the Pollatsek-Tversky model expresses risk as a convex combination of mean and variance. These may be viewed as two attributes of a prospect. The Luce-Weber model represents each prospect by five numerical attributes and establishes an additive-multiplicative representation. However, the exact relation between attributes of perceived *risk* and attributes used to construct *preferences* among prospects is made most clear and explicit in the model of Jia et al. (1999). This model represents preferences among prospects by the following function:

$$\begin{aligned} EU(X) &= V(m) - \phi(m)R(X) \qquad \text{(Perceived risk model)} \\ &= \text{expected value} - \text{perceived risk} \end{aligned}$$

where

- $V(m)$ = value function for the mean of the prospect
- $R(X) = R(X') - R(0)$
- $R(X')$ = standard measure of risk for the prospect (see Chapter 6)
- X' = the zero-mean prospect defined by: $X' = X - E(X)$
- $\phi(m)$ = a positive weight that decreases as mean return, m, increases
- $R(0) = -u(0)$ is a constant.

The perceived risk model $R(X) = \phi(m)[R(X') - R(0)]$ has the following qualitative properties, which are well supported by experimental evidence.

1. Adding a zero-mean lottery to a prospect increases its risk. This is consistent with the concept of mean-preserving spreads as increasing risk, previously discussed in the context of SSD for SAUT prospects.

Table 3: Axiomatic Models of Perceived Risk (Jia, Dyer, and Butler, 2001)

Authors	Mathematical form and Proposed Attributes	Comments
Pollatsek and Tversky (1970)	Mean-variance model: $R(X) = \theta Var(X) - (1-\theta)E(X),\ 0 \leq \theta < 1$ *Risk attributes:* Mean, variance. (Coombs and co-workers used mean, range, and skewness or positive and negative outcomes and probabilities, instead.)	Coombs and Bowen, 1971, showed that skewness and other factors also affect perceived risk. Pollatsek and Tversky assume *additive independence* of risks (e.g., risk rankings of prospects are unaffected by adding the same constant or random variable to each prospect). This contradicts empirical evidence.
Luce, 1980	Expected logarithmic or expected power functions	These models, based on axioms, are contradicted by experimental evidence (Weber, 1984, Keller et al., 1986).
Luce and Weber, 1986	Conjoint Expected Risk model. This is a weighted sum of: $Pr(X = 0)$, $Pr(X > 0)$, $Pr(X < 0)$, $E(X^k \mid X > 0)Pr(X > 0)$, $E(X^{k'} \mid X < 0)Pr(X < 0)$. *Risk attributes:* Probabilities of loss, gain, and zero; positive and negative outcomes	The Conjoint Expected Risk (CER) model has seven parameters (i.e., weights on each of the five terms and the exponents k and k'). Although it can describe risk judgments reasonably well, it is not very parsimonious.
Fishburn, 1982, 1984	No specific functional form suggested. *Risk attributes:* Distribution of loss, loss probability , gain probability, distribution of gain	Fishburn's axioms treat gains and losses separately and assume that risk is zero if there is no chance of loss.
Jia, Dyer, and Butler, 1999	Preferences for prospects can be represented by a *risk-value model:* $V(m) - \phi(m)[R(X') - R(0)]$, where • $m = E(X)$ • $V(m)$ is a value function for the mean of X • $\phi(m)$ is a positive, decreasing function of m • $X' = X - m$, $R(0) = -u(0)$ • $R(X') = -E[u(X - m)] = -E[u(X')]$ for some NM utility function u is the standard measure of risk.	The perceived risk model $R(m, X') = E[u(m, X')]$ can be represented in the form $\phi(m)[R(X') - R(0)]$ if and only if the following four axioms* hold for any two zero-mean prospects X' and Y': 2. If $(m_0, X')\ R_p\ (m_0, Y')$ for some m_0, then $(m, X')\ R_p\ (m, Y')$ for all values of m. 3. $(0, X')\ R_p\ (0, Y')$ if and only if $X'\ R_p\ Y'$. 4. $(m, X')\ R_p\ (m + k, X')\ \forall k > 0$ 5. For all m, $(m, 0) \sim_p (0, 0)$

*Here, R_p is the binary relation "is at least as risky as" (and $\sim_p$ means "is as risky as") between prospects. R is a function that represents R_p.

2. Multiplying all consequences by a factor greater than 1 (e.g., gambling for dollars instead of pennies) increases the risk of any zero-mean prospect.
3. Repeating a zero-mean lottery many times increases risk.
4. Adding a positive constant to any prospect decreases its risk (if more of the consequence attribute is preferred to less.)

There is also some evidence to support a fifth implication: that adding or subtracting the same constant to each of two prospects with the same mean will not change their perceived risk ordering.

In summary, MAUT theory – specifically, for two attributes – has been applied to develop a risk-value model of perceived risks for prospects with a single consequence attribute. The model has considerable empirical support. It includes many earlier models of perceived risk as special cases, but without requiring features that have been refuted by experimental evidence (Jia et al., 1999).

4. RISKS TO MULTIPLE PEOPLE: RISK EQUITY

Early in the history of probabilistic risk assessments (PRAs) and quantitative risk assessments (QRAs) of the safety of industrial facilities, it became common practice to present risk profiles showing the probability of at least x people being killed in any one year by an identified hazard, with x ranging from 1 to the maximum plausible number of fatalities. The corresponding risk profiles for fatalities (sometimes called "F-N curves" or "Farmer curves", especially in England) are part of the analytic methods taught in many introductions to QRA for engineers. However, when multiple people are exposed to a hazard, such a fatality risk profile does not necessarily summarize all of the information about fatality risks that is relevant for choosing among alternative risk management options. Specifically, it omits information about the *equity* of the distribution of individual risks.

Example: Fatality Risk Profiles Ignore Risk Equity

Suppose that a risk manager must choose between two acts, A and B, each of which will produce between 1 and 5 fatalities, depending on which of five equally-likely states of the world occurs. The fatality risk profiles are determined by the data in the following table. Clearly, acts A and B have identical fatality risk profiles. But the distributions of *individual* risks corresponding to these profiles may be entirely different.

	States				
Acts	S1	S2	S3	S4	S5
A	1	2	3	4	5
B	2	3	4	5	1
State Probs:	1/5	1/5	1/5	1/5	1/5

For example, suppose that the five states S1 through S5 correspond to five different accident scenarios, each with occurrence probability 0.2. Suppose that under act A, the individuals killed if each accident scenario occurs are as follows.

	Individuals				
Accident	*1*	*2*	*3*	*4*	*5*
S1	x				
S2	x	x			
S3	x	x	x		
S4	x	x	x	x	
S5	x	x	x	x	x
Individual Risks:	1.0	0.8	0.6	0.4	0.2

Key: x indicates individual in column is killed if scenario in row occurs.

Thus, if accident scenario j occurs, it kills individuals 1 through j, for each j = 1 through 5. (Thus, S1 through S5 may indicate accidents of increasing sizes, with individual 1 located closest and individual 5 furthest from the accident site.) In this case, the individual risks are very unequal, ranging from 100% probability of death for individual 1 to 0.2 probability of death for individual 5. Now, under act B (which may involve relocating the individuals), suppose that the individual fatality risks are as follows:

	Individuals				
Accident	*1*	*2*	*3*	*4*	*5*
S1	x				
S2	x				x
S3		x	x	x	
S4		x	x	x	x
S5	x	x	x	x	x
Individual Risks:	0.6	0.6	0.6	0.6	0.6

Key: x indicates individual in column is killed if scenario in row occurs.

Then, even though the fatality risk profile is unchanged (with equal probabilities of 0.2 for killing 1, 2, 3, 4, or 5 individuals, depending on which scenario occurs), the individual risks are no longer unequal. Each individual now has a 0.6 probability of death. In choosing between A and B, a decision-maker may care about the distribution of individual risks, as well as about the probability distribution for the total number of fatalities.

To model risk equity, Keeney (1980), followed by Sarin (1985) and Fishburn (1984), applied single-attribute and multiattribute utility theory to population fatality risks. Keeney interpreted individual probabilities of death as "attributes", and applied MAUT independence axioms to derive utility functions over the single composite attribute "number of fatalities." He interpreted preferences for risk equity in terms of the shape (convexity) of utility functions over this attribute. However, as suggested by the preceding example and the following one, there are fundamental limitations to any approach that treats "number of fatalities" as an attribute.

Example: Non-Existence of a Utility Function for Lives Lost

The following example is adapted from a discussion by Keeney and Winkler (1985) of a "compatibility theorem" given by Fishburn (1984). Suppose that the target population B consists of two individuals, 1 and 2, and that a decision-maker must choose between the following acts:

- A assigns a probability of 1 to the death of individual 1.
- B assigns a probability of 1 to the death of individual 2.
- C assigns a probability of 0.5 to the death of individual 1 and a probability of 0.5 to the death of individual 2: exactly one of these events will occur.
- D chooses act A with probability 50% and chooses act B with probability 50%.

If an N-M utility function exists over the attribute "number of fatalities', then the utility of acts A and B will be the same. Hence, act D must have the same expected utility as each of them, since it is a probabilistic mixture of them. Finally, D must have the same expected utility as C, since both give a 50% chance of the consequence in which 1 dies and a 50% chance of the consequence in which 2 dies, and therefore must have the same expected utility. Thus, a preference for act C or act D over act A or B is inconsistent with the hypothesis that an N-M utility function over "number of fatalities" exists. In other words, a preference for "risk equity" is incompatible with the existence of a utility function over this attribute.

Keeney and Winkler conclude from this that "Basically, *ex ante* equity is a characteristic associated with the decision strategy that should not change as uncertainties are resolved over time." However, the following argument suggests that the opposite conclusion may be more appropriate. The difficulty raised by the example is that EU theory implies that a d.m. who is indifferent between two consequences must be indifferent among all probability mixtures of them. If the condition of *being at risk* is itself part of the consequence, however, then this implication must be refined: uncertainties that resolve at different times are parts of different consequences. From this perspective the resolution of the above example is that act D will in general not be indifferent to acts A and B, and will have different values depending on how soon it is resolved. Only if it is resolved at the moment death of 1 or 2 arrives will it be indifferent to C. Earlier resolution implies that one of the individuals will live under the shadow of certain (as opposed to probable) death for a longer time, and this is a different consequence than would be implied by later resolution.

Example: Equity-Preserving Transformations

Fishburn (1984) suggested that risks should be represented not only by risk profiles for the probable number of deaths, but also by an individual risk vector summarizing the fatality risk borne by each individual, sorted from largest to smallest. However, even this expanded summary may not capture all risk information of interest.

Consider the following two fatality risk matrices, D1 and D2.

Individuals

	Accident	Pr(i)	*1*	*2*	*3*
D1 =	1	0.3	1	1	0
	2	0.4	0	1	1
	3	0.3	0	0	1
			0.3	0.7	0.7

Individuals

	Accident	Pr(i)	*1*	*2*	*3*
D2 =	1	0.3	0	0	1
	2	0.7	1	1	0
			0.7	0.7	0.3

In the matrix (or hypergraph) represented by D1, three individuals are at risk (labeled 1 to 3 across the column headings). Three accidents compete to happen, and the individuals killed by each accident if it occurs are indicated by the 1's in the matrix. The first accident has a 0.3 probability of being the one to occur first. If it occurs, it will kill individuals 1 and 2. The second accident has a 0.4 probability of occurring, and will kill individuals 2 and 3. The third accident has probability 0.3 and a fatality set consisting of individual 3 alone. The individual risk vector, (0.3, 0.7, 0.7) is shown beneath D1. The unconditional individual risk, or fatality probability, for any individual i is the sum of the occurrence probabilities of the fatality sets (rows of the fatality matrix) that include i as a member. The population risk profile for D1 evidently puts a 0.7 probability on two fatalities and a 0.3 probability on one fatality. It is found by assigning to each number x the sum of the probabilities of fatality sets that contain exactly x fatalities.

D1 may be thought of as being derived from D2 by transferring a probability mass of 0.4 from individual 1 to individual 3. Thus, individual 1 goes from having an individual risk probability of 0.7 in D2 to a probability of 0.3 in D1, while individual 3 goes from having a probability of 0.3 to having a probability of 0.7. Individual 2 has a 0.7 fatality probability in both cases. By symmetry, this transfer might be considered an 'equity-preserving" transformation, since in effect individuals 1 and 3 have just traded places as far as individual fatality probabilities are concerned. It remains an open question (Fishburn, 1984, p. 907) whether all fatality risk matrices that have the same risk curve and individual fatality probabilities (apart from a permutation) are derivable from each other by sequences of equity-preserving transformations. Of course, if some individuals belong to the same families or have other affiliations, then the fatality risk matrix will not contain all of the information about fatality risks that might matter to a decision-maker.

Rather than focusing on risk profiles for number of fatalities, Sarin (1985) considered the numerical measurement of equity as one component of a multiattribute utility function for prospects involving fatality risks. This approach was continued and refined in Keeney and Winkler (1985) and subsequently. In particular, Boiney (1995) proposed treating individual and group *envy*, as well as a measure of efficiency, as attributes for use in societal decision analyses for difficult choices such as where to locate a facility that produces a mix of risks and benefits to its neighbors.

The *envy* (or utility difference) that individual i has in relation to individual j for a consequence that gives a vector of attributes (e.g., costs, risks, and benefits) x^i to i and a vector x^j to j is defined as the difference in i's utilities for his own vector of outcomes compared to j's:

$$\text{i's envy relative to j} = u_i(x^i) - u_i(x^j). \qquad \text{(Envy)}$$

Here, $u_i()$ is i's utility function, normalized to run from 0 to 1, with the same extreme outcomes used to anchor the scales for all individuals. Economic theory has established interesting relations between envy-free allocations, Pareto-efficient allocations, and equilibrium allocations in various social decision-making processes (see Chapter 8). Boiney proposes several measures of group envy (e.g., based on sums of individual envies) to use as attributes in social utility functions for making decisions involving risks, costs, and benefits.

Attempts to treat risk equity via EU theory and MAUT have helped to clarify limitations in traditional displays of population risks and to highlight challenges for defining satisfactory definitions of risk equity. In practice, as discussed in Chapter 6, perceptions about fairness and equity can strongly affect judgments about the acceptability of risks to life and health. But it is most often the fairness and equity of the *process* by which decisions are made that is judged, including how individual rights and preferences are taken into account in setting agendas and resolving conflicts. An outcome may be perceived as fair or not according as the political, legal, and other decision processes that led to it are perceived as legitimate and as having been open and fair to those affected. From this perspective, it seems unlikely that MAUT will ever give a fully satisfactory theory of risk equity as a basis for decision-making. Instead, models of group decision processes such as those in Chapter 8 are required to fully address issues of equity and fairness in decision-making about risks.

5. BEYOND MAUT: MCDM APPROACHES

A pragmatic response to the challenges of multiattribute preference theory is to abandon some of the usual normative requirements for rational decision-making, such as weak ordering of consequences by preferences, and to focus instead on using *partial* information about preferences to eliminate bad choices. Eliminating inferior alternatives, rather than optimizing a function that completely orders all alternatives, is a central theme of *multi-criteria decision-making* (MCDM) methodology (Vincke, 1992).

MCDM methods typically allow for incomplete, ambiguous, and uncertain preferences. They recognize that, while some pairs of alternatives may not be comparable (at least, not easily), it may be unnecessary to compare them to identify the best one. For example, if alternative *a* outranks both *b* and *c*, then both *b* and *c* can both be eliminated without ever having to compare them to each other. The term *outranking* is a generic one for a binary relation used to compare pairs of alternatives in such a way that if alternative *a* outranks alternative *b*, then *b* would not be selected in preference to *a*. Many MCDM algorithms seek a minimal subset of alternatives such that every alternative not in the subset is outranked by at least one alternative in the subset. Such a subset is called a *kernel* of the outranking relation in graph-theoretic terminology, when nodes represent alternatives and directed arcs represent a binary outranking relation. For acyclic outranking relations, a unique kernel exists and it can be computed using known graph algorithms.

One important class of MCDM methods first scores or ranks each alternative action in the choice set *A* according to each of several criteria, yielding a multivariate payoff table. *Efficient actions* are defined as those that generate *undominated* points in the criteria space. If there are *m* criteria, then one *m*-tuple dominates another if it is at least as desirable according to each of the *m* criteria and is more desirable according to one or more of them. An action that maximizes a convex combination or weighted average (with non-negative weights) of the *m* criteria is efficient. Various generalizations of this approach have been explored, including allowing non-numerical scores (e.g., ranges or probability distributions) instead of precise numerical values in the multivariate payoff table. Several specific outranking relations have been proposed and built into MCDM decision support software (e.g., the PROMETHEE program and the ELECTRE family of programs). They use various measures for the importance of different criteria, the concordance and/or discordance among criteria, the confidence level or degree of outranking of one alternative by another, and preference or indifference relations based on the difference in criterion values for pairs of alternatives (Vincke, 1992). For current information on the state of the art in MCDM techniques, see the *Journal of Multi-Criteria Decision Analysis* (http://www3.interscience.wiley.com/cgi-bin/jtoc?ID=5725).

Another set of MCDM methods emphasizes *iterative interaction* between a decision-maker and a software program (e.g., Koksalan and Sagala, 1995). The program queries the user to both (a) Elicit and gradually refine (or, if necessary, help to construct) needed preference information; and (b) Eliminate inferior alternatives and/or move the set of options being evaluated toward better alternatives. These iterative approaches often use ideas from iterative (e.g., hill-climbing) numerical optimization programs, with the user being asked to make local comparisons, preference trade-offs, or choices (e.g., of gradient direction and/or step size) at each step. Unlike numerical optimization programs, however, most MCDM methods are designed to reduce burden on the user by limiting the number and/or difficulty of questions that must be answered. Some of the older iterative MCDM methods assumed that the user's preferences could be represented by a well-defined, coherent value function, and then sought to use iterative questioning to identify choices that would maximize that initially unknown function. However, the assumption of consistent, detailed preferences is often unrealistic and can lead to inconsistencies or cycles in such straight-forward optimization approaches.

Finally, an MCDM method that has attracted widespread user acceptance and enthusiasm, but also much justified academic criticism, is the Analytic Hierarchy Process (AHP). AHP has several attractive features. It asks users to estimate ratios of importance weights for pairs of criteria or attributes and then estimates a final set of weights that best represents (based on the eigenvalues of the matrix of ratios) these multiple, potentially inconsistent, pair-wise ratio estimates. It uses a hierarchical decomposition of objectives and/or actions for achieving them to break the choice problem into pieces. These and other features make the method plausible and comfortable to many decision-makers that have tried it. Yet, there are some severe conceptual difficulties with AHP. It is not clear exactly what the importance weights and their manipulations mean, why or whether the resulting recommendations should be trusted, or how some of the simplifications made distort the results. (For example, use of 1-9 scales to express relative importance makes it impossible to express weights that are more than one order of magnitude apart.) The method requires such strong assumptions about preferences and independence conditions that it may have no advantages over simple additive value functions. French (1988) describes these and other criticisms of AHP and rejects it as a normative framework for MCDM. Nonetheless, the popularity of the method and many efforts to modify it and to find a mathematical framework to justify it continue to make it too prominent to ignore.

Formal MCDM methods are not yet as widely used in health risk management as they probably will be in the future. But, applications have been made in both public health and safety and in medical decision-making. For example, Papazoglou et al., (2000) describe an MCDM decision support system for identifying efficient (undominated) decisions about land use around an oil refinery using potential loss of life, noise levels and economic benefits as criteria. Papazoglou and Kollas (1997) present an MCDM approach to choosing among possible protective measures in contaminated areas of the former USSR after the Chernobyl accident using population life-time dose and cost as criteria. This application focused on identifying efficient decisions, with value tradeoffs postponed until the last stage of the decision process. Yu (1997) applied a machine-guided iterative optimization approach to MCDM in radiation oncology, where competing treatment plans were ranked based on the decision maker's preference tradeoffs among several dimensions. A genetic algorithm accomplished the optimization. This optimization-based MCDM procedure provided an alternative to both complete MAUT-type utility assessment and more traditional human-guided MCDM procedures for exploring the efficient solution set.

In summary, MCDM methods differ from the MAUT framework in the following main ways:

a) They focus on eliciting the least amount of information from the user that is needed to make a specific decision with high confidence. By contrast, MAUT models represent entire preference structures, potentially allowing many more problems to be solved than the specific one at hand.
b) They relax normative axioms (e.g., the assumption that any two alternatives can be compared) built into many MAUT models.
c) They tend to be more constructive than deductive, helping the user to create, discover, or clarify preferences rather than assuming that a coherent preference ordering already exists.
d) Many MCDM methods are highly algorithmic, presenting a defined sequence of steps for moving from an initial set of options to an undominated subset, outranking kernel, or other "good" subset. MCDM procedures typically take into account realistic limitations on the user's ability to assess his or her own preferences, choose among pairs of alternatives, or assign weights to goals or criteria.

The use of MCDM methods in risk management seems likely to expand, as many risk-cost-benefit comparisons involve multiple criteria with difficult value trade-offs and conflicting, incomplete, or uncertain preferences that are not easily addressed in the MAUT framework.

6. CHOOSING AMONG TEMPORAL PROSPECTS

Risks involve time. Actions taken now affect consequences received later. Consequences of past actions may be learned only gradually, as initial uncertainties are resolved, or they may not be learned at all. Although the consequence function ($c: A \times S \rightarrow C$) framework does not emphasize it, time plays a key role in the definition and evaluation of many health-related consequences, as well as in other areas of applied risk analysis.

This section examines the following aspects of time in health risk analysis: Evaluation of *delayed consequences,* including discounting; *sequential choices* and consistency of plans over time; *repeated gambles* and consistency of choices when given the opportunity to make the same decision repeatedly; *dynamic utility theory* for acts with consequences distributed over time; and *stochastic utility theory* for acts with stochastic transition processes on a set of health states as consequences.

6.1 Discounting of Delayed and Gradual Consequences

Choices that affect health and safety often have delayed consequences. For example, a teen's decision about whether to smoke may affect his or her probable health in old age. Young workers may have to decide whether to purchase insurance, disability, and benefit plans that cover possible illnesses or disabilities much later in life. Diet, exercise, and other lifestyle choices often have delayed consequences, or gradual ones to which one may adapt.

These create conflicts between the satisfaction of present and future selves. How people make such choices may depend in part on their attitudes toward current *vs.* future satisfaction. In practice, such choices frequently violate the predictions from simple discounting models.

Example: Personal Discount Rates and Smoking (Bickel et al., 1999)

Current cigarette smokers, never-smokers, and ex-smokers were asked to indicate preferences for different amounts of immediate versus delayed money, for various delays. Current smokers discounted the value of delayed money significantly more than did the other two groups. (Delayed cigarettes lost their subjective value even to smokers even faster than delayed money.) This finding held for both exponential and hyperbolic models of discounting, discussed next. (The hyperbolic equation fit the data better than the exponential equation for 74 out of 89 comparisons, but both models showed that current cigarette smokers discount money and cigarette rewards more than people who do not currently smoke.)

Example: Time Insensitivity for Protective Investments

A class of choices that often violate traditional economic models of discounting involves *investments in risk reduction measures for property loss,* such as deadbolt locks or earthquake protection measures (e.g., bracing the foundations of a house for a cost of $1500.) Analysis of questionnaire data (Kunreuther et al., 1998) suggests that many individuals do not consider the additional benefits from more years of service life when making decisions about whether to purchase a property protection measure or how much they are willing to pay for it.

Exponential/geometric discounting

The traditional way to make financial decisions with known but delayed monetary consequences is to discount future costs and benefits and to select actions that maximize net present value (NPV). In finance and economics, this procedure is typically justified by appealing to the operation of credit markets, which bring about a constant geometric discount rate when borrowers and lenders are in equilibrium.

In decision analysis, four axioms for multiattribute value functions imply geometric discounting and the use of NPV when cash flows received in different periods are treated as attributes of the consequence. The axioms are: *dominance* (more in each period is preferred to less*), successive pairwise preferential independence* (preferences between two time streams that differ only in periods t and t + 1 do not depend on the common values received in other periods), and *constant* pairwise substitution rates (indifference curves for amounts received in periods t and t + 1 are parallel straight lines) that are *time-invariant* (the slopes are the same for all t.) (Keeney and Raiffa, 1976, Chapter 9.)

An alternative set of axioms that is also necessary and sufficient for NPV with a constant discount rate consists of: *weak ordering* of time streams (i.e., attribute vectors); *constant* relative trade-offs between cash flows in all pairs of years; *equal* trade-off rates from one period to the next for all periods, and a *monotonicity* condition generalizing the idea that more is preferred to less to apply to time sequences (French, 1988). Of course, in the absence of a market mechanism to allow trades, the discount rates used by different people may differ. Also, the same person may discount different types of consequences at different rates.

Personal discount rates for both cash flows and health effects typically appear to decrease with time and for larger amounts (Thaler, 1981; Albrecht and Weber, 1997; Henderson and Langford, 1998). For cash flows, discount rates are usually smaller for losses than for gains ("gain-loss asymmetry"). Moreover, implicit discount rates for postponed rewards are greater than for expedited payments, and the reverse is true for losses (Albrecht and Weber, 1997). Even if individual discount rates are constant (i.e., exponential discounting), the aggregate discount rate based on a sample of individuals with different constant discount rates may be hyperbolic (discussed below) (Henderson and Langford, 1998). In practice, different studies have found implied discount rates ranging from 2% for wages and health risks to about 13% for auto safety measures to over 20% for energy-saving durables (Kunreuther et al., 1998, Table 10.)

In health applications, individual discount rates for delayed health effects differ widely. Many people appear to display negative discounting, while others exhibit various degrees of positive discounting (Dolan and Gudex, 1995). For an intervention taken at the age of 75 (but paid for years in advance) that will extend life expectancy from 10 years to 11, econometric modeling of survey responses suggests a discount rate of 0.3% to 1.3% based on a standard willingness to pay (WTP) estimate. (This estimate increases to 1.6% to 3.4% using a different estimation method) (Johannesson and Johansson, 1996.)

Geometric discounting can be extended to finite or infinite time streams of consumption, rather than only of cash flow. The goal is to represent preferences among time streams of the form $(c_0, c_1, c_2, \ldots)$, where c_t denotes the consumption in period t. The following axioms due to Koopmans (1972) (Keeney and Raiffa, 1976, Chapter 9) accomplish this extension.

1. *Preferential independence of consumptions in periods 1 and 2 from consumption in other periods.* Consumption in each of periods 1 and 2 is preferentially independent of consumption in other periods. Moreover, consumption in periods 1 and 2 is pairwise preferentially independent of consumption in other periods, meaning that preference comparisons of time sequences that differ only in these first two periods do not depend on consumption levels in subsequent periods.
2. *Stationarity.* If two time streams have the same first-period consumption, then their preference ordering is the same as for the modified streams obtained by deleting the first period consumption and advancing all subsequent consumptions by one period (so that the period 2 consumptions are now received in period 1, etc.)

When these axioms hold, preferences for consumption streams can be represented by a net present value (NPV) function:

$$NPV(c_0, c_1, c_2, \ldots) = v(c_0) + av(c_1) + a^2v(c_2) + \ldots$$

where $0 < a < 1$ and v is interpreted as a value function for consumption. One time stream of consumption is preferred to another, under these conditions, if and only if it has a greater NPV.

Hyperbolic discounting

Psychologists and social scientists have found that real personal and social decisions involving delayed gains or losses often seem to exhibit *declining* discount rates. Far-future payoffs are discounted less than near-term payoffs, violating the stationarity axiom of geometric discounting. Time preferences are better described by a hyperbolic discounting model that assigns a present value at time 0 of

$$V_0(c_t) = v(c_t)/(1 + t)^r \qquad \text{(Hyperbolic discounting function)}$$

to a payoff of c_t received t periods in the future (Ainslie, 1975). Many studies have confirmed the prevalence of hyperbolic discounting in human decision-making, although it may reflect sub-additive evaluations of delays divided into sub-intervals rather than declining impatience (Read, 2001).

Example: Hyperbolic Discounting of Health Effects and of Lives Saved

Surveys and experimental results suggest that hyperbolic discounting often provides a psychologically realistic description of individual time trade-off preferences for health consequences as well as for consumption and cash flows. For example:

- A postal survey of 5,120 randomly selected adults in England, Scotland, and Wales contained six choices between ill health in the near future and ill health in the further future. Respondents were asked to indicate the maximum duration of ill health they would accept in return for a specified delay in the onset of the period of ill health. The responses were modeled via regression analysis. The results suggest that the median discount rate for one's own health is approximately 5.0% for a 5-year delay, 4.6% for an 8-year delay, and 3.8% for a 13-year delay. For evaluations of others' health, the rates were 6.4%, 5.7%, and 3.8%, respectively. Thus, the stationarity axiom required for constant (exponential) discounting is violated, but the hyperbolic model fits these data quite well (Cairns and van der Pol, 2000).
- Given a choice between two hypothetical pollution control programs for saving lives now *vs.* saving more lives later, the responses from 3,200 households surveyed in the United States suggest that discount factors, $a(t)$, decline as follows: 0.46 for lives saved in year 5, 0.346 for year 10, 0.168 for year 25, 0.096 for year 50, and 0.024 for year 100 (Henderson and Langford, 1998, citing studies by Cropper et al., 1992). The best-fitting hyperbolic discount function is approximately:

$$a(t) = 1/(1 + t/5) \qquad \text{(Hyperbolic discount function for lives saved)}$$

with an adjusted R^2 of 0.996.

Interestingly, animal behavior is also often consistent with hyperbolic discounting of rewards, although the time scales used in experiments are typically short, e.g., seconds or minutes rather than years (Richards et al., 1997). However, even hyperbolic discounting does not describe all aspects of time preferences. For example, having a negative experience at all (even for a short time) may receive a weight that is out of proportion to the apparent discount rate for experiences with longer durations (Dolan and Gudex, 1995). How an unpleasant experience (e.g., a visit to a dentist) starts or ends may receive disproportionate weight when the overall quality of the experience is recalled and evaluated (Camerer, 1995). Discounting does not describe the details of how experiences on short time scales are assembled, edited, and integrated into a perceived flow of value on longer time scales.

Turning from descriptive to normative principles, Harvey (1995) showed that, under certain preferential independence and regularity conditions, preferences among time streams can be represented by a general additive value function of the form:

$$V(c_0, c_1, c_2, \dots) = v_0(c_0) + v_1(c_1) + v_2(c_2) + \dots$$

This specializes to the *linear-value form:*

$$V(c_0, c_1, c_2, \dots) = a(t_0)v(c_0) + a(t_1)v(c_1) + a(t_2)v(c_2) + \dots$$

if the outcome variable c is an extensive variable. This means that receiving two amounts c and then c' sufficiently close together in time has approximately the same value to the recipient as receiving $c + c'$ at the earlier time (*ibid,* Theorem 2.1). Here, c_j denotes the amount received at discrete time t_j. The discounting sequence $a(t)$, regarded as a function of time, may be increasing, decreasing, or constant. Harvey (1995) further established that:

a) If $a(t)$ is constant, then the linear value model reduces to the geometric discounting NPV model, $a(t) = a^t$ for $t \geq 0$, and Koopmans' stationarity axiom is satisfied.
b) Preferences for time streams are called *timing regular* if, whenever (c, t_0) is indifferent to $(c', t_0 + s_0)$ for *some* s_0, then there are constants α and β such that $(c, t_0 + s)$ is also indifferent to $(c', t_0 + \alpha s + \beta)$ for *all* $s > 0$. (Here, the notation (c, t) denotes the one-time outcome is which c is received at time t.) If preferences are timing regular, then either $a(t)$ is constant or else

$a(t) = [b/(b + t)]^r$ for $t \geq 0$. $b > 0$ and r are time preference parameters.

c) Preferences are said to be *timing proportional* if, whenever (c_0, t_0) is indifferent to (c_1, t_1) for benefits $c_0 > 0$ and $c_1 > 0$, then $(c_0, t_0 + \Delta_0)$ is also indifferent to $(c_1, t_1 + \Delta_1)$ for any Δ_0 and Δ_1 such that $\Delta_0/\Delta_1 = c_0/c_1$. (A rough interpretation is that one is willing to wait more extra time to get more, and is willing to wait in proportion to the amount extra received.) In the above linear value model, preferences are timing proportional if and only if either $a(t) = 1$ or $a(t) = b/(b + t)$, for $t > 0$.

Thus, the hyperbolic discounting model,

$$a(t) = b/(b + t) = 1/(1 + t/b)\text{ , for } t > 0 \qquad \text{(Hyperbolic discounting)}$$

can be derived from preference conditions such as timing proportionality. It has been proposed for possible use in social cost-benefit analyses with long-delayed consequences that seem too important to be discounted using the usual geometric discounting formula (Harvey, 1995). Decisions affecting nuclear waste repositories, global climate, non-renewable resources, and so forth might be candidates for hyperbolic discounting models. In general, when geometric discounting appears to drastically under-weight far future consequences, hyperbolic discounting may be considered instead.

Chichilnisky's criterion for long-term social utility

An alternative approach for avoiding excessive discounting of far-future events relies on axioms for comparing infinite time sequences of utilities, $\{u_1, u_2, \ldots, u_t, \ldots\}$. These are interpreted as the total social utilities of successive generations, where the *social utility* of each generation is a (possibly weighted) sum of the utilities of the people in it (see Chapter 6 and Dhillon and Mertens, 1999 for justification). Examples of informally stated axioms that may be useful in ranking such utility sequences include the following (Heal, 1998, p. 71):

- The ranking should be sensitive to the long-run future (e.g., changing future utilities can change preferences among sequences even if the changes are far in the future)
- The ranking should be sensitive to the present and near future.
- The social utility or total welfare of a sequence varies continuously (in the sup norm) with changes in the sequence.
- Pareto efficiency (i.e., if one sequence makes one generation better off and no other worse off than a second sequence, then the first is preferred)
- Linearity (total welfare is linear in the welfare of the generations, as implied by the axioms of social utility theory; see Chapter 8). A possible ethical motivation for this axiom is given by assuming that each generation makes decisions as if it were equally likely to be born at any

position in the time sequence, not knowing its own position *a priori*. (This is a form of the veil *of ignorance* described in Chapter 8.)

- Different generations rank alternatives the same way (i.e., the utility function for current consumption, resources, etc. is the same in each generation).

Assuming that utilities are bounded, these axioms, formalized by Chichilnisky (1996, see Heal, 1998) imply that the ranking of utility sequences can be represented by a weighted average of (a) Discounted utility with an arbitrary (not necessarily geometric) set of discounting weights; and (b) Eventual (as t goes to ∞) per-generation utility. In symbols, this representation is:

$$\alpha \int_{[0,\infty]} u(t)w(t)dt + (1-\alpha)\lim_{t\to\infty} u(t),\ 0 < \alpha < 1 \quad \text{(Intergenerational Utility)},$$

where w is any countably additive measure, i.e., a weighting function with $\int_{[0,\infty]} w(t)dt = 1$. Suggestive as such results are, they ignore the realistic complexities introduced by *uncertainties* e.g., about the preferences, sizes, and existence of future generations and the potential effects of current actions on future generations and their preferences.

Uncertainty about the preferences and values of future generations for environmental amenities, nonrenewable natural resources, and other consequences of current policy decisions complicate ethical decision-making. Concern for the preferences and well-being of future generations, combined with uncertainty about how well their preferences will be adapted to the conditions they inherit (and whether allowing for such adaptation is morally desirable) extend to the sphere of policy-making many difficulties and ambiguities arising in the foundations of intertemporal choice for individuals, as discussed in Section 7.5. It also motivates considerations of inter-generational risk equity for decisions having prolonged but uncertain consequences (e.g., Hammitt and Harvey, 2000). Initial attempts to model inter-generational welfare in the presence of uncertain future preferences suggest that uncertainty about the preferences of future generations does not by itself imply a motive for current conservation efforts. If it is suspected that future generations will probably value non-renewable resources more than current ones, however, then economic risk-cost-benefit analysis with an inter-generational welfare objective function and uncertain preferences for future generations does motivate current conservation (Heal et al., 1996).

6.2 Sequential Choices and Effects of Event Sequencing

One of the most interesting ways in which the Reduction Axiom of rational (EU) decision-making fails to describe real decisions and behaviors is that, for many people, the *sequence* of events and actions leading to a consequence matters. In fact, decision-makers often act as if the utilities for final consequences resulting from particular sequences of gains and losses depend on what has happened and how recently. Moreover, the opportunity to take actions as part of a sequence of events that determines final outcomes may itself affect how the outcomes are valued (Patt and Zeckhauser, 2000).

The psychology of aspiration levels, reference points, and asymmetric weighting of gains and losses, as in Cumulative Prospect Theory, suggests that how news about a multi-part consequence is broken may affect its perceived utility. In other words, the path used to resolve a conjunction of uncertainties may affect the perceived value of the final consequence. The familiar question "Which do you want first, the good news or the bad news?" reflects this possibility.

The following examples deal with how sequences of decisions affect *actions,* rather than with how they affect *evaluation* of consequences. Both may be important in risk communication and risk management deliberations.

Example: Consolation Effects in Insurance Decisions

Many people are more willing to buy insurance – and more willing to go to some trouble to collect insurance compensation following a loss – on a highly valued object than on a less valued object, even when the amount of compensation is held fixed (Hee and Kunreuther, 2000). This suggests that they attach positive value to *consolation,* i.e., to receiving some positive reward (an insurance payment) following a loss, with the value of a given consolation payment being higher after a more grievous loss. In deciding whether to insure a beloved object, for example, people may evaluate a *sequence* (loss of the object followed by the inconvenience of collecting compensation followed by the receipt of the compensation amount) differently from how they would evaluate an unsequenced triple of attributes representing the loss, inconvenience, and compensation amounts.

Example: The "House Money Effect": Changing Reference Levels and Aspirations in a Sequence of Decisions

Gamblers and game-show participants, as well as participants in controlled experiments and in the stock market, often make bets that they would not ordinarily make if they first gain or lose more than initially expected. Specifically, the *house money effect* (Thaler, 1992) refers to a tendency of subjects who have just experienced an unanticipated monetary gain to become risk-seeking for monetary gambles that are small compared to it. For example, individuals may be willing to bet more of their recent winnings (which they may perceive as the "gambling house's money") than they would of their "own money", starting from the same financial position. This indicates a failure or delay in *asset integration,* i.e., in perceiving newly acquired gains as part of one's base level of assets. Instead, people often perceive the *sequence* of decisions as an entity and attribute changes in asset position to the sequence as a whole, rather than analyzing and optimizing each decision separately.

Example: The Break-Even Effect in a Sequence of Decisions

Individuals who have just lost money from a sequence of decisions may be unwilling to accept fair bets whose payoffs are small relative to the amount already lost, and yet be willing to take fair bets that allow the possibility of recouping the losses. This is sometimes called the *break-even effect* (Thaler, 1992). Both the house money effect and the break-even effect can be interpreted in terms of a three-argument utility function. The arguments are: a reference level of wealth, the cumulative gain or loss from the sequence of decisions taken so far, and the incremental gain or loss from the current decision (Neilson, 1998). The utility function allows for the commonly observed pattern of risk-aversion for gains and risk-seeking for losses for the first choice in the sequence, while accounting for house money and break-even effects in subsequent choices.

6.3 Repeated Choices and Iterated Prospects

Suppose that a monetary gamble is too risky for a decision-maker to accept, i.e., its certainty equivalent is negative. If the d.m. is offered a finite sequence of N such gambles, might it be rational to accept the sequence (at least for large enough N), even though no one of them would be acceptable in isolation? If each individual gamble is unacceptable at all levels of wealth, then in principle the answer must be no, as can be proved by induction (starting from the last gamble and working backward). Yet, in practice, many individuals will accept a finite sequence of individually unacceptable gambles (Benartzi and Thaler, 1999). Under some conditions involving low-probability, high-loss gambles, many individuals will also reject a finite sequence of individually acceptable gambles (Langer and Weber, 2001). These phenomena appear to be more rooted in cognitive biases in decision-making (such as "narrow framing" of a decision problem to focus on its isolated parts) than in temporal aspects of decision-making.

6.4 Preferences for Timing of Uncertainty Resolution

When future consequences are uncertain, discounting alone no longer suffices to describe how intertemporal trade-offs are made or should be made. New considerations arise, such as anxiety while waiting to learn what will happen or what is likely to happen and preferences for earlier *vs.* later resolution of uncertainties. Moreover, when the prospect being discounted is risky, qualitative patterns of discounting seem to be affected in unexpected ways. For example, in matching tasks, a larger discount rate holds for the first period than for subsequent ones for both risky and certain payoffs, but this does not appear to be true in choice tasks (Albrecht and Weber, 1997).

The objects for which preferences must be defined are also more complex. Instead of deterministic time streams of consequences, it is natural to use recursively defined objects, such as decision trees with time-stamped nodes in which the first chance node in the tree is a lottery yielding sub-trees as prizes. Rewards or penalties may be collected as one traverses the tree. Together with the times at which they are received and the probabilities of different branches, these rewards or penalties contribute to the value of the temporal prospect represented by the tree. But anxiety about up-coming uncertainties, disappointment or relief about recently resolved ones, variations in expectations and reference points, and other psychological adjustments and reactions may also strongly affect the perceived value and risk of the tree process.

Preferences for time streams of economic consequences

The theory of preferences for uncertain time streams of *consumption* has been extensively developed in economics. Substitution and complement effects among consumption in different periods, formation of habits and acquired tastes over time, and other dynamic aspects of preferences are now routinely included in economic and econometric models of consumer preferences (e.g., Deaton, 1992). Indirect preferences for earlier resolution of uncertainty based on ability to *use* this information to take advantageous actions have also been extensively modeled in the context of consumption and saving decisions, borrowing opportunities, and life-cycle consumption planning.

Preferences for *income streams* induced by (a) Preferences for consumption streams and (b) Availability of borrowing and lending opportunities in general *cannot* be represented by EU theory. Applying standard decision analysis methods to uncertain income streams, such as those generated by a decision tree with delayed resolution of uncertainties, implicitly assumes that uncertainties are resolved sooner than they really are. They can therefore substantially overestimate the true values of risky financial projects (Smith, 1998). Correct "present certainty equivalent values" (PCEVs) for such temporal prospects over *income* streams can be accomplished using a simple dynamic programming "roll-back" procedure if the d.m.'s preferences for *consumption* streams can be represented by additive-exponential utility functions. These have the form:

$U(c_0, c_1, ..., c_T) = -\Sigma^T_{t=0} k_t \exp(-c_t / \rho_t)$ (Additive-exponential utility function) where the utility weights k_t and the risk tolerances ρ_t are assumed to be positive. They express the d.m.'s time preferences and risk preferences for consumption streams, respectively. If utility for consumption does not have this special form, then correct calculation of PCEVs may be computationally intractable (*ibid,* p. 1696).

Preferences for temporal prospects and decision trees

Despite these developments in the economic theory of preferences for consumption and income time streams, temporal prospects have played only a minor role in the development of most modern utility theory and decision and risk analysis. By far the dominant approach represents decisions as choices among static probability distributions (or probability measures) for consequences. For example, although decision trees intrinsically represent the dynamic refinement of probability measures over time as uncertainties are resolved and choices are made (i.e., a controlled filtration), that is not how they are typically analyzed. Instead, trees are usually reduced to algebraically equivalent conditional probability distributions (or their certainty equivalents) for final "terminal asset positions" or other indicators of the values of consequences at the tips of the tree. Expected utility theory is then invoked to justify choosing decisions that select the EU-maximizing probability distribution or the most preferred certainty equivalent.

The famous "averaging out and rolling back" dynamic programming procedure for decision trees (Raiffa, 1968) rests solidly on EU foundations. [Indeed, the roll-back procedure *requires* the Independence axiom of EU theory (LaValle and Wapman, 1985). Without this axiom, the computational tractability of dynamic programming procedures for sequential decisions is lost.] However, the averaging out and rolling back procedure does not address preferences for the order and timing of decisions, incremental payoffs, and resolutions of uncertainty within the tree. EU theory treats all trees that give the same probability distribution over final outcomes as being equivalent, in that any d.m. should be indifferent among them. Similarly, if the d.m. has only one possible course of action available, then EU theory implies that she must be indifferent to when (or whether) information is received about its probable consequences.

By contrast, recent preference models for stochastic consumption streams, starting with two seminal papers by Kreps and Porteus (1978, 1979), allow for the possibility that decision-makers may prefer earlier to later resolution of uncertainties (or conversely) on purely psychological grounds. Aversion to anxiety or sustained uncertainty may, in these theories, create different preferences for trees that would be considered equivalent under EU theory. Descriptive models of preferences have similar implications for sequential decision-making even when the consequences of interest are not necessarily consumption streams. For example, Gul's disappointment model (Gul,1991, see Table 3.1) implies that *disappointment-averse individuals will strictly prefer uncertainties to be resolved simultaneously rather than sequentially* (Palacios-Huerta, 1999). This may help to explain the apparent preference of many at-risk individuals to remain ignorant about their own health status, as has sometimes been observed in HIV testing (*ibid,* 261-262).

However, possible deviations from EU theory are sharply limited if other attractive properties for preferences are assumed. For example:

- Any d.m. whose preferences do not satisfy the Independence axiom of EU theory *must* exhibit dynamic inconsistency for some decision trees, and must also assign a negative value to free information in some situations (e.g., Quiggin, 1993, Chapter 9). "Dynamic inconsistency" means that the d.m. will choose an action at the first stage of a two-stage tree based on an assumption or plan about what he will do at the next stage – but will then not do it when that stage is reached. A key theoretical result is the following. *If temporal preferences for acts in decision trees satisfy dynamic consistency, then the Independence axiom (and hence EU theory, if the remaining axioms hold) must also hold.* Dynamic consistency is often referred to as *recursivity* of temporal preferences. Versions of this requirement appear in many recent temporal utility theories.
- Suppose that: (a) Preferences have an RDEU representation; (b) Preferences are dynamically consistent; and (c) *Sequential consistency* holds, i.e., a decision-maker evaluates the stages of a temporal lottery and the lotteries induced by each strategy in a decision tree using the same family of preferences. Then preferences for temporal lotteries can be represented by EU except possibly at the last stage (Sarin and Wakker, 1998).
- Suppose that preferences for multi-stage lotteries satisfy (a) dynamic consistency, (b) preference for early resolution of uncertainty, and (c) first-order stochastic dominance (FSD) conditions. If these preferences, when restricted to certain classes of lotteries with one-shot resolution of uncertainties, also satisfy RDEU and risk-aversion, then they can be represented by EU (Grant et al., 2000).

The main theme of these results is that, even when some forms of anxiety and intrinsic preferences for early or late receipt of information are considered, several normative models of preferences for temporal prospects lead back to an EU representation, except perhaps in one or two periods.

6.5 Changeable and Uncertain Preferences

People are often uncertain about their own future preferences. One who has not tried alcohol, cigarettes, coffee, or regular exercise may be uncertain about how much he might eventually enjoy them. One's own subjective reactions to a range of experiences, including acquired tastes, aversions, habits or addictions may also be hard to predict or evaluate in advance. Indeed, even without uncertainty, current and future preferences for a habit, addiction, or experience may be very different (Pollak 1970; Orphanides and Zervos, 1995). This has led to the concept of a divided self or a "team" of selves distributed over time and having different preferences at different times for the same activities and their consequences (Schelling, 1984).

Moreover, people often are uncertain about their own characters, capabilities, and future choice behaviors. For example, in contrast to the implications of most models of rational decision-making, some people may prefer to select from reduced choice sets (e.g., to dine at a restaurant with a restricted menu of healthy items instead of at a comparable restaurant with a superset of selections) if the larger choice set includes temptations that they fear succumbing to and then regretting (Dekel et al., 2001). Uncertainty about one's own reactions may be born of experience. For example, there is evidence that people tend to mis-predict their own responses to information such as the outcome of HIV tests, fearing that a positive test result will lead to more distress than it actually does (Sieff et al., 1999). Many people also exhibit *projection bias,* i.e., a tendency to underestimate how greatly future preferences may differ from current ones and the degree to which their own current behaviors may affect their future utilities (Loewenstein et al., 2000).

Uncertainty about one's own (current or future) preferences and reactions may prompt decision-makers to take actions designed to learn about or signal to themselves their own characters, likely future behaviors and preferences (Bodner and Prelec, 2001). A jogger may persist in jogging in the rain because she believes that doing so signals desirable characteristics, such as strength of will or commitment to a regimen, that is expected to bring future benefits. Analogously, in an experiment in which tolerance for pain from hot or cold water was represented as being diagnostic of a healthy heart (more specifically, absence of a certain heart defect), subjects systematically showed greater tolerance for the condition (hot or cold) that they had been told was associated with a healthy heart (Quattrone and Tversky, 1984). This was not because tolerating the painful stimulus would *cause* a healthy heart, but because it might *signal* a healthy condition. People are willing to tolerate some additional pain to signal "good news" to themselves. This propensity applies to health-related behaviors and choices. For example, part of the utility from following a healthy diet or exercise regimen may be based on the indirect "diagnostic utility" derived from the fact that the behavior suggests good news about one's resolution, character, or future behavior, as opposed to the direct "consequence utility" from the behavior (Bodner and Prelec, 2001).

Several extensions of EU theory have been made to model decision-making with uncertain future preferences. A traditional approach is to allow for *state-dependent preferences,* in which the utility function for consequences may depend on the state of the world in which the consequences occur (Karni, 1985; Wakker and Zank, 1999). State-dependent utilities have long been used to model health insurance-purchasing decisions and other decisions in which future preferences for financial or other consequences may depend on the state of the decision-maker (e.g., healthy, ill, or dead).

The extent of subjective uncertainty about future preferences and/or future contingencies can be represented mathematically in terms of *preferences for flexibility* in future choices (Dekel et al., 2001; Nehring, 1999). For someone with no uncertainty about future preferences or contingencies, the expected utility of a choice set is the expected utility of its most-preferred member, and hence all other choices in the set are irrelevant for determining this value. But suppose that preferences over members of the choice set at the time of decision (i.e., when a selection from the choice set must be made) are uncertain at the (presumably earlier) time when the choice set itself (or a lottery over choice sets) must be selected. Then, at the earlier date, the decision-maker may have a preference for (future) choice sets that offer more options, and hence a preference for current actions that lead to such future choice sets. Formalizing this idea leads to generalizations of stochastic dominance comparisons for acts that determine probable future choice sets (Nehring, 1999). It also leads to mathematical generalizations of expected utility representations, now defined both *ex ante* and *ex post* for a decision, that express the preferences for flexibility in future actions induced by uncertainties about future preferences (Dekel et al., 2001).

The well-documented malleability and inconsistency of preferences over time (Strotz, 1956; Thaler, 1991) and uncertainty about future (and perhaps present) preferences, character, and behavior raise fundamental challenges for defining what rational choice really means. Among the issues needing clarification are (a) Whether it makes sense to have (and act on) preferences for future preferences and/or character traits ("meta-preferences" in the terminology of Bodner and Prelec, 2001); (b) Whether one can have well-defined preferences among alternative acts whose consequences have never been experienced at the time of decision (and so are not truly known, in the sense of not having been directly experienced before); and (c) What kinds of self-limiting behavior (such as deliberate rejection of larger choice sets or binding precommitments to courses of action that one knows she may dislike later) are truly "rational", given the divided nature of the self over time (Schelling, 1984; Bodner and Prelec, 2001).

These foundational philosophical issues for temporal preferences may be very relevant for individual decisions affecting health and safety, such as decisions to smoke or to experiment with recreational drugs. But a well-developed normative theory that embraces these aspects of temporal decision-making and preferences, as well as the psychological processes by which experiences over time are integrated into an overall assessment of retrospective satisfaction or preference, has yet to be created.

6.6 Choosing Among Stochastic Processes for Health States

Health risks affect the rates at which people enter and leave different health states over time. For example, high air pollution levels may increase mortality and morbidity rates among the elderly and the frequency or duration of asthma attacks among asthmatics. Exposure-related health effects may occur more than once, with age and other covariates affecting the probable time between occurrences for any given exposure history. Thus, exposure and other risk factors can influence the average amount of time that people spend in different health states and the quality of life associated with health state durations and transition rates among them.

How does quality of life vary with transition rates among health states? This is hard to answer (or perhaps even to think about usefully) in the abstract. Even for small examples, the issues involved are daunting. They include possible mental adjustments and adaptations to long-lasting health states, anxiety about potential future adverse transitions, dread or anticipation of future health states, aversion to risks that include death or extreme debilitation as possible outcomes, relief at unexpected improvements in health prospects, and the integration of such reactions into an overall evaluation of quality of life.

A slight simplification may be achieved by replacing the criterion "quality of life" with the more definite criterion of preference between alternative transition processes. But how to construct such preferences from (a) more basic preferences for consequences – in this case, process histories; and (b) their probabilities (which can be calculated from transition rates for the stochastic process) is still challenging.

Example: Valuing Health State Transition Processes

Consider the following health state transition process:

Well ⇒ Not Well ⇒ Dead

Let λ_{01} and λ_{12} denote the transition rates from Well to Not Well and from Not Well to Dead, respectively, measured in units of expected transitions per unit time per person in the Well state and the Not Well state, respectively. For simplicity, suppose these rates are constant and known. Assume also that the Not Well state and the transition from Not Well to Dead have been vividly and clearly defined, so that the prospects involved are well understood by a decision-maker. Then what principles should be used to construct indifference curves for parameter pairs (λ_{01}, λ_{12})? For example, to what extent should someone be willing to trade-off decreased expected time in the Well state (i.e., higher values of λ_{01}) against longer expected life (i.e., lower values of λ_{12})?

Now, consider the following more general process:

Well $\Leftrightarrow$ *Slightly Ill* $\Rightarrow$ *Very Ill* $\Rightarrow$ *Dead*

Again, for simplicity, assume that this is a Markov process with states 0 = Well, 1 = Slightly Ill, 2 = Very Ill, 3 = Dead, with all states and transitions defined well enough to be clearly understood by the decision-maker. The vector of transition rate parameters describing the process is:

$$(\lambda_{01}, \lambda_{10}, \lambda_{12}, \lambda_{23}).$$

It is reasonable to suppose that preferences are increasing in λ_{10} (i.e., the *Slightly Ill* $\Rightarrow$ *Well* transition rate) and decreasing in the remaining three parameters. The trade-offs among these parameters might be approached by applying multiattribute utility theory directly to such parameter vectors, rather than to the corresponding stochastic processes of times spent in different health studies. However, such applications of MAUT to evaluate stochastic processes are not common.

Mean Quality-Adjusted Duration

A simplified approach to evaluating stochastic processes for health states is to use the quality-adjusted duration formula:

$$Q = \Sigma_x v(x) T(x) \qquad \text{(Quality-Adjusted Duration).}$$

Here:

- $v(x)$ = quality factor for time spent in state x.
- Typically, $v(x)$ is normalized so that $v(Well) = 1$, $v(Dead) = 0$. (This must be modified if states worse than death are to be considered.)
- $T(x)$ is the time spent in state x.

For general stochastic process over health states, including non-Markov models, the mean value of Q can be estimated by Monte-Carlo simulation of the process. For the special case of Markov and semi-Markov processes over health states, special algorithms (roll-back and value iteration algorithms from dynamic programming) can also be used to calculate mean Q values (Hazen and Pellissier, 1996). Different processes induced by different risk management actions – e.g., public health measures that change the transition rates among states – can therefore be compared by their estimated mean Q values. However, this simple, tractable approach is generally unsatisfactory because it ignores risk aversion for time spent in different states.

Example: Mean Quality-Adjusted Duration is Not Satisfactory for Decision-Making (Hazen and Pellissier, 1996)

Should a 40-year old white male with a 0.025 annual mortality rate accept an intervention that gives a 50% chance of immediate death in exchange for a 50% reduction in the annual mortality rate if immediate death does not occur? Most people would say no. Yet, if mean quality-adjusted duration of life were the sole decision criterion, these two prospects would be equally valued. For, without the intervention, the expected remaining lifetime is:

$$1/0.025 = 40 \text{ years.}$$

With the intervention, the expected remaining lifetime is:

$$(0.5)(0) + (0.5)(1/0.0125) = 40 \text{ years.}$$

Thus, a decision-maker whose preferences depend only on the mean Q value would be indifferent between these two products. But many risk-averse decision-makers would strongly reject the chance to play "double or nothing" (at even odds) for remaining duration of life. Such examples show the need to incorporate risk aversion into the comparison of health state stochastic processes.

Recursive Utility Models for Stochastic Processes

A growing literature in mathematical economics has started to address the challenge of developing value functions to represent preferences for stochastic processes having consumption levels at different time as outcomes. For evaluation of such processes to be computationally tractable, it is necessary to make assumptions such as that future consumption is utility independent (UI) of past consumption. This is necessary and sufficient for the existence of "memoryless preference summaries" (Meyer, 1976). A *preference summary* (called a *state descriptor* by Meyer) is a quantity calculated from the history of a process such that preferences depend on the past only through the preference summary computed from its past history. A preference summary is *memoryless* if it depends only on the current state and not on the history preceding it.

The assumption that preference summaries exist leads to *recursive value functions* in which the current value of a stochastic process depends only on: (a) the current state; (b) the preference summary calculated from the history of the process so far; (c) the current (state-dependent) probability distribution over next states; and (d) the value of the stochastic process starting from each next state (Kreps and Porteus, 1978). The recursion comes from the fact that the current value of the process depends, via part (d), on the values of its future possible sub-processes. Mathematically, this recursion is represented by a difference equation. The economics literature has also modeled recursive utilities using *aggregators*. An aggregator is a bivariate function that takes as its two input arguments the current state (i.e., consumption level) and the certainty equivalent of the value of the possible future sub-processes, conditioned on the history of the process so far and on the next state. As output, it calculates the current value of the process. Some important aspects of temporal utility, especially intrinsic preferences for earlier *vs.* later resolutions of uncertainty, have been greatly clarified by these models. However, economic models of recursive utilities have typically focused on consumption trajectories rather than on health state trajectories with death as a possible outcome.

Recursive utility models have more recently been developed for a broad class of health state transition processes that include continuous-time Markov and semi-Markov processes. Under mild regularity conditions (e.g., continuity of the utility function in its duration arguments), it can be shown that *a memoryless preference summary for a continuous-time semi-Markov process exists if and only if preferences among stochastic processes can be represented by quality-adjusted duration with state-dependent discount rates* (Hazen and Pellissier, 1996). Mathematically, the discounted quality-adjusted utility of being in state x for time $t \geq 0$ is expressed as:

$$u(x, t) = [v(x)/a(x)][1 - e^{-a(x)t}] \text{ for } a(x) \neq 0 \quad \text{(Memoryless Utility)},$$

where

- $v(x)$ = quality factor for state x
- $a(x)$ = discount factor for state x.

In effect, such preferences can be represented by a utility function with a constant risk attitude, possibly different for each state, for durations spent in different states.

In the special case of a Markov transition processes with the rate of departure from state x being λ, the utility of being in health state x and facing departure rate λ is just:

$$u(x, \lambda) = v(x)/[a(x) + \lambda] \text{ for } a(x) + \lambda > 0. \quad \text{(State utility in Markov process)}$$

This provides the needed generalization of quality-adjusted duration to account for risk-aversion for duration-in-state, represented by the factor $a(x)$. Assessing the two numbers $v(x)$ and $a(x)$ for each health state x in a Markov transition process provides all the information required to compare alternative processes described by different sets of transition rates, given the initial state (or its probability distribution). In particular, the problems posed at the beginning of this section, dealing with trade-offs among the transition rate parameters for Markov processes such as:

Well $\Leftrightarrow$ *Slightly Ill* $\Rightarrow$ *Very Ill* $\Rightarrow$ *Dead*

with parameter vector (λ_{01}, λ_{10}, λ_{12}, λ_{23}) can now be solved, once $v(x)$ and $a(x)$ have been assessed for each of the four health states.

The memoryless property for preference summaries is a strong requirement. It implies the restricted representation of preferences for Markov transition processes just described. This representation seems over-specialized for many purposes. For example, it seems plausible that the utility associated with a particular health state might depend on the previous state (is the current state better or worse?) A transition might itself carry some value, either positive or negative, e.g., because of the disappointment or elation when an unexpected change takes place to a new, probably long-lasting health state. These modifications can be accommodated if the preference summary allows memory of the most recent state visited. Then the recursive utility representation of preferences allows a much more flexible functional form in which transitions between states may incur rewards or penalties and trigger a multiplier (that can depend on the start and end states of the transition) that amplifies or diminishes the utility of the future process.

Utility function assessment procedures are fairly straightforward for both of these classes of preferences, and also for the still more general class in which the preference summary is allowed to include the time spent in the most recent state (*ibid*). In each case, one is required to evaluate a small set of numbers [such as $v(x)$ and $a(x)$, in the memoryless case] for each state and perhaps for each transition. Then, the values of different health state transition processes can be calculated (via the roll back and other algorithms described by Hazen and Pellissier, 1996) or estimated by Monte Carlo simulation. The resulting values indicate normative preferences for alternative transition processes among health states, just as NM utilities indicate normative preferences for probability distributions over consequences. Thus, recursive utility models with succinct preference summaries (e.g., memoryless, previous-state only, or previous-state and duration in it only) supply a computationally practical technique for deriving preferences among Markov and semi-Markov processes. (Moreover, Hazen and Pellissier, 1996, show that these are the *only* models that lead to computationally practical techniques for comparing such processes.)

In summary, preference models have been developed for an important class of stochastic processes (continuous-time Semi-Markov health state transition processes, and hence Markov models) and for an important class of temporal preferences (those that can be represented using low-dimensional preference summaries to capture the dependence of preferences for future processes on past experience). This modeling technology covers many interesting examples. As yet, no comparable approach exists for more general stochastic processes and preferences.

7. CONCLUSIONS

The topics introduced in this chapter, especially attempts to extend expected utility theory to represent and prescribe preferences for risk equity, multi-person consequences, multi-criteria consequences, intergenerational welfare, and a variety of temporal prospects, are at the forefront of current research in decision analysis and economics. Yet, these issues are central to many current applied risk analyses and risk management decisions. No well-developed prescriptive theory, analogous in scope and normative power to SAUT but applicable to the more complex prospects and decisions addressed in this chapter, is yet available. But practical risk management decisions cannot wait. Further development of the issues and conceptual frameworks introduced in this chapter may be among the most practically important challenges for operations researchers and risk analysts working on the foundations and technologies of effective decision-making and selection among risky prospects that represent realistically complex choices.

Meanwhile, in the absence of fully satisfying normative theories for choosing among these complex prospects, much can still be accomplished within the framework of traditional EU theory and SAUT. Several decades of successful applications attest to the power and practical value of using formal representations (e.g., influence diagrams, decision trees, or simulation models) to help structure and make complex decisions. Unresolved foundational issues, such as how to deal with changing preferences can be bypassed, as suggested in Chapter 6, by using current preferences for risk profiles over a single "value" attribute. "Value" represents current preferences for various outcomes as of a specific future date. The potential subtleties and conceptual complexities involved in fully defining this concept may simply be ignored in favor of developing operational elicitation procedures that ask people for their preferences without dwelling on precise definitions.

The current state-of-the-art in applied decision and risk analysis perforce favors such a simplified approach. It is useful. Many decision-makers have felt that their decisions have been clarified and improved with the help of traditional EU-based SAUT or MAUT decision theory (Clemen, 1996). Meanwhile, there are many opportunities to extend the EU framework (or perhaps replace it) to more explicitly address temporal prospects and preferences, risk equity, uncertain (current or future) preferences, and preferences of multiple parties. These opportunities present an ongoing challenge to decision and risk theorists. They highlight areas where further improvements are both needed and expected in the normative theory and practical application of decision analysis to problems involving health risks.

Chapter 8

Multi-Party Risk Management Decision Processes

1. INTRODUCTION: RISK MANAGEMENT DECISION PROCESSES

Societal risk management decisions are usually made by multiple participants and reflect the interests of multiple stakeholders with partially conflicting interests and beliefs. The participants interact through *decision processes* in which individual proposals, choices, offers, commitments, and actions or behaviors are iteratively modified until an outcome is reached. Examples of such decision processes include contracts, negotiation (with or without arbitration), auctions, voting, tort liability systems, regulation, market processes and legislative politics.

In general, *risk management decision processes* refer to procedures, typically with multiple stages or steps, by which multiple participants jointly determine how risks are to be managed. Each participant uses information about what others have done, claimed, or offered to decide what to do next. Their interacting decisions determine how risks are managed. This chapter introduces methods to analyze and model such processes.

Table 1 list various risk management decision processes organized by (a) The types of decisions being made, constrained, informed, or facilitated by the risk management decision process (shown in the rows); and (b) The type of decision process used – centralized, decentralized, or a blend labeled "participatory" (shown in the columns). The entries in the table are different approaches or decision processes for managing health and safety risks. The conceptual boundaries between columns are somewhat fuzzy, although the distinctions among prototypal cases seem clear.

Table 1: Decision processes for managing risks

Process:	Centralized	Decentralized	Participatory
Risk Management Decision Problem			
Restrict private individual choices and activities: *What to allow?*	Paternalistic regulation and laws (e.g., speed limits)	Warning labels. liability rules (e.g., comparative negligence rules)	Legislation, safety laws
Restrict industrial activities and products: *What to permit?*	Regulation (EPA, OSHA, CPSC) Inspection (USDA)	Tort litigation for negligence, nuisance, etc.	Standards and grades, labels; environmental laws
Restrict voluntary economic transactions	Bans, regulation, licenses; control of substances		Laws (drugs, pornography)
Determine due care and prudent business practice: *How much care is enough?*	Set test /labeling requirements; set an approval processes (FDA)	Product liability; product testing, decision analysis (Chapter 5)	Standards, grades, labels, inspections, grades (e.g., USDA)
Site investigation and remediation: *What to do next?*	Optimize resource allocation (e.g., set priority indices)	Real-estate and environmental laws; RCRA	Risk communicatio n with community
Pollution abatement: *What to require?*	Optimize mix of measures; Set regulated standards	Minimize costs of complying with laws, regulations, and standards	Environmental laws
Compensation, assigning liability	Fiat	Contracts, tort law, litigation processes	Shapley value
Provision of public goods	Social utility theory; risk-cost-benefit analysis		Voting, political processes
Siting hazardous facilities	MAUT, MCDM, optimization (e.g., minimize average or maximum risk or exposure)	Spatial location games and processes (players move over time)	Fair division Auctions and bidding
Methods and models	MAUT, social utility theory, risk-cost-benefit analysis, MCDM, optimization	Contracts, games, arbitration, EU theory, markets , negotiation, bargaining	Collective choice, voting, bidding and auctions, social contract

A highly *centralized* decision process emphasizes command-control. A decision-making entity, e.g., a regulatory agency, collects information about risks (i.e., probabilities of consequences given actions) and tells agents in the economy what they must do. Compliance with these mandates is usually intended to assure that risks stay at levels that the decision-maker considers acceptable. Social utility theory, discussed in the next section, attempts to provide a formal framework for deciding on behalf of a set of individuals what mandated risk-cost-benefit trade-offs are in their best collective interests.

Decentralized approaches to risk management emphasize choices made by individuals, either privately (perhaps using EU theory) or through transactions and contracts with each other. A fully decentralized approach allows individuals to freely choose what risks to create or accept and to enter into voluntary contracts and transactions with each other – but it also sets rules (e.g., for liability and litigation) that the individuals may use in managing the risks arising from their transactions. The rules will in general result in a distribution of decentralized decisions and outcomes for the conduct of hazardous activities and subsequent allocation of responsibility, compensation, and penalties. Game-theoretic models and methods discussed in this chapter seek to predict the likely behaviors and outcomes resulting when self-interested rational players act within a framework of rules.

Participatory processes let those affected participate in decisions or in designing the rules used to make decisions. These processes may collect information about preferences for consequences, and perhaps about beliefs and preferences for actions, from participants and stakeholders to inform the risk management process. (A *stakeholder* is defined as anyone who cares about the outcome.) Collective choices that cannot be made via bilateral agreements, e.g., social decisions such as whether to have a mandatory school vaccination program, are commonly also made through participatory processes. Of course, in a republic, participation may be indirect, in that the direct participants are representatives of larger constituencies. Models of collective choice as well as highly idealized models of political and legislative processes, ranging from social contract theory to more detailed models of agenda-setting and political processes, provide analytic frameworks for understanding many of the philosophical and some of the practical aspects of participatory processes. Alternatively, the study of *risk communication* as a sub-discipline within risk analysis has yielded a number of pragmatic insights into what works (and what doesn't) for achieving various goals by sharing risk information with individuals or communities.

The suggested partitioning of processes into centralized, decentralized, and participatory is inexact. For example, a centralized planning body might well design a process for liability and litigation – or, for that matter, a process for arbitration and adjudication of contracts – in order to promote effective decentralized risk management via tort and contract law. Indeed, a major challenge explored in this chapter is how a centralized body or a single "social decision-make" should design a set of rules to benefit individuals as they engage in various economic transactions such as hazardous production, consumption, and occupational activities. On the other hand, centralized decision processes may also be justified by appealing to (perhaps hypothetical) participatory processes, such as a hypothetical social contract or an actual legislative process.

Despite such overlaps, Table 1 shows that the expected utility (EU) and optimization approaches to single-person decision-making studied in chapters 5-7 constitute only a small subset of the models and methods useful for understanding risk management decision processes. They are primarily relevant for individual decision-making (e.g., how much insurance to buy, what activities to engage in, and how much care to take, possibly in light of warnings and other risk information provided by a centralized authority) and perhaps for centralized decision-making on behalf of individuals in society (the "social decision-maker" perspective explored in Chapters 1 and 7.) For understanding true multi-person decision-making, it is necessary to use different models such as those from game theory and collective choice theory, introduced in this chapter.

A risk management decision process may be characterized by the following elements.

Defining Characteristics of Decision Processes

1. *Who makes the decision?* For example, is a single *social decision-maker* (SDM), perhaps an elected official, trying to act in society's best interest? Do the affected parties (or *stakeholders*) themselves decide, perhaps via bidding procedures, voluntary transactions, arbitration, or political negotiations? Is the decision constrained by law? How is it enforced (e.g., through litigation, regulatory monitoring, etc.)
2. *What is its scope?* Does it address a specific case or set of cases, as in many retrospective tort liability, and compensation decisions? Or, does it establish general rules that will govern future cases and/or provide incentives that shape future actions and that may prevent future cases? Is the decision intended to be lasting, or is it only an interim measure, designed to be replaced or updated when better QRA information becomes available? Does it contain explicit conditions that would trigger an update or change in the current decision?
3. *Who bears its consequences?* Who is exposed to what risks before and after the decision has been implemented? Who receives the benefits and who pays the costs of implementation and of compliance?
4. *How is the decision made?* Is the decision process centralized or decentralized, collective or distributed, hierarchical or single-level? Who

provides what inputs and receives what outputs? Is it made by a single decision-maker, by a team with shared goals, by a syndicate with mixed interests, by a small group of decision makers, or by many economic agents responding to a mixed market economy constrained by laws, regulations, and litigation activities?

5. *How is it implemented?* Who implements what, how, using what resources? How is the execution of the decision monitored and enforced?
6. *How is it evaluated, and by whom?* Are the consequences of the decision identified and attributed to it and used to improve future decisions or decision processes? Is the decision based on preferences for actions or on preferences for their consequences (or on something else, such as the perceived fairness and legitimacy of the decision process)? How are the legitimacy, efficiency, effectiveness, and fairness of particular decisions and of the decision processes and institutions that generate them judged and modified over time?

The following sections show how to apply ideas from social utility theory, game theory, collective choice theory, and risk communication to address these questions for a range of risk management decision processes.

The theories and models introduced in the following sections provide frameworks for understanding both how risk management decisions should be made and how ideally rational people might make them. Applying them to various risk management decision problems suggests several distinct potential roles for a social decision-maker in improving risk management. Risk management problems often spark hotly contested policy debates because they involve conflicting opinions and intuitions about the proper roles of rights, duties, private ethics, collective choice rules, and the fairness and equity of outcomes and processes. These factors, too, will be modeled and analyzed. Finally, some practical aspects of multi-party risk management will be addressed in the context of insights and guidelines for effective risk communication.

Motivating Applications and Examples

A fully adequate theory of multi-party risk management should be able to address a range of challenging philosophical and policy questions. This section lists representative examples to explain and illustrate the categories in Table 1. Although formal theories of multi-party risk management, such as the game-theory models developed later in this chapter, do not fully answer all of these questions, they do provide useful frameworks for addressing most of them, as shown subsequently.

Key examples of the risk management questions that useful normative theory should inform are as follows.

- *Restricting individual activities.* Should cigarette smoking be banned in public places to protect against the health effects of environmental tobacco smoke (ETS)? Should parents be forced to vaccinate their children against measles and other infectious diseases? Which, if any, of the following should be made illegal in order to protect health: smoking cigarettes in private, drinking alcohol, eating junk food (or fatty foods, for people at risk of heart attacks), recreational drug use, riding a motorcycle (or skiing or bicycling or roller-blading) without a helmet, driving a car without a seat belt or without air bags?
- *Individual choices and private decisions.* How much care should individuals take in using products and what choices should they make about diet, exercise, exposure to sun, and other self-controlled risk factors? Should the levels of care taken affect eligibility for compensation for health outcomes to which these factors might contribute? What insurance policies should individuals buy (and what policies should insurance companies or the government offer) to mitigate the consequences of health risks? What medical procedures should individuals choose to have (or should physicians recommend or be required to use) when the outcomes are uncertain?
- *Restricting industrial activities and products.* Should employers be required to keep eight-hour average concentrations of benzene in air inhaled by workers less than 10 parts per million (ppm) to protect worker health? Should the oxygenate additive MTBE be required in gasoline to promote clean air, or should it be banned because it creates more environmental harm than good? Should phthalate ester plasticizers be banned from children's pacifiers and toys and from surgical tubing to prevent cancer risks, or are they safe in these products?
- *Restrictions and requirements for private transactions.* Should manufacturers be prevented from making and retailers be prevented from selling silicone breast implants, cigarettes, beer, or hand guns? More generally, what products should be banned, and should the ban apply equally to manufacture, sale, purchase, and use of hazardous products?
- *Due care and prudent business practices.* How much care should product manufacturers take to inspect, test, and research products to assure that they are reasonably safe for use? (In legal terminology, how far does the "duty to know" extend?) How much information about dangers from unintended or inappropriate uses of products should they be required to post on warning labels (a) In order to mitigate potential product liability (legal duty to warn); and (b) To serve the public interest (a possible regulatory goal)? Similarly, how much research about workplace safety, how much effort to enhance it, and how much information about it should an employer provide to reduce potential liability, serve the public interest?

- *Site investigation and remediation decisions.* How should resources be allocated to investigate and clean known or suspected hazardous waste sites? How should priorities be set for Superfund cleanups, new site investigations, and environmental monitoring and search activities? How much and how far should one clean up around a contaminated site or after an oil spill? How much money should be spent to assure that unexploded ordnance has been removed from former army testing grounds where civilians now jog and play? Should friable asbestos be removed from schools, or left in place? Should lead paint be removed from walls, asbestos from furnaces, and radon gas from basements before houses can be sold?
- *Pollution abatement.* Should diesel exhaust emissions be eliminated or severely curtailed compared to their 1997 levels in the state of California to protect public health? Should power plants and other industrial facilities be forced to prevent or to substantially reduce emissions of fine particulate matter (e.g., 2.5 microns or less in diameter) to protect public health? By how much should each factory in a region that fails to meet air pollution standards be forced to curtail its own emissions?
- *Compensation decisions.* Should women with silicone breast implants have been paid over three billion dollars by Dow Corning to compensate for possible adverse health effects? To what extent does the answer depend on how well, if at all, studies ever establish a link between silicone implants and health risks? Should tobacco companies pay hundreds of millions of dollars in punitive damages to plaintiffs who feel that they were misled or not adequately warned of the dangers of tobacco use and addiction until it was too late? How, if at all, should the U.S. government compensate cancer victims who may have been exposed to fallout from nuclear weapons tests in Utah and Nevada, or veterans who complain of Gulf War Syndrome or ill effects from Agent Orange? More generally, what compensation rules should be used for ill health effects when causation is uncertain?
- *Retrospective allocation of financial liability.* How much, if any, should each defendant pay to the plaintiff in a tort law case where the defendants' behaviors jointly caused or contributed to the injury of the plaintiff? How should compensation, remediation, and investigation costs be allocated among multiple defendants in a tort law case, a Superfund site cleanup, or an oil spill?
- *Provision of public goods.* Which public health and safety risk reduction projects should be implemented? How should their costs be allocated and how should their benefits be defined and estimated?
- *Facility-siting decisions.* Where should land fills, incinerators, hazardous production facilities, hazardous material (HAZMAT) transportation routes, and the like be located? The answer typically requires balancing

or adjudicating among the perceived interests of multiple stakeholder groups. These in turn depend on economic and health risk factors (actual or perceived), as well as on the perceived fairness and legitimacy of the decision process. The usual political answer to noxious facility location problems is "Not In My Back Yard" (NIMBY). Can risk analysis help to create more useful answers, and if so, how?

First-Party, Second-Party, and Third-Party Risks

To clarify the central issues, it is helpful to distinguish among three broad categories of risks based on the roles played by participants in hazardous activities. We will call them first-party, second-party, and third-party risks, respectively. Table 1 illustrates all three.

First-party risks are those that an individual imposes on herself by voluntarily participating in hazardous activities. Examples include skiing, hang-gliding, private cigarette-smoking, and choices about diet, exercise, and other lifestyle factors. Decisions about how to manage first-party risks are often left to the individual. Theoretical and behavioral models of how individuals make decisions about their own risk, covered in earlier chapters, address first-party risks. Governments may also impose paternalistic bans and restrictions, e.g., prohibiting use of recreational drugs, especially if they believe that individuals do not correctly understand or cannot correctly perceive the risks involved. Alternatively, risk management may emphasize providing individuals with adequate information about hazards, e.g., through warning labels and product hazard disclosures. Medical insurance markets can also help individuals to manage some of the potential financial and health consequences of risky activities.

Second-party risks are created when two or more parties voluntarily agree to a transaction (e.g., an employment contract, purchase of real-estate, sale of a hazardous product) that exposes one or both of them to a hazard. Typically, all the parties involved believe they will benefit from the transaction, since it is voluntary. However, some of them might change their minds if they had better information about the hazards involved. Conversely, some transactions that all potential participants would agree to if they were well-informed about the risks involved may not be consummated if they have insufficient or incorrect information about the risks. In such cases, risk management may focus on producing and communicating relevant information to help improve the quality of private transactions.

Paternalistic management of second-party risks arises when the government prevents sales of certain dangerous products to susceptible populations even though buyers and sellers would otherwise voluntarily agree to the transactions. Concern for risk equity may also prompt paternalistic intervention that prohibits people from making voluntary and informed choices that they otherwise would make. For example, a landowner might wish to sell inexpensive housing to poor families on land near hazardous facilities. The families might be willing to accept the risks of the location in return for affordable housing. However, to prevent exploitation of the poor, such transactions may be forbidden, leaving both parties in situations that they may find less desirable (but perhaps consider more just) than the one that would result if the transaction were allowed.

Third-party risks arise when one or more individuals engage in activities that expose third parties to hazards. For example, pollution from an industrial facility may create hazards for its down-wind neighbors. Failing to vaccinate some children may put other children at risk. Such *risk externalities* invite coordination and agreement before decisions are made and adverse effects occur, perhaps using the format of an analysis-deliberation framework. Both ex ante and ex post compensation payments may be considered, in addition to mitigation measures, as possible approaches to risk management decision-making. If explicit agreement is impracticable, litigation or regulation or both may ensue instead.

Example: Risk Management Decisions for a Production Plant

Consider a production process or manufacturing plant that creates potentially hazardous exposures for workers, consumers (who are "exposed" to the product), neighbors of the production facility, and the broader environment. If production is regarded as a source of hazards (as well as of goods or services), then the flows of exposures from the process to stakeholders might appear as in the following diagram.

		neighbors (Bhopal)		
		⇑		
workers (benzene)	⇐	PRODUCTION	⇒	consumers (Ford Pinto, Alar)
		⇓		
		public (acid rain)		

The items in parentheses are intended to suggest examples of the types of exposures of concern to health and safety risk managers. Workers and consumers are involved in second-party risks, while the general public is involved in a third-party (externality) risk from long-range transport of plant emissions. Neighbors who had lived in the area before the plant was introduced may also be exposed to third-party risks, while neighbors who move in after the plant is operating (perhaps attracted by depreciated home prices) may have an element of first-party risk (voluntary and informed assumption of risk) as well. Thus, the first-party, second-party, third-party distinction is not pure.

Managing the health risks from the production process requires at least the following types of decisions.

- *Acceptable risk decisions:* 1. Should the plant be allowed to operate? Who should decide (e.g., plant owners and managers only, taking into account relevant liability laws and estimated risks; plant managers together with community members; or plant managers together with regulators; or regulators only?) How should the decision be made? What input-output characteristics (including resources consumed and beneficial products produced, as well as health and safety risks) are or should be acceptable to the public, the community, and regulators? 2. Is an offered employment contract at the plant acceptable to a prospective worker? Is it acceptable to a social decision-maker? (Should it be allowed? What information about workplace hazards must be shared to make the offer legal?) 3. Is a proposed home near the plant worth buying? Do the convenience and price discount (if any) of living near the plant outweigh the potential hazards?
- *Optimization decisions:* How much care should be taken in operating the plant, assuming that care is expensive? (This question may be answered differently by owners neighbors, regulators, and courts.) Under what conditions should it be allowed to operate? Should plant owners pay to relocate neighbors? Should quality of product be enforced by warranties, by regulation, by insurance, or by liability?
- *Rule design decisions:* What compensation should the plant owners make to an injured neighbor or employee if the cause of injury is uncertain? If it is known to be plant-related? Should prospective compensation to the community (e.g., building parks or libraries) affect the amounts of compensation awarded *ex post* to members of the community who suffer illnesses that may be attributable (or partly attributable) to the plant?

The game-theoretic models and methods introduced later in this chapter provide ways to address most of these questions. Expected utility (EU) theory provides a framework for answering the rest, e.g., questions about the acceptability of risk-cost-benefit trade-offs to consumers, workers, or neighbors once information about the risks (and costs and benefits) has been provided.

General Risk-Management Options

Risk management options generally include *prevention, mitigation, transfer, sharing, spreading,* and *compensation.* This section briefly reviews these approaches to risk management.

Participants in hazardous transactions sometimes manage risks by ex ante contracts that specify explicit, agreed-on risk-sharing arrangements. Negotiation of such agreements typically takes place in an institutional setting that determines outcomes if agreement is not reached, e.g., by litigation or arbitration. Models of bargaining and negotiation with and without arbitration (but with a specified no-agreement outcome) are given

Rather than negotiating risk-sharing contracts, participants in hazardous activities and transactions may instead rely on insurance plans to *spread* the financial component of risks among members of an insurance pool. Such spreading exploits the fact that each risk-averse individual participating in an insurance pool is willing to pay a risk premium over and above its expected (i.e., actuarial) value to avoid an uncertain, potentially large, individual financial loss. As a result, two or more risk-averse individuals may each find that *shares* in the uncertain prospect representing the stochastic gains or losses from operation of the insurance pool over some amount of time are desirable prospects for each of them (certainty equivalents greater than zero) even if none of them would accept the undivided prospect. The economic theory of risk-sharing and syndicates (e.g., Raiffa, 1968; Kobayashi, 1980) explains in detail how multiple risk averse decision-makers can mutually benefit by sharing in the initial investment costs, stochastic gains or losses, and management control of prospects with uncertain returns. Insurance is sometimes said to *transfer* risk from the insured to the insured in exchange for a premium payment. However, this description omits the spreading of risk that is essential to viable insurance.

If private agreements and insurance arrangements fail, risk management may be undertaken on behalf of participants in risky activities and transactions by lawmakers and by government regulators and inspectors. This may be cost-effective when the costs of producing and assimilating relevant risk information are lower for a centralized group (e.g., an agency) than for the individuals who need that information to make mutually beneficial decisions and agreements. In such cases, individuals may prefer to have the centralized agency collect and understand risk information and impose constraints that reflect it, rather than doing this work themselves.

Regulators and lawmakers can reduce or prevent some risks by banning or restricting certain activities, as in Table 1. They may also promulgate and enforce standards for the conduct of activities and for the design, manufacture, and sale of products. A less coercive option is to inform or warn the individuals involved so that they can avoid hazards if they choose to, or take precautions to mitigate (i.e., reduce or avoid) potential adverse consequences. Indirect approaches to prevention include taxing hazardous activities or products (such as cigarettes) and attaching crushing liability and compensation payments for known or hypothesized health damages to product manufacturers.

Retrospective decision processes (e.g., liability-allocation and compensation rules) can help to overcome market imperfections by internalizing the economic and risk externalities imposed on others by an individual's actions. Liability and compensation payments may also help to mitigate or offset the injured party's damage. Compensation rules known in advance of an injury transfer some risk from potential injured parties to potential injurers. Often, lawmakers and regulators seek to design rules and processes that give individuals incentives to discover and engage in only those activities and transactions whose net social benefits are positive.

Approaches for making this concept operational and useful are explored in the next section.

In summary, generic risk management options include the following:

GENERIC RISK-MANAGEMENT OPTIONS

1. **Warning:** Inform or warn potential participants about risks of activities and transactions.
2. **Agreement:** Facilitate voluntary risk management agreements by verifying and publicizing relevant risk information, enforcing risk-sharing contracts, and specifying no-agreement outcomes.
3. **Insurance:** Insure hazardous activities in exchange for premium payments. Underwrite health care costs for specific illnesses to mitigate the financial component of losses.
4. **Regulation:** Restrict voluntary activities or transactions by imposing constraints, standards, and regulatory requirements based on risk information.
5. **Litigation and process design**: Design and enforce processes and rules (e.g., tort liability rules, inspection and labeling or licensing programs, workers' compensation) that will induce desired behavior in the selection of activities and transactions by assigning penalties to and/or allocating costs of adverse outcomes.
6. **Compensation:** Compensate known or suspected victims of hazardous activities, or force others (e.g., their known or suspected injurers, or tax payers) to pay compensation.

Deciding what specific rules members of society would most want a regulator to enforce is largely an unsolved problem. For example, what mix of liability rules, standards, and regulations will best serve the public interest in balancing product cost against product safety? How much of society's resources should be allocated to remediating environmental contaminants? When can paternalistic restriction of individual liberties and of voluntary transactions be justified by concern for health and for risk equity and justice? It is easy to imagine that processes and institutions designed to manage health and safety risks might come to impose different restrictions and consume a different level of resources than their constituents want. How, if at all, can and should individual preferences and risk attitudes be incorporated into social risk management decision processes? These questions make risk management difficult and worthwhile.

Which approach or mix of approaches is most effective for particular applications depends in part on whether the risks to be managed are perceived by participants as first-party, second-party, or third-party risks. It also depends on how well-informed the parties are (and believe that they are) about relevant risks, on the costs of administering each approach, and on the credibility and reputation of those who would do the administering.

Conflict over how to manage risks arises partly from different perceptions about these factors. Conflict may also arise from disagreements about what outcomes are most desirable – who should engage in or be prohibited from what activities and transactions, who should compensate whom how much for various injuries, and so forth. Differences in risk attitudes and in assumptions about the initial allocation of rights may also drive different evaluations of risk management options. Risk management therefore faces a substantial challenge: to identify and apply normative principles that command wide agreement and that clarify what outcomes are desirable and what methods can best be used to obtain them. The following sections survey possible approaches to meeting this challenge.

2. SOCIAL UTILITY: MODERN UTILITARIANISM

Expected utility (EU) theory was developed for individual decision-making. However, it can be extended to groups for purposes of societal decision-making. This may be done by creating a *social utility function* (SUF) for consequences that is a weighted sum of the utility functions of individuals affected by the decision. The socially preferred risk management option is then defined as the one that maximizes expected social utility among all available options.

The following set of axioms provides one normative justification for this essentially utilitarian procedure. There are other justifications, too, i.e., other sets of axioms that imply that decisions should be made to maximize expected social utility. An alternative justification given by Harsanyi and others for a SUF with equal weights is that it maximizes the expected utility of someone born into a random position in society, i.e., who is equally likely to have the preferences of any member of society. This is related to the *veil of ignorance* concept for defining just outcomes, discussed later.

Axiomatic Social Utility Theory

Assume that individual and societal preferences obey these three axioms:

- *Rationality of individual preferences:* Each individual's preferences for consequences satisfy the N-M utility axioms and can therefore be represented by an N-M utility function (see Chapter 5).

- *Rationality of societal preferences:* There is a social utility function (SUF) for consequences that the social decision maker (SDM) uses to make decisions. It also satisfies the N-M axioms.
- *Relation between individual and social preferences:* If every individual prefers one consequence to another, then so does the social utility function. More generally, the social utility function is a strictly increasing function of the individual utilities.

Social Utility Representation Theorem: If the preceding three conditions hold, then the social utility function is a weighted sum of the individual utility functions for consequences, with positive weights. The SDM should evaluate and compare alternative risk management options by the expected value of the SUF.

- *Comment 1:* If individual utilities are considered to be measured in the same utility units and if the SUF treats individuals symmetrically, then the weights of the individual utility functions in the SUF are equal. The theorem implicitly assumes that the individual utility functions are linearly independent, i.e., their weighted sum is not identically equal to zero.
- *Comment 2:* The above version of the theorem applies when the probabilities of different consequences from different risk management options are known. Otherwise, a version can be used in which the SDM uses his or her subjective probabilities to calculate expected social utilities of different options. In this case, individual preferences for consequences are aggregated into the SUF and the SDM's beliefs are then used to calculate expected social utilities of different actions.

Reference: Harsanyi, 1992

Several alternative axiom sets lead to similar conclusions (e.g., Karni, 1998; Dhillon and Mertens, 1999). In particular, Dhillon and Mertens provide an axiomatization for individual and social preferences over lotteries in which social preferences can be represented by the (unweighted) sum of individual utilities, all normalized to the interval from 0 to 1. Importantly, each individual's utility function in the sum is independent of others' preferences. This representation is called *relative utilitarianism* (RU). In the special case of only two alternatives, RU reduces to majority rule.

The necessary and sufficient conditions for RU include those in th ebox below. If the set of alternatives is finite, then these axioms already imply RU generically, i.e., if any three individuals have linearly independent preferences. Otherwise, additional axioms (of Continuity and Independence of Redundant Alternatives) are required. The theorem holds when there are at least 3 individuals and at least 5 alternatives. For smaller problems, it still holds if it can be assumed that duplicating an alternative will not change social preferences, or that adding an individual who is indifferent among all alternatives will not change social preferences.

Axioms for Relative Utilitarianism

- *Individualism* (If all individuals are indifferent, so is society.)
- *Nontriviality* (Not all alternatives are always indifferent, and at least two individuals do not have equal or opposite preferences.)
- *No Ill Will* (If all but one individual are indifferent among alternatives, then society's preferences are not opposite the individual's.)
- *Anonymity* (A permutation of individuals leaves social preferences unchanged.)
- *Monotonicity* (This is implied by the "Good will" axiom that if all but one individual are indifferent between lotteries A and B and that individual strictly prefers A to B, then so does society.)

Despite such theoretical justifications for a utilitarian procedure, there are practical and conceptual difficulties with any sum-of-utilities approach. The most critical are: (1) Individual utilities may not be well-defined (see Chapter 5); (2) Individuals may not reveal truthfully their utility functions for consequences; (3) Some of the affected individuals may not be available to interview (e.g., future generations) and/or may have uncertain future preferences; (4) The probabilities of consequences may be uncertain, making direct application of EU and societal EU theory problematic; (5) Use of social utility does not imply a specific *process* used to collect individual preference data and make collective decisions. But, it may well be that people care strongly about the justice (including fairness, equity, equal access, and so forth), legitimacy, openness, and efficiency of the decision process, rather than the aggregate social utility of particular decisions.

In practice, many processes and institutional frameworks for societal risk management decision-making, including bargaining, negotiation, bidding, and litigation, may produce outcomes very different from what social utility theory would prescribe. More generally, as emphasized by Dhillon and Mertens (1999), non-utilitarian considerations of feasibility, justice, and right and wrong may be introduced in the form of constraints on the set of alternatives that are to be evaluated. Despite these limitations, social utility theory makes a useful starting point for analyzing societal risk management decisions, just as EU theory makes a useful starting point and baseline for discussions of individual decision-making.

3. GAME THEORY: BASIC IDEAS

A limitation of the social utility and social decision maker (SDM) frameworks for risk management is that they deal only with centralized decisions. The SDM is, in effect, a dictator, though perhaps a benevolent one. He or she ultimately decides what will be done, perhaps after considering information about the utilities of various stakeholders.

In many cases, however, it is more realistic to model risks as being managed by the joint decisions or interactions of multiple parties. For example, a manufacturer may choose a level of care to use in manufacturing a consumer product, but its consumers must also choose how much care to take in using it. The risk to a consumer of the product will typically depend on both decisions. Should an injury result from (perhaps careless or improper) use of the product, the injured consumer must decide whether to sue and, if so, the manufacturer and consumer must jointly decide whether to settle prior to trial. Similarly, communities or their political representatives may negotiate over (or bid for) the location and compensation for a shared hazardous facility; polluters in the same area may sell emissions permits to each other; voters may respond to surveys or participate in public hearings to express preferences that affect political decision processes, and so forth. In all of these cases, normal-form decision analysis and EU theory do not describe how decisions should be made. Instead of a single-person decision theory, an approach is needed that explicitly considers the likely actions of other decision-makers.

Game theory extends normal form decision analysis to apply to multiple decision-makers. An *N-person game in normal form* (also called normal strategic form) generalizes the mapping for a single d.m. from act-state pairs to utilities. It provides a *payoff function* that maps *N*-vectors of individual actions, one for each individual or "player", into *N*-vectors of individual expected utilities – again, one for each player. The general form of such a function is thus:

$$\pi: A_1 \times A_2 \times \ldots \times A_N \rightarrow [0, 1]^N \qquad \text{(payoff function for N-person game)}$$

Here, A_i denotes the set of acts available to player i, for $i = 1, 2, \ldots, N$. If other (perhaps random) variables affect the relation between *N*-tuples of acts and resulting consequences, then either of the following two modeling options may be used. First, "Nature" may be treated as one of the players. Nature is modeled as selecting an act (namely, the values of all random variables) from an appropriate joint distribution. Alternatively, the payoffs resulting from player acts may be expressed as *N*-vectors of *expected* utilities, rather than as utilities of deterministic consequences. The expected values of the utilities corresponding to player acts are evaluated with respect to the joint probability distribution of other (non-act) variables. Either approach allows for chance variables, as well as player acts, to be included in the game theory model.

Given a payoff function π, each player calculates an optimal strategy, namely, probabilities (or a probability measure) of choosing each available act, to maximize her own expected utility. The expected utility of each act for a player is calculated with respect to all available information about the strategies of other players, taking into account that they, too, will seek to maximize their own expected utilities. (Only Nature plays strategies without regard for maximization.) Any N-tuple of strategies having the property that no player can increase her own expected utility, given the other strategies, by unilaterally changing her own strategy is called a *Nash equilibrium* of the game. Every game in which the players have finite act sets A_i has at least one Nash equilibrium. However, there may not be any Nash equilibrium in which each player uses a pure (i.e., non-randomized) strategy.

Example: Locating Noxious-Facilities: A Game with No Equilibrium

Four contiguous locations lie along a river in which the water flows from left to right. The locations may be diagrammed as follows:

0	1	2	3

river current →

(Alternatively, the locations may be on a strip of land with predominant wind direction blowing form left to right.) Suppose that a polluter already occupies cell 0 and that two recently arrived players, A and B, are trying to decide where to locate themselves among sites 1, 2, and 3. Each player prefers to be as far away from an upstream polluter as possible and prefers to be downstream from as few polluters as possible. Only one player at a time can occupy a location. With these preferences and constraints, no assignment of players to locations is stable, meaning that, starting from any configuration, some player will be able to move to a preferred location. If player A chooses among rows (representing feasible locations) and player B chooses among columns (also representing locations) in the following configuration table, then the arrows indicate the changes that each player can make to achieve a preferred position, starting from each possible configuration.

Configuration table (A chooses rows, B chooses columns)

	1		2		3
1	--		f	→	a
			↑		↓
2	c	←	--	←	b
	↓		↑		
3	d	→	e		--

For example, starting from (A location, B location) = (1, 3) in the upper right-hand cell (configuration "a"), player A can move from 1 to 2 (producing configuration "b"), thus moving further away from the upstream polluter in location 0. Player B can then move from 3 to 1 (indicated by the two leftward arrows in row 2 and resulting in configuration "c"), thus still having an adjacent upstream polluter (at 0 now instead of at 2) but reducing the number of upstream polluters from 2 to 1. But then A can move to 3, increasing the distance from the upstream polluters. Similarly for all other configurations: either A or B can always move to a preferred position. Thus, players A and B may end up chasing each other through a cycle of moves in an attempt to escape each other's negative externalities.

Note that in this game, even if the three players were to adopt a constitution that permits collective decision-making via majority rule, there would still be no obvious equilibrium outcome. For any proposed assignment of the three players to locations, there is a different assignment that two of the three players will prefer. (Just take the two most downstream players and shift their locations upstream, displacing the third player into the most downstream location just vacated.) Considered as a cooperative three-person game, this one is said to have an *empty core*. This means that every outcome is dominated by another that can be achieved by some coalition of players, all of whom prefer it.

This contrived example illustrates some of the issues that might make real siting decisions for noxious facilities difficult. *Any* proposed process for assigning players to locations raises questions of fairness and equity (as outcomes are inherently asymmetric), individual rights to choose locations, stability of the proposed assignment under the operation of political processes, rights and duties of those who arrive first with respect to those who arrive subsequently, and economic efficiency of the outcomes. These same issues arise if, instead of using an explicit decision process, one simply assigns property rights and lets voluntary buying and selling determine the final allocation of properties to players. If there is no acceptable way to initially assign rights to different locations, then they might be auctioned off. Game theory allows the equity, efficiency, and other properties of procedures for allocating locations (or other scarce resources) to players to be studied analytically within a well-developed conceptual and theoretical framework.

In general, a player's *strategy* defines a mapping from what she knows at any moment (an information set or filtration) to a choice from the set of acts available to her then. Thus, dynamic games in which information is acquired over time as other players act and as values of random variables are revealed, can be analyzed within the framework of game theory. (Games represented explicitly this way are said to be in *extensive form* rather than in strategic form. Extensive forms generalize decision trees just as strategic forms generalize payoff matrices.) However, defining (a) exactly what probabilistic beliefs rational players should hold about each other; (b) what starting assumptions or facts (e.g., about player preferences, resources, and beliefs) should be considered common knowledge; (c) how beliefs should be revised if other players are observed to choose acts that seem to be irrational (non-Nash equilibrium); and what strategies are optimal when there are multiple Nash equilibria, requires a great deal of relatively sophisticated theory (Aumann, 1992; Myerson, 1991).

3.1 Evolutionary Game Theory and Learning

Traditional game theory has been criticized as requiring unrealistically subtle and abstract reasoning by the players. Super-human rationality and computational prowess are needed to calculate equilibrium strategies in many interesting games. An alternative, called *evolutionary game theory,* allows Nash equilibria to be reached by trial-and-error adjustment processes rather than being derived simultaneously by all players via pure reason. Such "evolutionary" adjustment processes depend on gradual adaptation, with more successful strategies tending to be played more frequently in the population until they drive out other strategies.

For example, companies that engage in a variety of risky activities, some of which lead to litigation and financial losses, may either learn to avoid those behaviors or be driven out of business. Or, they may adapt to the current penalty structure so that behaviors are tried with frequencies at which their benefits exceed their costs. If the costs of litigation and resulting penalties depend in part on the frequency of risky behaviors, then the behaviors of companies and potential litigants may co-evolve to an equilibrium in which the frequencies of law suits with various outcomes and of behaviors that trigger them are in balance.

In an evolutionary game, a Nash equilibrium that confers a survival or reproduction advantage to those who use it, even if a small subset of other players adopts a different (non-equilibrium) strategy, is said to be an *Evolutionarily Stable Strategy* (ESS). Key research questions are (a) Under what conditions will plausible adaptive processes converge to a unique Nash equilibrium? (b) What conditions allow players to learn (or evolve to) desirable behavior patterns? For example, in the evolutionary game of potential injurers and potential litigants, will the two populations evolve to an equilibrium in which behaviors are practiced at frequencies such that their average social benefits outweigh their average social costs? Samuelson (1998) presents methods for analyzing the dynamics of evolutionary games and answering such questions.

3.2 Mechanism Design

A part of game theory that is especially relevant to many social risk management problems is called *mechanism design theory,* also known as *implementation theory.* The framework involves a social decision-maker – the mechanism designer – who defines a set of rules mapping player acts to deterministic or probable consequences (and hence indirectly to utilities or to expected utilities). In effect, the SDM can choose a payoff function π from a set Π of feasible payoff functions. The other players are then assumed to act strategically within the game created by π. The SDM's problem is to choose the rules so that, when the other players choose their strategies within the resulting game, the outcome maximizes social utility or has other desirable properties. For example, the designer may seek to create a set of rules that will elicit a fair and efficient outcome as a Nash equilibrium.

Potentially important applications of game theory and mechanism design theory to health and safety risk management include the following:

- *Design of liability rules.* An ideal application of mechanism design theory is designing product liability and tort liability rules that will elicit socially desirable levels of before-the-fact ("ex ante") care-taking, and also of *ex post* litigation in the event of injury, by different participants in hazardous transactions is. What constitutes "socially desirable" levels of *ex-ante* risk-avoidance and hazard-reduction effort and ex-post litigation effort may be decided by appeals to expected social utility. An approximation often used is that whoever can reduce or avoid risk at least cost should be held responsible for doing so as long as the expected cost of care is less than the expected benefit (to someone) from that care. (This echoes the Learned Hand rule from tort theory.) Game theory and mechanism design theory are also used to study incentives for litigation and the probability of settlement *vs.* trial under different liability, compensation, and cost-allocation rules.
- *Design of risk management contracts.* One way to manage risks is via contracts. In the simplest case, participants explicitly agree ex ante on how the financial costs of an accident or injury will be allocated and on how compensation for non-financial costs shall be paid. Power may be symmetric, in that any participant may refuse to agree to the contract, resulting in well-defined no-agreement outcomes (e.g., the status quo arrangements). This setting leads to a rich positive theory of bargaining and negotiation that predicts what contractual agreements will be reached and how long it will take to reach them. There is also a normative theory that provides prescriptive solution concepts for what contracts *should* be reached to comply with various principles of efficiency, fairness, and practicality. As explained later, a practical solution approach must give participants incentives to participate and to reveal the private information (e.g., about preferences) needed to make the approach work.
- *Insurance policies:* The theory of mechanism design has been applied to the problem of designing insurance policies and regulations that will encourage desired participation, risk pooling and sharing, and care-taking by policy-holders.
- *Risk-sharing syndicates.* How, if at all, should several risk-averse players agree to share the costs, potential gains, and potential losses of a

risky undertaking? A detailed theory of syndicates exists for financial risks (Kobayashi, 1980). No comparable theory exists for health and safety risks, of course, since health consequences are generally not divisible or transferable.

- *Regulation:* Deciding what constraints, reporting requirements, monitoring and enforcement procedures, and penalties for detected violations to impose on a regulated industry to elicit socially desirable risk management behavior raises many of the same mechanism-design issues as the design of liability rules and contracts.
- *Comparing risk management institutions.* The theory of mechanism design, information, and incentives has many other applications within the field of law-and-economics, e.g., in deciding whether regulation, tort liability, insurance markets, or a mix will be most effective in managing specific risks to human health and safety. The *Rand Journal of Economics* (formerly the *Bell Journal of Economics)* as well as the *Journal of Legal Studies* and occasionally *the American Economic Review,* have helped to lead the widespread application of game-theoretic methods and models to questions of health and safety risk management.

The following sections show how to apply game-theoretic concepts to risk management, using relatively simple examples to motivate and explain the most important ideas.

4. TWO-PERSON GAMES OF RISK MANAGEMENT

Game theory has produced many models and methods useful for addressing risk management issues. The following sections survey important game theory models for risk management applications, focusing on two-player games modeling second-party risks.

4.1 Prisoner's Dilemma: Individual *vs.* Group Rationality

Suppose that players A and B share an environment in which the activities of one player (e.g., pollution, reckless driving, hunting, etc.) may increase the risk of harm to the other. What level of care should each player choose, if care is costly to the one taking it?

The following simple game models some key aspects of this basic ethical problem. Player A chooses one of the two rows (top or bottom), while B chooses one of the two columns (left or right). Each cell records the expected utilities received by A and B when the corresponding choices are made. A pair of numbers (x, y) means that A gets x (i.e., A's expected utility from the outcome in that cell is x) and that B gets y. The utilities have been normalized to run from -1 to 1, instead of from 0 to 1.

Payoff Matrix for Two-Person Prisoner's Dilemma

	B takes much care	B takes little care
A takes much care	(1, 1)	(-1, 1)
A takes little care	(1, -1)	(0, 0)

Games with payoffs having the rank ordering shown are instances of the notorious *Prisoner's Dilemma* (e.g., Myerson, 1991). A striking feature of such games is that each player has a dominant strategy, but the outcome when the dominant strategies are used is Pareto-inferior (i.e., less preferred by both payers) to some other outcome. A *dominant strategy* for a player is defined as one that returns at least as high a reward as any other strategy, no matter what the other player does. In this example, each player always does at least as well, and sometimes does better (i.e., receives greater expected utility) by choosing "little care" as by choosing "much care", no matter what level of care the other player chooses. Hence, to be rational, each should choose to take "little care". But then, each achieves 0 payoff (lower right-hand corner), compared to the payoff of 1 achieved if each chooses "much care" (upper left-hand corner). The outcome (0, 0) is the game's unique Nash equilibrium outcome.

The possible objection that the behaviors implied by the strategically dominant choices are unethical misses the point. All aspects of preference, including concerns for altruism and ethics, are considered already in forming the utility payoffs. Given *these* preferences, the conflict between individual and collective rationality seems unavoidable.

The basic idea of the two-person Prisoner's Dilemma has been extended to more than two players and to allow for repeated plays, player uncertainties about the population frequency distribution of other player "types" (summarizing player beliefs, information, and preferences), asymmetric preferences, survival of successful (high-payoff) strategies and extinction of unsuccessful ones, and so forth. Nonetheless, the simple two-person single version provides a powerful metaphor and model for the need for coordination and constraints when one player's choice can affect another's utility as well as his own.

Example: N-Person Prisoner's Dilemma and Free Riders

Suppose that in a community of many people, each person must decide whether to be vaccinated against a contagious disease. For simplicity, the expected utility to any person will be modeled as follows:

$$EU(x, a) = x - 0.5a,$$

where

- x = fraction of the population that is vaccinated (between 0 and 1);
- a = 1 for a person who is vaccinated; a = 0 otherwise.

A rough interpretation is that being vaccinated incurs a utility cost of –0.5, but when fraction x of the population is vaccinated, each member of the population receives a benefit of x. (Utilities are not normalized to lie between 0 and 1. The interpretation just suggested assumes linear utility.) For a large population, x is not significantly affected by any individual's choice.

In this game, as in two-person prisoner's dilemma, there is a unique dominant strategy for each player: don't be vaccinated. This is the unique Nash equilibrium. A player's expected utility is always higher by 0.5 if he or she avoids vaccination, no matter what the other players choose to do, i.e., for any value of x. This Nash equilibrium outcome gives a value of EU(0, 0) = 0 to each player. All players would be better off if each chose to be vaccinated (EU(1, 1) = 0.5), or if at least half chose to be vaccinated, with the rest being "free riders" (i.e., receiving the benefits but not bearing the costs.) The problem is that the incentive to free ride always outweighs the incentive to be vaccinated. When everyone elects to be a free rider, the potential net benefits of vaccination are lost entirely. Mandatory vaccination overcomes the free rider problem, but at the cost of a reduction in freedom of choice.

Prisoner's Dilemma has been used to explain why people in a Hobbesian "state of nature" might voluntarily agree to a binding social contract, enforced by a government, that restricts individual choices (Mueller, 1996). To prevent uncoordinated, individually rational actions from leading to a Pareto-dominated outcome, players might agree to any of the following institutional solutions:

Some Possible Solutions to the Prisoner's Dilemma

- *Criminal law solution:* Make the "little care" choice illegal, i.e., restrict the choice set available to each player. The loss of freedom buys an increase in expected utility.
- *Tort law solution:* Change the payoffs, e.g., by making players liable for any harm that their actions cause to others (if this can be determined). The costs of liability are sometimes said to *internalize* the "external" costs (or "negative externality") that exists when one player's action creates costs or harms that are borne by another.
- *Contract law solution:* Have the players negotiate and voluntarily agree on the level of care that each will take. Enable each to make a binding commitment (to be enforced by a third agent, e.g., the government, the court, etc.) to the level of care that he or she will choose. If credible commitments can be made, then the players may voluntarily agree to adopt the (1, 1) outcome.

- *Incentives solution:* Change the payoffs by taxing the "little care" choice and/or subsidizing the "much care" choice.
- *Regulatory solution:* Make the "much care" choice mandatory.

Players in a hypothetical state of nature might voluntarily agree to be bound by such institutions in order to achieve greater payoffs than they can achieve by rational but uncoordinated actions. This provides a possible social contract justification for these institutions.

4.2 Moral Hazard in Insurance

The following table shows a game in which both an insurer and a potential insured party can gain, compared to the status quo payoff of (0, 0), by having the insurer issue a policy and having the insured take "much care".

Payoff Matrix for Moral Hazard in Insurance

	Insurer underwrites	Insurer refuses to underwrite
Insured takes much care	(1, 1)	(0, 0)
Insured takes little care	(2, -1)	(0, 0)

Then, the outcome will be (1, 1), i.e., each party gains 1. But these gains are not attainable without outside help. For, after a policy has been written, the insured no longer has an incentive to choose "much care". Choosing "little care" gives a higher utility, i.e., 2 instead of 1, and reduces the insurer's payoff from 1 to –1. Knowing this, the insurer should refuse to underwrite the policy. Thus, even though both parties know that gains are possible in principle, both also know that the incentives are structured so that the insured will take actions after securing insurance that will make the insurance contract undesirable for the insurer. This illustrates the problem of *moral hazard* in insurance.

The moral hazard problem can be solved if the level of care taken by the insured is observed freely by both parties. In this case, the payment of insurance claims can be made contingent on the insured having taken "much care" instead of "little care". If monitoring is costly or noisy, however, then potential gains from trade may be left unachieved because the insurer cannot trust that the insured will use enough effort. This phenomenon generalizes to an entire class of important games called Principal-Agent (PA) problems, discussed later. Deductibles and experience-based premiums can reduce the incentive problem, but they leave some risk uncovered and some potential benefits unrealized.

Although the payoff matrices for Prisoner's Dilemma and Moral Hazard look similar, i.e., the normal forms of the games are both 2 x 2 tables, their extensive forms are entirely different. In Prisoner's Dilemma, the players are assumed to choose their strategies simultaneously. In Principal-Agent games, the Principal moves first and the Agent, after observing what the Principal has done, moves second. The Principal considers this in deciding what to do.

Example: Private Information and Adverse Selection in Insurance

Suppose that a risk-neutral health insurance company is required by public policy to offer insurance policies at the same price (i.e., premium) for all people of the same age or in the same group. For a price of \$p per year, set by the insurer, any individual in the group can be insured against the full costs of illness. The individual average annual costs of illness are modeled as independent random variables, uniformly distributed between 0 and 1 for all members of the group. Each member of the group is assumed to know his or her own average annual cost of illness. Each is risk-averse and is willing to pay up to 1.5 times his or her expected annual cost of illness to protect against possible very large costs. That is, an individual with average annual claims of \$c will be willing to pay up to \$1.5c per year for insurance. Thus, the insurer can potentially make a profit. What price, p, should the insurer charge, and what benefits will the insurance offering create for the population?

If the insurer charges a price \$p per year, then each individual analyzes his or her own insurance-buying decision as follow. It is worth paying \$p per year for a health insurance policy if and only if the expected annual cost of illness, say, \$c, satisfies:

$$1.5c > p \text{ or } c > (2/3)p.$$

Hence, all and only those individuals with $c > (2/3)p$ will buy insurance policies if the price of a policy is set equal to p. The fraction of the population that will buy at price p is:

$$1 - (2/3)p, \text{ for } 0 \leq p \leq 1.5.$$

Moreover, the average annual cost of illness in this sub-population of insurance buyers is:

$$E(c \mid c > 2p/3) = [1 + (2/3)p]/2 = 0.5 + p/3, \text{ for } 0 < p < 1.5.$$

The expected profit at price p is therefore:

$$\begin{aligned}\text{Profit}(p) &= [1 - (2/3)p][p - (0.5 + p/3)] = [1 - (2/3)p][(2/3)p - 0.5], \text{ for } 0 < p < 1.5\\ &= p - (2p/3)^2 - 0.5\end{aligned}$$

This is positive if $1.5 > p > 0.75$, i.e., the insurer must charge at least 0.75 in order to break even. It is maximized at $p = 1.125$, with profit of 0.16. Most of the population will choose to remain uninsured.

Notice that, if no one knew his or her own average medical costs, then the expected cost per person would be E(c) = 0.5 and everyone would be willing to pay up to 0.75 to buy insurance. Setting p = 0.67, for example, would result in everyone buying insurance and in the insurance company making an average profit of 0.17, while the total consumer surplus (i.e., total willingness to pay minus total amount charged) would be 0.75 – 0.67 = 0.08. Thus, prices could be lower, coverage could be greater, and the insurer's profits could be increased if no one knew his or her own true costs. When people know their own risks, those who are most at risk will tend to self-select into insurance programs, driving up the average costs – and, if premiums follow, potentially driving away lower-cost customers. In the extreme case, although not in this example, such *adverse selection* can eventually make it impossible to offer insurance, because the average cost of care for those who accept high premiums will exceed the premium payments. Then, the potential benefits of insurance will be lost to all.

In general, in insurance settings, discovering more information about individual risks can undermine the basis for risk-pooling and reduce the net benefits from insurance. Even, in the simplest model of a mutual insurance company, e.g., where farmers insure each other against possible fires, revealing in advance whose barn will burn down would make insurance impossible, reducing everyone's expected utility. A similar argument suggests why it may be undesirable to use genetic screening or other detailed individual information about uncontrollable risk factors in developing insurance plans.

In the above example, the insurer could only choose the price, p, at which to offer insurance. Now, suppose that the insurer can choose both the price and the amount of coverage purchased for that price. More than one insurance product may be offered, each specifying a different amount of compensation to be paid in the event of an illness and each costing a different amount. In this case, relatively high-risk customers may end up paying more to purchase full insurance against possible losses, while lower-risk customers spend less but are only partially insured. Such an outcome is called a *separating equilibrium,* since the choices made by different customers reveal their types (high-risk or low-risk), separating them according to which policies they purchase.

Alternatively, if the difference in average costs is small enough, both high-risk and low-risk customers may prefer the same insurance policies, creating a *pooling equilibrium* in which different types of players behave the same, and hence cannot be distinguished by their behaviors. If there are two or more rival insurance companies, then there may be no equilibrium set of policies. Whenever one offers policies that create a separating equilibrium, the other can profit by introducing rival policies that lead to a pooling equilibrium. But, whenever a pooling set of contracts is offered, the other company can introduce separating contracts to attract some customers away, making the remaining pool unprofitable. In this situation, no set of offered policies is proof against the introduction of other policies: the only Nash equilibrium requires the rival insurance companies to randomly mix their offers (Dasgupta and Maskin, 1986; Rasmusen, 1989, Section 8.4.)

4.3 Inefficiencies Due to Asymmetric Private Information

The previous discussion showed that inefficiencies may arise in insurance markets when each player's risk "type" (e.g., high risk or low risk) is his private information. Such inefficiencies have parallels in non-insurance settings when player types are private information. For example, if there are multiple types of manufacturers making products of different qualities, with each manufacturer's type (e.g., care taken in making a product) being private information, then free-riding by low-quality manufacturers may deprive consumers and high-quality manufacturers of the potential benefits from trade.

Example: Private Information Can Block Beneficial Product Sales

Suppose that a manufacturer offers to sell a product to a consumer. The product may be defective and cause damage to the consumer. The manufacturer knows, but the consumer does not, that the probability of a product defect is $p = 0.01$. The consumer knows only that half of all manufacturers provide low-quality products with $p = 0.09$, while half provide high-quality products with $p = 0.01$. The customer's maximum acceptable probability of defect in exchange for the product benefits is 0.03. What inefficiencies might arise in these circumstances, and how can regulatory or legal measures be used to improve economic efficiency?

To the consumer, given his information about product qualities, the probability that this manufacturer's product will prove defective is $(0.5)(0.01) + (0.5)(0.09) = 0.05$, which is unacceptable. There is thus an economic inefficiency: the consumer will refuse to buy the manufacturer's product even though she would do so if she knew what the manufacturer knows. Both parties might expect to gain from the sale if they could credibly share their private information.

If manufacturers are allowed to assume and are able to afford high liability payments for product defects, then economic efficiency (i.e., consummation of mutually beneficial transactions) may be restored. Manufacturers of low-risk products can afford to bear greater product liability than can manufacturers of high-risk products, thus sending a market signal to the customer that reveals that they are low-risk producers. This is one way in which legal liability can be used to promote economic efficiency—by driving high-risk producers out of the market. Alternatively, a regulatory standard that is enforceable by the government and that prevents low-quality products from being sold can give customers the assurance they need to be willing to buy the product. In practice, institutional constraints may limit the ability of manufacturers to use liability to signal their private information about product quality to consumers (Shavell, 1982). For example, limited corporate liability, bankruptcy and wealth constraints, and the fact that liability may be determined retrospectively by the courts rather than prospectively through agreements between buyers and sellers limit the signaling role of private liability agreements.

As an aside, this example illustrates some of the limitations of the much-discussed concept of acceptable risk. Objective aspects of risk (viz., that the true probability of defect is $p = 0.01$) may interact with who knows, or can find out, what in determining the "acceptability" of a risky transaction to those who participate in it – and perhaps to those who regulate it. Thus, suppose that it becomes technically feasible for manufacturers to screen their products more accurately for defects at little or no additional cost. Then a product whose risk was formerly considered acceptable by regulators and consumers on the grounds that the expected net benefit of each sale is positive for all parties might now become unacceptable. The ability of one party to take additional care at reasonable cost, rather than simply the probable magnitude and severity of adverse consequences, helps to determine the acceptability of a risky transaction. More generally, transactions create distributions of risks, costs, and benefits among those participating in them. Arguably, the risk of a transaction should be considered acceptable (with respect to the set of all feasible transactions) only if no other feasible transaction would create a preferred distribution of risk, costs, and benefits.

4.4 Designing Product Liability Standards

The mechanism-design perspective is well illustrated by the problem of designing liability rules that will promote socially desirable outcomes.

Example: Comparing Alternative Product Liability Rules

Consider the following simplified model of a liability rule design problem. A manufacturer produces and sells a product to a consumer who may take much care or little care in using it. The manufacturer may also take much care or little care in the design and manufacture of the product. Taking little care costs 0 for each party, while taking much care is expensive. The parameters of the problem are as follows:

- Cost of manufacturer's care = b for much care, 0 for little care. $b > 0$.
- Cost of consumer's care = c for much care, 0 for little care. $c > 0$.
- Pr(Consumer is harmed) = $1 - 0.5*M\_care - 0.5*C\_care$. Here, M_care = 1 if manufacturer takes much care and 0 otherwise. C_care = 1 if consumer takes much care and 0 otherwise.
- Amount of harm, if it occurs = -A
- Value of product sale = 1 to consumer, 1 to manufacturer.

The following payoff matrix summarizes the model.

A Parametric Payoff Matrix for Product Liability

	Manufacturer takes much care	Manufacturer takes little care
Consumer takes much care	$(1 - c, 1 - b)$	$(1 - c + 0.5(L_2 - A), 1 - 0.5 L_2)$
Consumer takes little care	$(1 + .5(L_1 - A), 1 - b - 0.5L_1)$	$(1 + L_3 - A, 1 - L_3)$

Given these parameters, a mechanism designer, whom we will call the *social decision-maker (SDM),* wishes to create liability rules that will induce a Pareto-efficient outcome. To do so, she may choose liability payment amounts that are contingent on available information. A choice of liability rules creates a specific game that the manufacturer and consumer then play. The liability rules are specified by three numbers, L_1, L_2, and L_3, showing how much the manufacturer must pay the consumer if an accident occurs, for each of the three cells with non-zero accident probabilities.

As a specific example, the following payoff matrix shows the game induced by the *strict-liability rule* in which the manufacturer always compensates the consumer fully for any harm. This rule can be implemented even if the players' levels of care are not observed. The costs of taking care are assumed to be equal for the two players: $b = c = 0.5$.

A Payoff Matrix for Strict Manufacturer Liability, b = c = 0.5

	Manufacturer takes much care	Manufacturer takes little care	Manufacturer exits
Consumer takes much care	(0.5, 0.5)	$(0.5, 1 - 0.5A)$	(0, 0)
Consumer takes little care	$(1, 0.5 - 0.5A)$	$(1, 1 - A)$	(0, 0)

In this case, the consumer's dominant strategy is always to take little care. If $A > 1$, then the manufacturer is unable to make a profit and he will prefer to exit the business, creating a joint payoff of (0, 0). This is Pareto-inferior to the (0.5, 0.5) outcome that would be achieved if both players would take much care. (The strategy of exiting, or refusing to play the game is assumed to be always available to both players.) If $0 < A < 1$, then the manufacturer will choose to take little care and the outcome is $(1, 1 - A)$.

Even if $A > 1$, the (0.5, 0.5) outcome can still be achieved by setting appropriate liability rules, if the SDM can observe the manufacturer's levels of care. Suppose that the manufacturer is held liable for any harm to the consumer if and only if the manufacturer did not use much care, a condition that may be called "negligence" liability. If $L = A$ (so that the consumer is compensated either fully or not at all) and if $A > 1$, then the manufacturer's dominant strategy is to take much care. Given that the manufacturer will choose much care, the consumer's best response is also to take much care, leading to an outcome of (0.5, 0.5) with no risk of harm and no liability payments. In this case, the mere threat of liability is sufficient to assure that both players will take care, so that both obtain the benefits of having the product produced and sold.

If A < 1, then the consumer will always choose to take little care under the negligence liability standard. The manufacturer will choose much care if A > 0.5 and will choose little care if A < 0.5.

Payoff Matrix for Negligence Liability

	Manufacturer takes much care	Manufacturer takes little care
Consumer takes much care	(0.5, 0.5)	(0.5, 1 – 0.5A)
Consumer takes little care	(1 - 0.5A, 0.5)	(1, 1 – A)

Finally, if there is no liability, then the manufacturer will always choose little care and the consumer will either buy the product and choose much care (if A < 2) or will not buy the product at all (if A > 2). For A < 1, the zero-liability outcome (1 - 0.5A, 1) is Pareto-superior to the outcome of (0.5, 0.5) obtained under negligence liability.

Payoff Matrix with No Product Liability

	Manufacturer takes much care	Manufacturer takes little care
Consumer takes much care	(0.5, 0.5)	(1 - 0.5A, 1)
Consumer takes little care	(1 - .5A, 0.5)	(1 – A, 1)
Consumer exits market	(0, 0)	(0, 0)

In summary, the players' payoffs for different values of the harm parameter A and for different liability rules are as follows:

A > 2 & no liability: (0, 0)
A < 2 & no liability: (1 - 0.5A, 1)
A > 1 & strict liability of manufacturer for any harm: (0, 0)
A > 1 & negligence liability only: (0.5, 0.5)
A < 1 & strict liability of manufacturer for any harm: (1, 1 – A)
A < 1 & negligence liability only: (1, 1 – A)

Thus, to maximize the sum of manufacturer and consumer values, the SDM should enforce no liability if A < 2 and negligence liability if A > 2. To maximize consumer utility only, the SDM should enforce either strict or negligence liability if A < 1 and negligence liability if A > 1. However, if negligence cannot be determined because the manufacturer's level of care is private information, then negligence liability is not an option. In this case, consumer utility is maximized by strict liability if A < 1 and by zero liability otherwise.

This type of liability rule modeling and analysis can be made more realistic in several ways. For example, Daughety and Reinganum (1995) considered a two-stage model in which:

1. The manufacturer first chooses a level of care from a continuous range of possible values (interpreted as R&D to make its product safe).
2. After observing the outcome of the R&D process, the manufacturer chooses a price to charge for the product.

The consumer cannot observe the level of safety of the product prior to use, but may draw inferences about its safety based on its price (and on knowledge of applicable liability laws.) If the customer decides to buy the product at the offered price, then injuries occur with a probability determined by the product's safety level and liability payments are made as requited by the liability law.

Some of the main findings from this model are as follows. If product safety is known to all for free, then safer products command higher prices if there is no manufacturer liability, but lower prices if consumer injuries from the product are largely compensated by the manufacturer. By contrast, when product safety is the firm's private information, safer products may be less expensive, even for liability rules that would make safer products more costly if product safety were common knowledge. In addition, the manufacturer may perform less R&D, sell less safe products, and make less profit than would be the case if there were a regulation that specified and enforced a minimum safety standard.

4.5 Principal-Agent (PA) Models

Principal-Agent (PA) games provide a unifying framework for modeling many risk management decision processes for second-party (i.e., transaction-generated) risks in which there is *asymmetry* in the information and power of the players. In a typical PA game, one player, the Principal, hires a second player, the Agent, to perform a task whose outcome depends on both the Agent's level of effort and on random events. The Agent has private information about his own level of effort and about what events occurred. He can observe them, but the Principal cannot. Thus, the Agent has more information than the Principal. However, the Principal typically has more power, in that he moves first. He determines the terms of an offer or contract to present to the Agent.

The Principal wishes to design a payment scheme, perhaps incorporating some form of risk-sharing, that will induce the Agent to provide a desired level of effort. The payment scheme, also called a *sharing rule,* can make the Agent's payment depend on the outcome. However, it cannot depend on the Agent's level of effort because that is not observed by the Principal, and hence is non-contractible. Typically, the payment scheme maps outcomes (observed by both players) into a payment from the Principal to the Agent. Thus, in PA games, instead of an external SDM designing a mechanism to elicit socially desirable outcomes, one of the players – the Principal – designs the mechanism to achieve outcomes that he prefers. The Agent then chooses actions to maximize his own expected utility, given the mechanism established by the Principal.

In the class of PA games just described, the Agent's action (e.g., his level of risk-avoiding effort) is hidden from the Principal. In another class of PA games, it is the Agent's "type" (e.g., high-risk or low-risk, in an insurance setting) that is hidden from the Principal. Both types of hidden information lead to similar results in terms of the types of offers and compensation rules that the Principal will adopt. However, adverse selection and signaling issues arise primarily when agents have hidden types, while moral hazard based on insufficient effort by the Agent after accepting the Principal's offer arises when agent actions are hidden.

Key results in the PA literature establish the impossibility, under many conditions, of designing a payment scheme that will make the players as well off as they would be if the Agent's true level of effort (or true type) were observed by the Principal instead of being private information. Instead, the Agent puts in less effort and gets paid less than he would if the Principal could observe his effort (Rasmusen, 1989).

Table 2 summarizes several potential applications of the PA framework to risk management issues, including the following.

- *Insurance contracts and moral hazard.* The Principal is the insurer, the Agent is the insured, and the Agent pays a higher deductible and takes less care than is optimal. In other words, both players would prefer a smaller deductible and a greater level of care by the insured. But they cannot rationally achieve this outcome when the insured's true level of care is private information – the problem of "moral hazard" in insurance.
- *Product safety.* The Principal is a consumer and the Agent is a product manufacturer. The manufacturer spends less on product safety (assumed to be private information) and is paid less for it than is socially optimal.
- *Product use.* The Principal is a seller and the Agent is a buyer. How much care the buyer takes in using the purchased product is her private information. The buyer takes less care and is charged more for the product (to help cover warranty or liability costs) than would be the case if care could be observed by the seller.
- *Regulated firm.* The Principal is a regulator. The Agent is a company that must choose how much care to take, e.g., in controlling emissions. The firm takes less care and the regulator exacts higher penalties than they would if the company's true level of care could be observed.
- *Regulator performance.* The Principal is a tax-payer and the Agent is a regulator. The Regulator's level of competence, effort, or performance is lower, and the tax-payer pays less (if able to elect how much to pay), than

would be optimal if the tax-payer could directly observe the regulator's level of performance.

- *Hazardous employment.* The Principal employs the Agent, whose level of care affects accident probabilities. The Agent takes less care and receives less worker's compensation in the event of an accident than would be the case if the Principal knew the Agent's care level.

Table 2: Adverse Effects of Private Information in PA Models

Application	**Principal/ Agent**	**Private Info.**	**Predictions from PA Model**
Insurance contracts and moral hazard	Insurer/ Insured	Insured's level of care	Insured pays higher deductible and takes less care than is Pareto-efficient. (Both parties would gain if agent's care level were common knowledge.)
Health and safety regulation	Consumer tax payer/ Regulator	Regulator's effort	Consumer pays less and regulator is less competent than if regulator's competence could be costlessly observed.
Health and safety regulation of a firm	Regulator (OSHA, EPA, FDA, USDA, CPSC)/ Firm	Firm's safety efforts	Firm takes less care and pays more penalties than if its level of effort were free public knowledge.
Occupational safety	Employee / Firm	Firm's safety efforts	Workplace is more hazardous and firm pays higher wages than if firm's occupational safety care level were free public knowledge.
Occupational safety	Firm/ Employee	Employee care level	Employee takes less care and firm pays more in worker compensation claims than if employee care levels could be costlessly monitored
Product safety	Consumer / Firm	Firm's care in making products	Products are more hazardous and sell for less than they would if manufacturer's care in making them were common knowledge.
Product safety	Firm/ Consumer	Consumer care in using products	Consumer takes less care in use and pays a higher cost than is socially optimal.
Food safety	Producer/ Inspector	Inspector's effort	Inspectors are less effective and are paid less than if their efforts were publicly observed.

Of course, these pessimistic predictions are based on simplified PA models that may not adequately describe these situations. The PA model is a severe abstraction. It ignores multiple layers of management and review, other players, operation of competitive markets, and other realistic complexities. It gives the Principal unilateral power to select a sharing rule. It assumes that the Agent's private information really is – that there is no practical way for the Principal to monitor or observe it. Nonetheless, despite its limitations and simplifications, the PA model is a powerful paradigm for understanding potential inefficiencies in multi-party risk management.

PA theory also suggests why some insurance, regulation, and liability rules for allocating the consequences of risky activities have the forms that they do. For example, in many insurance PA models with a risk-neutral insurer and a risk-averse insured, the optimal contract entails a fixed deductible (or coinsurance), with complete compensation paid except for this deductible. More exactly, this is the form of a *"second-best contract"*, meaning that it is Pareto-optimal given the constraints of the problem. These constraints include asymmetric information (i.e., the insurer cannot observe the insured's level of care), incentive-compatibility (i.e., the insured will act to maximize his own expected utility once the contract is assigned), and voluntary participation (i.e., both the insurer and the insured expect non-negative gains from the contract). If the insurer is one among many competitors, then an additional zero-profit constraint may hold, reflecting the operation of competitive market forces. The second-best solution reflects optimization of the sharing rule (i.e., mechanism) by the Principal subject to these constraints. By contrast, a *first-best contract*, meaning the one that would obtain if all information were freely and publicly available, would typically entail full insurance of a risk-averse insured.

PA theory (Rasmusen, 1989) shows that first-best contracts are usually not achievable when agents are risk-averse. The need to give them incentives to take care generally requires leaving them exposed to some risk that they would otherwise be willing to pay to avoid. However, small deductibles may be possible that lead to outcomes close to those from first-best contracts. Moreover, in repeated games, players may invest in building reputation (e.g., a risk-averse insured may seek to establish a low-risk experience rating), and this can help to close the gap between first-best and second-best outcomes.

4.6 Bargaining and Contracts for Allocating Liability

In the PA model, the players have asymmetric bargaining power and information. Only the Principal has the power to design a sharing rule while only the Agent has access to private information. In many settings, however, bargaining power and information are more evenly distributed among the participants and negotiation of contracts, rather than design of contracts by the Principal, becomes the key theme. Game theory models that predict, explain, and/or prescribe the outcomes of bargaining and negotiation games have been extensively developed and applied to many examples.

Perhaps the most famous example is the "Divide a dollar" game, in which two players will be given a dollar, split between them in any way that they agree on, provided that they agree. If no agreement is reached (perhaps within a certain amount of time), they forfeit the dollar and neither player receives any benefit. More generally, a two-person bargaining game consists of an *opportunity set* of possible allocations of the gains from bargaining, i.e., of the gains that the players can achieve by agreement. There is also a *disagreement outcome,* often interpreted as the status quo, specifying what each player will receive if no agreement is reached. Power is symmetrically distributed, in that either player can block agreement. This description extends immediately to $N > 2$ players. Variations include settings in which player utilities for outcomes in the opportunity set are common knowledge (i.e., all players know them, know that everyone else knows them, etc.) and cases in which they are private information. Opportunities for pre-play communication and for binding pre-commitments (e.g., to accept no less than a certain amount) enrich the possible specifications of bargaining games.

There are many solution concepts for bargaining games. Each specifies a rule that maps pairs consisting of an *opportunity set* (the set of possible player utilities from different outcomes) and a *no-agreement outcome* (often interpreted as the status quo) to a single outcome in the opportunity set. The best known solution concept is the *Nash bargaining solution,* defined as an outcome that maximizes the product of the players' utility gains over the no-agreement outcome. This is the unique solution concept satisfying the following four normative axioms (Myerson, 1991; Rasmusen, 1989).

Axiomatic Theory for the Nash Bargaining Solution

- **Invariance:** The solution outcome does not depend on the units used to measure utility. (No interpersonal comparisons of utilities are required.)
- **Pareto Efficiency:** The solution outcome is Pareto-undominated, meaning that no other outcome is preferred by all players.
- **Anonymity:** Re-labeling or re-numbering the players does not change the solution.
- **Independence of Irrelevant Alternatives:** If one bargaining game differs from another only by having some previously possible utility outcomes removed, and if the original solution was not removed, then it is still the solution of the new bargaining game with the reduced

opportunity set. (Thus, the solution does not change if only non-solution outcomes are eliminated.)

THEOREM (Nash, 1950): The one and only solution concept satisfying these four axioms maximizes the product of the players' utility gains from the outcome, compared to the utilities that they receive from the no-agreement outcome.

The Nash bargaining solution is simple, elegant, and intuitive in many applications. For example, in the divide-a-dollar game, it prescribes that two players who have utilities linear in money should divide the dollar equally. It incorporates a principle of justice in the anonymity axiom, i.e., different individuals are treated symmetrically by the procedure. However, it requires the opportunity set and the no-agreement outcome to be known before the solution can be calculated, and this will be unrealistic in many settings.

In practice, attempting to identify the utilities of different players for different outcomes may be impossible. Among the likely problems are that (a) Well-defined utilities for outcomes involving health and safety risks as well as economic benefits and costs may not exist and participants may not have constructed stable, useful preferences for risk-cost-benefit combinations. (b) Answers to surveys about hypothetical outcomes may be deliberately distorted or unintentionally differ from what real decisions would reveal. (c) Answers to questions that have real consequences (e.g., questions on a referendum) may not be answered truthfully if participants perceive that strategically manipulating their responses can increase the likelihood of achieving outcomes that they want. In addition, the axiomatic approach does not specify any process by which the Nash solution will be reached. Unlike the previous games analyzed in this chapter, no analysis of equilibrium strategies has been offered to justify the outcome in terms of plausible strategic behaviors. This deficiency is repaired below, where it is shown that the Nash solution also results from Nash equilibrium strategies in a more explicit (alternating-offer) specification of the bargaining process.

In experiments with bargaining games, outcomes often diverge from the Nash solution. Salience of solutions (e.g., equal division of a dollar even when utility functions are likely to be quite different), perceptions of fairness (including unwillingness to accept manifestly unfair outcomes no matter how weak one's bargaining position), and expectations about what is customary often drive the agreed-to outcomes far more than rational (i.e., game-theoretic) strategic thinking (Young, 1994).

Negotiating private risk-management contracts

The simplest case of joint decision-making about risk is a voluntary transaction between two economic agents, with each party free and able to withhold agreement if she so chooses. The transaction will not be consummated unless both parties agree to it, so both must expect to gain for it to occur. If they complete the transaction, then one or both agents incur some hazard. Examples of such transactions include sale of a potentially hazardous product or property to a willing (though perhaps ignorant) buyer; and employment of a willing (though perhaps not fully informed) worker in a hazardous occupation.

Suppose that, in contrast to the Principal-Agent (PA) model, the parties are equally powerful and reach agreement (or fail to do so) by making alternating offers until an offer is accepted or until the parties agree not to complete the transaction. Even in this relatively simple case, asymmetries in the private information available to the different parties about risks can undermine efficiency and create ambiguity about the degree to which the less-informed party "voluntarily" accepts a risk.

Neither party directly control risk, except by choosing whether to agree to the terms of an offered transaction. Risk to the potential victim (consumer or employee) is characterized by a pair (p, D), where p is the probability that he will be injured if the transaction is consummated and D is the magnitude of the damage if injury occurs. For the moment, suppose that both players are risk neutral and that D is measured in dollar terms. (This is oversimplified since we are concerned with health and safety risks, but is warranted by the fact that most of our results are negative; extensions to more general utility functions can easily be made.)

A *contract* in this context is a pair (w, L). Here, w is the ex ante monetary transfer (the wage or selling price, i.e., increment in wealth) transferred from the buyer to the seller of the hazardous product or labor. L is the ex post payment to be made by the producer to the victim in the event of an accident. It may represent provider insurance, worker's compensation, or producer liability. By risk neutrality, a contract (w, L) is equivalent (indifferent), for both parties, to the contract (w - pL, 0) with compensation being paid ex ante rather than ex post. The two parties may thus be seen as actually negotiating over the single quantity w - pL. This is an example of *single-issue distributive bargaining,* as the players bargain over the distribution of the potential gains from trade.

Let the dollar value of the consumer product be v to the buyer and c to the seller. (Henceforth, the producer-consumer example will be used to discuss second-party transactions. All results apply equally well to the employer-employee relationship.) The possible expected gains from trade, evaluated at the time of contracting, are thus v - c - pD, which must be positive for trade to occur. The terms of the contract, w - pL, allocate this potential social surplus between the two agents. The producer's surplus is w - pL - c and the consumer's surplus is v - pD - (w - pL).

To understand the effectiveness of private contracting as a mode of risk management, it is useful to model the mechanism of negotiation as alternating offers and counteroffers of proposed w - pL amounts. The potential consumer is assumed to open the negotiations, and the first offer made by one party that is accepted by the other is implemented. Modern game-theoretic models can predict the outcome of this contract negotiation process. A player's strategy is a (possibly randomized) decision rule specifying the offer he should make on any turn as a function of the sequence of offers and counteroffers observed so far. A pair of strategies (f, g) for the producer and consumer, respectively, is a Nash equilibrium pair if neither player can increase his expected payoff by unilateral deviation from his strategy. Thus, f is a "best response" to g, and conversely. For example, any division of the social surplus could be a Nash equilibrium, if each player could credibly commit himself to accepting nothing else. (f, g) is a *(subgame) perfect equilibrium* (SPE) if, after any sequence of offers generated during the bargaining, f and g are still in Nash equilibrium from that point on. Strategies in an SPE do not include non-credible threats. Thus, a player cannot threaten to take an action such as permanently walking away if his initial demands aren't met if it is common knowledge that he would regret actually having to carry out.

The SPE solution concept as a model for bargaining between rational players can be applied to the w - pL negotiation game. If each player's preferences are stationary, continuous, and exhibit a weak form of impatience – three concepts made precise in Rubinstein, 1982 (see Myerson, 1991) – and if these features are common knowledge, then a unique SPE solution exists to this negotiation problem. Specifically, suppose that the producer and consumer have discount rates such that the payoff to player j from receiving a benefit of \$x from a transaction in period t is $\delta_j^t x$. Here, j = 1 indicates the consumer and j = 2 indicates the producer and $0 \leq$ is $\delta_j < 1$. Then in the discounted infinite-period alternating-offers bargaining game, the consumer receives the following share of the total gains from trade:

$$\text{Consumer's share} = (1 - \delta_2)/(1 - \delta_1\delta_2) \qquad \text{(Solution to bargaining game)}$$

When offers can be exchanged arbitrarily quickly and the players' discount rates reflect their different degrees of risk aversion, *this SPE solution becomes the Nash bargaining solution!* Thus, while the axiomatic approach leading to the Nash solution provided no description of a bargaining mechanism by which rational strategic players might arrive at that outcome, the alternating offers mechanism does provide such an explanation.

When there is no discounting but it costs each player a certain amount to make an offer, the SPE solution allocates all of the gains from trade to the consumer if it is common knowledge that the consumer has a lower cost per move than the producer. If it is common knowledge that the producer has a lower cost per move than the consumer, than all the gains from trade go to the producer, except for an amount equal to the producer's cost of an offer. Specifically, under these assumptions, the producer and consumer will immediately agree to a contract. The player with the lower bargaining costs captures essentially all the expected gains from trade. For example, if price w is fixed by market conditions and negotiation is over liability L alone, then the unique SPE outcome requires the producer to immediately accept either full liability or (almost) no liability, depending on whether the producer's bargaining costs are greater than or less than the consumer's respectively. (For formal and precise statements and proofs of results, see the references.)

The Rubinstein model implies that the producer and consumer will agree to a lopsided allocation of risk reflecting relative bargaining strength. (If the producer has lower bargaining costs, he will accept his one-period bargaining cost for liability, but we assume that this is trivial.) Although possibly inequitable, this outcome is Pareto-efficient. It is achieved without costly haggling. The potential social surplus is actually obtained and allocated between the two players. There is no role for a social decision maker (SDM) to improve *ex ante* efficiency, although she may which to intervene to prevent one party from appropriating a disproportionate share of the expected gains from trade, or to prevent inequitable *ex post* allocations of risk.

However, when there is an opportunity cost of time, with each player deciding whether to pay the cost of participation at the start of each period, there exists an equilibrium (SPE) in which neither player pays the participation cost in any period and therefore agreement is never reached, so that Pareto-efficiency is lost (Anderlini and Felli, 2001).

Contract Negotiation with Private Information

Now suppose that one or both players have private information. Then, even if potential gains from trade exist, there may be no way for the two parties to achieve them without outside assistance. A very general result is *that private information creates inefficiency in private negotiations and distributive bargaining* (Samuelson, 1984; Chatterjee, 1985). This inefficiency extends to many contexts, including competitive markets in which producers have private information about product safety. If safer products are more costly to produce but impossible to signal, for example, then bad producers will drive good ones out of the market. The entire market may collapse as consumers cease buying products which are no longer high-safety on average (Shapiro, 1982).

Private information can make it impossible for players to reach a mutually beneficial contract even if one exists, or can impose dead-weight costs on the players during the negotiation process. To avoid these costs, players may find it advantageous ex ante to commit themselves to negotiation procedures (such as first-and-final offers) that put positive probability on an impasse even when gains from trade exist in principle. The "efficiency" of any particular contract negotiation process must be judged separately from the standpoint of each player's ex ante expectations based only on his private information at that time, and from the standpoint of the information that he ends up having after the process has concluded.

Example: One-Sided Risk Control and Negligence Standards

Similar conclusions about inefficiency emerge when the producer has a control technology $[p(x), c(x)]$ available to him and can choose from an entire range of x values ("care levels"). For example, instead of only knowing the probability of a product defect, the producer can choose it. Lower defect probabilities cost the producer more to achieve. In this setting, the producer has an incentive in any one-shot transaction to choose a low x value, and so the consumer, knowing this, may be unwilling to buy from him.

Even if producer liability L is not exogenously constrained, efficient contracts can not in general be designed when x is private information. For let (w, L) by any contract. The expected payoff to the producer is then

$$A(x) = w - c(x) - p(x)L, \qquad \text{(Producer's payoff)}$$

while the expected payoff to the consumer is

$$B(x) = v - w - p(x)(D - L). \qquad \text{(Consumer's payoff)}$$

In principle, a Pareto-efficient outcome requires that the producer choose x to maximize the sum

$$T(x) = A(x) + B(X) = v - c(x) - p(x)D \qquad \text{(Total surplus)}$$

This total surplus (which is independent of w and L) should then be allocated between the two players. However, in fact, the producer seeks to maximize A(x) given the allocation determined by the contract (w, L). That is, he will choose x such that $A'(x) = 0$, rather than so that $T'(x) = 0$ (assuming that $[p(x), c(x)]$ is such that each of these first-order conditions identifies a unique interior maximum.)

Only if $L = D$ will these first-order conditions coincide, so that in solving his own problem – maximization of A(x) – the producer will also be maximizing social surplus T(x). In summary, *a Pareto-optimal contract requires that the risk neutral producer fully "insure" the consumer by paying him an amount $L = D$ equal to the damage that he suffers if an accident occurs.* Any other choice of L will, under appropriate assumptions about the control technology $[p(x), c(x)]$, (e.g., that first-order conditions determine a unique internal maximum) induce the producer to choose a value of x that is either too great or too small. Whenever the producer is able to negotiate for himself a value of L less than D, he will also end up putting in too little care to maximize total social surplus, T(x).

Now suppose that D, the amount of damage to the consumer if an accident occurs, is an ex post jointly observable random variable whose distribution F(D | x) is stochastically decreasing in the producer's care level, x. Consider a Principal-Agent game in which the consumer can propose any contract terms (w, L(D)), that the producer may reject if it gives him negative expected utility. Given any acceptable contract, the producer will choose x to maximize his own expected utility. Assume that c(x) displays increasing marginal costs and is continuous in x, that both players are risk averse, and that the control technology (F, c) is common knowledge. For any liability rule L = L(D), the consumer can calculate what value of x, say x*(L), the producer would choose in response to L to maximize his own expected utility. The consumer can also calculate what his own expected utility would be, given the producer's "best-response" level of care, x*(L). He will therefore choose a liability rule L* such that x*(L*) gives him a higher expected utility than x*(L), for any other liability rule L. The pair (L*, x*(L*)) is a Nash equilibrium: each player's choice (of L for the consumer, and of x for the producer) is a "best response" to the other's.

In this setting, *if x and D are both jointly observable,* than an ex ante Pareto-efficient liability rule has the following form, called a *pure negligence standard* in tort law (Harris and Raviv, 1978; Shavell, 1982):

$$\begin{aligned} &L^*(x, D) = 0 \text{ for } x > x', \\ &L^*(x, D) = L^*(D) \text{ for } x \text{ less than } x'. \end{aligned} \qquad \text{(Pure negligence standard)}$$

The payment made from the producer to the consumer is 0 if he has not been "negligent" (used a level of care less than some level x' specified in the contract), and depends only on-the amount of damage, but not on his care level, if he has been negligent. However, under this negligence standard, *the producer takes less care when x is his private information than when x is publicly observable.* Both the producer and the consumer have lower ex ante expected payoffs (utilities) in the Nash equilibrium outcome when x is private information than when x is jointly observable. This implies that having x private information unambiguously damages welfare. It is another illustration in the PA context of the inefficiency theme: that ex ante Pareto-efficient outcomes can not in general be implemented as strategic equilibria of contracts arrived at by agents with private information. If x is "partially" jointly observable, in that some variable s (such as expenditure on safety research) that is correlated with x is jointly observable, then both players' expected payoffs can be improved by making the liability rule depend on s.

Example: Two-Sided Control (Holmstrom, 1982; Kambhu, 1982)

Often both players have some control over risk. For example, the producer can take care in the design and manufacture of a product, but the consumer can also take care in using the product safely and appropriately. In this case, a pure negligence standard may be less effective than a *comparative negligence standard,* such as one in which the producer must pay the consumer for damages only if the producer was negligent but the consumer was not.

To model two-sided risk control, suppose that both players are risk neutral and that each player's level of care is his private information. The producer's payoff from a contract (w, L) when he uses care level x and the consumer uses care level y is

$$A(w, L, x, y) = w - c(x) - p(x, y)L.$$

The consumer's payoff is

$$B(w, L, x, y) = v - w - b(y) - p(x, y)(D - L),$$

where b(y) is the consumer's cost function for taking care. b(x) and c(x) are assumed to be increasing and convex (i.e., increasing marginal costs) with b(0) = c(0) = 0. The risk function p(x, y) is assumed to be smoothly decreasing and convex (diminishing returns to care) in each argument.

The Nash equilibria corresponding to different liability rules L can be analyzed as above by considering the "best response" functions x*(y; L) and y*(x; L). These specify each player's optimal (expected payoff-maximizing) level of care as a function of the other's level of care and the liability rule being used. Where they intersect, the players' care levels, say x* and y*, are in Nash equilibrium. Under appropriate assumptions about the control technology, [p(x, y), c(x), b(y)], e.g., that x choice and y have complementary effects on reducing p(x, y), it can be shown that the response curves x*(y) and y*(x) will intersect exactly once, establishing the existence of a unique Nash equilibrium. However, *there is in general no value of L that will induce both players to choose jointly Pareto-outcome optimal levels of care as a Nash equilibrium.* In other words, as in the principal-agent context, it is impossible to implement a Pareto-efficient outcome as a Nash equilibrium of some liability rule when x and y are private information. This result holds no matter how L is determined – whether by negotiation, by the consumer alone, or by an outside SDM. (The conclusion depends on specific technical assumptions, however, such as that x and y are continuous control variables, rather than having only a few levels.)

In the above situation, ex ante Pareto efficiency can sometimes be improved by allowing each player to commit himself to paying a fine if an accident occurs (where by a "fine" is meant any payment that goes outside the system. For example, destruction of one's property would constitute a "fine" in this sense, but a payment by one party to the other in the event of an accident would not. By committing themselves to pay fines if an accident occurs, the players can signal to each other their intentions to take enough care so that payment will very probably not be necessary. As in the case of one-sided information, budget constraints or other limitations on contingent payments can make such signaling impossible and reduce Pareto efficiency, either by preventing a contract from taking place, or by causing to few or too many resources to be spent on risk avoidance. Too much defensive expenditure occurs when one party spends resources on risk reduction, even though the other has lower marginal costs of risk reduction, because the cheaper risk averter can't be trusted to spend to reduce risk.

The use of fines to signal intentions well illustrates the conflict between ex ante and ex post efficiency when players have private information. Clearly, it is inefficient for players to pay a fine after an accident occurs, when they already taken due care to reduce its risks. Yet unless they can commit themselves ex ante to pay such penalties in the event of an accident, they will not be able to trust each other to take due care (and in fact will have incentives not to), so that a contract may become impossible or inefficient. Keeping payments within the system by redistributing the fines between the players would undermine their incentives to take care and the signaling value of the commitment to pay penalties. Similar results apply to any number of players.

In summary, risk management via direct negotiation among stakeholders often looks attractive because each participant is free to reject unattractive proposals. This suggests that the outcome will be satisfactory to all. However, game-theoretic analysis shows that outcomes of such negotiations may be inefficient (some other feasible outcome would be preferred by all stakeholders), unfair (participants with weak bargaining positions or initial endowments may fail to share equitably in the gains from trade), or both. Applications of game-theoretic bargaining theory to risk management issues such as negotiating pay for hazardous occupation, negotiating the site for a hazardous waste facility (Samuelson, 1985), and negotiating product price and safety confirm these limitations of bargaining and negotiation for managing second-party risks.

4.7 Litigation and Bargaining Under Arbitration

In practice, bargaining over the resolution of disputes arising from product defects, occupational injuries, or risk externalities such as feared health effects of pollution, usually takes place in the context of litigation and/or alternative dispute resolution. A judge, jury, or arbiter can affect the payoffs to both parties if they do not reach a voluntary agreement or settlement. This raises the challenge of *how best to design arbitration and litigation mechanisms* to promote socially desirable outcomes. A substantial literature on law-and-economics has arisen that addresses this design challenge from both theoretical and empirical perspectives. It characterizes and compares a wide variety of litigation and arbitration systems with respect to various normative criteria, using both game-theoretic models and empirical analyses.

The following box summarizes some criteria widely used to compare alternative litigation and arbitration systems, both real and proposed. Most theoretical evaluation and comparisons use game theory models to predict how players would (or should) respond to the incentives created by different systems or rules. These predicted behaviors and the outcomes they generate are then evaluated by means of the following criteria.

Criteria for Evaluating a Litigation or Arbitration System

a) *Selection of disputes to litigate:* The incentives that it provides to members of society to select disputes to submit to the process for resolution (Eisenberg and Farber, 1997). For example, are frivolous suits deterred and worthwhile ones pressed? Interestingly, the parties to a dispute are often subject to a self-serving "fairness bias" that tends to make each side believe that settlements *should* be in its favor. This perception can increase the number of disputes that go to litigation over what a purely fact-based Nash equilibrium analysis would suggest (DelRossi and Phillips, 1999, citing earlier work of Loewenstein et al., 1993.)
b) *Value of information produced:* The incentives that the system provides to litigants to produce, use, and disclose information about risks. There may be social benefits to such information production.
c) *Risk management incentives:* The incentives the system provides to non-litigants to choose actions that appropriately manage risks. A good litigation system should provide incentives for potential litigants to take appropriate levels of care, but without enforcing a standard of care and liability amounts so high as to deter or prevent activities with net social benefits. Unfortunately, there appear to be many examples of products with net social benefits, ranging from pediatric vaccines to football helmets, for which fear of liability has led to product withdrawals and/or the collapse of competition among manufacturers – arguably, a Pareto-inefficient outcome (Daughety and Reinganum,1995).
d) *Fairness and efficiency of outcomes:* The fairness and Pareto-efficiency of outcomes achieved for participants both in the litigation or arbitration process and in pre-litigation or pre-arbitration settlements. The mere existence of a litigation or costly arbitration system for resolution of disputes can affect out-of-court settlement agreements, and the impact of a resolution process on cases that do *not* to use it must be considered to understand its total effects. For example, a costless court system that randomly assigns a disputed property right to its original owner or to a plaintiff if the two fail to negotiate a mutually acceptable sales price can actually *reduce* the efficiency of the bargaining process. This has been established both by analysis of Nash-equilibrium bargaining strategies and by experimental results from laboratory games in which each player's true valuation of the disputed property right is his private information and a first-and-final offer bargaining mechanism is used (DelRossi and Phillips, 1999). The presence of the court can reduce total gains from trade by 12% to 20% in the absence of any litigation expenses or court costs.
e) *Fairness, legitimacy, and practicality of the process:* The fairness, legitimacy, and incentive compatibility of the process used to determine outcomes. For example, does the process give rational participants reason to participate? Does it induce them to reveal their private information well enough to obtain a fair and efficient outcome? Does it protect their rights and promote just outcomes, both as perceived by the participants and as judged by more absolute standards?

Examples of insights and propositions from this literature include the following.

- *Negligence standards.* The *Judge Learned Hand rule* for determining negligence requires *both* the buyer and the seller to take actions to prevent damage whenever the cost of taking action is less than the expected damage prevented. However, this glosses over the difficulty that the expected damage prevented by the customer's action may depend on the producer's action – which the customer may not know.
- *Defining and promoting due care.* Suppose that damages to the consumer (if any), probabilities of damage as a function of care taken by buyer and seller, and costs of taking care are all common knowledge and costless to observe. Define the *due care* levels as the levels of care by buyer and seller that minimize the expected total social cost of an accident – i.e., $p(x, y)D + A(x) + B(y)$ in previous notation. Then, if due care levels are positive, both a negligence standard and a strict liability standard with a defense of contributory negligence can create incentives for both parties to take due care (Landes and Posner, 1984).
- *Use of risk information in litigation.* Regulatory information about the risks from exposure to pollutants or other hazards, produced at public expense, can lead to better-informed victim decisions about whether to pursue litigation *if* the victims' private information about the probable cause of harm can be verified by the court. (Such private information may include a victim's knowledge about his exposure to the specific hazard and his own behaviors that might have caused the harm.) When this private information is unverifiable, however, withholding regulatory information until after a lawsuit is brought may help to deter frivolous suits (Che and Earnhart, 1997).
- *Settling out of court.* The probability of out-of-court settlement decreases with the size of the stakes and with the plaintiff's uncertainty about the defendant's true behavior and potential liability. It increases with bargaining and litigation costs (e.g., Fournier and Zuehlke, 1996).
- *Should losing plaintiffs compensate defendants?* To minimize the litigation costs of deterring potential injurers by a specified amount, losing plaintiffs should pay a penalty to defendants and winning plaintiffs should receive increased awards, in contrast to current widespread practice in the United States (Polinsky and Rubinfeld, 1996).

These results fairly illustrate the scope and breadth of past and on-going inquiries into the implications of game theory for risk management via arbitration and litigation.

Similar models and methods have been used to compare litigation, insurance, regulation, and market forces as possible mechanisms for managing second- and third-party risks (e.g., Shavell, 1987). In general, none of these approaches dominates the others, and determining which combination to use in managing any specific risk often requires detailed analysis of incentives effects and costs of information acquisition by the different parties involved, e.g., producers, consumers, and regulators, insurers, litigators and arbiters or courts.

The theoretical models have had some success in explaining empirical data and the forms of legal and other institutional decision rules. Deductibles in insurance, legal standards for negligence in tort law, and several forms of contracts have all been interpreted and explained by game-theoretic models. [The optimality of deductible policies from the standpoint of the insured also follows from single-person decision-analysis (EU) models and from non-EU models with SSD (Meyer and Ormiston, 1999.)]

However, there are also psychological and attitudinal determinants of decisions to litigate, as well as of jury award decisions, that do not fit well within the framework of economic rationality. For example, the fact that jury awards are usually expressed in dollar terms can introduce large response mode biases. Final award sizes are notoriously difficult to predict, even though punitive intent can often be predicted quite well from (a) The extent of victim's harm, (b) How well victims can be identified and (c) Shared outrage among jurors (based primarily on defendant's perceived malice or recklessness). The mapping of punitive intent to final dollar awards by juries is highly variable, with award amounts being very sensitive to anchoring effects and typically increasing dramatically with the size of corporate defendants (Kahneman et al., 1998). Moreover, jury awards often appear to be based more on intent to punish defendants – to exact retribution – rather than on any calculus of incentives, deterrence, and optimization of societal incentives and care-taking (*ibid,* p. 77).

4.8 Potential Roles for a Social Decision Maker (SDM) in Managing Second-Party Risks

Understanding the limitations of voluntary contracting, with or without arbitration, as a method of managing second-party risks suggests several potential roles for the SDM in improving the efficiency of risk management. The SDM may intervene to accomplish any or all of the following.

1. *Promote equity:* As already suggested, the SDM may wish to intervene in private negotiations to prevent one party from exploiting its superior bargaining strength or information about risks to take unfair advantage of the other. OSHA standards for workplace safety, CPSC and FDA standards for consumer products, as well as right-to-know and duty-to-disclose legislation and regulations may be seen in this light as devices that prevent exploitative behavior in second-party transactions.
2. *Provide information:* When the producer has control technology [p(x), c(x)], the SDM can establish a minimum level of care x by fiat, i.e., through (enforced) regulation. Knowing that this minimum level of x is

guaranteed by the SDM may give potential consumers the assurance they need to engage in mutually beneficial transactions with vendors of risky products. In addition to outright regulation (setting levels for x), the SDM can also signal information through certification, licensing, and grading programs – a role played by government agencies such as USDA and FDA. A third way in which the SDM can fill the informational gap between agents that might otherwise prevent mutually beneficial transactions is by providing for ex post determination of care level after an accident occurs. By making care levels publicly observable after the fact, the SDM can allow ex ante compensation rules with improved efficiency, i.e., rules of the form L(x, D) (or possibly L(s, d) if the SDM's ex post assessment is imperfect), inducing Nash equilibria that are Pareto-superior to any achievable through private contracts of the form L(D). This informational role may, in practice be filled to some extent by the courts, by insurance companies, and by agencies such as the FAA that conduct inquiries into who was "at fault", and to what extent, in causing an accident.

3. *Set terms:* Potentially prohibitive private negotiation costs can be alleviated by having the SDM set the terms of "implicit contracts" that agents are required to abide by. Much of tort law can be interpreted as a set of implicit contracts that avoids the necessity of costly explicit contracting in everyday activities and transactions. Rather than having the parties negotiate compensation rule L, the SDM (or the courts, in practice), can set L based on the equilibrium strategic behavior (care levels) that it is expected to induce. (If x is not observable to the SDM, however, then PA theory implies that there may be no rule that will induce a Pareto-efficient level of care as a Nash equilibrium.)
4. *Underwrite/signal good bets:* The effects of individual budget constraints on limiting an individual's ability to signal his high care levels has been mentioned several times. If x is observable to the SDM, then he can signal high levels of x to potential consumers by providing extensive, publicly observable, liability insurance to the producer, if it is known that he acts as an expected profit-maximizing insurer.

The SDM can also serve as a sink for fines or penalties. The producer, for example, can pre-commit himself to pay a penalty (possibly in excess of the damage to the consumer) if an accident occurs, and the SDM can collect these payments and use them elsewhere in society – perhaps to operate the insurance role just described.

In summary, because voluntary contracting tends to be inefficient when individuals have private information, the SDM has a chance to improve on the resource allocations that pairs of individuals could obtain by themselves. Three primary avenues for intervention are (i) the regulatory approach, in which the SDM directly sets and enforces levels of care, x and y, to be used by the parties; (ii) the tort law approach, in which the SDM clarifies property rights and sets liability or compensation rules, e.g., L(x, y) (depending on what can be observed) that will motivate the parties to take due care in strategic equilibrium; and (iii) the administrative approach, including social insurance, in which the SDM sets or collects fines and makes payments or provides insurance, both to give agents appropriate risk-avoidance incentives to allow them to signal high care levels, and to maximize their expected welfare from participation in an economy of risky transactions.

5. PROPERTY RIGHTS AND RISK EXTERNALITIES

Many socially important risks arise not out of voluntary transactions between a potential injurer and a potential victim, but out of production or sale risk externalities. For example, the health and safety risks to local populations from the operation of a chemical plant or a nuclear reactor may be imposed on neighbors without their consent and perhaps without their knowledge. Even if it is imposed as the result of a public collective choice process, the questions of how such risk externalities should be managed and how victims or potential victims should be compensated remain.

As another example, the risks to sensitive individuals from a new drug that is safe and beneficial to most consumers, but dangerous to a few, create a different type of risk externality. Here, protecting the few sensitive members of the population, if there is no way to screen them, may deprive others of the benefits of the drug. As a third example of a risk externality, the risk created by multiple producers, or by the interaction of multiple products or risk factors (e.g., the health risks of the combined pollutants from multiple sources) is an example of a distributed hazard. For this type of hazard, it may not be possible even in principle for the victim to identify a single party as a probable cause of or contributor to his injury, either before or after it occurs.

In second-party negotiations, failure of the parties to reach a mutually beneficial agreement implies that the risk-producing transaction will not take place. Each participant's right (if there is one) not to have unacceptable risk-cost-benefit combinations imposed upon him is protected by his ability to veto any proposed agreement. The third-party case is more complicated, as the right of the producer to produce (and perhaps of his consumers to have him produce) must be balanced against neighbors' rights to be protected from imposed risks.

To cut through some of this complexity, the SDM can allow the producer and neighbor to negotiate privately over the amount of compensation that the producer (and perhaps consumers through increases in product price) will pay to neighbors to secure the right to produce. Payment may be made directly to neighbors or may be made indirectly but on their behalf by having the producer fund pollution controls, safety systems, and so forth. Only if no negotiated settlement is achieved will the SDM impose a solution. Then the problem of compensation rule design for the SDM reduces to the problem of choosing a no-agreement point for third-party negotiations. (A similar role is played by courts when they provide the no-agreement outcome for out-of-court settlement negotiations after an accident or injury has occurred.)

An SDM or judge typically has at least the following goals in designing default compensation rules:

- *Economic efficiency in individual cases.* If the producer and neighbor in a specific case are thought of as negotiating over "ownership" of a production right, then economic efficiency requires that the one who values it more highly should end up with it. For example, production should continue only if its value to the producer (and customers) exceeds the value of non-production to the producer's neighbors.
- *Economic efficiency in the population of potential litigants (e.g., producers and neighbors).* The court wishes to design compensation rules that will induce efficient ex ante risk avoidance behavior and efficient out-of-court settlements in cases that never come to court. Since most cases in the United States are settled out of court, this is an extremely important objective in practice.
- *Protection of individual rights.* The compensation rule used should respect the initial allocation of rights to produce or to be protected from production externalities, respectively. Pre-assigned or court-determined rights automatically establish the no-agreement point for private third-party negotiations, and the court merely has to enforce them. How rights should initially be assigned is a separate, difficult subject, involving entitlement dynamics such as differences in the allocation of production rights according to whether the producer arrived before or after his current neighbors.

These goals are in general incompatible. When the producer and neighbor have private information about the value of the production right to themselves, *no initial pre-assignment of rights can guarantee economically efficient negotiated outcomes.* Economic inefficiency resulting from private information extends to situations with externalities, and hence to efforts to manage third-party risks through private agreements. The reason is that ownership of a production right is assigned before learning who values it most highly. If private values for production rights were common knowledge and there were no effective budget constraints or transaction costs, then the initial assignment of rights would not compromise long-run economic efficiency. Whichever party, producer or neighbor, valued the production right most highly would end up owning it or purchasing it ("*Coase's Theorem*", discussed next.) In a world of private information, however, private bargaining is not efficient, and Coase's Theorem no longer applies.

Coases's Theorem states that, "In the absence of transactions and bargaining costs, affected parties to an externality will agree on an allocation of resources which is both Pareto optimal and independent of any prior assignment of property rights" (Mueller, 1996, citing Coase, 1960). Merely assigning transferable rights to use resources or to own property enables the exchange of these rights for cash payments until, in equilibrium, the resources are used in a Pareto-efficient manner. Of course, the initial allocation of property rights may be perceived as more or less fair or desirable, but the efficiency of the final outcome is not in doubt. This type of argument has motivated the EPA's efforts to establish transferable pollution rights and has suggested that corrective taxes and regulatory interventions may be unnecessary to control externalities when clearly identified property rights can be established and enforced by the courts.

Unfortunately, the Coase theorem was not stated and proved as a formal theorem with clearly specified necessary and sufficient conditions. In fact, if individuals are uncertain about each others' utility functions or have other incomplete information, formal analysis (Samuelson, 1985) shows that:

a) The parties to an externality are in general *not* necessarily able to negotiate a Pareto-efficient agreement.
b) Even if an efficient agreement is eventually reached, it may require costly delays to reach it, during which the players test and draw inferences about each others' true preferences.
c) The efficiency of the outcome in general *does* depend on the initial allocation of property rights, as well as on the specific bargaining mechanism used. (Indeed, allocating the initial property rights by a competitive bid will often improve the Pareto efficiency of the final outcome.) Thus, private information or other forms of incomplete information that induce the players to engage in costly experimentation during the bargaining process destroy the efficiency suggested by the Coase argument.

Example: *Ex Ante* Compensation Agreements are Inefficient

Suppose that a community must decide whether or not to permit a pollution-generating plant to be built. Members of the community and the prospective plant's owner decide to negotiate with each other to determine whether the plant will be built. Specifically, the firm first announces a decision rule that it will use to make compensation payments to members of the community if the plant is built. Payments to residents in the community are to be based on information about estimated monetary damages that they will furnish (not necessarily truthfully) to the firm. Once the firm has announced the decision rule (i.e., mechanism) used to determine compensation payments, members of the community may accept or reject the proposal. Property rights are such that any resident may veto the firm's proposal and block the building of the plant if he or she does not feel adequately compensated under the firm's proposal. The true damage to each individual is his or her private information.

Detailed game-theoretic analysis of this situation (Rob, 1989) shows that the *equilibrium strategies can produce Pareto-inefficient outcomes.* Specifically, the plant may not be built even when it would have been better for all players for it to have been built and for the proposed compensation payments to have been made. Intuitively, the reason is that residents have an incentive to over-state their true losses to increase their compensation if the plant is built. This can make the total compensation payment required greater than the potential profit from the plant, so that building the plant becomes unacceptable to the firm – even though the plant would have been built had the residents' private information been common knowledge. This inefficient outcome becomes likely if the number of residents in the community is large, with probability approaching 1 as the number of residents increases. The efficient outcome predicted by the Coase argument fails when specific bargaining mechanisms and realistic limitations on public information are considered.

When there are several different types of residents, the probability that the plant will be built if it is socially optimal to do so can be greater than zero, even for large populations. Pesenforfer (1998) describes mechanisms that give positive probability of an efficient outcome. These mechanisms give residents an incentive to voluntarily participate and to provide the information needed, while allowing each resident veto power and non-negative transfer payments.

Instead of passively pre-assigning rights and allowing private negotiations to take care of third-party risks thereafter, the SDM can actively serve as an arbiter in cases where out-of-court settlement negotiations fail to lead to agreement. Suppose that the SDM hears many such cases and enforces a consistent rule or set of rules in each. In effect, these rules map statements of preferences, beliefs, abilities, and so forth by the producer and neighbor into an amount of compensation (possibly zero) to be paid by the producer to or on behalf of the neighbor for production to continue. When players have private information about their preferences and beliefs (including beliefs about each other's preferences and beliefs), *there may be no compensation rule mapping player statements into enforced settlements that will induce the players to make statements that will produce a Pareto-efficient settlement* (Myerson and Satterthwaite 1983). This negative result can be proved if it is assumed that players are strategically rational, i.e., make statements that are in Bayesian Nash equilibrium with each other. In other words, neither player believes that he can increase his own expected payoff by changing his own statement to the SDM (judge), given his beliefs about the judge's compensation rule and about the other player's statement.

To increase economic efficiency averaged over all cases, the SDM may have to enforce settlements in particular cases that violate case-specific efficiency. For example, if the SDM expects the neighbor to exaggerate the value of non-production to him, he may allow the producer to make a compensation payment that is less than the neighbor's stated minimum acceptable amount. But this risks having production continue even when the loss to the neighbor really does exceed the benefit to the producer. Conversely, the SDM may sometimes shut down activities that would have been left to operate if all private information had been made public.

6. AGREEING ON RULES: SOCIAL CONTRACTS

The idea of a *social contract* has long played a leading role in Western political theory and philosophy. It is often useful in considering how third-party risks (i.e., risk externalities) should be managed. Social contract theory gained new precision and power with the introduction and application of game-theoretic methods, mainly since the 1960s (Binmore, 1998). Key ideas and developments are as follows.

1. *Bargaining over rules instead of outcomes.* Social contract theory considers what rules individuals might agree to if they were making this decision outside the context of any specific case in which the interests of the different players are already known and in which their support for different rules might therefore be biased by their preferences for outcomes.
2. *Hypothetical bargaining.* A social contract specifies a set of rules ("mechanisms") for resolving conflicts, making collective decisions, and constraining individual behaviors in particular situations. It is justified by the claim that members of society *would* voluntarily agree to or choose it *if* they could participate in a hypothetical idealized bargaining procedure that provided adequate time and resources to propose, discuss, understand, and accept or reject agreements on different rules.
3. *Veil of ignorance.* To promote justice in the social contract, the hypothetical multi-player bargaining process is further idealized by assuming a Rawlsian "veil of ignorance" behind which the social contract is supposed to be negotiated. Each (hypothetical) player is assumed not to know, while participating in selection of the social contract, which role or position he or she will occupy in society. In this sense, the contract is designed to be good for everyone, rather than reflecting the bargaining power of different special interests. (It is also extremely hypothetical.) The veil of ignorance assumption is sometimes formulated by supposing that each participant in the social contract bargaining process is equally likely, after the contract is agreed to, to end up in any position in society.

If she ends up in position j, then she is assumed to have the values, preferences, risk attitudes, and so forth – in short, the utility function – that goes with position j.

4. *No strategic manipulation.* The idealized bargaining process is assumed to be free of strategic misrepresentations of preference, manipulation of the agenda, time pressures on the participants, incomplete information, and so forth. (Indeed, behind the veil of ignorance, there may be no incentive to misrepresent preferences.)
5. *Costless transactions and enforcement.* The participants in the hypothetical contracting process can credibly commit themselves to any subsequent behaviors (e.g., to always cooperate in instances of Prisoner's Dilemma) simply by agreeing to it. There are no costs involved in the bargaining process or in making, keeping, or enforcing any agreements reached.

These idealizations delineate the gulf between the philosophical construct of a social contract and any realistic political process. Nonetheless, it has often been proposed in Western political philosophy that societies and their members should act according to the rules that would be selected if a social contract were actually executed.

If this framework is applied to final *outcomes,* such as societal distributions of costs, risks, benefits, and opportunities, the conclusion is that each individual will agree to the alternative that maximizes the sum of the individual position-specific utilities. (Problems of interpersonal comparisons of utility and identification of utilities from statements or from behaviors are glossed over by assuming that each individual can imagine being in each position well enough to assess the same utility function for it.) This provides a possible justification for using social utility functions, equal to sums of individual utility functions, to choose among outcomes (Harsanyi, 1992).

When the social contract framework is applied to decision *mechanisms* or risk management *processes* by which members of society may interact to determine outcomes, strong theoretical and practical conclusions follow. Because the framework is so abstract and idealized, it is sometimes viewed as having little relevance for policy making and institution design. The following examples suggest otherwise.

Example: Should Insurance Companies Use Genetic Information?

Suppose that a new genetic test is introduced that can identify people who are at high risk of early death from a specific cause, such as heart attack or lung cancer. Should health insurance companies be allowed to use this information is setting individual health premiums? An argument in favor of doing so might be that not using relevant information to better estimate individual risks and to adjust individual premiums accordingly unfairly forces lower-risk individuals to subsidize higher-risk individuals. This seems (at least superficially) analogous to the widely accepted argument that smokers and non-smokers should not pay the same health insurance premiums, since information about smoking status helps to predict future medical expenses and since non-smokers should not have to subsidize the health costs of smokers.

However, the social contract framework implies that genetic test information should *not* be used in setting insurance premiums if people would reject such use if the decision had to be made before learning who was at risk and who was not. In other words, suppose that everyone would have greater *ex ante* expected utility by agreeing not to use such information if the decision had to be made behind the veil of ignorance, i.e., with no one knowing which person with what risk characteristics he or she would become. Then the social contract framework for ethical decision-making implies that the information still should not be used *ex post,* when people have learned something about their own roles in society. [This example assumes that individuals do not use genetic information in making insurance-purchasing decisions. If individuals use such information but insurance companies do not, then adverse selection can create bankrupting costs (Subramanian, 1999).]

Modifying this example suggests some limitations of the social contract construct as a guide to ethical action. For example, should automobile insurance companies be able to use driving records (e.g., history of DUI convictions) to help deermine individual premiums? Many people would say yes. But suppose that a genetic test could provide statistically useful predictive information about tendency toward alcoholism (or drug abuse). Would it be fair for health (or automobile) insurers to use such information to help determine individual premiums? Many people would say no, on the grounds that statistical tendencies identified in groups do not predestine the individual choices and behaviors of group members.

The answer to what information insurers should be allowed to use must depend, even in a social contract theory, on unverified hypotheses about free will and personal responsibility. The idea of abstract "positions in society" used in social contract theory abstracts away from the choices that, in part, lead individuals into certain positions. If some choices lead to high risks, then higher premiums may be appropriate for those who make them. To the extent that such choices cannot usefully be determined by the "positions" of the social contract theory, the theory cannot prescribe how best to manage the risks arising from choices.

Example: A Social Contract Justification for Majority Rule

Suppose that the hypothetical participants in a social contract formation process are given the task of selecting a rule to use to settle conflicts among subsets of members of society about which proposed agreements or collective measures to adopt. For example, a subset of members may find themselves disagreeing about whether to use a certain body of water for fishing or boating; about whether to allow or to prohibit plasticizers in children's toys or fluoride in drinking water, and so forth. Suppose that, *a priori*, any randomly selected member of society is equally likely to favor a proposed measure as to oppose it. The expected gain, to a member who favors a measure, from having it passed is assumed to equal the expected loss to a member who opposes it. In other words, all members expect to care about issues with equal intensity (at least to an adequate approximation.) Under these assumptions, what decision rule would a social contract pick for resolving the contested issue? The answer is that the selected rule should minimize the probability that any member supports a defeated measure or opposes an accepted one. This maximizes the *a priori* expected utility for each participant in each conflict situation. Majority rule is the unique rule that accomplishes this goal. This result is the Rae-Taylor theorem (1969) (Mueller, 1996).

There are other routes to this conclusion, not all requiring the social contract. For example, May (1952) proved that majority rule is the only procedure satisfying the following four axioms for choosing between two alternatives, A and B:

1. *Decisiveness:* The procedure must declare A preferred to B, B preferred to A, or A indifferent to B.
2. *Anonymity:* The procedure does not distinguish among individuals.
3. *Neutrality:* If one person prefers A to B and everyone else is indifferent, then the procedure selects A over B.
4. *Positive responsiveness:* If one member's preferences change in favor of A over B and A was already preferred to B according to the procedure, then A is still preferred to B for the new preferences.

6.1 Bargaining from Behind a "Veil of Ignorance"

Private information undermines the Pareto-efficiency of bargaining over rules as well as of bargaining over outcomes. If private information could somehow be removed before bargaining begins, then perhaps more efficient results might be achieved. The "veil of ignorance" provides one way to pursue this idea (Binmore, 1998).

Suppose that agents must agree on a set of compensation rules to be implemented by the SDM. These rules specify how second-party and third-party risks will be managed – e.g., via liability standards, voting procedures, and so forth. A suitably general formulation is that a *compensation rule* will map claims and evidence – or, more generally, statements – presented by disputants in any particular case into a decision about who will pay how much to whom and whether activities will be permitted or prohibited. The payments may include compensation paid to victims by known or suspected injurers; purchases of use-rights made to owners of resources or property; fines and penalties paid to the government, and so forth. Compensation rules are to be used to settle particular conflicts as they arise.

In this context, a version of the veil of ignorance requires players to agree on what compensation rules should be used *before* learning their private information (e.g., before learning who will play what roles in future cases).

It can be shown that rules agreed to this way can give greater expected payoffs to all participants than any set of rules that they would be able to agree on *after* learning their private information (Crawford, 1985). For example, agents may all be able to gain on average by agreeing to accept inefficiently great (e.g., bankrupting) punitive damages for certain types of accidents, if the deterrent effects of the agreement allow everyone to be assured that such accidents will be very rare. Or, they may expect to gain on average by agreeing to let the preferences of a sufficient majority over-rule the preferences of the remaining minority in certain circumstances, as in the Rae-Taylor result. Such an agreement might never be reached on a particular issue once the members of the minority have been identified, lifting the veil of ignorance. Thus, although the veil of ignorance is most often used in discussions of justice, analyzing the rules that would be selected from behind one form of the veil of ignorance – in which people do not yet know their private information – can also promote ex ante Pareto efficiency.

Whether these expected gains in efficiency translate into actual gains for each individual over his lifetime, or whether some individuals suffer ex post inequitable losses, depends on the transactions that each individual participates in and on how they are affected by the compensation rules. Therefore, cost-risk-benefit analysis of one case or one decision at a time may not always provide the most fruitful perspective for public risk management decisions. From a social contract perspective, it is important to judge the efficiency and equity of the *decision process* averaged over all cases, rather than of each outcome that it generates.

The veil of ignorance can help to clarify, and perhaps to justify, rules used to govern liability and compensation for private transactions. Thus, it may provide a partial foundation for tort law theory. The concept can also be applied fruitfully to *collective choice problems,* in which entire groups or societies must decide on something, e.g., whether to require a school vaccination program, whether to adopt a proposed environmental regulation, what level of a public good to produce, or how much of a scarce resource to consume. Collective choice theory is discussed next.

6.2 Collective Choice and Social Choice Functions (SCFs)

Suppose that several players must jointly decide whether to undertake a public project – or, more generally, which of several competing alternative actions to take. The choice can affect each of their utilities. If the project is undertaken, or if a particular alternative is selected, then the players must decide how to share its costs. Moreover, those who favor the selected option (or who claim that they do) may have to compensate those who don't (or who say that they don't). How should the players decide what to do, how to pay for it, and how to allocate compensation? This is the defining problem of *collective choice.*

Several potential answers have already been suggested. One is to generalize two-person bargaining solution concepts to N players. This is easy for the Nash bargaining solution, which selects the outcome that maximizes the product of the players' utility gains. It is harder for more practical mechanisms. Key issues include: who controls the agenda and how, strategic misrepresentation of preferences and other private information, how agreements should be monitored and enforced, what to do (if anything) about strategic collusion among subsets of players, and how previous agreements should be re-negotiated as new information is gained or as preferences change. Such complexities, along with the costs and delays resulting from the combinatorial complexity of the bargaining process, can make direct bargaining among large numbers of people impractical.

A second possible approach is to invoke the veil of ignorance again to ask how players ideally *would* agree to make collective choices if they had to agree on methods *before* obtaining their private information and initial endowments. If this analysis led to a clear procedure that could be implemented in practice, then the result might be an ethically attractive collective choice procedure. However, it is far from clear how collective choices ideally should be made, even when this meta-decision is considered from behind the veil of ignorance.

A third approach, developed in this section, grows out of the ideas of *mechanism design.* Whenever a collective choice must be made, each player is assumed to submit information (not necessarily truthfully) to a centralized procedure. This procedure then determines the solution to the collective choice problem based on the submitted information. It specifies which collective action will be selected, how its costs will be allocated, and who will receive how much compensation. Thus, it allocates risk-cost-benefit combinations to all participants. Such a procedure is called a *social choice function* (SCF). Because it is centralized, it avoids many of the combinatorial difficulties of multi-person bargaining. Because it is generic, the principles that any SCF should satisfy can be discussed in the abstract rather than in the context of particular interests and information. Thus, the problem of how to define a "good" SCF allows scope for ethical principles, as well as for practical implementation concerns.

An even more ambitious strand of collective choice theory deals with *Social Welfare Functions* (SWFs) instead of Social Choice Functions (SCFs). The difference is that an SCF maps individual statements (of preferences or other information) into a decision about what to do, while an SWF maps individual preference orderings for alternatives into an entire social preference ordering for alternatives. Thus, a SWF produces far more information as output than is needed to make a choice. It also specifies the runner(s)-up and the ranking of all rejected alternatives. On the other hand, it allows for less information as input than an SCF. SWFs only consider preference orderings rather than player utilities, expressions of preference intensities, or other information about preferences . (The IIR property in Table 3 below explicitly prohibits expressions of preference intensity, e.g., allocations of a fixed total number of points among alternatives to reflect strength of preference.) Thus, an SWF attempts to produce more detailed output while using less detailed input than many SCFs.

Moreover, an SWF deals with preferences for alternative collective *choices,* i.e., actions, rather than with preferences for *consequences* of collective actions. By contrast, generalizations of EU theory to groups, such as the theory of syndicates (Raiffa, 1968), use preferences for consequences (and beliefs about the probable consequences of alternative choices). In general, SWFs demand both too much information as output and too little information as inputs to provide a useful basis for collective decision-making. In fact, the most famous result about SWFs is Arrow's Impossibility Theorem, which states that no SWF satisfying the desirable properties of non-dictatorship and Pareto-efficiency for arbitrary (i.e., unrestricted) individual preferences exists if IIR is assumed (see Tables 3 and 4 for definitions). For these reasons, SCFs are more interesting and potentially more useful than SWFs.

Many principles have been proposed for constructing SCFs. Table 3 summarizes principles in several major groups, as follows.

- *Ethical properties* impose restrictions on the SCF motivated by concerns for fairness, equity, efficiency, and justice. Many of the fairness-related properties, such as Anonymity, Symmetry, and Non-dictatorial reflect the idea that how people are treated by an SCF should not be affected by their identities. All those who behave the same way, express the same preferences, accept the same offers, and so forth should be treated equally by the SCF. The efficiency-related concepts (Pareto-efficiency, Envy-free, Individualism, No Ill Will, Monotonicity) require that the SCF choose an outcome that reflects individual preferences and that makes some attempt to accommodate them. They stipulate that social preferences or choices should be positively responsive, in some sense, to individual preferences. Conditions such as non-bossiness and minimal liberalism (not shown) attempt to capture the intuition that some things are or should be left to individuals to decide for themselves, not subject to the preferences of others.
- *Structural properties* such as weak-ordering describe the type of result (e.g., rank ordering, unique choice, etc.) to be produced.
- *Incentive properties* deal with the pragmatic consideration that individuals in a free society can be expected to only participate in processes and provide information that they expect will benefit them.

Table 3: Properties for SCFs and Welfare Functions SWFs

Axiom or Property	Interpretation
Ethical Properties	
Non-dictatorial	The SCF or SWF depends on more than one person's preferences (i.e., it is not just the preference ordering or choice function for one individual).
Non-bossiness	If a change in one agent's preferences does not change his/her assigned allocation, then it does not change the selected outcome (Fleurbaey and Maniquet, 1999)
Anonymity/impartiality	If the preferences of two agents are exchanged, then the allocations assigned to them by the SCF are exchanged. A permutation of individuals leaves social preferences among alternatives (i.e., collective choices) unchanged (Dhillon and Mertens, 1999).
Symmetry	Agents with the same preferences are treated equally.
Pareto-efficiency	The SCF chooses a Pareto-efficient outcome (i.e., no other feasible outcome is preferred by all players)
Envy-free	No one strictly prefers another's allocation to his own.
Individualism	If all individuals are indifferent among alternatives, then so is society.
No ill will	If all but one individual are indifferent between two alternatives, the SCF agrees with his preferences.
Monotonicity	An allocation of bundles to agents remains selected if one agent's preferences change, but she still (weakly) prefers her assigned bundle to all those that she weakly preferred it to before the change (Fleurbaey and Maniquet, 1997).
Structural Properties	
Weak ordering	The SWF provides a rank ordering (with ties allowed) of all alternatives.
Independence of irrelevant alternatives (IIA)	The ranking of any two collective actions by the SWF (a function mapping individual preference orderings to social preference orderings) depends only on their ranking by the individuals.
Incentive Properties	
Strategy-proof	Truthful revelation of preferences is a dominant strategy (at least in the equivalent direct revelation game, having announcements of preferences as the only possible strategies, associated with the SCF)
Individual rationality	An SCF should not make any individual worse off than he would be if he did not participate (e.g., paying no costs and receiving no compensation or benefits)
Domain Constraints	
Unrestricted domain	Players may have any preferences for the alternatives (e.g., any preference orderings, for a SWF).
Budget-balancing	The total amount collected from all agents equals the cost of producing the public good.

Table 4: Theoretical Results on Possible SCFs and SWFs

Assumed Axioms/Properties	Result	Reference
Pareto-efficient SWF *IIA* *Unrestricted domain* *Non-dictatorial*	No such SWF exists satisfying all four properties ("Arrow Impossibility Theorem")	Arrow, 1951
Strategy-proof SCF *Non-dictatorial* *Unrestricted domain* Collective choice set has at least 3 alternatives	No such SCF exists (Gibbard-Satterthwaite Theorem)	Gibbard, 1973; Satterthwaite, 1975
Strategy-proof SCF *Pareto-efficient* *Unrestricted domain*	No such SCF exists	Serizawa, 2000
Preferences are restricted so the SWF satisfies the other three Arrow conditions (IIA, Pareto-efficient, Non-dictatorial))	These three Arrow conditions hold if and only if the Gibbard-Satterthwaite conditions (strategy-proof SCF) holds	
Nash equilibria of players' revelation strategies give *Pareto efficient* choices for players with preferences linear in money.	Only *Groves-Ledyard mechanisms* (in which each individual pays a tax sufficient to compensate others for the loss his choice causes them) satisfy these conditions. They fail to balance budgets, in general	Green and Laffont, 1979
Strategy-proof *Budget-balancing* *Symmetry* *Individually rational.* Cost function for public good is continuous and increasing.	In a Public Good game, all agents pay equal cost shares and the minimum amount demanded for the public good is produced. (This may not be Pareto-optimal.)	Serizawa, 2000
Strategy-proof *Non-bossiness* *Symmetry* *More of each attribute is preferred to less*	An allocation function (over multiattribute risk-cost-benefit vectors) that satisfies these conditions will also be envy-free.	Fleubaey and Maniquet, 1999

- *Domain of applicability* describes the types of situations to which an SWF or SCF is to be applied.

Various subsets of these principles are mutually inconsistent, leading to impossibility results such as the Arrow and Gibbard-Satterthwaite theorems (see Table 4). Other subsets imply a unique process for mapping individual submissions to collective decisions.

From a game-theory perspective, an SCF is a mechanism that maps players' announcements to final decisions. In most applications, the final decisions determine *allocations* of costs, taxes, and compensation payments to individuals and a *decision* about which collective choice to make (from a set of alternatives) and/or what level of a public good (e.g., an expensive public health or safety measure) to provide. A player's *strategy* is a (possibly randomized) mapping of all of her information (including private information, such as preferences, true willingness-to-pay and willingness-to-accept compensation for public goods or projects, and so forth) into an announcement submitted to the SCF. This mapping is also called a *revelation strategy,* since it determines what each player reveals. A player's full set of private information defines her *type*. The revelation strategies used by different types of players in Nash equilibrium when (a) The prior frequency distribution of player types is common knowledge; and (b) Players draw inferences about each others' probable types based on observed behaviors, define a *Bayesian Nash equilibrium* (BNE) solution concept for N-person games. It has been applied to analyze what revelation strategies rational players will use for different SCFs, and therefore the degree to which different SCFs can be expected to elicit fair, efficient, or otherwise desirable behavior based on the criteria in Table 3 (Myerson, 1995). Table 4 summarizes representative theoretical results on design of SCFs (and SWFs) satisfying different combinations of properties.

The theory of SCFs consists of three main types of results. *Impossibility theorems* assert that certain combinations of properties cannot be satisfied by any SCF. The Gibbard-Satterthwaite theorem is of this type. *Possibility theorems* establish that some combinations of properties are consistent. Sometimes, they imply a unique function for mapping individual submissions to outcomes, at least in some types of games. The Groves-Ledyard mechanism is an example. Finally, *implication theorems* show that some conjunctions of properties imply others. These can be surprising and informative when the implied properties seem to be conceptually very different from the ones that imply them. For example, the property of producing envy-free allocations is implied by a wide variety of other conditions, many reflecting concern for fairness, efficiency, equity, and logical consistency in allocation of limited goods (Young, 1994).

6.3 Fair Allocation, Fair Division, and Fair Auctions

Two related collective choice problems of especial interest in risk management are (a) Deciding in which of N communities to place a hazardous or noxious facility; and (b) Deciding how to allocate costs and compensation payments for noxious facilities among the N communities. Choosing where to put a facility can be cast as a problem of *fair allocation of an indivisible public good.* Deciding who should pay how much of its costs (which may include environmental, health, and safety losses as well as financial costs) and to whom, especially when costs depend on total use is a problem of *fair cost allocation.*

Instead of locating one facility at a time, it may be more sensible to plan the location and financing of several undesirable facilities at once. This may allow different communities to exploit differences in their preferences to help allocate facilities (and payments) to communities in a way that will produce a Pareto-efficient, envy-free assignment. The joint problem of designing bundles of facility location/taxation/compensation plans and choosing one to implement may be viewed as a problem of *fair division of multiple indivisible goods.*

Many normative axioms and resulting procedures have been developed for solving fair division and fair allocation problems (Young, 1994). For example, the Shapley value, already discussed in the context of the Average Marginal Contribution (AMC) solution concept for risk attribution problems in Chapter 4, exemplifies the axiomatic approach to fair allocation and the types of insights that may result into fair allocation rules. Different axioms lead to different procedures. Most of the axioms considered fall into the same categories as in Table 3 (ethical, structural, incentives, and domain), but additional axioms consider consistency (e.g., do the allocations made to every subset of players satisfy the same properties?) A host of ingenious combinatorial algorithms have also been developed that generalize the "cake-cutting" or divide-and-choose" methods for obtaining outcomes to fair allocation games with divisible goods that all participants will regard as "fair", or at least as envy-free (Robertson and Webb, 1998).

In the SCF paradigm, each player makes just one submission to a centralized procedure. By contrast, fair-division algorithms, whether for divisible or indivisible goods, typically involve a sequence of stages, with players making multiple offers and interim decisions on the way to a final outcome. For example, one player (perhaps selected at random or by auctioning the position of initiator) may begin by proposing a set of divisions of goods into bundles that all participants will then choose among. The other players then take turns in which they either (a) Reject the current proposal and replace it (perhaps with a known default allocation of bundles to players); (b) Refine the current proposal (e.g., by subdividing one of the originally proposed bundles); or (c) Accept the proposed bundles. Finally, players choose from among the proposed bundles in some (perhaps randomized) order. Possible variations include allowing new proposal after each player's choice and allowing for post-allocation trades among the participants.

Games of fair division and fair allocation have been analyzed using both the axiomatic approach for defining solution concepts (as illustrated by the Shapley value and various alternative formulas for fair allocation) and via strategic analysis of how ideally rational players would conduct themselves in seeking their most-preferred outcomes. A striking finding is that, at least in theory, appropriately designed auction mechanisms can satisfy many criteria for equity and efficiency in allocating indivisible goods – and hence in solving facility-location problems.

Example: Siting a Locally Undesirable Facility

Problem: Suppose that a locally undesirable land use (LULU) facility such as an incinerator, prison, landfill, or hazardous production facility must be placed in exactly one of N communities. No community wants to host it, but all will benefit from it. Community j assigns a nuisance cost of c_j dollars to being the host and puts an equivalent monetary value of $(s - c_j)$ dollars on hosting the facility and being paid \$s to do so by the other communities. These preferences are common knowledge. Where should the facility be put and what payments should be made by the communities to compensate the one that hosts the facility?

Solution: Kleindorfer and Sertel (1994) propose the following simple solution. Place the LULU facility in the community with the lowest nuisance cost (breaking ties arbitrarily if there is more than one least-cost provider). If that community's nuisance cost is C, then each of the N communities (including the host) should pay an equal share of this cost to the host community, i.e., each community pays an amount C/N to the host community.

Despite its simplicity, this solution has the following desirable properties:

- *Efficient:* No other solution would be preferred by all players.
- *Balanced budget:* There is no surplus or deficit payments, i.e., no money enters or leave the system of N players to make the solution work.
- *Fair:* The solution is envy-free (no player would prefer to trade places with any other). It is also equitable insofar as it entails equal payments by all.
- *Incentive to participate:* If each player weakly prefers someone else to host the facility in the absence of compensation and if the default option in the absence of an agreement is for each community to host its own facility, then all expect to gain by participating in the process.
- *Implementable as Nash equilibrium of an auction:* The proposed solution can be achieved as the Nash equilibrium of a "k^{th} lowest bidder wins" auction in which the players (communities) are invited to submit bids for the amounts they would have to be paid as compensation for hosting the facility.

This is an impressive list for such a simple mechanism. However, it depends on the fact that the players hold no private information and have no uncertainty about their own or others' nuisance costs, c_j. When auctions are used to allocate indivisible goods in the presence of private information, uncertain values, or many complementary goods, efficiency is in general impossible to guarantee. One reason is the notorious "Winner's Curse", where imperfect information leads those who most over-estimate an item's true value to bid the most for it (Rasmusen, 1989).

The mathematical game theory of SCFs, auctions, and collective choice mechanisms is rich and insightful. Yet, it is perhaps most useful as an aid for thinking about how people *should* make collective decisions. Its practical implications for the design of risk management processes are less clear. Actual behaviors in collective choice processes, as well as in bargaining processes and other games, are often driven less by purely strategic considerations than by customs, shared expectations based on cultural norms, perceptions of legitimacy and fairness in process and outcomes, and beliefs or speculations about the motivations and behaviors of other participants than (Thaler, 1992; Young, 1994). Practical techniques for designing and managing multi-party risk management processes to promote efficient, fair, and politically acceptable outcomes focus largely on lessons learned about effective *risk communication,* rather than on formal game-theoretic models.

7. INTRODUCTION TO RISK COMMUNICATION

The field of risk communication provides guidelines, derived mainly from experience, survey data, interviews, and experiments, for how to accomplish risk management goals by sharing risk information among stakeholders and decision-makers. The most common goals for risk communication programs are *informing* individuals about risks so that they can make better-informed decisions or seek more information; *influencing* people to change their behaviors, their attitudes and beliefs about hazards, and their acceptance of risk management decisions and policy recommendations; *involving* affected parties in the decision process; and *facilitating* their participation in conflict-resolution, consensus-building, and collective decision-making about risk management. Effective facilitation enables stakeholders to participate more effectively in risk management decision processes by structuring how their beliefs, values, and concerns are elicited, shared, used to create and evaluate decision options, and acted on. It may also enable the facilitator to pursue policy goals in setting the agenda and managing the process to promote certain ends. This section introduces lessons learned about how to present risk information and structure participatory risk management processes to achieve these various objectives.

7.1 Toward More Effective Risk Presentations

There is a tension in many risk communication efforts between *informing* and *influencing* or manipulating target audiences in presenting risk information (Ng and Hamby, 1997). Risk communication programs are often designed and evaluated based on their success in changing individual behaviors, e.g., by persuading people to stop eating fish from polluted lakes, to start using sun block, to participate in vaccination programs, to wear seatbelts, or to refrain from smoking. Other risk presentations have as their main goals to make decisions that have already been reached palatable to those affected (often a lost cause if those affected did not participate in the decision) and to confer legitimacy on decision processes by holding open meetings and sharing information.

Communication and presentation styles that are most effective in changing behaviors typically differ in structure, content, and emphasis from those that best express the technical content of risk assessment findings or that invite and elicit fruitful participation and interaction. For example, accurate communication of technical findings about risks and uncertainties to technically trained decision makers, and effective internal communication about facts, assumptions, conclusions, and uncertainties among expert members of a risk-assessment or risk management team, can greatly benefit from the types of technical methods and displays discussed in Chapters 1-5. Causal graph models, simulation-based what-if analyses, sensitivity analyses, risk profiles, and Bayesian posterior distributions can convey precisely what is known, how it is known, and what remains unknown or assumed – to audiences well trained in such methods. But technically accurate risk communication does *not* address other key goals, such as telling people what has been decided or what they should do so persuasively and credibly that they will agree. It may not even give non-specialists the information they need to make improved decisions. It does not address the need to elicit stakeholder concerns and values or to address them in risk assessment and decision making.

By contrast, persuasive communication about risks and risk management decisions to stakeholders, media, and the public requires different tools and skills. Building trust, gaining and maintaining credibility, and preparing effective summaries of decision-relevant information using appropriate framing techniques (Chapters 5-7) are important to winning acceptance and changing behavior. Brevity, clarity, focus, candor, cogent examples, and deliberate attempts to distance one's self from negative stereotypes of risk communicators may be crucial for communicating technical risks to non-specialist audiences so that the message is listened to instead of being tuned out or dismissed (Peters et al., 1997; Byrd and Cothern, 2000, Chapter 12.) (For example, industry presenters should exhibit concern and care, while government presenters should emphasize commitment and involvement, to overcome possible audience preconceptions and negative stereotypes.) These factors help to establish an audience's perception of knowledge and expertise, openness and honesty, and concern and care – all of which, in turn, tend to promote trust in the speaker and acceptance of his or her risk messages. More generally, audience members consider the source of information, emotional style, framing, and imputed motives of the speaker in assessing the credibility of the message and in responding to it.

Successful facilitation requires yet different skills, including managing group dynamics to ensure that concerns and preferences are elicited, acknowledged, and responded to in ways that confer legitimacy on the process and its outcomes. At a pragmatic level, models of learning styles and cognitive information processing suggest whether logical, visual, example-based, or emotional presentations may work best for different audiences and purposes. The most appropriate level of detail and use of examples will also vary with the audience. Other theoretical frameworks for predicting what types of messages and presentation styles will be most effective in getting risk messages listened to and acted on by an audience emphasize their knowledge and beliefs, represented by facts and mental models (*knowledge theories*); attitudes, ideologies and values (*personality* and *cultural theories*), and socioeconomic status (*context theories*).

Each of these theories has considerable supporting research and conjecture. Yet, no single framework for understanding and practicing risk communication has yet provided dependable, generally applicable guidance on what to say and how to say it to improve risk management processes and/or to streamline the complex concepts, conflicts, and discussions that surround many risk and risk management efforts (Slovic, 1999). Thus, risk communication as an applied discipline for effective risk management is still in its infancy. Many of its best-established recommendations are perhaps not very surprising. Moreover, many prescriptions for effective risk communication and facilitation are not very specific to risk, but reflect general principles for effective communication and presentation. For example, knowing one's target audience(s), having clear messages in mind, communicating clearly and credibly, and eliciting effective participation all draw on skills and preparation that are useful for most areas of business and technical communication.

The following guidelines for communicating regulatory agency risk analyses and risk management decisions to the public are representative of much prescriptive literature that helps to structure risk communication and management efforts (see, e.g., Ng and Hamby, 1997).

Elements of a Successful Agency Risk Management Plan

1. *Be clear on the roles and goals* of your risk management program (e.g., is the goal to inform, influence, or involve the audience?)
   a. What is the agency's desired role with respect to each audience?
   b. Articulate clear agency values and mission for the communication.
   c. What defines success for the communication effort?

2. *Know your target audiences* and stakeholders and their concerns. What knowledge, beliefs, values, attitudes, cultures, and contextual factors shape their concerns and motivate their actions?
3. *Study/understand risk perceptions*, concerns, and most effective communication styles for each audience. Prepare materials that address the concerns and answer key questions efficiently and convincingly.
4. *Involve stakeholders.* Successful risk communication should be interactive and participatory, not a one-way broadcast.
5. Develop technical risk assessment content to support effective risk communication by answering specific questions/addressing concerns.
   - *Emphasize decisions and consequences,* not pure science
   - Be sure to *address key issues* and stakeholder concerns
6. *Organize risk assessments to facilitate effective presentation* of content Identify outcomes of interest or concern to stakeholders, identify decision options, show how they affect outcome probabilities, and quantify trade-offs among likely consequences of different options.
7. *Organize risk management decision processes* to eliminate outrage, accomplish goals, serve chosen roles, and reflect Agency values.

To this generally good advice, risk analysis adds some more specific principles and content, especially in the areas of decision-analytic methodology, framing of results, and risk perception.

Decision-Analytic Structuring of Discussions and Analyses

Analytic discussions among experts, as well as analysis and deliberation within broader groups, can often benefit from decision-analytic structuring of information and issues (Keeney and von Winterfeldt, 1986). The decision analysis framework explicitly considers value trade-offs and individual preferences for consequences – as well as individual attitudes toward risk-bearing for uncertain prospects, as quantified by risk aversion – via value and utility functions. It carefully separates values and preferences for consequences from beliefs about probable consequences and preferences for actions. This separation can help to clarify analysis and to isolate sources of disagreement about what should be done. Although it is impossible to directly extend to group decision making the concepts of Bayesian decision analysis with separate aggregate subjective probabilities and aggregate utility functions (Hylland and Zeckhauser, 1979), still maintaining the distinction between beliefs (which may be informed and clarified by risk assessment models) and values, which must inform risk management decisions, can help to organize and clarify analyses and deliberation.

Understanding whether conflicts arise from differences in preferences for consequences, beliefs about the probable consequences of alternative actions, or willingness to accept risk in order to achieve potential gains, can also identify where (and whether) better technical information might help to resolve conflicts. The separation of facts and values also helps in making appropriate use of technical expertise in teams. Expertise is usually best used to explain the probabilities of various consequences for alternative decision options, rather than to directly recommend actions, which may require conflating expertise about facts with values and risk attitudes.

Framing Results to Have Impact on Behaviors

As discussed in Chapters 5 to 7, how risk information is formatted and presented can greatly affect how recipients assimilate and act on it. For example, in medical decisions, people are more likely to elect a medical procedure when it is described as "99% safe" then when it is described as having "1% chance of complications" (Gurm and Litaker, 2000). Presenting relative risks rather than absolute risks and using loss framing instead of gain framing make it more likely that patients will adopt screening procedures (Edwards et al., 2001). In presenting chemical risks, the language used to describe risks may trigger speculations about the presenter's motives and undermine his or her credibility with the target audience (MacGregor et al., 1999). Understanding such effects can help in preparing the presentation of factual information in ways that are likely to elicit desired responses.

A striking insight from the framing literature is that there may be no neutral way to present risk information. Any presentation carries with it potential presentation and framing effects and biases that may affect the recipients' attention, interpretation, and actions. Presenting the same information in different ways and emphasizing fact-rich displays (e.g., cumulative risk profiles) that are not strongly associated with known presentation biases may come as close as possible to providing the information needed for rational decision-making without influencing the decision. Such displays often lack the brevity and focus that are most effective in action-oriented presentations.

Understanding Risk Perceptions and Concerns

Psychological research on risk perceptions and risk communication has shown that effective risk communication must be sensitive to the nature of the hazard (e.g., natural vs. man-made, voluntary vs. imposed, etc.) as well as to the perceptual filters or motivating concerns of the audience. As discussed in Chapters 5 and 6, individual risk perceptions can often be predicted surprisingly well based on political attitudes and affiliations and correlates such as socioeconomic indicators. Cultural values (e.g., hierarchical, egalitarian, or individualist) (Douglas and Wildavsky, 1983) and contextual factors such as voluntariness, familiarity, control, dread, fairness, man-made vs. natural, child-affecting, and so forth (Slovic et al., 1982; Slovic, 1987) also help to predict who will worry about what and how much. Such political, cultural, attitudinal and contextual factors affect how risks are perceived and how regulatory risk management agendas are set as much or more than the frequency-severity attributes emphasized in risk profiles and technical presentations of risk (Slovic, 1999).

Especially for third-party risks, such as chemical releases from a facility, effective risk communication presentations must be sure to address what people worry about – e.g., whether it is safe to eat garden vegetables and to have children play outside, rather than focusing solely on risk profiles and other technical and analytic descriptions of risk. Effective communication must be concerned with process as well as with outcome. If people believe that identifiable groups are having risks imposed on them unfairly by identified others having superior power, authority, or information, the result is likely to be outrage (Ng and Hamby, 1997). Unresolved outrage can quickly destroy the chances for joint problem-solving as an approach to risk management decision-making and conflict resolution. To resolve such situations, it is important to acknowledge and address the perceived unfair situation, either by correcting it or by discussing how decisions *should* be made when values and interests genuinely conflict and then demonstrating willingness to abide by agreed-to principles of fairness in deciding and communicating what will be done.

7.2 Toward Designs for Better Risk-Management Processes

A hard lesson learned in the 1980s and 1990s is that, in many settings, *no* risk communication effort can succeed in building trust and credibility, changing attitudes and behaviors, or gaining acceptance for proposed risk management actions unless well-designed risk management decision processes have first prepared the way. Any decision reached without the participation of those affected, or by decision processes perceived as lacking credibility and legitimacy, may be unable to win community approval or acceptance no matter how much careful preparation is brought to the risk communication stage.

To lead to politically acceptable solutions for risk management problems, a decision process should be perceived as satisfying informal versions of several principles similar to many of those in Table 3. It should be widely perceived to protect, or at least not obviously violate, individual rights, It should promote economic and political fairness and Pareto-efficient, envy-free outcomes, while satisfying additional constraints of procedural legitimacy. Perceived procedural legitimacy typically requires that a collective decision process have visible, formal mechanisms to elicit, acknowledge, document, and respond to the concerns of stakeholders. It should reach decisions based on principles (e.g., analysis and deliberation followed by voting) that most people consider fair and reasonable in general, even though they may produce outcomes that work against the interests of specific groups in particular cases.

The process should be open and participatory, with those affected or their representatives involved and empowered to provide information and suggestions that can affect the outcome. Many of these characteristics are similar to principles in Table 3, but with the key distinction that they now refer to *perceived* properties of the decision procedure rather than to mathematically exact, rigorously defined properties. As previously mentioned, perceptions of risks, as well as of proposed risk management decisions and processes, may be affected more by political and ideological attitudes and other non-technical factors than by technical aspects alone.

The following box summarizes properties of a decision process that are often associated with its perceived legitimacy and hence with effectiveness in changing people's attitudes and behaviors (e.g., Slovic, 1999)

Elements of Legitimate Multi-Party Decision Processes

A legitimate decision process for risk management should be perceived to:

- Identify and involve key players (or "stakeholders") whose participation, assent or consent will later be needed.
- Give each stakeholder both the opportunity and a positive incentive to participate (e.g., an expectation of obtaining collective choices that he prefers thereby).
- Allow individual concerns, preferences and values to be surfaced, acknowledged, and responded to.
- Confront and resolve conflicts among individual beliefs and/or preferences using shared, accepted principles for how decisions *should* be made when individuals disagree.
- Enable group decisions to be made, implemented, monitored and enforced using agreed-to procedures, even when individual differences and disagreements remain.
- Produce fair and efficient decisions for each participant, at least on average. (There should be no identifiable, systematic losers.)
- Partner with stakeholders to build trust in the process, get it used, and improve it over time.

Not all of these properties are easy (or perhaps even possible) to achieve simultaneously in precise game-theoretic models (see Table 4). However, relatively straightforward techniques for managing group dynamics and for organizing and running effective meetings and hearings can often create a broadly shared perception that most of these elements have been accomplished – even if they have in truth been addressed only superficially. For example, making sure that all stakeholders are given opportunity to comment; recording and systematically responding to (or at least noting) points raised, and actively encouraging participation are simple methods that go far toward making a process look and feel legitimate – even if, in truth, little or nothing is done with the suggestions and concerns of participants beyond noting them. Allowing participants to take turns speaking, keeping and publishing careful notes and written responses to questions and issues raised, and providing multiple opportunities to review and comment before a final decision is made are all now standard methods for creating legitimacy for public risk management processes.

Evolution of Risk Communication

In an influential review of the history of risk communication, Fischhoff (1995) characterized the field as having evolved through a sequence of dominant paradigms or major emphases. These may be caricatured by a succession of slogans: "All we have to do is... Get the numbers right; tell them (i.e., stakeholders) the numbers; explain what we (i.e., risk analysts and communicators) mean by the numbers; show them they've accepted similar risks in the past; show them it's a good deal for them; treat them nicely; make them partners; and do all of the above". Many multi-party decision processes for risk management, including some organized and administered by state and Federal regulatory agencies, have carefully considered this history and its lessons in designing their current risk communication strategies. The need for participatory risk management and early partnering with stakeholders are now widely regarded as essential components for successful risk management and risk communication, indicating a late stage in Fischhoff's proposed evolutionary ladder.

Indeed, approaching risk management decisions as exercises in joint problem-solving by the participants, backed by a commitment to use mutually agreed-on principles and procedures (e.g., of fairness or voting) to resolve conflicts when necessary, provides a powerful approach for creating consensus and acceptance of outcomes. (A similar approach to principled negotiation has also long been advocated as effective for interpersonal negotiations; see Fisher and Ury, 1981). However, as suggested by some of the theoretical results summarized in Table 4, decision processes with multiple players are subject to various types of manipulation and inefficiency, even if they are successful in creating a shared perception of legitimacy. One problem is that, those who set the agendas for decision processes (and process the results) may be able to manipulate the probable outcomes even for decision processes (e.g., voting) that are widely perceived as fair and legitimate. If there is private information, then strategic misrepresentation of interests, preferences, and beliefs may hamper the success of decision processes in obtaining fair, efficient outcomes with high probability.

Increasing expertise in building perceived legitimacy and acceptance for risk management decisions and decision processes raises the threat that agendas, processes and perceptions will come to dominate risk management decisions at the expense of relevant facts. The result may be effective political processes that make ineffective allocations of resources to address health risks. Some professional risk analysts fear that this is already happening (Slovic, 1999). To obtain results that truly serve the public interest – at least as it would be defined for a fully-informed public – the earliest lessons of risk communication (i.e., "get the numbers right") must be revisited. The right numbers are still important for guiding public risk management decision-making to achieve goals such as addressing the worst and most easily remedied problems first. But, it is more important than ever to find ways to incorporate relevant technical information about risks into decision processes that are generally accepted as legitimate.

8. CONCLUSION

In summary, the relatively young disciplines of risk communication and design of risk management decision processes can transform applied risk analysis. They can make it much more useful by introducing its quantitative results into analysis, deliberation, and decision processes where trust, credibility, and perceived legitimacy have been established by other means. Doing so magnifies the chances that the technical information about risk will be heeded and used. But, if sound risk assessment modeling and technical information are neglected, these same communication and process design techniques can potentially be used to achieve consensus and win political acceptance for actions based on inadequate factual analysis. Then, societal risk management resources could be diverted to relatively low-yield endeavors and the field of risk analysis would ultimately lose credibility and value to those participating in risk management.

To a large extent, therefore, the future success and long-run practical impact of the field of risk analysis may depend on how well risk communication and decision process designers insist on obtaining and using high-quality risk assessment information to inform risk management decisions. The causal, statistical, and simulation-based models and methods of analysis in Chapters 2 to 4 of this book and the decision-analytic models and methods of Chapters 5 to 7 can produce and structure risk and uncertainty information to inform and improve decisions. The risk communication and decision-process design approaches surveyed in this chapter can bring such technical information to bear at points in decision processes where it can most clarify and improve analysis, deliberation, and risk management decisions. Doing so will maximize the policy and economic value of high-quality risk assessment modeling. Such a successful link between production of dependable risk and uncertainty information and its application in risk management decision-making will put a premium on methods for more accurately modeling the probabilistic causal relations between alternative decision options and their consequences.

References

Aaerustrup FM, Nielsen EM, Madsen M, Engberg J. Antimicrobial susceptibility patterns of thermophilic Campylobacter spp. from humans, pigs, cattle, and broilers in Denmark. *Journal of Antimicrobial Chemotherapy* 1997;41:2244-2250.

Aardema MJ, Albertini S, Arni P, Henderson LM, Kirsch-Volders M, Mackay JM, Sarrif AM, Stringer DA, Taalman RD. Aneuploidy: a report of an ECETOC task force. *Mutat Res* 1998 Feb;410(1):3-79.

Abdellaoui M. Parameter-free elicitation of utility and probability weighting functions. *Management Science* 2000;46:1497-1512.

Adak GK, Cowden JM, Nicholas S, Evans HS. The Public Health Laboratory Service national case-control study of primary indigenous sporadic cases of Campylobacter infection. *Epidemiol Infect* 1995 Aug;115(1):15-22.

Adler-Mosca H, Luthy-Hottenstein J, Martinetti Lucchini G, Burnens A, Altwegg M. Development of resistance to quinolones in five patients with Campylobacteriosis treated with norfloxacin or ciprofloxacin. *Eur J Clin Microbiol Infect Dis* 1991 Nov;10(11):953-957.

Ahn WK, Bailenson J. Causal attribution as a search for underlying mechanisms: an explanation of the conjunction fallacy and the discounting principle. *Cognit Psychol* 1996 Aug;31(1):82-123.

Ahn H, Chen JJ. Tree-structured logistic model for over-dispersed binomial data with application to modeling developmental effects. *Biometrics* 1997;53:435-455.

Aickin M, Gensler H. Adjusting for multiple testing when reporting research results: the Bonferroni vs Holm methods. *Am J Public Health* 1996 May;86(5):726-728.

AIHC Delivered Dose Work Group, Cox LA Jr. Managing Statistical Uncertainties in PBPK Modeling. Washington (DC):American Industrial Health Council (AIHC);1993.

Ainslie G. Specious reward: a behavioral theory of impulsiveness and impulse control. *Psychological Bulletin* 1975;82:463-496.

Albrecht M, Weber M. An empirical study on intertemporal decision making under risk. *Management Science* 1997 June;43(6):813-826.

Allen LM, Creaven PJ, Nelson RL. Studies on the human pharmacokinetics of isophosphamide (NSC-109724). *Cancer Treatment Reports* 1976;60(4):451-458.

Altekruse SF, Swerdlow DL, Stern NJ. Microbial food borne pathogens. Campylobacter jejuni. *Veterinary Clinics of North America Food Animal Practice* 1998 Mar;14(1):31-40.

Anderlini L, Felli L. Costly bargaining and renegotiation. *Econometrica* 2001 Mar;69(2):377-411.

Andersen PK, Borgan O, Gill RD, Keiding N. *Statistical Models Based on Counting Processes.* New York:Springer;1993.

Anderson KE, Carmella SG, Ye M, Bliss RL, Le C, Murphy L, Hecht SS. Metabolites of a tobacco-specific lung carcinogen in nonsmoking women exposed to environmental tobacco smoke. *J Natl Cancer Inst* 2001 Mar 7;93(5):378-381.

Andersson H, Britton T. *Stochastic Epidemic Models and their Statistical Analysis.* New York:Springer;2001.

Andradóttir S. Optimization of transient and steady-state behavior of discrete event systems. *Management Science* 1996;42:717-737.

Antoniadis A, Gijbels I, Gregoire G. Model selection using wavelet decomposition and applications. *Biometrika* 1997;751-763.

Armstrong BG, Oakes D. Effects of approximation in exposure assessments on estimates of exposure-response relationships. *Scand J Work Environ Health* 1982;8 Suppl 1:20-23.

Armstrong BG, Whittemore AS, Howe GR. Analysis of case-control data with covariate measurement error: application to diet and colon cancer. *Stat Med* 1989;8(9):1151-1163.

Armstrong BG. The effects of measurement errors on relative risk regressions. *Am J Epidemiol* 1990;132(6):1176-1184.

Arrow KJ. *Social Choice and Individual Values.* New York:Wiley;1951.

Aubry-Damon H, Courvalin P. Bacterial resistance to antimicrobial agents: selected problems in France, 1996 to 1998. *Emerging Infectious Diseases* 1999;5(3):315-320.

Aumann, RJ, Hart S, editors. *Handbook of Game Theory with Economic Applications.* Volume 1. Amsterdam:North-Holland;1992.

Babcock L, Loewenstein G, Issacharoff S, Camerer C. Biased judgments of fairness in bargaining. *American Economic Review* 1995:1337.

Bacchetti P, Segal MR. Survival trees with time-dependent covariates: application to estimating changes in the incubation period of AIDS. *Lifetime Data Analysis* 1995;1(1):35-48.

Baker RC, Paredes MDC, Qureshi RQ. Prevalence of Campylobacter jejuni in eggs and poultry meat in New York State. *Poultry Science* 1987;66:1766-1770.

Barlow RE. *Engineering Reliability.* ASA-SIAM Series on Statistics and Applied Probability. Philadelphia (PA):Society for Industrial and Applied Mathematics;1998.

Barron FH, Barrett BE. Decision quality using ranked attribute weights. *Management Science* 1996 Nov;42(11):1515-1523.

Barry D. Differential recall bias and spurious associations in case/control studies. *Stat Med* 1996 Dec 15;15(23):2603-2616.

Bashir SA, Duffy SW. The correction of risk estimates for measurement error. *Ann Epidemiol* 1997 Feb;7(2):154-164.

Basu A, Blanning RW. The analysis of assumptions in model bases using metagraphs. *Management Science* 1998;44(7):982-995.

Beattie J, Covey J, Dolan P, Hopkins L, Jones-Lee M, Loomes G, Pidgeon N, Robinson A, Spencer A. On the contingent valuation of safety and the safety of contingent valuation: Part I – caveat investigator. *Journal of Risk and Uncertainty* 1998 Oct;17(1):5-25.

Beck-Bornholdt HP, Dubben HH. Potential pitfalls in the use of p-values and in interpretation of significance levels. *Radiother Oncol* 1994 Nov;33(2):171-176.

Becker NG. Uses of the EM algorithm in the analysis of data on HIV/AIDS and other infectious diseases. *Stat Methods Med Research* 1997;6(1):24-37.

Bell DE. Disappointment in decision making under uncertainty. *Operations Research* 1985;33:1-27.

Bell DE. A contextual uncertainty condition for behavior under risk. *Management Science* 1995;34:1416-1424.

Ben Abdelaziz F, Lang P, Nadeau R. Dominance and efficiency in multicriteria decision under uncertainty. *Theory and Decision* 1999 Dec;47(3):191-211.

Ben-Shakhar G, Bar-Hilliel M, Bilu Y, Shefler G. Seek and ye shall find: test results are what you hypothesize they are. *J Behavioral Decision Making* 1998;11(4):235-250.

Benartzi S, Thaler RH. Risk aversion or myopia? Choices in repeated gambles and retirement investments. *Management Science* 1999;45(3):354-381.

Benjamin DK, Dougan WR, Buschena D. Individuals. estimates of the risks of death: Part II – new evidence. *Journal of Risk and Uncertainty* 2001 Jan 22;(1):35-58.

Bentler PM, Stein JA. Structural equation models in medical research. *Stat Methods Med Res* 1992;1(2):159-181.

Berger JO, Pericchi LR. The intrinsic Bayes factor for model selection and prediction. *Journal of the American Statistical Association* 1996;91(433):109-122.

Bertsekas DP, Tsitsiklis JN. *Neuro-Dynamic Programming.* Cambridge (MA):Athena Scientific;1996.

Bertsimas D, Nino-Mora J. Restless bandits, linear programming relaxations, and a primal-dual index heuristic. *Operations Research* 2000 Jan-Feb;48(1):80-90.

Bickel WK, Odum AL, Madden GJ. Impulsivity and cigarette smoking: delay discounting in current, never, and ex-smokers. *Psychopharmacology* (Berl) 1999 Oct;146(4):447-454.

Bielza C, Muller P, Insua DR. Decision analysis by augmented probability simulation. *Management Science* 1999;45(7):995-1007.

Biggs D, de Ville B, Suen E. A method of choosing multiway partitions for classification and decision trees. *Journal Applied Statistics* 1991;18(1):49-62.

Billera LJ, Heath DC. Allocation of shared costs: a set of axioms yielding a unique procedure. *Math Oper Res* 1982;7:1.

Binmore K. *Game Theory and the Social Contract: Volume 1 – Playing Fair.* Cambridge (MA):MIT Press;1998.

Birnbaum MH, Navarette JB. Testing descriptive utility theories: violations of stochastic dominance and cumulative independence. *Journal of Risk and Uncertainty* 1998 Oct;17(1):49-78.

Black RE, Levine MM, Clements M, Hughes TP, Blaser MJ. Experimental Campylobacter jejuni infection in humans. *The Journal of Infectious Diseases* 1988 Mar;157(3):472-479.

Blair RC, Troendle JF, Beck RW. Control of familywise errors in multiple endpoint assessments via stepwise permutation tests. *Stat Med* 1996 Jun 15;15(11):1107-1121.

Blalock HM. *Causal inferences in nonexperimental research.* Chapel Hill (NC):University North Carolina Press;1961.

Bleichrodt H, Pinto JL. A parameter-free elicitation of the probability-weighting function in medical decision analysis. *Management Science* 2000 Nov;46(11):1485-1496.

Bodner R, Prelec D. Self-signaling and diagnostic utility in everyday decision making. 2001. Available from: URL: http://www.ecare.ulb.ac.be/ecare/Juan/pdf-chapters/04.prelec_F.pdf

Boffey B, Karkazis J. "Location, Routing, and the Environment." In: Drezner Z, editor. *Facility Location: A Survey of Applications and Methods.* New York:Springer-Verlag;1995.

Bogen KT, McKone TE. Linking indoor air and pharmacokinetic models to assess tetrachloroethylene risk. *Risk Analysis* 1988;8:509-520.

Boiney LG. When efficient is insufficient: fairness in decisions affecting a group. *Management Science* 1995 Sep;41(9):1523-1537.

Borch K. A note on uncertainty and indifference curves. *Review of Economic Studies* 1969;36(1):1-4.

Borcherding K, Eppel T, von Winterfeldt D. Comparison of weighting judgments in multiattribute utility measurement. *Management Science* 1991 Dec;37(12).

Boudjellaba H, Dufour JM, Roy R. Testing causality between two vectors in multivariate autoregressive moving average models. *Journal of the American Statistical Association* 1992;87:1082-1090.

Box GEP, Draper NR. *Evolutionary Operations.* Wiley;1969.

Box MJ, Davies D, Swann WH. *Non-linear Optimization Techniques.* Edinburgh:Oliver and Boyd;1969.

Bramwell VH, Williams CJ. Do authors of review articles use systematic methods to identify, assess and synthesize information? *Ann Oncol* 1997 Dec;8(12):1185-1195.

Breiman L, Friedman J, Olshen R, Stone C. *Classification and Regression Trees.* Wadsworth Publishing;1984.

Brenner H, Savitz DA, Gefeller O. The effects of joint misclassification of exposure and disease on epidemiologic measures of association. *J Clin Epidemiol* 1993 Oct;46(10):1195-1202.

Brenner H, Loomis D. Varied forms of bias due to nondifferential error in measuring exposure. *Epidemiology* 1994 Sep;5(5):510-517.

Brenner H, Blettner M. Controlling for continuous confounders in epidemiologic research. *Epidemiology* 1997 Jul;8(4):429-434.

Breslow NE. Statistics in *Epidemiology*: the case control study. *Journal of the American Statistical Association* 1996;91(433):14-28.

Brockett PL, Kahane Y. Risk, return, and skewness. *Management Science* 1992;38(6):851-866.

Brown PJ, Vannucci M, Fearn T. Multivariate Bayesian variable selection and prediction. *Journal of the Royal Statistical Society Series B* 1998;60:627-642.

Buckland ST, Burnham KP, Augustin NH. Model selection: an integral part of inference. *Biometrics* 1997;53:603-618.

Buckley JJ. Entropy principles in decision-making under risk. *Risk Analysis* 1985;5:303-313.

Buede D. Decision analysis software survey: aiding insight IV. *ORMS Today* 1998 August:56-64.

Burmaster DE. A lognormal distribution for time spent showering. *Risk Analysis* 1998 Feb; 18(1):27-32.

Bushway S, Brame R, Paternoster R. Assessing stability and change in criminal offending: a comparison of random effects, semiparametric, and fixed effects modeling strategies. *Journal Quantitative Criminology* 1999;15(1):23-61.

Buzas JS, Stefanski LA. Instrumental variable estimation in generalized linear measurement error models. *J American Statistical Association* 1996;91(435):999-1006.

Byers S, Raftery AE. Nearest-neighbor clutter sampling for estimating features in spatial point processes. *Journal of the American Statistical Association* 1998 Jun;93(442):577-584.

Byrd DM, Cothern R. *Introduction to Risk Analysis.* Rockville (MD):Government Institutes;2000.

Cairns J, van der Pol M. The estimation of marginal time preference in a UK-wide sample (TEMPUS) project. *Health Technol Assess* 2000;4(1):1-83.

Calabrese EJ, Baldwin LA. Can the concept of hormesis be generalized to *Carcinogenesis*? *Regul Toxicol Pharmacol* 1998 Dec;28(3):230-241.

Calabrese EJ, Baldwin LA, Holland CD. Hormesis: a highly generalizable and reproducible phenomenon with important implications for risk assessment. *Risk Analysis* 1999;19(2):261-282.

Camerer C. "Individual Decision Making." In: Kagel JH, Roth AE, editors. *The Handbook of Experimental Economics.* Princeton (NJ):Princeton University Press;1995.

Campbell DT, Stanley JC. *Experimental and Quasi-Experimental Designs for Research.* Chicago:Rand McNally;1963.

Carlin BP, Louis TA. *Bayes and Empirical Bayes Methods for Data Analysis.* 2nd ed. New York:Chapman and Hall/CRC;2000.

Carpenter CL, Morgenstern H, London SJ. Alcoholic beverage consumption and lung cancer risk among residents of *Los Angeles County. J Nutr* 1998 Apr;128(4):694-700.

Carroll RJ. Covariance analysis in generalized linear measurement error models. *Stat Med* 1989 Sep;8(9):1075-1093.

Carroll RJ. Surprising effects of measurement error on an aggregate data estimator. *Biometrika* 1997;84(1):231-134.

Casadeus-Masanell R, Klibanoff P, Ozdenoren E. Maxmin expected utility over Savage acts with a set of priors. *Journal of Economic Theory* 2000 May;92:35-65.

Casey JT. Predicting buyer-seller pricing disparities. *Management Science* 1995 June;41(6):979-999.

CDC-FoodNet. 1999 Surveillance results preliminary report. Centers for Disease Control and Prevention, National Center for Infectious Diseases, Division of Bacterial and Mycotic Diseases, Foodborne and Diarrheal Diseases Branch. Atlanta:CDC;2000 March. Available from: URL: http://www.cdc.gov/ncidod/dbmd/foodnet/

CDC-FoodNet. Population survey atlas of exposures: 1998 – 1999. Centers for Disease Control and Prevention. Foodborne Diseases Active Surveillance Network (FoodNet) Atlanta:CDC;1999a.

CDC-FoodNet. 1998 Surveillance results final report. Centers for Disease Control and Prevention, National Center for Infectious Diseases, Division of Bacterial and Mycotic Diseases, Foodborne and Diarrheal Diseases Branch. Atlanta:CDC;1999b. Available from: URL: http://www.cdc.gov/ncidod/dbmd/foodnet/

Chapman K, Hariharan G. Controlling for causality in the link from income to mortality. *Journal of Risk and Uncertainty* 1994 Jan;8(1):85-93.

Charnes JM, Shenoy PP. A forward Monte-Carlo method for solving influence diagrams using local computation. Paper presented at the Sixth International Workshop on *Artificial Intelligence* and Statistics. Fort Lauderdale (FL);1997 January 4-7.

Chateauneuf A, Wakker P. An axiomatization of cumulative prospect theory for decision under risk. *Journal of Risk and Uncertainty* 1999;18(2):137-145.

Chatfield C. *Statistics for Technology.* 3rd ed. New York:Chapman and Hall;1983.

Chatterjee K. "Disagreement in Bargaining: Models with Incomplete Information". In: Roth A, editor. *Game-Theoretic Models of Bargaining.* Cambridge (MA):Cambridge University Press;1985.

Chatterjee P. Fault Tree Analysis: Reliability Theory and Systems Safety Analysis. ORC 74-34. University of California, Berkeley:Operations Research Center;1974.

Che YK, Earnhart D. Optimal use of information in litigation: should regulatory information be withheld to deter frivolous suits? *RAND Journal of Economics* 1997 Spring;28(1):120-134.

Checkoway H, Rice CH. Time-weighted averages, peaks, and other indices of exposure in occupational *Epidemiology. Am J Ind Med* 1992;21(1):25-33.

Cheeseman P, Olford RW, editors. *Selecting Models from Data.* In series: Lecture Notes in Statistics New York:Springer-Verlag;1994;89:339-350.

Chen H, Rupa DS, Tomar R, Eastmond DA. Chromosomal loss and breakage in mouse bone marrow and spleen cells exposed to benzene in vivo. *Cancer Research* 1994;54:3533-3539.

Chen H, Eastmond DA. Synergistic increase in chromosomal breakage within the euchromatin induced by an interaction of the benzene metabolites phenol and hydroquinone in mice. *Carcinogenesis* 1995;16(8):1963-1969.

Chen J, Shao J. Jackknife variance estimation for nearest-neighbor imputation. *Journal of the American Statistical Association* 2001;96(453):260-269.

Chen VCP, Ruppert D, Shoemaker CA. Applying experimental design and regression splines to high-dimensional continuous-state stochastic dynamic programming. *Operations Research* 1999 Jan-Feb;47(1):38-53.

Cherkassky V, Mulier F. *Learning from Data: Concepts, Theories, and Methods.* New York:Wiley;1998.

Chevalier S, Roberts RA. Perturbation of rodent hepatocyte growth control by nongenotoxic hepatocarcinogens: mechanisms and lack of relevance for human health. *Oncol Rep* 1998 Nov-Dec;5(6):1319-1327.

Chew, SH. "Axiomatic Utility Theories with the Betweenness Property." In: Fishburn PC and LaValle IH, editors. *Annals of Operations Research* 19, 1-4: *Choice Under Uncertainty.* Basel, Switzerland:JC Baltzer;1989.

Chib S, Jeliazkov I. Marginal likelihood from the Metropolis-Hastings output. *Journal of the American Statistical Association* 2001;96(453):270-281.

Chiu WA, Hassenzahl DM, Kammen DM. A comparison of regulatory implications of traditional and exact two-stage dose-response models. *Risk Analysis* 1999 Feb;19(1):15-22.

Choi BC, Noseworthy AL. Classification, direction, and prevention of bias in epidemiologic research. *J Occup Med* 1992 Mar;34(3):265-271.

Chuma T, Makino K, Okamoto K, and Yugi H. Analysis of distribution of Campylobacter jejuni and Campylobacter coli in broilers by using restriction fragment length polymorphism of flagellin gene. *J Vet Med Sci* 1997;59(11):1011-1015.

Ciranni R, Barale R, Adler ID. Dose-related clastogenic effects induced by benzene in bone marrow cells and in differentiating spermatogonia of Swiss CD1 mice. *Mutagenesis* 1991;6(5):417-421.

Clemen RT. *Making Hard Decisions: An Introduction to Decision Analysis.* 2nd ed. Boston:PWS-Kent;1996.

Coase RH. The Problem of Social Cost. *J Law and Econom*;1960.

Cobelli C, Lepschy A, Jacur GR. "Identification Experiments and Identifiability Criteria for Compartmental Systems." In: Matis JH, Patten BC, White GC, editors. *Compartmental Analysis of Ecosystem Models.* Fairfield (MD):International Cooperative Publishing House;1979.

Cohen LL, Jao JY, Jusko WJ. Pharmacokinetics of cyclophosphamide in man. British *Journal of Pharmacology* 1971;43:677-680.

Cohen PR, Gregory DE, Ballesteros L, Amant RS. "Two Algorithms for Inducing Structural Equation Models from Data." In: Fisher D, Lenz H-L, editors. *Learning from Data: Artificial Intelligence and Statistics V.* New York:Springer-Verlag;1996.

Concato J, Feinstein AR, Holford TR. The risk of determining risk with multivariable models. *Ann Intern Med* 1993 Feb 1;118(3):201-210.

Cooper R, Ross TW. Product warranties and double moral hazard. *RAND Journal of Economics* 1985 Spring;16(1):103-113.

Cover TM, Thomas JA. *Elements of Information Theory.* New York:Wiley;1991.

Cowell R. "Advanced Inference in Bayesian Networks." In: Jordan M, editor. *Learning in Graphical Models*. MIT Press;1999. p. 27-50.

Cox LA Jr. A new measure of attributable risk for public health applications. *Management Science* 1985;31(7):800-814.

Cox LA Jr. "Managing Uncertain Risks Through 'Intelligent' Classification: A Combined Artificial Intelligence/ Decision Analysis Approach." In: Bonin JJ, Stevenson DE, editors. *Risk Assessment in Setting National Priorities.* Plenum Press;1989.

Cox LA Jr, Ricci PF. Reassessing benzene cancer risks using internal doses. *Risk Analysis* 1992;12(3):401-410.

Cox LA Jr. An exact analysis of the multistage model explaining dose-response concavity. *Risk Analysis* 1995;15(3):359-368.

Cox LA Jr. Reassessing benzene risks using internal doses and Monte-Carlo uncertainty analysis. *Environmental Health Perspectives* 1996;104 Supplement 6:1413-1429.

Cox LA Jr. "Using Causal Knowledge To Learn More Useful Decision Rules from Data." In: Fisher D, Lenz H-J, editors. *Learning from Data: Artificial Intelligence and Statistics V.* New York:Springer-Verlag;1996.

Cox LA Jr. Does diesel exhaust cause human lung cancer? *Risk Analysis* 1997;17(6):807-829.

Cox LA Jr. A biomathematical model of hematotoxicity. *Environment International* 1999 Sep;25(6/7):805-817.

Cox LA Jr. Internal dose, uncertainty analysis, and complexity of risk models. *Environment International 1999b* Sep;25(6/7):841-852.
Crawford VP. "The Role of Arbitration and the Theory of Incentives." In: Roth A, editor. *Game-Theoretic Models of Bargaining.* Cambridge (MA):Cambridge University Press;1985.
Crosbie JH, Glazebrook KD. Index policies and a novel performance space structure for a class of generalised branching bandit problems. *Mathematics of Operations Research* 2000 May;25(2):281-297.
Cubitt RP, Sugden R. Dynamic decision-making under uncertainty: an experimental investigation of choices between accumulator gambles. *Journal of Risk and Uncertainty* 2001;22(2):103-128.
Cummins DJ, Filloon TG, Nychka D. Confidence intervals for nonparametric curve estimates: toward more uniform pointwise coverage. *Journal of the American Statistical Association* 2001;96(453):233-246.
Currim IS, Sarin RK. A comparative evaluation of multiattribute consumer preference models. *Management Science* 1984;30(5):543-561.
Curtis LM, Patrick M, Blackburn CD. Survival of Campylobacter jejuni in foods and comparison with a predictive model. *Letters in Applied Microbiology* 1995;21:194-197.
Das TK, Gosavi A, Mahedevan S, Marchalleck N. Solving semi-Markov decision problems using average reward reinforcement learning. *Management Science* 1999 Apr;45(4):560-574.
Dasgupta A, Raftery AE. Detecting features in spatial point processes with clutter via model-based clustering, *Journal of the American Statistical Association* 1998 Mar;93(441):294-302.
Dasgupta P, Maskin E. The existence of equilibrium in discontinuous economic games, II: Applications. *Review of Economic Studies* 1986 Jan;53(1):27-41.
Daughety AF, Reinganum JF. Product safety: liability, R&D, and signaling. *American Economic Review* December 1995;85(2):1187-2010.
De Flora S. Threshold mechanisms and site specificity in chromium(VI) *Carcinogenesis* 2000 Apr;21(4):533-541.
Deaton A. *Understanding Consumption.* New York:Oxford University Press;1992.
Degraeve Z, Koopman GJ. Achieving air quality standards in the European Union. *Operations Research* 1998 Sep-Oct;46(5):643-654.
DeGroot MH. *Optimal Statistical Decisions.* McGraw-Hill;1970.
DeGroot MH. *Probability and Statistics.* Reading (MA):Addison-Wesley;1975.
Dekel E, Lipman BL, Rustichini A. Representing preferences with a unique subjective probability state space. *Econometrica* 2001 Jul;6(4):891-934.
DelRossi AF, Phillips OR. Pretrial bargaining in the face of a random court decision: evidence from laboratory games. *Journal of Risk and Uncertainty* 1999 Oct;18(3):271-294.
Deming M, Tauxe R, Blake P. Campylobacter enteritis at a University: transmission from eating chicken and from cats. *Am J of Epidemiol* 1987;126:526-534.
Denison DGT, Mallick BK, Smith AFM. A Bayesian cart algorithm. *Biometrika* 1998.
Detweiler JB, Bedell BT, Salovey P, Pronin E, Rothman AJ. Message framing and sunscreen use: gain-framed messages motivate beach-goers. *Health Psychology* 1999 Mar;18(2):186-196.
Dhillon A, Mertens JF. Relative utilitarianism. *Econometrica* 1999;67(3):471-498.
Dolan P, Gudex C. Time preference, duration and health state valuations. *Health Econ* 1995 Jul-Aug;4(4):289-299.
Doucet P, Sloep PB. *Mathematical Modeling in the Life Sciences.* New York:Ellis-Horwood;1992.
Douglas MT, Wildavsky AB. *Risk and Culture: an Essay on the Selection of Technical and Environmental Dangers.* Berkeley and Los Angeles:University of California Press;1983.

Doull J, Cattley R, Elcombe C, Lake BG, Swenberg J, Wilkinson C, Williams G, van Gemert M. A Cancer Risk Assessment of Di(2-ethylhexyl)phthalate: application of the new U.S. EPA Risk Assessment Guidelines. *Regul Toxicol Pharmacol* 1999 Jun;29(3):327-357.

Doye DG, Berry JG, Green PR, Norris PE. Broiler production: considerations for potential growers. Oklahoma Cooperative Extension Service, Division of Agricultural Sciences and Natural Resources;1992. Report F-202. Available from: URL: http://www.ansi.okstate.edu/EXTEN/POULTRY/

DPI. How Broilers are Marketed, Delmarva Poultry Industry, Inc;2000. Available from: URL: http://www.dpichicken.org

Dubeau CE, Voytovich AE, Rippey RM. Premature conclusions in the diagnosis of iron-deficiency anemia: cause and effect. *Med Decis Making* 1986 Jul-Sep;6(3):169-173.

Dubois D, Wellman MP, D'Ambrosio B, Smets P, editors. *Uncertainty in Artificial Intelligence: Proceedings of the Eighth Conference.* San Mateo (CA):Morgan Kaufmann;1992.

Dyer JS, Sarin RK. Group preference aggregation rules based on strength of preference. *Management Science* 1979;25(9).

Dyer JS, Jia J. Preference conditions for utility models: a risk-value perspective. Risk Value Study Series Working Paper 5. Graduate School of Business, University of Texas at Austin. *Annals of Operations Research* 1997. Available from: URL: http://www.bus.utexas.edu/~dyerj/prefcond.pdf

Easley D, Rustichini A. Choice without beliefs. *Econometrica* 1999 Sep;67(5):1157-1184.

Eberhart-Phillips J, Walker N, Garrett N, Bell D, Sinclair D, Rainger W, Bates M. Campylobacteriosis in New Zealand: results of a case-control study. *J Epidemiol Community Health* 1997 Dec;51(6):686-691.

Eckel C. Commentary on The effects of financial incentives in experiments: a review and capital-labor production framework. *Journal of Risk and Uncertainty* 1999 Dec;19(1-3):47-48.

Eisenberg T, Farber HS. The litigious plaintiff hypothesis: case selection and resolution. *RAND Journal of Economics* 1997;28(0):92-112.

Eframovich S. *Nonparametric Curve Estimation.* New York:Springer;1999.

Efron B. The length heuristic for simultaneous hypothesis tests. *Biometrika* 1997;84(1):143-157.

Elandt-Johnson RC, Johnson NL. *Survival Models and Data Analysis.* New York:Wiley;1980.

Elder J, Pregibon D. "A Statistical Perspective on Knowledge Discovery in Databases." In: Fayyad UM, Piatetsky-Shapiro G, Smyth P, Uthurusamy R, editors. *Advances in Knowledge Discovery and Data Mining.* AAAI Press/The MIT Press;1996.

Ellwein LB, Cohen SM. The health risks of saccharin revisited. *Crit Rev Toxicol* 1990;20(5):311-326.

Elting LS, Martin CG, Cantor SB, Rubenstein EB. Influence of data display formats on physician investigators' decisions to stop clinical trials: prospective trial with repeated measures. *British Medical Journal* 1999 Jun 5;318(7197):1527-1531.

Endtz HP, Ruijs GJ, van Klingeren B, Jansen WH, van der Reyden T, Mouton RP. Quinolone resistance in Campylobacter isolated from man and poultry following the introduction of fluoroquinolones in veterinary medicine. *J Antimicrob Chemother* 1991;27:199-208.

Engberg J, Aarestrup FM, Taylor DE, Gerner-Smidt P, Nachamkin I. Quinolone and macrolide resistance in Campylobacter jejuni and C. coli: resistance mechanisms and trends in human isolates. *Emerg Infect Dis* 2001 Jan-Feb;7(1):24-34.

Epstein LG, Le Breton M. Dynamically consistent beliefs must be Bayesian. *Journal of Economic Theory* 1993;61:11-22.

Ericsson NR, Irons JS, editors. *Testing Exogeneity.* Oxford, England:Oxford University Press;1994.

Erkut E, Verter V. "Hazardous Materials Logistics." In: Drezner Z, editor. *Facility Location: A Survey of Applications and Methods.* New York:Springer-Verlag;1995.

Erkut E, Verter V. Modeling of the transport risk from hazardous materials. *Operations Research* 1998 Sep-Oct;46(5):625-642.

Farber JM, Medeiros D, Sanders G, Austin J, Graham C. Quinolone resistance among clinical and food isolates of Campylobacter spp. Poster presented at the 43rd Annual Meeting of the Canadian Federation of Biological Societies;2000 June; Ottawa, Ontario, Canada.

Farris GM, Wong VA, Wong BA, Janszen DB, Shah RS. Benzene-induced micronuclei in erythrocytes: an inhalation concentration-response study in B6C3F1 mice. *Mutagenesis* 1996 Sep;11(5):455-462.

Farris GM, Robinson SN, Gaido KW, Wong BA, Wong VA, Hahn WP, Shah RS. Benzene-induced hematotoxicity and bone marrow compensation in B6C3F1 mice. *Fundam Appl Toxicol* 1997;36(2):119-129.

Fayyad UM, Piatetsky-Shapiro G, Smyth P, Uthurusamy R, editors. *Advances in Knowledge Discovery and Data Mining.* AAAI Press/The MIT Press;1996.

FDA-CVM. Risk assessment on the human health impact of fluoroquinolone resistant Campylobacter associated with the consumption of chicken, US Food and Drug Administration, Center for Veterinary Medicine;2000a. Available from: URL: http://www.fda.gov/cvm/fda/mappgs/ra/risk.html

FDA-CVM. Human health impact of fluoroquinolone resistant Campyblocter attributed to the consumption of chicken, US Food and Drug Administration, Center for Veterinary Medicine;2000b.

Fedorka-Cray PJ, Petersen LE, Tollefson L, Dargatz DA, Muhmed W, Hollinger K, Wineland N, Headrick M, Ferris K. Significance of antibiotic resistant Campylobacter. Presented at the 88th *Poultry Science* Association Annual Meeting, Campylobacter Symposium;1999 Aug 8-11; Springdale (AR). (Abstract available through the Poultry Science Association.)

Feeny D, Furlong W, Boyle M, Torrance GW. Multi-attribute health status classification systems. Health Utilities Index. *Pharmacoeconomics* 1995 Jun;7(6):490-502.

Field RW, Smith BJ, Brus CP, Lynch CF, Neuberger JS, Steck DJ. Retrospective temporal and spatial mobility of adult Iowa women. *Risk Analysis* 1998 Oct;18(5):575-584.

Feinendegen LE, Bond VP, Sondhaus CA, Altman KI. Cellular signal adaptation with damage control at low doses versus the predominance of DNA damage at high doses. *C R Acad Sci* III 1999 Feb-Mar;322(2-3):245-251.

Feinstein AR, Savitz DA, Greenland S, Stolley PD, Kelsey JL. Scientific standards of criticism: a reaction to "Scientific Standards In Epidemiologic Studies of the Menace of Daily Life." *Epidemiology* 1990 Jan;1(1):78-83.

Feller W. *An Introduction to Probability Theory and Its Applications. Volume 1, 3rd ed.* New York:Wiley;1968.

Finkelstein MM. Potential pitfall in using cumulative exposure in exposure-response relationships: demonstration and discussion. *Am J Ind Med* 1995 Jul;28(1):41-47.

Fischer GW. "Prescriptive Decision Science: Problems and Opportunities." In: Fishburn PC, LaValle IH, editors. *Annals of Operations Research* 1989;19:1-4: *Choice Under Uncertainty.* Basel, Switzerland:JC Baltzer;1989.

Fischer GW, Jia J, Luce MF. Attribute conflict and preference uncertainty: the RandMAU model. *Management Science* 2000 May;46(5):669-684.

Fisher R, Ury W. *Getting to Yes.* New York:Penguin Books;1981.

Fischhoff B, Cox LA. "Conceptual Framework for Regulatory Benefits Assessment." In: Bentkover JD, et al., editors. *Benefits Assessment: The State of the Art.* Boston:Reidel;1986.

Fischhoff B. Risk perception and communication unplugged: twenty years of process. *Risk Anal* 1995;15(2):137-145.

Fishburn PC. Equity axioms for public risks. *Operations Research* 1984;32:901-908.

Fishburn PC. Foundations of risk measurement I: Risk as probable loss. *Management Science* 1984;30:296-306.
Fishburn PC. Foundations of risk measurement II: Effects of gain on risk. *Journal Mathematical Psychology* 1984;25:226-242.
Fishburn PC. "Generalizations of Expected Utility Theories: A Survey of Recent Proposals." In: Fishburn PC, LaValle IH, editors. *Annals of Operations Research* 1989;19:1-4: Choice Under Uncertainty. Basel, Switzerland:JC Baltzer;1989.
Fishburn PC, Sarin RK. Dispersive equity and social risk. *Management Science* 1991:751-769.
Fishburn PC, LaValle IH. Stochastic dominance on unidimensional grids. *Mathematics of Operations Research* 1995 August;20(3):513-525.
Fisher D, Lenz HJ, editors. *Learning from Data: Artificial Intelligence and Statistics V*. New York:Springer-Verlag;1996.
Fleurbaey M, Maniquet F. Implementability and horizontal equity imply no-envy. *Econometrica* 1997 Sep;65(5):1215-1219.
Fournier GM, Zuehlke TW. The timing of out-of-court settlements. *RAND Journal of Economics* 1996 Summer;27(2):310-321.
French SA. Decision Theory: An Introduction to the *Mathematics of* Rationality. Ellis Horwood, Ltd;1988.
Friedman C, Reddy S, Samuel M, Marcus R, Bender J, Desai S, Shiferaw B, Helfrick D, Carter M, Anderson B, Hoekstra M, and the EIP Working Group. Risk factors for sporadic Campylobacter infections in the United States: a case-control study in FoodNet sites, International Conference on Emerging Infectious Diseases;2000a July; Atlanta, GA. Available from: URL:http://www.cdc.gov/ncidod/dbmd/foodnet/mce/ICEID/C_frie.htm
Friedman CR, Neimann J, Wegener HC, Tauxe RV. "Epidemiology of *Campylobacter jejuni* Infection in the United States and Other Industrialized Nations." In: *Campylobacter*, Nachamkin I, Blaser MJ, editors. 2nd ed. Washington:ASM Press;2000b. p. 121-138.
Friedman N, Goldszmidt M. "Learning Bayesian Networks with Local Structure." In: Jordan M, editor. *Learning in Graphical Models*. MIT Press;1999. p. 421-459.
Funk LM, Sedman R, Beals JA, Fountain R. Quantifying the distribution of inhalation exposure in human populations: 2. Distributions of time spent by adults, adolescents, and children at home, at work, and at school. *Risk Anal* 1998 Feb;18(1):47-56.
Gafni A, Torrance GW. Risk attitude and time preference in health. *Management Science* 1984;30(4):440-451.
Gail MH. A bibliography and comments on the use of statistical models in *Epidemiology* in the 1980s. *Stat Med* 1991 Dec;10(12):1819-1885.
Galles D, Pearl J. Axioms of causal relevance. Artificial Intelligence, 1997;97(1-2):9-43.
Gamerman D. *Markov Chain Monte Carlo: Stochastic Simulation for Bayesian Inference*. New York:Chapman and Hall;1997.
Gao S, Hui SL. Logistic regression models with missing covariate values for complex survey data. *Stat Med* 1997 Nov 15;16(21):2419-2428.
Garshick E, Schenker MB, Munoz A, Segal M, Smith TJ, Woskie SR, Hammond SK, Speizer FE. A retrospective cohort study of lung cancer and diesel exhaust exposure in railroad workers. *Am Rev Respir Dis* 1988 Apr;137(4):820-825.
Gasser T, Kneip A. Searching for structure in curve samples. *Journal of the American Statistical Association* 1995;90(432):1179-1187.
Gaudreau C, Gilbert H. Antimicrobial resistance of clinical strains of Campylobacter jejuni subsp. jejuni isolated from 1985 to 1997 in Quebec, Canada. *Antimicrobial Agents and Chemotherapy* 1998;42(8):2106-2108.
Gaunt PN, Piddock LJ. Ciprofloxacin resistant Campylobacter spp. in humans: an epidemiological and laboratory study. *Journal of Antimicrobial Chemotherapy* 1996;37:747-757.

Gelfand AE, Mallick BK. Bayesian analysis of proportional hazards models built from monotone functions. *Biometrics* 1995 Sep;51(3):843-852.

Gelles GM, Mitchell DW. Broadly decreasing risk aversion. *Management Science* 1999;45(10):1432-1439.

Genigeorgis C, Hassuneh M, Collins P. Campylobacter jejuni infection on poultry farms and its effect on poultry meat contamination during slaughter. *Journal of Food Protection* 1986;49:895-903.

Geoffrion AM. Structured modeling: survey and future directions. *ORSA CSTS Newsletter,* 1994;15:1. An update of this paper and references to other work are at URL: http://personal.anderson.ucla.edu/~art.geoffrion/home/docs/csts/index.htm

Gerber HU. An Introduction to Mathematical Risk Theory. Huebner Foundation Monograph 8. Wharton School, University of Pennsylvania;1979.

Geweke J. "Inference and Causality." In: Griliches Z, Intriligator MD, editors. *Handbook of Econometrics, Vol. 2.* Amsterdam:North-Holland;1984.

Gibbard A. Manipulation of voting schemes. A general result. *Econometrica* 1973;41:587-601.

Gittins JC. Multi-armed Bandit Allocation Indices. Wiley;1989.

Glymour C, Cooper GF. *Computation, Causation, and Discovery.* Menlo Park (CA):AAAI Press;1999.

Goering PL, Aposhian HV, Mass MJ, Cebrian M, Beck BD, Waalkes MP. The enigma of arsenic *Carcinogenesis*: role of metabolism. Toxicol Sci 1999 May;49(1):5-14.

Gold LS, Stone TH, Ames BN. What do animal cancer tests tell us about human cancer risk?: Overview of analyses of the Carcinogenic Potency Database. *Drug Metabolism Reviews* 1998;30(2):359-404.

Golden RJ, Holm SE, Robinson DE, Julkunen PH, Reese EA. Chloroform mode of action: implications for cancer risk assessment. *Regul Toxicol Pharmacol* 1997;26(2):142-155.

Goldsworthy TL, Lyght O, Burnett VL, Popp JA. Potential role of alpha-2 mu-globulin, protein droplet accumulation, and cell replication in the renal carcinogenicity of rats exposed to trichloroethylene, perchloroethylene, and pentachloroethane. *Toxicol Appl Pharmacol* 1988 Nov;96(2):367-379.

Gollier C. Repeated optional gambles and risk aversion. *Management Science* 1996 Nov;42(11):1524-1530.

Gonzalez-Mantega W, Cao R, Marron JS. Bootstrap selection of the smoothing parameter in nonparametric hazard rate estimation. *Journal of the American Statistical Association* 1996;91(435)1130-1140.

Gonzalez-Vallejo C, Wallsten TS. Effects of probability mode on preference reversal. J. Exp *Psychol Learn Mem Cogn* 1992 Jul;18(4):855-864.

Goodman J. A comment on the maximum entropy principle. *Risk Analysis* 1987;7(2):269-272.

Gori GB. *Epidemiology* and the concept of causation in multifactorial diseases. *Regul Toxicol Pharmacol* 1989 Jun;9(3):263-272. Bethesda (MD):Health Policy Center.

Gotoh J, Konno H. Third degree stochastic dominance and mean-risk analysis. *Management Science* 2000 Feb;46(2):289-301.

Goutis C, Casella G, Wells MT. Assessing evidence in multiple hypotheses. *Journal of the American Statistical Association* 1996;91(435):1268-1277.

Goutis C, Robert CP. *Model Choice In Generalised Linear Models: A Bayesian Approach Via Kullback-Leibler Projections.* 1998.

Granger CWJ, Newbold P. Spurious regression in econometrics. *Journal of Econometrics* 1974;2(2):111-120.

Granger CWJ. Tests for causation: a personal viewpoint. Journal of Economic Dynamics and Control 1980;2:329-352.

Granger CWJ. "Spurious Regression." In: Eatwell J. *The New Palgrave: Econometrics.* San Francisco:WW Norton and Company;1990.

Grant S, Kajil A, Polak B. Temporal resolution of uncertainty and recursive non-expected utility models. *Econometrica* 2000 Mar;68(2):425-434.

Green JD, Snyder CA, LoBue J, Goldstein BD, Albert RE. Acute and chronic dose/response effect of benzene inhalation on the peripheral blood, bone marrow, and spleen cells of CD-1 male mice. *Toxicology and Applied Pharmacology* 1981a;59:204-214.

Green JD, Snyder CA, LoBue J, Goldstein BD, Albert RE. Acute and chronic dose/response effect of inhaled benzene on multipotential hematopoietic stem (CFU-S) and granulocyte/macrophage progenitor (GM-CFU-C) cells of CD-1 male mice. *Toxicology and Applied Pharmacology* 1981b;59:204-214.

Green PJ. Reversible jump Markov chain Monte-Carlo computation and Bayesian model determination. *Biometrika* 1995:82:711-732.

Greenland S. Modeling and variable selection in epidemiologic analysis. *Am J Public Health* 1989 Mar;79(3):340-349.

Greenland S, Robins JM. Empirical-Bayes adjustments for multiple comparisons are sometimes useful. *Epidemiology* 1991 Jul;2(4):244-251.

Greenland S. Methods for epidemiologic analyses of multiple exposures: a review and comparative study of maximum-likelihood, preliminary-testing, and empirical-Bayes regression. *Stat Med* 1993 Apr 30;12(8):717-736.

Greenland S, Finkle WD. A critical look at methods for handling missing covariates in epidemiologic regression analyses. *Am J Epidemiol* 1995 Dec 15;142(12):1255-1264.

Greenland S. Basic methods for sensitivity analysis of biases. *Int J Epidemiol* 1996 Dec;25(6):1107-1116.

Grenfell BT, Kleczkowski A, Gilligan CA, Bolker BM. Spatial heterogeneity, nonlinear dynamics and chaos in infectious diseases. *Stat Methods Med Res* 1995 Jun;4(2):160-183.

Grisham JW. Interspecies comparison of liver *Carcinogenesis*: implications for cancer risk assessment. *Carcinogenesis* 1997 Jan;18(1):59-81.

Gul F. A theory of disappointment aversion. *Econometrica* 1991;59:667-686.

Gurland J, Sethuraman J. How pooling failure data may reverse increasing failure rates. *Journal of the American Statistical Association* 1995;90(432):1416-1423.

Gurm HS, Litaker DG. Framing procedural risks to patients: is 99% safe the same as a risk of 1 in 100? *Acad Med* 2000 Aug;75(8):840-842.

Guy RL, Dimitriadis EA, Hu PD, Cooper KR, Snyder R. Interactive inhibition of erythroid 59Fe utilization by benzene metabolites in female mice. *Chem Biol Interact* 1990;74(1-2):55-62.

Haas CN, Rose JB, Gerba CP. *Quantitative Microbial Risk Assessment.* New York:John Wiley and Sons;1999.

Hale J, Kocak H. *Dynamics and Bifurcations.* New York:Springer-Verlag;1991.

Hall P, Turlach BA. Interpolation methods for adapting to sparse designs in nonparametric regression. *Journal of the American Statistical Association* 1997;92(438):466-472.

Hammitt JK, Graham JD. Willingness to pay for health protection: inadequate sensitivity to probability. *Journal of Risk and Uncertainty* 1999 April;18(1):33-62.

Hammitt JK, Harvey CM. Equity, efficiency, uncertainty, and the mitigation of global climate change. *Risk Analysis* 2000 Dec;20(6):851-860.

Harber P, Shusterman D. Medical causation analysis heuristics. *J Occup Environ Med* 1996 Jun;38(6):577-586.

Hard GC, Rodgers IS, Baetcke KP, Richards WL, McGaughy RE, Valcovic LR. Hazard evaluation of chemicals that cause accumulation of alpha 2u-globulin, hyaline droplet nephropathy, and tubule neoplasia in the kidneys of male rats. *Environ Health Perspect* 1993 Mar;99:313-349.

Harrell FE Jr, Lee KL, Pollock BG. Review. Regression models in clinical studies: determining relationships between predictors and response. *J Natl Cancer Inst* 1988 Oct 5;80(15):1198-1202.

Harrell FE Jr, Lee KL, Mark DB. Multivariable prognostic models: issues in developing models, evaluating assumptions and adequacy, and measuring and reducing errors. *Stat Med* 1996 Feb 28;15(4):361-387.

Harris M, Raviv A. Some results on incentive contracts with applications to education and employment, health insurance, and law enforcement. *American Economic Review* 1978;68(1):20-30.

Harris N, Thompson D, Martin D and Nolan C. A survey of Campylobacter and other bacterial contaminants of premarket chicken and retail poultry and meats, King County, Washington. *American Journal of Public Health* 1986;76(4):401-406.

Harris N, Weiss N, Nolan C. The role of poultry and meats in the etiology of Campylobacter jejuni/coli enteritis. *Am J Pub Health* 1986b;76:407-411.

Harsanyi JC. "Game and Decision Theoretic Models in Ethics." In: Aumann RJ, Hart S, editors. *Handbook of Game Theory with Economic Applications: Volume 1.* North-Holland;1992.

Harvey C. Proportional discounting of future costs and benefits. *Mathematics of Operations Research* 1995 May;20(2):381-399.

Haukka JK. Correction for covariate measurement error in generalized linear models--a bootstrap approach. *Biometrics* 1995 Sep;51(3):1127-1132.

Hawkes AP, Wilkins JR 3rd. Assessing agreement between two job-exposure matrices. *Scand J Work Environ Health* 1997 Apr;23(2):140-148.

Hazen GB, Pellissier JM. Recursive utility for stochastic trees. *Operations Research* 1996;44(5):788-809.

Hazen GB, Sounderpandian J. Lottery acquisition versus information acquisition: prices and preference reversals. *Journal of Risk and Uncertainty* 1999;18(2):125-136.

Heal G, Chichilnisky G, Beltratti A. "Uncertain Future Preferences and Conservation." Working Paper Series In: Money, Economics and Finance. Columbia Business School, Columbia University;1996. Available from: URL: http://www.google.com/search?q=cache:4UqnasSyPqE:www.columbia.edu/cu/business/wp/96/pw-96-03.pdf+utility+uncertain+future+preferences&hl=en

Heal G. *Valuing the Future: Economic Theory and Sustainability.* New York:Columbia University Press;1998.

Heidenreich WF, Hoogenveen R. Limits of applicability for the deterministic approximation of the two-step clonal expansion model. *Risk Analysis* 2001 Feb;21(1):103-106.

Heise DR. *Causal Analysis.* New York:Wiley;1975.

Heller WD, Scherer G, Sennewald E, Adlkofer F. Misclassification of Smoking in a Follow-up Population Study in Southern Germany. *J Clin Epidemiol.* 1998 Mar;51(3):211-8.

Henderson N, Langford I. Cross-disciplinary evidence for hyperbolic social discount rates. *Management Science* 1998 Nov;44(11):1493-1500.

Herring AH, Ibrahim JG. Likelihood-based methods for missing covariates in the Cox proportional hazards model. *Journal of the American Statistical Association* 2001;96(453):292-302.

Hethcote HW. The Mathematics of infectious diseases. *SIAM Review* 2000 Dec;42(4):599-654.

Hetrick DM, Jarabeck AM, Travis CC. Sensitivity analysis for phsiologically based pharmacokinetic models. *Journal of Pharmacokinetics and Biopharmaceutics* 1991;19(1):1-20.

Hjorth JS. *Computer Intensive Statistical Methods: Validation, Model Selection, and Bootstrap.* Chapman & Hall;1994.

Hoel DG, Li P. Threshold models in radiation *Carcinogenesis. Health Phys* 1998 Sep;75(3):241-250.

Holmstrom B. Moral hazard in teams. *Bell Journal of Economics* 1982 Autumn;13(2):324-340.

Hoogenveen RT, Clewell HJ, Andersen ME, Slob W. An alternative exact solution of the two-stage clonal growth model of cancer. *Risk Analysis* 1999;19(1):9-14.
Hooper DC, Wolfson JS. Bacterial resistance to the quinolone antimicrobial agents. *Am J Med* 1989 Dec 29;87(6C):17S-23S.
Hopkins RS, Olmsted R, Istre GR. Endemic Campylobacter jejuni infection in Colorado: identified risk factors. *Am J Public Health* 1984 Mar;74(3):249-250.
Hoppensteadt FC, Peskin CS. *Mathematics in Medicine and the Life Sciences.* New York:Springer-Verlag;1992.
Hosoya Y. "Causal Analysis and Statistical Inference on Possibly Non-Stationary Time Series." In: Kreps DM, Wallis KF, editors. *Advances in Economics and Econometrics: Theory and Applications. Volume III.* Cambridge University Press;1997.
Hougaard P. *Analysis of Multivariate Survival Data.* New York:Springer;2001.
Hsee CK, Kunreuther HC. The affection effect in insurance decisions. *Journal of Risk and Uncertainty* 2000 March;20(2):141-159.
Huchzermeier A, Loch CH, Project management under risk: using the real options approach to evaluate flexibility in R&D. *Management Science* 2001 Jan;47(1):85-101.
Huff JE. Chemical toxicity and chemical Carcinogenesis. Is there a causal connection? A comparative morphological evaluation of 1500 experiments. IARC Sci Publ 1992;(116):437-475.
Hylland A, Zeckhauser R. The impossibility of Bayesian group decision making with separate aggregation of beliefs and values. *Econometrica* 1979;47(6):1321-1336.
Ibrahim JG, Chen MH. Predictive variable selection for the multivariate linear model. *Biometrics* 1997 Jun;53(2):465-478.
Ikram R, Chambers S, Mitchell P, Brieseman MA, Ikam OH. A case control study to determine risk factors for Campylobacter infection in Christchurch in the summer of 1992-1993. *N Z Med J* 1994 Oct 26;107(988):430-432.
Inman JJ, Dyer JS, Jianmin J. A generalized utility model of disappointment and regret effects on post-choice valuation. *Marketing Science gy* 1997;16:2.
Irons RD. Studies on the mechanism of 1,3-butadiene-induced leukemogenesis: the potential role of endogenous murine leukemia virus. *Environ Health Perspect* 1990 Jun;86:49-55.
Isaaks EH, Srivastava RM. *Applied Geostatistics.* Oxford University Press;1989.
Iwasaki Y, Simon HA. Causality and model abstraction. *Artificial Intelligence* 1994;67(1):143-194.
Izat AL, Gardner FA, Denton JH, Golan FA. Incidence and level of Campylobacter jejuni in broiler processing. *Poultry Science* 1988;67(11):1568-1572.
Jaakkola MS, Samet JM. 1.2 Summary: Workshop on Health Risks Attributable to ETS Exposure in the Workplace *Environmental Health Perspectives* 1999 Dec;107 Suppl 6.
Jacob-Reitsma WF, Kan CA, Bolder NM. The induction of quinolone resistance in Campylobacter bacteria in broilers by quinolone treatment. *Letters in Applied Microbiology* 1994a;19:228-231.
Jacob-Reitsma WF, Bolder NM, Mulder RW. Cecal carriage of Campylobacter and Salmonella in Dutch broiler flocks at slaughter: a one-year study. *Poultry Science* 1994b;73(8):1260-1266.
Jacob-Reitsma WF. "Campylobacter in the Food Supply." In: Campylobacter. Nachamkin I, Blaser MJ, editors. 2nd ed. Washington:ASM Press;2000. p. 467-481.
JAMA rt J. Statistical tests in medical research. *Acta Oncol* 1992;31(7):723-727.
Jenkins-Smith H, Kunreuther H. Mitigation and benefits measures as policy tools for siting potentially hazardous facilities: determinants of effectiveness and appropriateness. *Risk Analysis* 2001;21(2):371-382.
Jensen FV. *An Introduction to Bayesian Networks.* New York:Springer-Verlag;1996.
Jenzarli A. "Solving Influence Diagrams Using Gibbs Sampling." In: Fisher D, Lenz H-J, editors. *Learning from Data: Artificial Intelligence and Statistics V.* New York:Springer-Verlag;1996.

Jessop A. *Informed assessments: An introduction to information, entropy and statistics.* New York:Ellis Horwood, 1995.

Jia J, Dyer JS. A standard measure of risk and risk-value models. *Management Science* 1996;42:1961-1705.

Jia J, Dyer JS, Butler JC. Measures of perceived risk. *Management Science* 1996;45(4):519-532.

Jia J, Dyer JS, Butler JC. Generalized disappointment models. Working Paper, Graduate School of Business, University of Texas at Austin;1997.

Jia J, Dyer JS, Butler JC. Generalized disappointment models. *Journal of Risk and Uncertainty* 2001 Jan;22(1):59-78.

Jiang R, Murthy DNP. Mixture of Weibull distributions – parametric characterization of failure rate functions. *Applied Stochastic Models and Data Analysis* 1998;14(1):47-65.

Johannesson M, Johansson P-O. The be or not to be, that is the question: an empirical study of WTP for increased life expectancy at an advanced age. *Journal of Risk and Uncertainty* 1996 Sep;13(2):163-174.

Johannesson M, Blomquist GC, Blumenschein K, Johannson P-O, Lijas B, O'Conner RM. Calibrating hypothetical willingness to pay responses. *Journal of Risk and Uncertainty* 1999 Apr;18(1):21-32.

Johansson P-O. *Evaluating Health Risks: An Economic Approach.* Cambridge University Press;1995.

Johnson LT. Cost perceptions and voter demand for environmental risk regulation: the double effect of hidden costs. *Journal of Risk and Uncertainty* 1999 Oct;18(3):271-294.

Johnson NL, Kotz S, Balakrishnan N. *Continuous Univariate Distributions, Volume 1. 2nd ed.* Wiley;1994.

Jones FT, Axtell RC, Rives DV, Scheideler FR, Tarver FR Jr, Walker RL, Wineland MJ. A survey of Campylobacter jejuni contamination in modern broiler production and processing systems. *Journal of Food Protection* 1991;54(4):259-262.

Jordan M, editor. *Learning in Graphical Models.* MIT Press;1999.

Josephson KL, Rubino JR, Pepper IL. Characterization and quantification of bacterial pathogens and indicator organisms in household kitchens with and without the use of a disinfectant cleanser. *Journal of Applied Microbiology* 1997;83:737-750.

Judge GC, Griffiths WE, Hill RC, Lutkepohl H, Lee T-C. *The Theory and Practice of Econometrics.* 2nd ed. New York:Wiley;1985.

Kagel JH, Roth AE, editors. *The Handbook of Experimental Economics.* Princton (NJ):Princeton University Press;1995.

Kahneman D, Tversky A. Prospect theory: an analysis of decisions under risk. *Econometrica* 1979;47:263-291.

Kahneman D, Slovic P, Tversky A. *Judgement Under Uncertainty: Heuristics and Biases.* Cambridge University Press;1981.

Kahneman D, Schkade D, Sunstein CR. Shared outrage and erratic awards: the psychology of punitive damages. *Journal of Risk and Uncertainty* 1998 Apr;16(1):49-86.

Kahneman D, Ritov I, Schkade D. Economic preferences or attitude expressions? An analysis of dollar responses to public issues. *Journal of Risk and Uncertainty* 1999 Dec;19(1-2):203-242.

Kalbfleisch JD, Prentice RL. *The Statistical Analysis of Failure Time Data.* New York:Wiley;1980.

Kambhu J. Optimal product quality under asymmetric information and moral hazard. *Bell Journal of Economics* 1982 Autumn;13:483-492.

Karni E. *Decision Making under Uncertainty: The Case of State-Dependent Preferences.* Cambridge (MA):Harvard University Press;1985.

Karni E. Impartiality: Definition and representation. *Econometrica* 1998;66(6):1405-1415.

Kassenborg H, Smith K, Vugia D, Fiorentino T, Bates M, Carter M, Davington A, Cassidy M, Marano N, Angulo F, and the EIP FoodNet Working Group. Eating chicken or turkey outside the home associated with domestically ascquired fluoroquinolone-resistant Campylobacter infections: a FoodNet case-control study. International Conference on Emerging Infectious Diseases;2000 July, Atlanta, GA.

Kawachi I, Colditz GA. Invited commentary: confounding, measurement error, and publication bias in studies of passive smoking. *Am J Epidemiol* 1996 Nov 15;144(10):909-915.

Keeney RL, Raiffa H. *Decisions with Multiple Objectives: Preference and Value Tradeoffs.* New York:Wiley;1976.

Keeney RL. Utility functions for equity and public risk. *Management Science* 1980 Apr:345-353.

Keeney RL, Winkler RL. Evaluating decision strategies for equity and public risks. *Operations Research* 1985 Sep-Oct:955-970.

Keeney RL, von Winterfeldt D. Improving risk communication. *Risk Anal* 1986 Dec;6(4):417-424.

Kenny DA. *Correlation and Causality.* New York:Wiley;1979.

Kim MY, Zeleniuch-Jacquotte A. Correcting for measurement error in the analysis of case-control data with repeated measurements of exposure. *Am J Epidemiol* 1997 Jun 1;145(11):1003-1010.

Klee V, "Sign Patterns and Stability." In: Roberts FS, editor. *Applications of Combinatorics and Graph Theory to the Biological and Social Sciences.* New York:Springer-Verlag;1989.

Klein JP, Moeschberger ML. *Survival Analysis: Techniques for Censored and Truncated Data.* New York:Springer-Verlag;1997.

Kleindorfer PR, Sertel MR. Auctioning the provision of an indivisible public good. *Journal of Economic Theory* 1994;64:20-34.

Knuiman MW, Divitini ML, Buzas JS, Fitzgerald PE. Adjustment for regression dilution in epidemiological regression analyses. *Ann Epidemiol* 1998 Jan;8(1):56-63.

Kobayashi T. Equilibrium contracts for syndicates with differential information. *Econometrica* 1980;48:1635-1665.

Koenraad PMFJ, Jacobs-Reitsma WF, van der Laan T, Beumer RR, Rombouts FM. Antibiotic susceptibility of Campylobacter isolates from sewage and poultry abattoir drain water. *Epidemiol Infect* 1995;115:475-483.

Koidis P, Doyle MP. Survival of Campylobacter jejuni in fresh and heated red meat. *Journal of Food Protection* 1983;46(9):771-774.

Koksalan MU, Sagala PNS. Interactive approaches for discrete alternative multiple criteria decision making with monotone utility functions. *Management Science* 1995 July;41(7):1158-1171.

Kolluru RV. "Risk Assessment and Management." In: Kolluru RV, editor. *Environmental Strategies Handbook: A Guide to Effective Policies and Practices.* New York:McGraw-Hill;1994.

Kornblatt AN, Barrett T, Morris GK, Tosh FE. Epidemiologic and laboratory investigation of Campylobacter enteritis associated with raw milk. American Journal of *Epidemiology* 1985;122(5):884-888.

Kreps DM, Porteus E. Temporal resolution of uncertainty and dynamic choice theory. *Econometrica* 1979;46:185-200.

Kreps DM, Porteus E. Temporal von Neuman-Morgenstern and induced preferences. *Journal of Economic Theory* 1979;20:81-109.

Kreps DM. Notes on the Theory of Choice. Boulder (CO):Westview Press;1988.

Krischer JP. An annotated bibliography of decision analytic applications to health care. *Operations Research* 1980;28(1).

Kuchenhoff H, Carroll RJ. Segmented regression with errors in predictors: semi-parametric and parametric methods. *Stat Med* 1997;16(1-3):169-188.

Kuha J. Corrections for exposure measurement error in logistic regression models with an application to nutritional data. *Stat Med* 1994 Jun 15;13(11):1135-1148.

Kuha J. Estimation by data augmentation in regression models with continuous and discrete covariates measured with error. *Stat Med* 1997;16(1-3):189-201.

Kunreuther H, Onculer A, Slovic P. Time insensitivity for protective investments. *Journal of Risk and Uncertainty* 1998;16(3):279-299.

Kushner HJ, Yin GG. Stochastic Approximation Algorithms and Applications. Springer;1997.

Laffont JJ. *The Economics of Uncertainty and Information.* Cambridge (MA):MIT Press;1989. Translated by JP Bonin and H Bonin.

Lam TH, He Y. Passive smoking and coronary heart disease: a brief review. *Clin Exp Pharmacol Physiol* 1997 Dec;24(12):993-996.

Lamm SH, Walters AS, Wilson R, Byrd DM, Grunwald H. Consistencies and inconsistencies underlying the quantitative assessment of leukemia risk from benzene exposure. *Environ Health Perspect* 1989 Jul;82:289-297.

Lancaster T. The Econometric Analysis of Transition Data. New York:Cambridge University Press;1990.

Landes WM, Posner RA. Tort law as a regulatory regime for catastrophic personal injuries. *J Legal Studies* 1984 Aug;13(3):417-432.

Langere T, Weber M. Prospect theory, mental accounting, and differences in aggregated and segregated evaluation of lottery portfolios. *Management Science* 2001 May;47(5):716-733.

Lauritzen SL, Spiegelhalter DJ. Local computations with probabilities on graphical structures and their application to expert systems. *Journal of the Royal Statistical Society Series B* 1998;50(2):157-224.

LaValle IH, Wapman K. Rolling back decision trees requires the Independence axiom! *Management Science* 1986;32(3):382-385.

LaValle IH, Fishburn PC. Structuring and assessing linear lexicographic utility. *Journal of Risk and Uncertainty* 1996 Sep;13(2):93-124.

Lawless JF. *Statistical Models and Methods for Lifetime Data.* New York:Wiley;1982.

Le CT. *Applied Categorical Data Analysis.* New York:Wiley;1998.

Lee A, Smith SC, Coloe PJ. Survival and growth of Campylobacter jejuni after artificial inoculation onto chicken skin as a function of temperature and packaging conditions. *J Food Protection* 1998;61(12):1609-1614.

Lee CC. On estimation for monotone dose-response curves. *Journal of the American Statistical Association* 1996;91(435):1110-1119.

Lenk PJ, Rao AG. Transition times: distributions arising from time heterogeneous Poisson processes. *Management Science* 1995;41(7):1117-1129.

Levois M, Switzer P. Differential exposure misclassification in case-control studies of environmental tobacco smoke and lung cancer. *J Clin Epidemiol* 1998 Jan;51(1):37-54.

Levy H, Wiener Z. Stochastic dominance and prospect dominance with subjective weighting functions. *Journal of Risk and Uncertainty* 1998 May/June;16(2):147-164.

Li J, Brainard D. Antibiotics a review of ED use, eMedicine World Medical Library;2000. Available from: URL: http://www.emedicine.com

Li Y, Slavik M. Electrical pasteurization method for reduction or elimination of bacteria on poultry carcasses and in processing water. In: Proceedings of the Food Safety Consortium Annual Meeting;1996 Oct 21-22; Kansas City, Missouri. Fayetteville:University of Arkansas;1996. p. 58-61.

Lichtenstein MJ, Mulrow CD, Elwood PC. Guidelines for reading case-control studies. *J Chronic Dis* 1987;40(9):893-903.

Lillard HS. The impact of commercial processing procedures on bacterial contamination and cross-contamination of broiler carcasses *Journal of Food Protection* 1990;53(3):202-204.

Lin CC, Segel LA. Mathematics Applied to Deterministic Problems in the Natural Sciences. Philadelphia (PA):Society for Industrial and Applied Mathematics;1988.

Lin DY. Non-parametric inference for cumulative incidence functions in competing risks studies. *Statistics in Medicine* 1997;16:901-910.

Lin DY, Psaty BM, Kronmal RA. Assessing the sensitivity of regression results to unmeasured confounders in observational studies. *Biometrics* 1998 Sep;54(3):948-963.

Lindsey JK, Jones B. Choosing among generalized linear models applied to medical data. *Stat Med* 1998 Jan 15;17(1):59-68.

Loch CH, Terwiesch C, Thomke S. Parallel and sequential testing of design alternatives. *Management Science* 2001 May;47(5):663-678.

Loewenstein G, O'Donoghue T, Rabin M. Projection Bias in Predicting Future Utility. 2000. Available from: URL: http://www.haas.berkeley.edu/groups/iber/wps/econ/E00-284.pdf

Loomes G, Sugden R. Disappointment and dynamic consistency in choice under uncertainty. *Review of Economic Studies* 1986;53:271-182.

Luce RD. Several possible measures of risk. *Theory and Decision* 1980;12:217-228.

Luce RD, Weber UK. An axiomatic theory of conjoint, expected risk. *Journal of Mathematical Psychology* 1986;30:188-205.

Luce RD, Narens L. "Measurement." In: Eatwell J, Milgate M, Newman P, editors. *The New Palgrave: Time Series and Statistics.* New York:WW Norton and Company;1990.

Luce RD. Coalescing, event commutativity, and theories of utility. *Journal of Risk and Uncertainty* 1998;16:87-114.

Ludbrook J. Multiple comparison procedures updated. *Clin Exp Pharmacol Physiol* 1998 Dec;25(12):1032-1037.

Luke CA, Tice RR, Drew RT. The effect of exposure regimen and duration on benzene-induced bone marrow damage in mice. I. Sex comparison on DBA/2 mice. *Mutation Research* 1988;203:251-272.

Luke CA, Tice RR, Drew RT. The effect of exposure regimen and duration on benzene-induced bone marrow damage in mice. II. Strain comparisons involving B6C3F1, C57Bl/6 and DBA male mice. *Mutation Research* 1988b;203:273-295.

Lutz WK. Dose-response relationships in chemical *Carcinogenesis*: superposition of different mechanisms of action, resulting in linear-nonlinear curves, practical thresholds, J-shapes. *Mutation Research* 1998 Sep 20;405(2):117-124.

Machina M. "Expected utility" analysis without the independence axiom. *Econometrica* 1982;50(2):277-323.

MacGregor DG, Slovic P, Malmfors T. How exposed is exposed enough? Lay inferences about chemical exposure. *Risk Anal* 1999 Aug;19(4):649-659.

Madigan D, Raftery A. Model selection and accounting for uncertainty in graphical models using Occam's window. *Journal of the American Statistical Association* 1994;89(428):1535-1546.

Magder LS, Zeiger SL. A smooth nonparametric estimate of a mixing distribution using mixtures of Gaussians. *Journal of the American Statistical Association* 1997;91(435):1141-1151.

Magnac T, Robin J-M. Dynamic stochastic dominance in bandit decision problems. *Theory and Decision* 1999 Dec;47(3):267-295.

Maldonado G, Greenland S. Impact of model-form selection on the accuracy of rate estimation. *Epidemiology* 1996 Jan;7(1):46-54.

Marais ML, Wecker WE. Correcting for omitted-variables and measurement-error bias in regression with an application to the effect of lead on IQ. *Journal of the American Statistical Association* 1998;93(442):494–504.

Marano N, Vugia D, Fiorentino T, Segler S, Carter M, Kassenborg H, Smith K, Sansky S, Hollinger K, Angulo F, and the EIP FoodNet Working Group. Fluoroquinolone-resistant Campylobacter causes longer duration of diarrhea than fluoroquinolone-susceptible Campylobacter strains in FoodNet sites. International Conference on Emerging Infectious Diseases;2000 July, Atlanta, GA. Synopsis available from: URL: http://www.cdc.gov/ncidod/dbmd/narms/pus/pub/presentations/2000/marano_n_3.htm

Marcus AH, Elias RW. Review: Some useful statistical methods for model validation. *Environ Health Perspect.* 1998 Dec;106 Suppl 6:1541-1550.
Marrazzini A, Chelotti L, Barrai I, Loprieno N, Barale R. In vivo genotoxic interactions among three phenolic benzene metabolites. *Mutation Research* 1994;341(1):29-46.
Marteau TM. Framing of information: its influence upon decisions of doctors and patients. *British Journal of Social Psychology* 1989 Mar;28 Suppl 1:89-94.
Marshall JR, Hastrup JL. Mismeasurement and the resonance of strong confounders: uncorrelated errors. *Am J Epidemiol* 1996 May 15;143(10):1069-1078.
Matis JH, Kiffe TR. *Stochastic Population Models: A Compartmental Perspective.* New York:Springer;2000.
Mattila L, Peltola H, Siitonen A, Kyronseppa H, Simula I, Kataja M. Short-term treatment of traveler's diarrhea with norfloxacin: a double-blind, placebo-controlled study during two seasons. *Clin Infect Dis* 1993 Oct;17(4):779-82.
Mazumdar J. *An Introduction to Mathematical Physiology and Biology.* New York:Cambridge University Press;1989.
McCullagh P, Nelder JA. *Generalized Linear Models.* New York:Chapman and Hall;1983.
McFadden D. Rationality for Economists? *Journal of Risk and Uncertainty* 1999 Dec;19(1-2):73-105.
Mead GC, Hudson WR, Hinton MH. Effects of changes in processing to improve hygiene control on contamination of poultry carcasses with Campylobacter. *Epidemiol Infect* 1995;115:495-500.
Mead PS, Slutsker L, Dietz V, McCaig LF, Bresee JS, Shapiro C, Griffin PM, Tauxe RV. Food-related illness and death in the United States. *Emerging Infectious Diseases* 1999;5(5). National Center for Infectious Diseases. Available from: URL: http://www.cdc.gov/ncidod/eid/vol5no5/mead.htm
Melnick RL, Huff JE, Roycroft JH, Chou BJ, Miller RA. Inhalation toxicology and carcinogenicity of 1,3-butadiene in B6C3F1 mice following 65 weeks of exposure. *Environmental Health Perspectives* 1990;86:27-36.
Melnick RL, Kohn MC, Portier CJ. Implications for risk assessment of suggested nongenotoxic mechanisms of chemical *Carcinogenesis. Environmental Health Perspectives* 1996 Mar;104 Suppl 1:123-134.
Messite J, Stellman SD. Accuracy of death certificate completion: the need for formalized physician training. *JAMA* 1996 Mar 13;275(10):794-796.
Meyer J, Ormiston MB. Analyzing the demand for deductible insurance. *Journal of Risk and Uncertainty* 1999 Oct;18(3):223-230.
Michie D, Spiegelhalter DJ, Taylor CC, editors. *Machine Learning, Neural and Statistical Classification.* Ellis Horwood Series in Artificial Intelligence. New York:Ellis Horwood;1994.
Mirman LJ, Samet D, Tauman Y. An axiomatic approach to the allocation of a fixed cost through prices. *Bell J Econom* 1983 Fall;141.
Miyamoto JM, Wakker P. Multiattribute utility theory with expected utility foundations. *Operations Research* 1998 Mar-Apr;44(2):313-326.
Miyamoto JM, Wakker P, Bleichrodt H, Peters HJM. The zero condition: a simplifying assumption in QALY measurement and multiattribute utility. *Management Science* 1998;44(6):839-849.
Moolgavkar SH, Dewanji A, Venzon DJ. A stochastic two-stage model for cancer risk assessment. I. The hazard function and the probability of tumor. *Risk Analysis* 1988;8(3):383-392.
Morgan DE. Levofloxacin to become UIHC's primary fluoroquinolone antimicrobial. Virtual Hospital, P&T News 1997 April. University of Iowa Health Care. Available from: URL: http://www.vh.org
Moskowitz H, Sarin RK. Improving the consistency of conditional probability assessments for forecasting and decision making. *Management Science* 1983;29(6):735-749.

Mueller DC. "Public Choice Theory." In: Greenway D, Bleaney M, Stewart I, editors. *A Guide to Modern Economics* . New York:Routledge;1996.

Müller P, Roeder K. A Bayesian semiparametric model for case-control studies with errors in variables. *Biometrika* 1997;84(3):523-527.

Musgrove MT, Cason JA, Fletcher DL, Stern NJ, Cox NA, Bailey JS. Effects of cloacal plugging on microbial quality of partially processed broilers. *Poultry Science* 1993;72(98).

Myerson R, Satterthwaite M. Efficient mechanisms for bilateral trading. *Journal of Economic Theory* 1993;29:265-281.

Myerson RB. *Game Theory: Analysis of Conflict.* Cambridge (MA):Harvard University Press;1991.

Nakamura T. Proportional hazards model with covariates subject to measurement error. *Biometrics* 1992;48:829-638.

Nakamura T, Akazawa K. Computer program for the proportional hazards measurement error model. *Comput Methods Programs Biomed* 1994 Nov;45(3):203-212.

Narens L, Luce RD. "Meaningfulness and Invariance." In: Eatwell J, Milgate M, Newman P, editors. *The New Palgrave: Time Series and Statistics.* New York:WW Norton and Company;1990.

NARMS. 1998 Annual Report of the National Anti-Microbial Resistance Monitoring System. 1999. Available from: URL: http://www.cdc.gov/ncidod/dbmd/narms

NARMS. 1999 Annual Report of the National Anti-Microbial Resistance Monitoring System. 2000. Available from: URL: http://www.cdc.gov/ncidod/dbmd/narms

Nehring K. Preferences for flexibility in a Savage framework. *Econometrica* 1999 Jan;67(1):101-120.

Neilson WS. Reference wealth effects in sequential choice. *Journal of Risk and Uncertainty* 1998 Oct;17(1):27-47.

Nelson CR, Schwert GW. Tests for predictive relationships between time series variables. *Journal of the American Statistical Association* 1982;77:11-18.

Nikula KJ, Avila KJ, Griffith WC, Mauderly JL. Lung tissue responses and sites of particle retention differ between rats and cynomolgus monkeys exposed chronically to diesel exhaust and coal dust. *Fundam Appl Toxicol* 1997 May;37(1):37-53.

Nilsson R. Environmental tobacco smoke and lung cancer: a reappraisal. *Ecotoxicol Environ Saf* 1996 Jun;34(1):2-17.

Ng KL, Hamby DM. Fundamentals for establishing a risk communication program. *Health Phys* 1997 Sep;73(3):473-482.

Norton JP. *An Introduction to Identification.* New York:Academic Press;1986.

Nurminen M. Some remarks on the operation of biases. *Scand J Work Environ Health* 1983 Oct;9(5):377-383.

Nylen G, Palmer SR, Bager F, Feierl G, Molbak K, Ruutu P, Sanchez-Serrano LP. Campylobacter seasonality in Europe. In: Proceedings of the 4th World Congress on Foodborne Infections and Intoxications;1998 Jun 7-12; Berlin, Germany;1998. p. 293-297.

O'Connor AM, Boyd NF, Tritchler DL, Kriukov Y, Sutherland H, Till JE. Eliciting preferences for alternative cancer drug treatments. The influence of framing, medium, and rater variables. *Med Decis Making* 1985 Winter;5(4):453-463.

Ogden MW, Morgan WT, Heavner DL, Davis RA, Steichen TJ. National incidence of smoking and misclassification among the U.S. married female population. *J Clin Epidemiol* 1997 Mar;50(3):253-263.

Ogryczak W, Ruszczynski A. From stochastic dominance to mean-risk models: semideviations as risk measures. *European Journal of Operations Research* 1999;116:33-50.

Okabe A, Brooks B, Sugihara K. *Spatial Tessellations: Concepts and Applications of Voronoi Diagrams.* Wiley;1992.

Orphanides A, Zervos D. Rational addiction with learning and regret. *Journal of Political Economy* 1995;103:739-758.

Ottenbacher KJ. Quantitative evaluation of multiplicity in *Epidemiology* and public health research. *Am J Epidemiol* 1998 Apr 1;147(7):615-619.
Owens DK, Shachter RD, Nease RF Jr. Representation and analysis of medical decision problems with influence diagrams. *Med Decis Making* 1997 Jul-Sep;17(3):241-262.
Paik MC, Tsai WY. On using the Cox proportional hazards model with missing covariates. *Biometrika* 1997;84(3).
Palacios-Huerta I. The aversion to the sequential resolution of uncertainty. *Journal of Risk and Uncertainty* 1999 October;18(3):249-269.
Pannell DJ. Sensitivity analysis of normative economic models: theoretical framework and practical strategies. *Agricultural Economics* 1997;16:139-152. Available from: URL: http://www.general.uwa.edu.au/u/dpannell/dpap971f.htm
Papazoglou IA, Kollas JG. Establishing protective long term measures after severe nuclear accidents using multiple criteria. *Health Phys* 1997 May;72(5):676-692.
Papazoglou IA, Bonanos GS, Nivolianitou ZS, Duijm NJ, Rasmussen B. Supporting decision makers in land use planning around chemical sites. Case study: expansion of an oil refinery. *J Hazard Mater 2000* Jan 7;71(1-3):343-373.
Parker RA, Rothenberg RB. Identifying important results from multiple statistical tests. *Stat Med* 1988 Oct;7(10):1031-1043.
Pastor R, Guallar E. Use of two-segmented logistic regression to estimate change-points in epidemiologic studies. *Am J Epidemiol* 1998 Oct 1;148(7):631-642.
Patt A, Zeckhauser R. Action bias and environmental decisions. *Journal of Risk and Uncertainty* 2000 July;21:145-172.
Payne JW, Schkade DA, Desvouges WH, Aultman C. Valuation of multiple environmental programs. *Journal of Risk and Uncertainty* 2000 July;21(1):95-115.
Pearl J. *Probabilistic Inference in Intelligent Systems: Networks of Plausible Inference.* San Mateo (CA):Morgan Kaufmann;1988.
Pearl J. "A Causal Calculus for Statistical Research." In: Fisher D, Lenz H-J, editors. *Learning from Data: Artificial Intelligence and Statistics V.* New York:Springer-Verlag;1996.
Pearl J. *Causality: Models, Reasoning, and Inference.* New York:Cambridge University Press;2000.
Peña EA. Smooth goodness-of-fit tests for the baseline hazard in Cox's proportional hazards model. *Journal of the American Statistical Association* 1998 Jun;93(442):673-692.
Peracchi F. *Econometrics.* New York:Wiley;2001.
Pesendorfer M. Pollution claim settlements under correlated information. *Journal of Economic Theory* 1997;79:72-105.
Peters RG, Covello VT, McCallum DB. The determinants of trust and credibility in environmental risk communication: an empirical study. *Risk Anal* 1997 Feb;17(1):43-54.
Philbrick CR, Kitanidas PK. Improved dynamic programming methods for optimal control of lumped-parameter stochastic systems. *Operations Research* 2001;49(3):398-412.
Phillips PCB. New tools for understanding spurious regressions. *Econometrica* 1998 Nov;66(6):1299-1325.
Piatelli-Palmarini M. *Inevitable Illusions: How Mistakes of Reason Rule Our Minds. New York:*Wiley;1994.
Piddock LJV. Implications for human health. *Journal of Antimicrobial Chemotherapy* 1999;44 Suppl A:17.
Piddock LJV, Ricci V, Stanley K, Jones K. Activity of antibiotics used in human medicine for Campylobacter jejuni isolated from farm animals and their environment in Lancashire, UK. *Journal of Antimicrobial Chemotherapy* 2000;46(2):303-306.
Pliskin JS, Beck CH Jr. A health index for patient screening: a value function approach with application to chronic renal failure patients. *Management Science* 1976;22(9):1009-1021.
Pliskin JS, Shepard DS, Weinstein MC. Utility functions for life-years and health status. *Operations Research* 1980;28(1):206-224.

Polinksky AM, Rubinfeld DL. *RAND Journal of Economics* 1996 Summer;27(2):269-280.
Pollak RA. Habit formation and dynamic demand functions. *Journal of Political Economy* 1970;78(4):745-763.
Pollatsek A, Tversky A. A theory of risk. *Journal of Mathematical Psychology 1970*;7:540-553.
Prakasa Rao BLS. *Identifiability in Stochastic Models.* Academic Press;1992.
Price PS, Su SH, Harrington JR, Keenan RE. Uncertainty and variation in indirect exposure assessments: an analysis of exposure to tetrachlorodibenzo-p-dioxin from a beef consumption pathway. *Risk Analysis* 1996;16(2):263-277.
Price PS, Scott PK, Wilson ND, Paustenbach DJ. An empirical approach for deriving information on total duration of exposure from information on historical exposure. *Risk Analysis* 1998 Oct;18(5):611-620.
Qin Biao Zhang J. A goodness-of-fit test for logistic regression models based on case-control data. *Biometrika* 1997;84(3).
Quattrone GA, Tversky A. Causal versus diagnostic contingencies: on self-deception and on the voter's illusion. Journal of Personality and Social Psychology 1984;46(2):237-248.
Quiggin J. *Generalized Expected Utility Theory: The Rank-Dependent Model.* Boston:Kluwer Academic Publishers;1993.
Rabin M. Risk aversion and expected-utility theory: a calibration theorem. *Econometrica* 2000 Sep;68(5):1281-1293.
Raftery AE, Madigan D, Hoeting JA. Bayesian model averaging for linear regression models. *Journal of the American Statistical Association* 1997 March;92(437).
Raiffa H. *Decision Analysis: Introductory Lectures on Choices Under Uncertainty.* Reading (MA):Addison-Wesley;1968.
Ramsey JR, Andersen ME. A physiologically-based description of the inhalation pharmacokinetics of styrene in rats and humans. *Toxicology and Applied Pharmacology* 1984;73:159-175.
Rasmusen E. *Games and Information: An Introduction to Game Theory.* Cambridge (MA):Basil Blackwell;1989.
Read D. Is time-discounting hyperbolic or subadditive? *Journal of Risk and Uncertainty* 2001 Jul;23(1):5-22.
Read D, Loewenstein G, Rabin M. Choice bracketing. *Journal of Risk and Uncertainty 1999* Dec;19(1-2):171-202.
Redelmeier DA, Koehler DJ, Liberman V, Tversky A. Probability judgement in medicine: discounting unspecified possibilities. *Med Decis Making* 1995 Jul-Sep;15(3):227-230.
Reed, G.H. 1994. Foodborne illness (Part 5): Foodborne Campylobacteriosis. Dairy, Food and *Environmental San.* 14(3):161-162.
Reina J, Borrell N, Serra A. Emergence of resistance to Erythromycin and fluoroquinolones in thermotolerant Campylobacter strains isolated from feces 1987-1991. *Eur J Clin Microbiol Infect Dis* 1992;11:1163-1166.
Reitz RH, Gargas ML, Andersen ME, Provan WM, Green TL. Predicting cancer risk from vinyl chloride exposure with a physiologically based pharmacokinetic model. *Toxicology and Applied Pharmacology* 1996 Apr;137(2):253-267.
Repace JL, Jinot J, Bayard S, Emmons K, Hammond SK. Air nicotine and saliva cotinine as indicators of workplace passive smoking exposure and risk. *Risk Analysis* 1998 Feb;18(1):71-84.
Richards JB, Mitchell SH, de Wit H, Seiden LS. Determination of discount functions in rats with an adjusting-amount procedure. *J Exp Anal Behav 1997* May;67(3):353-366.
Richardson S, Gilks WR. A Bayesian approach to measurement error problems in *Epidemiology* using conditional independence models. Unite 170, Institut National de la Sante et de la Recherche Medicale, Villejuif, France. *Am J Epidemiol* 1993 Sep 15;138(6):430-442.

Richardson S, Leblond L. Some comments on misspecification of priors in Bayesian modeling of measurement error problems. *Stat Med* 1997 Jan 15;16(1-3):203-213.
Riggs DS. *The Mathematical Approach to Physiological Problems.* Baltimore:Williams and Wilkens;1963.
Ripley BD. *Spatial Statistics.* Wiley;1981.
Rob R. Pollution claim settlements under private information. *J Economic Theory* 1989;47:307-333.
Robins J, Wasserman L. "On the Impossibility of Inferring Causation from Association Without Background Knowledge." In: Glymour C, Cooper G, editors. *Computation and Causality.* AAAI/MIT Press;1998. Available from: URL:http://www.sis.pitt.edu/~dsl/UAI/UAI97/Robins.UAI97.html
Robinson DA. Infective dose of Campylobacter jejuni in milk. *British Medical Journal* 1981;282:1584.
Robertson J, Webb W. *Cake-Cutting Algorithms: Be Fair if You Can.* Natick (MA):AK Peters;1998.
Robertson ML, Eastmond DA, Smith MT. Two benzene metabolites, catechol and hydroquinone, produce a synergistic induction of micronuclei and toxicity in cultured human lymphocytes. *Mutatation Research* 1991;249(1):201-209.
Ronchetti E, Field C, Blanchard W. Robust linear model selection by cross-validation. *Journal of the American Statistical Association* 1997;92(439):1017–1023.
Rosner B, Spiegelman D, Willett WC. Correction of logistic regression relative risk estimates and confidence intervals for measurement error: the case of multiple covariates measured with error. *Am J Epidemiol* 1990 Oct;132(4):734-745.
Ross SM. *Applied Probability Models with Optimization Applications.* San Francisco:Holden-Day;1970.
Ross SM. *Stochastic Processes.* 2nd ed. New York:Wiley;1996.
Rossiter S, Joyce K, Ray M, Benson J, Mackinson C, Gregg C, Sullivan M, Vought K, Leano F, Besser J, Marano N, Angulo F. High Prevalence of antimicrobial-resistant, including fluoroquinolone-resistant Campylobacter on chicken in US grocery stores. American Society for Microbiology 100th General Meeting;2000 May; Los Angeles, CA;2000.
Rottenstreich Y, Tversky A. Unpacking, repacking, and anchoring: advances in support theory. *Psychol Rev* 1997 Apr;104(2):406-415.
Ruanaidh JJK, Fitzgerald WJ. *Numerical Bayesian Methods Applied to Signal Processing.* Springer;1996.
Rubinstein A. Perfect equilibrium in a bargaining model. *Econometrica* 1982;53:1151-1172.
Russell S, Norvig P. *Artificial Intelligence: A Modern Approach.* Englewood Cliffs (NJ):Prentice-Hall;1995.
Saari DG. The sources of some paradoxes from social choice theory and probability. *Journal of Economic Theory* 1987;41:1-22.
Saenz Y, Zarazaga M, Lantero M, Gastanares MJ, Baquero F, Torres C. Antibiotic resistance in Campylobacter strains isolated from animals, foods, and humans in Spain in 1997–1998. *Antimicrobial Agents and Chemotherapy* 2000;44(2):267-271.
Samuelson L. *Evolutionary Games and Equilibrium Selection.* Cambridge (MA):MIT Press;1998.
Samuelson W. Bargaining under Asymmetric Information. *Econometrica* 1984;52:995-1005.
Samuelson W. "A Comment on the Coase Theorem." In: Roth AE, editor. *Game-Theoretic Models of Bargaining.* Cambridge (MA):Cambridge University Press;1985.
Sarin RK. Measuring equity in public risk. *Operations Research* 1985 Jan-Feb:210-217.
Sarin RK. Some extensions of Luce's measures of risk. *Theory and Decision* 1987;22:25-141.
Sarin RK, Wakker P. Dynamic choice and non-expected utility theory. *Journal of Risk and Uncertainty* 1998;17:87-119.

Sarkar SK, Chang CK. The Simes method for multiple hypothesis testing with positively dependent test statistics. *Journal of the American Statistical Association* 1997;92(440):1601–1608.

Satterthwaite M. Strategy-proofness and Arrow's conditions: existence and correspondence theorems for voting procedures and social welfare functions. *Journal of Economic Theory* 1975;10:187-217.

Sauerbrei W, Schumacher M. A bootstrap resampling procedure for model building: application to the Cox regression model. *Stat Med* 1992 Dec;11(16):2093-2109.

Schachter R. An ordered examination of influence diagrams. *Networks* 1990;20:535-563.

Schafer JL. *Analysis of Incomplete Multivariate Data.* New York:Chapman and Hall;1997.

Schell MJ, Leysieffer FW. An increasing failure rate approach to low-dose extrapolation. *Biometrics* 1989;45, 1117-1123.

Schelling TC. "The Initimate Contest for Self-Command." In: Schelling TC. *Choice and Consequence.* Cambridge (MA):Harvard University Press;1984.

Schmid CH, Rosner B. A Bayesian approach to logistic regression models having measurement error following a mixture distribution. *Stat Med* 1993 Jun 30;12(12):1141-1153.

Schmoyer RL. Sigmoidally constrained maximum-likelihood estimation in quantal bioassay. *Journal of the American Statistical Association* 1994;79, 448-453.

Seber GAF, Wild CJ. *Nonlinear Regression.* New York:Wiley;1989.

Serizawa S. Strategy-proof and symmetric social choice functions for public good economies. *Econometrica* 1999 Jan;67(1):121-145.

Shafer G. *The Art of Causal Conjecture.* Cambridge (MA):MIT Press;1996.

Shahabuddin P. Importance sampling for the simulation of highly reliable Markovian systems. *Management Science* 1994;40:333-352.

Shaked M, Shanthikumar JG. *Stochastic Orders and their Applications.* New York:Academic Press;1994.

Shanker S, Lee A, Sorrell T. Campylobacter jejuni in broilers, the role of vertical transmission. *J Hyg Camb* 1986;96:153-159.

Shao J. Bootstrap model selection. *Journal of the American Statistical Association* 1996;92(438):466-472.

Shao J. *Mathematical Statistics.* New York:Springer-Verlag;1999.

Shapiro C. Consumer information, product quality, and seller reputation. *Bell Journal of Economics* 1982 Spring;13(1):20-35.

Shavell S. On Liability and insurance. *Bell Journal of Economics* Spring 1982;13(1):120-132.

Shavell S. *Economic Analysis of Accident Law.* Cambridge (MA):Harvard University Press;1987.

Shenoy PP. Valuation-based systems for Bayesian decision analysis. *Operations Research* 1992;40(3):463-483.

Shore RE, Iyer V, Altshuler B, Pasternack BS. Use of human data in quantitative risk assessment of carcinogens: impact on epidemiologic practice and the regulatory process. *Regul Toxicol Pharmacol* 1992 Apr;15(2 Pt 1):180-221.

Sieff EM, Dawes RM, Loewenstein GF. Anticipated versus actual responses to HIV test results. *American Journal of Psychology* 1999;112(2):297-311.

Siegel S. Nonparametric Statistics. New York:McGraw-Hill;.

Simon HA. "Causes and Possible Worlds." In: Simon HA. *Models of Discovery.* Dordrecht:D. Reidel;1977.

Sims CA. "Multivariate Time Series Models." In: Eatwell J, Milgate M, Newman P editors. *The New Palgrave: Time Series and Statistics.* New York:WW Norton and Company;1990.

Silvey SD. *Optimal Design: An Introduction to the Theory for Parameter Estimation. New* York:Chapman and Hall;1981.

Sjoberg L. Factors in risk perception. *Risk Analysis* 2000 Feb;20(1):1-11.
Slovic P. Perception of risk. *Science* 1987;236:280-285.
Slovic P, Fischhoff B, Lichenstein S. Facts versus Fears: Understanding Perceived Risk. In Kahneman D, Slovic P, Tversky A, editors. *Judgment Under Uncertainty: Heuristics and Biases*. Cambridge:Cambridge University Press;1982.
Slovic P. Trust, emotion, sex, politics, and science: surveying the risk-assessment battlefield. *Risk Analysis* 1999 Aug;19(4):689-702.
Smidts A. The relationship between risk attitude and strength of preference: a test of intrinsic risk attitude. *Management Science* 1997 Mar;43(3):357-370.
Smith JE. Evaluating income streams. *Management Science* 1998 Dec;44(12 Pt 1):1690-1708.
Smith KE, Besser JM, Hedberg CW, et al. Quinolone-resistant Campylobacter jejuni infections in Minnesota, 1992-1998. *New England Journal of Medicine* 1999 May 20;340(20):1525-1582.
Sombe L. Reasoning under incomplete information in artificial intelligence. Special Issue of *International Journal of Intelligent Systems* 1990 Sep;5(4):323-472.
Soni MG, Ramaiah SK, Mumtaz MM, Clewell H, Mehendale HM. Toxicant-inflicted injury and stimulated tissue repair are opposing toxicodynamic forces in predictive toxicology. *Regul Toxicol Pharmacol* 1999 Apr;29(2 Pt 1):165-174.
Spearow J, Doemeny P, Sera R, Leffler R, Barkley M. Genetic variation in susceptibility to endocrine disruption by estrogen in mice. *Science* 1999;285:1259-1261.
Spiegelman D, Valanis B. Correcting for bias in relative risk estimates due to exposure measurement error: a case study of occupational exposure to antineoplastics in pharmacists. *Am J Public Health* 1998 Mar;88(3):406-412.
Srinivasan A, Muggleton SH, Sternberg MJE, King RD. Theories for mutagenicity: a study in first-order and feature-based induction. *Artificial Intelligence* 1996;85(1-2):277-299.
Stefanski LA, Cook JR. Simulation-extrapolation: the measurement error jackknife. *Journal of the American Statistical Association* 1995;90(432):1247-1256.
Sterling TD, Rosenbaum WL, Weinkam JJ. Bias in the attribution of lung cancer as cause of death and its possible consequences for calculating smoking-related risks. *Epidemiology* 1992 Jan;3(1):11-16.
Stern NJ, Clavero MRS, Bailey JS, Cox NA, Robach MC. Campylobacter spp. in broilers on the farm and after transport. *Poultry Science* 1995a;74:937-941.
Stern NJ. Influence of season and refrigerated storage on Campylobacter spp. contamination of broiler carcasses. *Journal of Applied Poultry Research* 1995b;4:235-238.
Strahilevitz, Michal, Loewenstein, George. The effect of ownership history on the valuation of objects. *Journal of Consumer Research* 1998;25:276-289.
Strotz R. Myopia and inconsistency in dynamic utility maximization. *Review of Economic Studies* 1956;23:165-180.
Stute W, Manteiga WG, Quindimil MP. Bootstrap approximations in model checks for regression. *Journal of the American Statistical Association* 1998 Mar;93(441):141-149.
Subramanian K, Lemaire J, Hershey JC, Pauly MV, Armstrong K, Ash DA. Estimating adverse selection costs from genetic testing for breast and ovarian cancers: the case of life insurance. *Journal of Risk and Insurance* 1999 Dec;66(4):531-550.
Subramanian U, Cropper M. Public choices between life saving programs: the tradeoff between qualitative factors and life saved. *Journal of Risk and Uncertainty* 2000 July;21(1):117-149.
Susser What is a cause and how do we know one? A grammar for pragmatic epidemiology. Susser M, Gertrude H. Sergievsky Center. *Am J Epidemiol* 1991 Apr 1;133(7):635-648.
Sutton RS, Barto AG. Reinforcement Learning: An Introduction. Cambridge (MA):MIT Press;1998.
Swanson NR, Granger CWJ. Impulse response functions based on a causal approach in residual orthogonalization in vector autoregressions. *Journal of the American Statistical Association* 1997;92(437):357-367.

Talsma E, Goettsch WG, Nieste HLJ, Schrijnemakers PM, Sprenger MJW. Resistance in Campylobacter species: increased resistance to fluoroquinolones and seasonal variation. *Clinical Infectious Diseases* 1999;29:845-848.

Tanner MA. *Tools for Statistical Inference, Third Edition.* Springer;1996.

Tarter ME, Lock MD. *Model-free curve estimation.* New York:Chapman and Hall, 993.

Tengs TO, Adams ME, Pliskin JS, Safran DG, Siegel JE, Weinstein MC, Graham JD. Five-hundred life-saving interventions and their cost-effectiveness. *Risk Anal* 1995 Jun;15(3):369-390.

Teunis P, Nagelkerke N, Haas C. Dose response for infectious gastroenteritis. *Risk Analysis* 1999 Dec;19(6):1251–1260.

Teunis PFM, Havelaar AH. The Beta Poisson dose-response model is not a single-hit model. *Risk Analysis* 2000 Aug;20(4):513-520.

Thayer DW. Use of irradiation to kill enteric pathogens on meat and poultry. *Food and Nutrition Press.* 1995;15(2):181-192.

Thaler R. *The Winner's Curse: Paradoxes and Anomalies of Economic Life.* New York:Macmillan;1992.

Thaler RH. Some Empirical Evidence on Dynamic Inconsistency. Quasi Rational Economics. Russell Sage Foundation;1991. p. 127-133.

Therneau T, Grambsch P. *Modeling Survival Data: Extending the Cox Model.* New York:Springer;2000.

Thomas DC. Pitfalls in the analysis of exposure-time-response relationships. *J Chronic Dis* 1987;40 Suppl 2:71S-78S.

Thompson SK, Seber GAF. *Adaptive Sampling.* Wiley;1996.

Threlfall EJ, Frost JA, Rowe B. Fluoroquinolone resistance in salmonellas and Campylobacters from humans. *British Medical Journal* 1999;318:943-944.

Thwaites RT, Frost JA. Drug resistance in Campylobacter jejuni, C coli, and C lari isolated from humans in north west England and Wales, 1997. *J Clin Pathol* 1999;52(11):812-814.

Tice RR, Luke CA, Drew RT. Effect of exposure route, regimen, and duration on benzene-induced genotoxic and cytotoxic bone marrow damage in mice. *Environ Health Perspect* 1989;82:65-74.

Titterington DM, Smith AFM, Makov UE. *Statistical Analysis of Finite Mixture Distributions.* New York:Wiley;1985.

Toft K, Olofsson T, Tunek A, Berlin M. Toxic effects on mouse bone marrow caused by inhalation of benzene. *Archives of Toxicology* 1982;51:295-302.

Torrance GW, Boyle MH, Horwood SP. Application of multiatiribute utility theory to measure social preferences for health states. *Operations Research* 1982;30(6):1043-1069.

Torrance GW, Feeny DH, Furlong WJ, Barr RD, Zhang Y, Wang Q. Multiattribute utility function for a comprehensive health status classification system. Health Utilities Index Mark 2. *Med Care* 1996 Jul;34(7):702-722.

Troendle JF. A permutational step-up method of testing multiple outcomes. *Biometrics* 1996 Sep;52(3):846-859.

Tucciarelli T, Karatzas GP, Pinder GF. A primal method for the solution of the groundwater quality management problem. *Operations Research* 1998July-Aug;46(4):463-473.

Tversky A. "On the Elicitation of Preferences: Descriptive and Prescriptive Considerations." In: Bell DE, Raiffa H, Tyersky A, editors. *Decision Making: Descriptive, Normative, and Prescriptive Interactions.* Cambridge University Press, 1988.

Tversky A, Kahneman D. The framing of decisions and the psychology of choice. *Science* 1981;211:453-458.

Tversky A, Kahneman D. Advances in Prospect Theory: cumulative representations of uncertainty. *Journal of Risk and Uncertainty* 1992;5:297-323.

Tversky A, Wakker P. Risk attitudes and decision weights. *Econometrica* 1995 Nov;63(6):1255-1280.

US Department of Health and Human Services (USDHHS). Determining Risks to Health: Federal Policy and Practice. Appendix 6. Dover (MA):Auburn House Publishing Company;1986.
USCB. Resident Population Estimates of the US by Sex, Race, and Hispanic Origin. Washington DC:US Census Bureau;2000a. Available from: URL: http://www.census.gov/population/estimates/nation/intfile3-1.txt
USCB. Monthly Estimates of the United States Population. Washington DC:US Census Bureau, Population Estimates Program;2000b. Available from: URL: http://www.census.gov/population/estimates/nation/intfile1-1.txt
USCB. Resident Population Estimates of the US by Age and Sex. Washington DC:US Census Bureau;2000c. Available from: URL: http://www.census.gov/population/estimates/nation/intfile2-1.txt
USCB. Enumerated and Adjusted Census Population by Age, Sex, Urban/Rural Residence. Washington DC:US Census Bureau;2000d. Available from: URL: http://www.census.gov/ipc/www/idbpyr.html
USDA. The US Broiler Industry, USDA Economic Research Service;1988. Available from: URL: http://usda.mannlib.cornell.edu/usda/
USDA. Changes in Food Consumption and Expenditures, USDA Economic Research Service;1993. Available from: URL: http://www.usda.mannlib.cornell.edu/data-sets/food/93004.
USDA. Broiler Chicken Microbiological Baseline Data Collection Program, USDA Food Safety and Inspection Service;1996. Available from: USDA Food Safety and Inspection Service, Science and Technology, Microbiology Division.
USDA. Statistical Highlights 1998-1999: Farm Economics, USDA National Agricultural Statistics Service;1999. Available from: URL: http://www.usda.gov/nass/pubs/stathigh/1999/ec-mc.htm
USDA. Poultry Slaughter, Figures released 2000 Jun 2, USDA National Agricultural Statistics Service;2000. Available from: URL: http://usda.mannlib.cornell.edu/reports/nassr/poultry/ppy-bb/2000
USDA-FAS. Per Capita Poultry Meat Consumption – Selected Countries, USDA Foreign Agricultural Service;1999.
USDA-FSIS. Broiler Chicken Microbiological Baseline Data Collection Program: July 1994 – June 1995, USDA Food Safety and Inspection Service, Science and Technology, Microbiology Division;1996.
Vacek PM. Assessing the effect of intensity when exposure varies over time. *Stat Med* 1997 Mar 15;16(5):505-513.
Vach W, Blettner M. Published erratum appears in *Am J Epidemiol* 1994 Jul 1;140(1):79. Biased estimation of the odds ratio in case-control studies due to the use of ad hoc methods of correcting for missing values for confounding variables. *Am J Epidemiol*
van Houwelingen HC. The future of biostatistics: expecting the unexpected. *Stat Med* 1997 Dec 30;16(24):2773-2784.
van Loon AJ, Goldbohm RA, van den Brandt PA. Lung cancer: is there an association with socioeconomic status in The Netherlands? *J Epidemiol Community Health* 1995 Feb;49(1):65-69.
Venables WN, Ripley BD. *Modern Applied Statistics with S-Plus.* Springer-Verlag;1994.
Verhoef LC, de Haan AF, van Daal WA. Risk attitude in gambles with years of life: empirical support for prospect theory. *Medical Decision Making* 1994 Apr-Jun;14(2):194-200.
Vincke P. *Multicriteria Decision-Aid.* New York:Wiley;1992.
Viscusi WK. *Fatal Tradeoffs: Public and Private Responsibilities for Risk.* New York:Oxford University Press;1992.
Viscusi WK, Magat WA, Huber J. An investigation of the rationality of consumer valuations of multiple health risks. *RAND Journal of Economics* 1997;18:465-479.

Vogel EW, Barbin A, Nivard MJ, Stack HF, Waters MD, Lohman PH. Heritable and cancer risks of exposures to anticancer drugs: inter-species comparisons of covalent deoxyribonucleic acid-binding agents. *Mutatation Research* 1998 May 25;400(1-2):509-540.

von Winterfeldt D, Edwards W. *Decision Analysis and Behavioral Research.* Cambridge University Press;1986.

Wacholder S. When measurement errors correlate with truth: surprising effects of nondifferential misclassification. *Epidemiology* 1995 Mar;6(2):157-61

Wagenknecht LE, Burke GL, Perkins LL, Haley NJ, Friedman GD. Misclassification of smoking status in the CARDIA study: a comparison of self-report with serum cotinine levels. *Am J Public Health* 1992 Jan;82(1):33-36.

Wakker PP, Deneffe D. Eliciting von Neumann-Morgenstern utilities when probabilities are distorted or unknown. *Management Science* 1996:1131-1150.

Wakker PP, Zank H. State-dependent expected utility for Savage's state space. *Mathematics of Operations Research* 1999 Feb;24(1):8-34.

Wei C. Multivariate adaptive stochastic approximation. *Annals of Statistics.* 1987;15:1115-1130.

Weinstein MC, Shepard DS, Pliskin JS. The economic value of changing mortality probabilities: a decision-theoretic approach. *Quarterly Journal of Economics* 1980 March:373-396.

Welch S, Salhi S. The obnoxious p facility network location problem with facility interaction. *Eur J Operational Research* 1997;102:302-319.

Wells AJ, English PB, Posner SF, Wagenknecht LE, Perez-Stable EJ. Misclassification rates for current smokers misclassified as nonsmokers. *Am J Public Health* 1998 Oct;88(10):1503-1509.

Wesley IV. Overview: Public health significance of Campylobacter in livestock and poultry. Proceedings of the 1998 USAHA (United States Animal Health Association) meeting;1998. Available from: URL: http://www.usaha.org/speeches/speech98/s98wesle.thml

Westfall PH, Young SS. *Resampling-Based Multiple Testing: Examples and Methods for p-Value Adjustment.* Wiley;1993.

Westfall PH. Multiple testing of general contrasts using logical constraints and correlations. *Journal of the American Statistical Association* 1997;92(437).

Wilbert C, Kallenberg M, Ledwina T. Data-driven smooth tests when the hypothesis is composite. *Journal of the American Statistical Association* 1997;92(439):1094-1104.

White PL, Baker AR, James WO. Strategies to control Salmonella and Campylobacter in raw poultry products. Revue scientifique et technique (International Office of Epizootics) 1997;16(2):525-541.

WHO. Use of Quinolones in Food Animals and Potential Impact on Human Health. Report of a WHO Meeting. Geneva, Switzerland;1998 Jun 2-5. Available from: URL: http://www.who.int/emc/ZDI/98.10

Whyte P, Collins JD, McGill K, Monahan C, O'Mahony H. Quantitative investigation of the effects of chemical decontamination procedures on the microbiological status of broiler carcasses during processing. *J Food Prot* 2001 Feb;64(2):179-183.

Willis WL, Murray C. Campylobacter jejuni seasonal recovery observations of retail market broilers. *Poultry Science* 1997;76:314-317.

Wolovich WA. *Automatic Control Systems: Basic Analysis and Design.* Holt, Rinehart and Winston;1994.

Wu G, Gonzalez R. Curvature of the weighting function. *Management Science* 1996 Dec;42(12):1676-1690.

Yang H, Ye J, Kim H, Li Y, Johnson M. Survival/Death of Salmonella typhimurium and Campylobacter jejuni on chicken skins and water during scalding and chilling processes. Poster presentation provided by authors. University of Arkansas, Fayetteville;2000.

Yao Q, Tritchler D. "Likelihood-Based Causal Inference." In: Fisher D, Lenz H-J, editors. *Learning from Data: Artificial Intelligence and Statistics V.* New York:Springer-Verlag;1996.

Ye J. On measuring and correcting the effects of data mining and model selection. *Journal of the American Statistical Association* 1998 March;93(441):120–131.

Young HP. *Equity in Theory and Practice.* Princeton (NJ):Princeton University Press;1994.

Yu Y. Multiobjective decision theory for computational optimization in radiation therapy. *Med Phys* 1997 Sep;24(9):1445-1454.

Zhang H. Classification trees for multiple binary responses. *Journal of the American Statistical Association* 1998;93(441):180-193.

Zhang H, Singer B. *Recursive Partitioning in the Health Sciences.* New York:Springer;1999.

Zhang J, Quan H, Ng J, Stepanavage ME. Some statistical methods for multiple endpoints in clinical trials. *Control Clin Trials* 1997 Jun;18(3):204-221.

Index of Topics

9 780792 376156

Batch number: 10620078

Printed by Printforce, the Netherlands